CONTROL OF FLOW SEPARATION

SERIES IN THERMAL AND FLUIDS ENGINEERING

EDITORS:

JAMES P. HARTNETT and THOMAS F. IRVINE, JR.

Chang	• **Control of Flow Separation: Energy Conservation, Operational Efficiency, and Safety**
Hsu and Graham	• **Transport Processes in Boiling and Two-Phase Systems, Including Near-Critical Fluids**

CONTROL OF FLOW SEPARATION

ENERGY CONSERVATION, OPERATIONAL EFFICIENCY, AND SAFETY

Paul K. Chang
Department of Mechanical Engineering
The Catholic University of America
Washington, D.C.

HEMISPHERE PUBLISHING CORPORATION
Washington London

McGRAW-HILL BOOK COMPANY
New York St. Louis San Francisco Auckland Düsseldorf
Johannesburg London Mexico Montreal New Delhi
Panama Paris São Paulo Singapore Sydney Tokyo Toronto

See Acknowledgments and Permissions on pages 505–509.

CONTROL OF FLOW SEPARATION: Energy Conservation, Operational Efficiency, and Safety

1 2 3 4 5 6 7 8 9 0 KPKP 7 8 4 3 2 1 0 9 8 7 6

This book was set in Press Roman by Hemisphere Publishing Corporation. The editors were Evelyn Walters Pettit and Martha M. Mahuran; the designer was Lilia N. Guerrero; the production supervisor was Rebekah McKinney; and the compositor was Wayne Hutchins. The printer and binder was The Kingsport Press.

Library of Congress Cataloging in Publication Data

Chang, Paul K.
Control of flow separation.

(Series in thermal and fluids engineering)
Includes bibliographies and indexes.
1. Fluid dynamics. 2. Boundary layer control.
I. Title.
TA357.047 620.1'064 75-17733
ISBN 0-07-010513-8

To my parents, Leo Kibin Chang and Lucia Huang Chang,
and other members of our family.

CONTENTS

PREFACE

Under certain conditions, fluid flowing over bodies or through passages separates from wall surfaces, causing pressure, velocity, and temperature conditions to change drastically from conditions of flow attached along the wall. The performances of fluid-handling machines (i.e., pumps, turbines, fans, compressors) and their parts (i.e., diffusers, channels, pipes), as well as airborne or seaborne vehicles (aircraft, missiles, ships, submerged undersea craft, etc.) are directly affected by the flow separation because these peak performances are at flow condition close to separation. If the flow separates, more power is required to compensate for energy loss, and detrimental surging, stalling, etc., may occur, destroying or damaging human bodies and machines. Thus, flow separation is an important problem in fluid mechanics and engineering.

On the other hand, the performance of machines and devices provoking flow separation may lead to a certain desirable efficiency. Extensive environmental destruction caused by wind flowing around buildings, dams, bridges, sea coasts, mountains, valleys, rivers, cities, etc., may be solved by the control of flow separation.

In this book an attempt is made to present the available control techniques of flow separation, after introducing briefly its specific features, so that it can be eliminated, reduced, or provoked in order to improve the performance. At the end of each chapter, quoted and unquoted references through 1975 have been listed, giving credit to the original sources wherever possible.

The fundamental aspects of theory, analysis and experiment, and a short introduction to flow separation control have been presented in my monograph, "Separation of Flow" (Pergamon Press, 1970), which will aid the reader who wishes to study the phenomena of flow separation in more detail. It may be of particular use to engineers, scientists, and students. I introduced a pioneering graduate course, "Separation of Flow and Its Control," at The Catholic University of America, Washington, D.C., in 1971.

U.S. Naval Air Systems Command, Washington, D.C., Contract No. N00019-71-C-0291 made it possible to compile and complete this book. I wish to thank G. L. Desmond and R. F. Broberg for their guidance. The following people from the U.S. Naval Ships Research and Development Center, Carderock, Maryland, also deserve credit: Dr. S. de los Santos and Dr. S. Sacks for reviewing the book, Ms. A. Cook and Mr. R. Hartley for editing it, and Ms. C. Applegate for typing it. Thanks are also due to the following persons from the Catholic University of America, Washington, D.C.: Dr. M. Casanella and Dr. R. Smith for proofreading, and Dr. J. LeNard for assisting in completion of the manuscript while he was at the university.

Paul K. Chang

GENERAL SYMBOLS

A	Wetted area or frontal area
a	Velocity of sound
C_D	Drag coefficient
C_L	Lift coefficient
C_p	Pressure coefficient
c_f	Skin friction coefficient
c_p	Specific heat at constant pressure
D	Drag
d	Diameter
H	Shape factor of boundary layer, $H = \delta^*/\theta$
h	Heat transfer coefficient, also height
Le	Lewis number
M	Mach number
Nu	Nusselt number
n	Index of power
Pr	Prandtl number
p	Pressure
q	Dynamic pressure
Re	Reynolds number
T	Temperature
t	Time
u	Streamwise velocity component
v	Velocity component in y-direction
w	Velocity component in z-direction
x	Coordinate, streamwise direction
y	Coordinate perpendicular to x-direction
z	Coordinate perpendicular to x and y-direction
α	Angle of attack
γ	Ratio of specific heats
δ	Thickness of boundary layer or shear layer
δ^*	Displacement thickness of boundary layer
ϵ	Eddy viscosity
θ	Momentum thickness of boundary layer
μ	Dynamic viscosity
ν	Kinematic viscosity
ρ	Density of fluid
τ	Shear stress
Ψ	Stream function

SUBSCRIPTS

e	Conditions at the outer edge of boundary layer
max	Maximum
min	Minimum
s	Separation
w	Wall
∞	Undisturbed free stream

CONTROL OF FLOW SEPARATION

chapter 1

CONTROL OF FLOW SEPARATION

INTRODUCTORY REMARKS

The effect of flow separation is frequently detrimental because it can cause loss of energy, instability, and so on. The term *stall* is used to express the unfavorable aspect of flow separation. Since stall usually occurs near peak operating performance, the control of stall is concerned with preventing or reducing flow separation in the interests of efficient operation. On the other hand, flow separation can be utilized to achieve a desired performance. This somewhat unconventional approach to separation control has not been investigated nearly as widely as has the prevention of stall.

This chapter attempts to outline the mechanics of flow separation and to describe the characteristics and effects of the phenomenon. Several examples of separation control techniques are also included. Because the phenomenon is complicated, it is first necessary to understand its physical aspects before proceeding to the principles for its control and the techniques whereby these principles are applied. Several methods are also presented for predicting the onset of steady flow separation, since it is necessary to determine whether or not separation occurs for given flow conditions. The problem of unsteady flow separation is not considered in this chapter.

For more details on the fundamental physical aspects of separation and methods for predicting its onset, the reader is referred to a monograph by the author (Chang, 1970).

1.1 MECHANICS, CHARACTERISTICS, AND EFFECTS OF FLOW SEPARATION

According to Maskell (1955), separation of flow over a body surface occurs if the surface streamline—the nearest streamline to the wall surface—leaves or breaks away from the body surface, as sketched in Fig. 1.1. The reversal of separation is then the reattachment as also shown in Fig. 1.1.

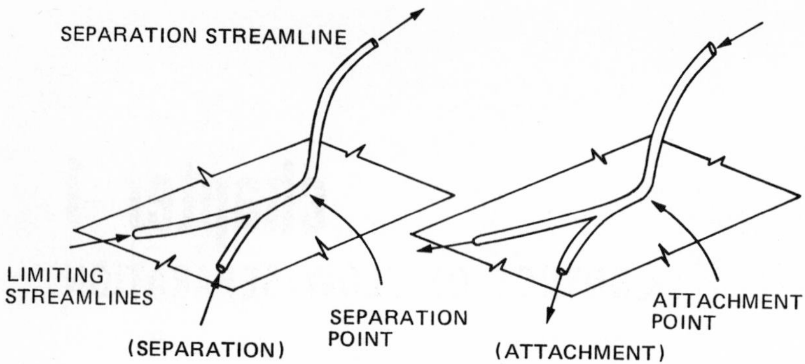

FIG. 1.1 The Maskell concept of ordinary separation and attachment (*from Mager, 1964*).

The surface streamline is given by

$$\frac{dz}{dx} = \lim_{y \to 0} \frac{w}{u} = \left(\frac{dw/dy}{du/dy}\right)_{y=0}$$

where u and w are velocity components in x, z coordinates, and y is the distance measured from the body surface and is perpendicular to x and z, as shown in Fig. 1.2.

The separation criterion may be given by the vanishing of the derivative of q with respect to y at the wall, i.e., by

$$\left.\frac{\partial q}{\partial y}\right|_{y=0} = 0$$

where q is the component of the velocity in the direction perpendicular to the separation line. It should be noted that the directions of the surface streamlines and the potential streamlines outside the shear layer are generally not the same; see Fig. 1.3. The milk streak in the lowest part of Fig. 1.3 clearly shows the different flow directions on the wall and above the wall; this indicates how potential streamlines differ from the surface streamlines. Thus it

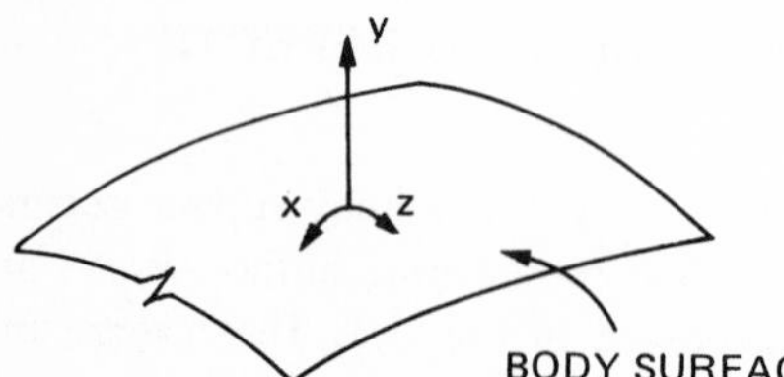

FIG. 1.2 Coordinate system.

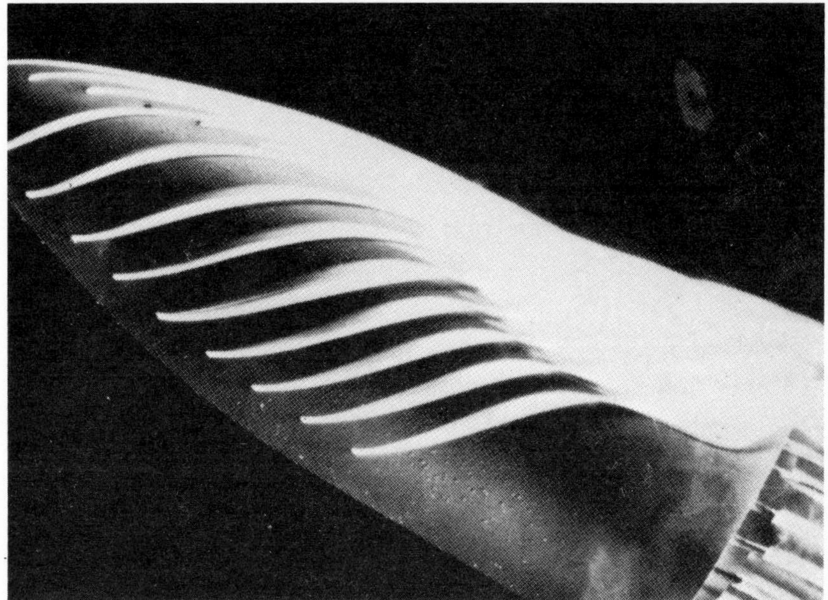

FIG. 1.3 Flow around an ogive at high angle of attack (*photograph courtesy of ONERA*).

is obvious that the separation is governed by the characteristics of flow along the wall where the viscosity effect is dominant. The interconnection between surface streamline and viscosity in flow separation may be found by the surface streamlines which define a kind of skeleton structure of the viscous region.

If the body surface is of finite dimension, then flow separation is inevitable because the flow expands over the downstream edge and flows away from the wall. Thus, flow separates at the trailing edge of a wing, around a corner of a rearward-facing step, and at a cavity; see Fig. 1.4.

Furthermore, flow also separates upstream of obstacles, such as a forward-facing step, a spoiler, and so on; see Fig. 1.5. Because the fluid is not capable of reaching an infinitely large velocity at the sharp corner, as the streamline along the wall approaches the obstacle, it leaves the wall.

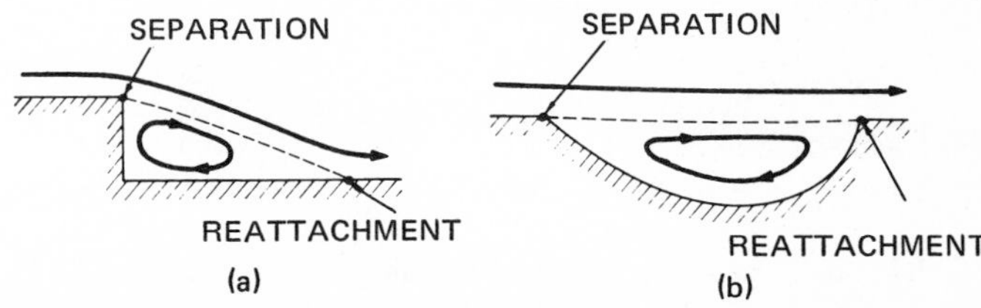

FIG. 1.4 Downstream separation: (*a*) rearward facing step; (*b*) cavity.

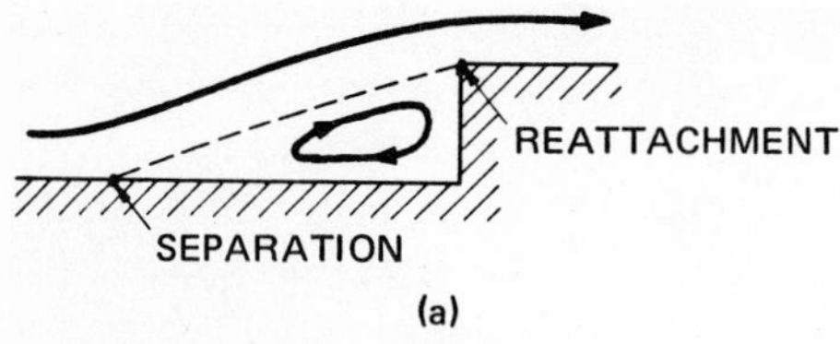

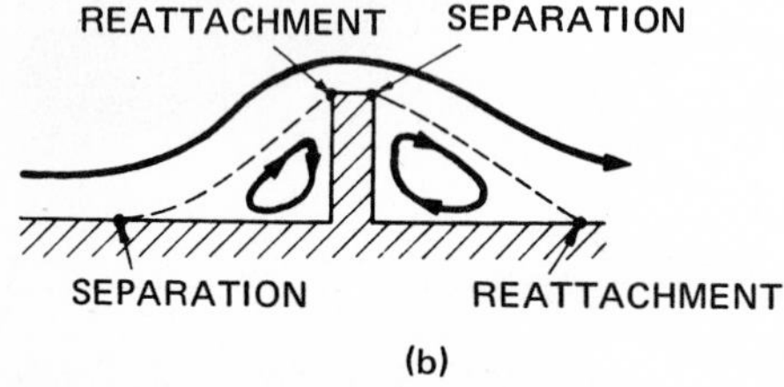

FIG. 1.5 Upstream separation: (*a*) forward facing step; (*b*) spoiler.

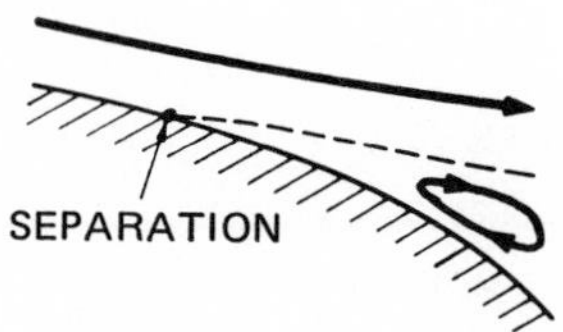

FIG. 1.6 Flow separation from a smooth surface.

The flow along a curved surface separates only if the streamwise pressure gradient is positive or adverse (as will be discussed later); this is shown in Fig. 1.6 for subsonic flow. Thus the separation point is located on the portion of the surface where the flow decelerates. Downstream of separation, reverse flow may occur.

As shown in Fig. 1.7, separation may occur at supersonic speeds if a shock wave strongly interacts with the boundary layer and creates a sufficiently large rise in pressure downstream from the shock interaction. Across a compression shock, the pressure rises, and the pressure difference spreads out in the lower level of the boundary layer. Thus, the pressure gradient that appears at the wall is determined by the properties of the boundary layer and by the strength of the shock wave. The pressure rise that is imposed on the boundary layer by the shock wave thickens the stream tubes of low velocity in the inner

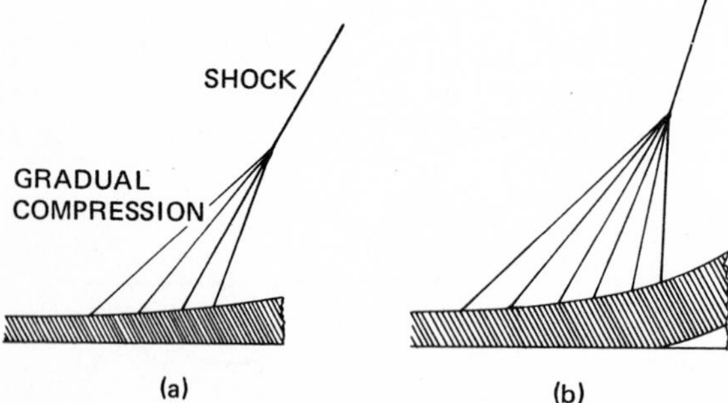

FIG. 1.7 Shock waves softened by the boundary layer (*from Pearcey, 1961*): (*a*) by boundary layer thickening; (*b*) by boundary layer thickening and separation.

(subsonic) part of the layer. This thickening deflects the outer part of the layer and the external flow outward, away from the wall. In turn, this generates a band of compression waves in the supersonic flow as indicated in Fig. 1.7. These waves start in the supersonic part of the boundary layer and propagate out into the external flow. If the shock is weak, the boundary layer converts the sharp, discontinuous pressure rise into a more gradual one so that it can negotiate the pressure gradient imposed by the shock, and the flow remains attached.

As the shock strength increases, however, the adverse pressure gradients increase, and the deceleration in the inner portion of the shear layer increases progressively until a stage is reached when the forward flow is no longer possible in the strata at the surface itself. These inner strata then leave the surface to continue the progressive increase in outward deflection in the outer parts of the boundary layer and in the external flow; see Fig. 1.7*b*. Thus, separation appears as a continuously developing flow pattern which unfolds further and further as the strength of the shock progressively increases.

Since flow separation is governed mainly by the pressure rise through the shock, a critical pressure rise can be defined as one that is just strong enough to cause separation.

The boundary layer separation of supersonic flow is also accompanied by a reversal of flow in the interaction region (Fig. 1.8). Note from Fig. 1.8 that during the interaction of incident shock wave and boundary layer, the characteristics of turbulent flow are different from those of laminar flow. The region of laminar interaction is spread out over a longer distance, many times the thickness of the boundary layer. In this interaction, there is a plateau

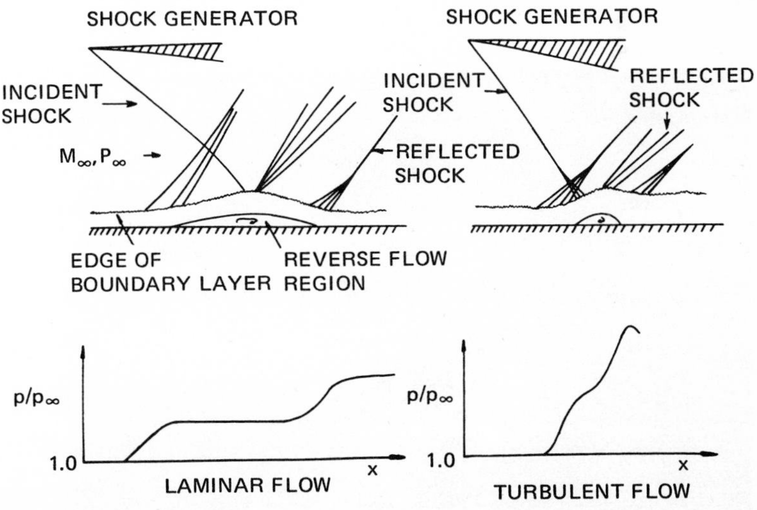

FIG. 1.8 Incident shock wave-boundary layer interaction (*from Torrillo & Savage, 1964*).

region where the static pressure remains constant. Because of this extended region, the external flow inclinations are small. For turbulent flow, however, the interaction takes place in a shorter distance and causes a strong normal pressure gradient. This behavior is similar to the step flow separation caused by a discontinuous compression shock, which will be described later in this chapter.

It is clear from these illustrations that the governing factors that cause separation for both subsonic and supersonic flow are the finite dimension of the body surface, the abrupt change of the geometrical shape of the body surface, the viscosity, and either a continuous or discontinuous adverse pressure gradient which reduces the streamwise velocity momentum of the flow near the body surface. This last factor occurs because the momentum of flow is used to overcome the pressure rise. The first two items (the finite dimension of the body surface and the abrupt change of the geometrical surface configuration) are associated purely with the body configuration. However, the flow that confronts the abrupt change of surface configuration may also be associated with the adverse pressure gradient caused (1) by the pressure of existing wall surface upstream of a corner (Fig. 1.5*a*) or (2) by a pressure gradient due to the formation of shock at a corner. In relation to fluid mechanics, these last two factors (adverse pressure gradient and viscosity) are essential for flow separation. If one of these two factors is missing, then flow cannot separate, regardless of whether it is external around a body or internal through a channel. For example, consider the subsonic channel shown in Figs. 1.9 and 1.10. Upstream of the throat, the channel converges in the flow direction; thus the streamwise pressure gradient is either negative or favorable. Hence, no separation takes place, and flow is completely attached

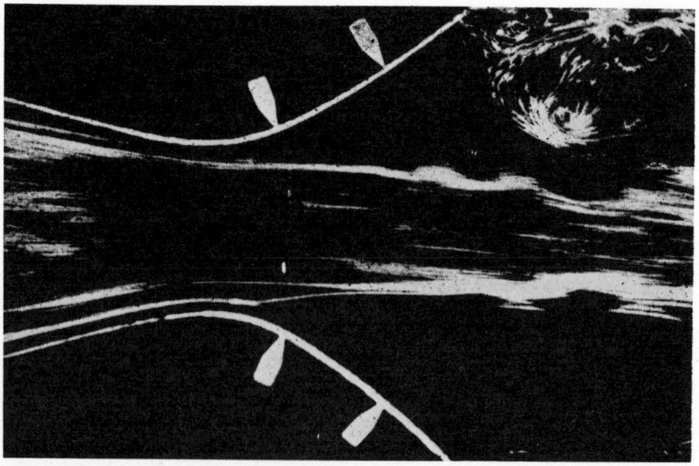

FIG. 1.9 Flow in a sharply diverging channel (*from Schlichting, 1968*).

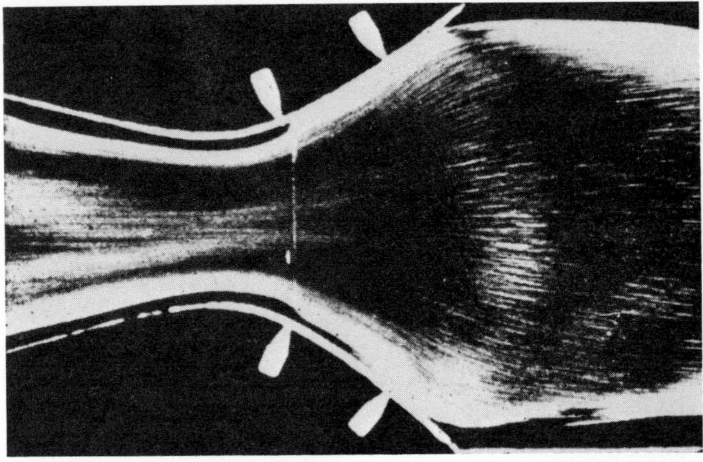

FIG. 1.10 A boundary layer sucked away on both walls (the flow is from left to right) (*from Schlichting, 1968*).

to the wall. Downstream from the throat, however, the channel is diverging and the pressure gradient is positive; thus the pressure rises downstream. Therefore, as shown in Fig. 1.9, the flow may separate if the adverse pressure gradient is large enough. The separation of flow can be prevented and the flow made to remain attached to the wall by suction of boundary layer flow. As shown in Fig. 1.10, the viscosity effect can be eliminated by removing the shear layer near the wall.

It is easily seen that the separation of flow can be prevented by designing the configuration of the body in such a way as to achieve a favorable streamwise pressure gradient or an adequately low streamwise adverse pressure gradient. Furthermore, it becomes apparent that the suction of the boundary layer can be used to control separation.

Figures 1.4 and 1.6 indicated that the vortices in the separated region are formed in such a way that the direction of the surface streamline of the vortices is toward the separation point and opposite to the surface streamline upstream from it. Hence, the separation streamline may be regarded as the line that divides all the surface streamlines into those coming from the separated and unseparated portions of the flow. Two sets of surface streamlines of opposite flow direction meet at the separation point and the loci of the separation points form a separation line. This phenomenon is the same for laminar as for turbulent flows.

As shown by analysis of three-dimensional shear layer, the necessary condition for separation of three-dimensional flow is that the derivative of a velocity component normal to the separation line at the wall must vanish, namely $\partial q/\partial y|_{y=0} = 0$ at the separation as previously mentioned (Oswatitsch, 1958). But this is not a sufficient condition for separation; the velocity of three-dimensional flows must be described by more than a single component. If the derivative of one of the velocity components vanishes, it may merely cause a change in the direction of the surface flow. The sufficient conditions are illustrated by the examples in the following discussion.

There are two types of separation, ordinary and singular. Ordinary separation is common in three-dimensional flow, and the singular type may be considered as a special case of the ordinary type, namely one that occurs in two-dimensional and axially symmetric flow. First consider ordinary separation.

Maskell (1955) reasoned that at the separation point, the values of α_0 defined by

$$\tan \alpha_0 = \lim_{y \to 0} \frac{\tau_x}{\tau_z}$$

are the same for both surface streamlines (upstream and downstream of the separation). This is so because both streamlines must be tangent to each other in the plane or wall, forming a cusp at the separation. Therefore, the separation points are located in cusps in the surface streamlines for three-dimensional flow. These streamlines must be tangent to the wall at the separation, in addition to the formation of tangency between these two streamlines, as shown in Fig. 1.11*a*. Moreover, because of these relations of

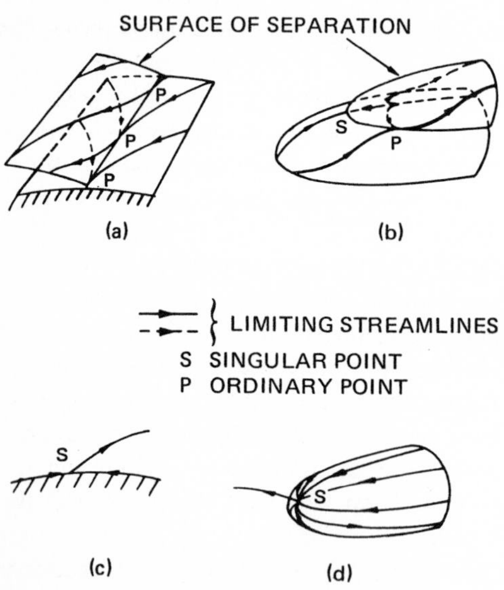

FIG. 1.11 Types of separation: (*a*) ordinary separation; (*b*) separation with an isolated singular point; (*c*) singular separation in two dimensions; (*d*) isolated separation point (*from Maskell, 1955*).

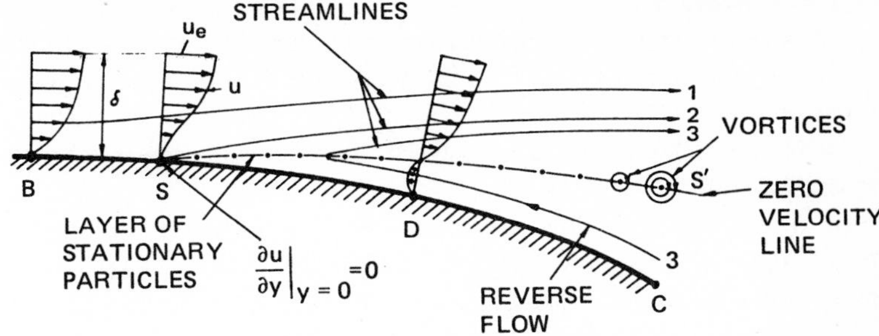

FIG. 1.12 Velocity profile near a separation point.

tangencies, the line of separation is an envelope of the separation streamlines.

Examples of singular separation which occur in two-dimensional and axisymmetric flows are shown in Figs. 1.11*b*–1.11*d*. For the singular separation, τ_x and τ_z become simultaneously zero; hence,

$$\tan \alpha_0 = \lim_{y \to 0} \frac{\tau_x}{\tau_z}$$

becomes indeterminate and the surface flow direction can be discontinuous and may have multiple values. In other words, at a singular point, the surface streamline can meet in other than a cusp. But at the separation point on the wall, the singular surface streamline must satisfy the condition of zero shear stress.

Prandtl, founder of boundary layer theory, explained the physical phenomenon of separation in this way: Consider the velocity profile of a two-dimensional or axially symmetric boundary layer upstream from separation where the pressure gradient is continuous and adverse (Fig. 1.12). Within the thin boundary layer of thickness δ, the effect of viscosity is such that a strong velocity gradient $\partial u/\partial y$ prevails near the wall (u is the streamwise velocity and y is the distance normal to the surface). With the exception of a rarefied gas, the flow velocity at the wall is zero and, with increasing distance y, u grows rapidly and gradually approaches the magnitude of u_e, the inviscid flow velocity at the outer edge of the boundary layer. Compared to the main stream, the retarded flow in the boundary layer suffers a relatively greater deceleration, and, since the momentum of the flow near the wall is small, the ability of the fluid to move forward against the pressure rise is also limited. This small amount of momentum and energy along the body surface is eventually used up downstream in order to overcome the pressure rise and friction. Hence the fluid particles are finally brought to rest. Because the main stream itself is decelerating, it is unable to energize the fluid in the boundary layer and accelerate. Thus, when the surface streamline reaches the point on the wall where $\partial u/\partial y|_{y=0} = 0$ (or zero stress position), it begins to break away from the wall and separates the flow from the surface. The zero stress condition also implies that τ_w changes at this point from a positive to a negative value. If τ_w is negative, then $\partial u/\partial y$ is negative at the wall, or u is negative near the wall. Thus reverse flow occurs downstream from the separation point.

It is evident that flow separation may be prevented or delayed if the momentum of the flow near the wall is increased. Thus, one means of separation control is to blow fluid into the shear layer.

The viscous mixing of fluid can augment the momentum near the wall. By its nature, turbulent flow provides such mixing of the slower fluid near the wall (the faster fluid is farther out). This increases momentum near the wall and results in a change in the velocity profile. Compare Fig. 1.13*b* to the laminar velocity profile shown in Fig. 1.13*a*. These figures also indicate that the velocity gradient is steeper in a turbulent than in a laminar boundary layer.

Hence, turbulent flow facilitates the streamwise movement of the fluid against the pressure rise and friction. Therefore, turbulent flow separates farther downstream than does laminar flow. If the level of mixing is raised artificially, for example, by a vortex generator, then the separation may be prevented or delayed.

The distinction between laminar and turbulent flow can be expressed numerically, for instance, by the shape parameter of the boundary layer H, defined by the ratio of the displacement thickness δ^* to the momentum thickness θ, namely $H = \delta^*/\theta$. The displacement thickness of the boundary layer is defined by

$$\delta^* = \int_0^\delta \left(1 - \frac{\rho u}{\rho_e u_e}\right) dy$$

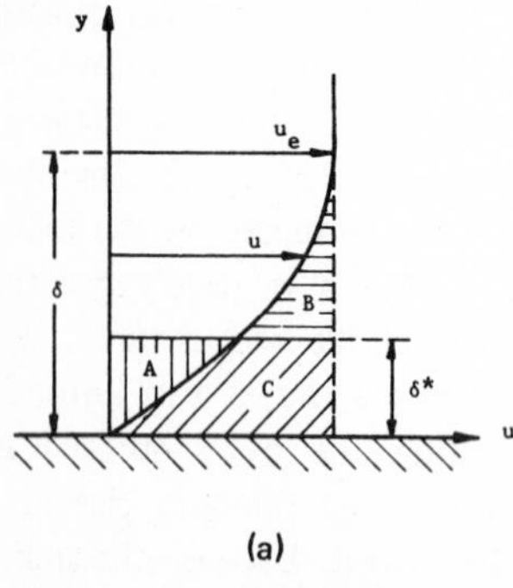

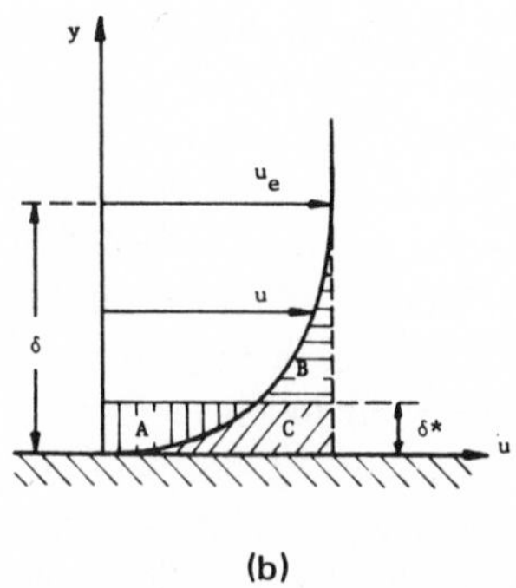

FIG. 1.13 Boundary layer velocity profiles: (*a*) laminar boundary layer; (*b*) turbulent boundary layer.

For incompressible flow it is

$$\delta^* = \int_0^\delta \left(1 - \frac{u}{u_e}\right) dy$$

In general, the momentum thickness is defined by

$$\theta = \int_0^\delta \frac{\rho u}{\rho_e u_e}\left(1 - \frac{u}{u_e}\right) dy$$

Also for incompressible flow, the definition is

$$\theta = \int_0^\delta \frac{u}{u_e}\left(1 - \frac{u}{u_e}\right) dy$$

The physical meaning of the displacement thickness is readily grasped by considering the mass flow in the boundary layer. As seen in Fig. 1.13 for incompressible flow, the magnitude of δ^* can be found by making Area A equal to Area B. Thus Area (A + C) equals Area (B + C). Therefore, δ^* is the thickness that a layer of fluid, represented by Area (A + C), would have if it had the same integrated velocity defect as the actual boundary layer, represented by Area (B + C). The boundary of the nonviscous flow is determined by displacing the boundary of the surface by a distance δ^*. Furthermore, it is apparent from Fig. 1.13 that the value of δ^* is smaller for turbulent than for laminar flow. In other words, if laminar flow becomes turbulent, the value of δ^* decreases abruptly. Similarly, if the momentum of the boundary layer is considered, the momentum thickness of the boundary layer may be explained as the thickness that displaces the boundary of the wall by the distance θ. However, the growth of momentum thickness through the transition region is negligible compared to displacement thickness. Hence, it is apparent that the value of $H = 2.6$ for incompressible laminar flow is suddenly reduced to about 1.3 when the flow becomes turbulent. Since this dimensionless parameter H characterizes the shape of the velocity profile, and since separation of flow is governed by the velocity profile, the value of H is often used to specify the onset of flow separation. The closer the flow is to separation, the larger the value of H becomes for attached turbulent flow.

The turbulent boundary layer has a thin laminar sublayer close to the wall, and a precise treatment should include consideration of this sublayer. However, its effect on the overall turbulent boundary layer is negligible.

The separated flow region is often called the "dead water" region. Actually, even though flow separation takes place over or behind the two-dimensional surfaces, this region consists of a complex vortical, unsteady three-dimensional flow that often has considerable velocity.

For convenience, separated flow may be classified into two different categories on the basis of the size of the separated region with respect to the body dimension. If the separated region is relatively small compared to the body and is enclosed by a separation streamline connecting the points of separation and reattachment, then this type of flow may be defined simply as *separated flow.* However, if the separated flow has a long streamwise dimension compared to the body and extends to an infinitely long distance downstream (rather than being enclosed behind the body), then this type of separated flow is referred to here as *wake flow.*

The static pressure is not necessarily constant in the separated flow region, and the pressure distribution at subsonic speeds is different from that at supersonic speeds. This is shown in Fig. 1.14 for flow over a flared flat plate. Note that the variation in Reynolds number at subsonic speeds $(0.4 \leqslant M \leqslant 0.8)$ causes only small changes in the pressure distribution and no measurable change in the pressure rise to separation. Moreover, there is almost

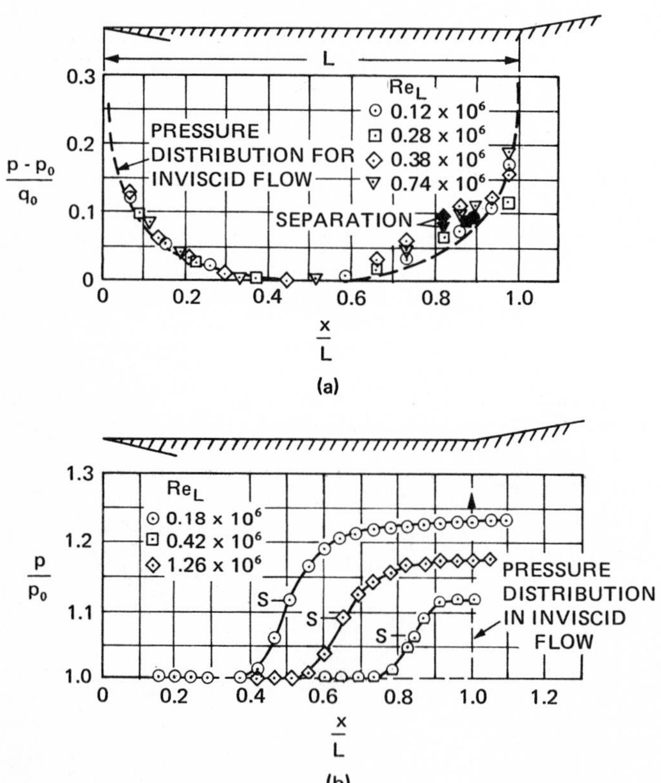

FIG. 1.14 Comparison of subsonic and supersonic flows at various Reynolds numbers: (*a*) subsonic (model CC10° – 3; $0.4 < M_\infty < 0.8$; $\alpha = 4°$); (*b*) supersonic (model CC10° – 2; $M_0 = 2.0$) (*from Chapman et al. 1957*).

no change in the position of the separation. The amount of the pressure rise given by $(p_s - p_0)/q_0$ is equal to 0.082 ± 0.005 for all Reynolds numbers. (Subscript 0 refers to the beginning of the interaction, that is, at the most downstream point upstream of which the pressure is essentially the same as the inviscid pressure.) Furthermore, the pressure distribution is roughly that which corresponds to the inviscid flow computed by using small-disturbance theory to superimpose the appropriate lift pressure distribution for an inclined plate; note the dotted line in Fig. 1.14*a*.

In contrast to the subsonic case, the pressure distribution at supersonic speeds ($M_\infty = 2$ at $0.18 \cdot 10^6 < Re_L < 1.26 \cdot 10^6$) depends on Reynolds number, and the position of the separation point changes. Further, there is a disparity between the measured pressure distribution and that calculated by inviscid flow theory. This calculated pressure distribution is constant upstream and then jumps discontinuously; note the dotted line in Fig. 1.14*b*. It may be concluded on the basis of these two cases that (1) for subsonic flow, the pressure distribution near and upstream of separation is determined primarily by the inviscid flow pressure distribution around the body shape and only secondarily by Reynolds number dependence on the interaction between the boundary layer and the external flow, whereas (2) for supersonic flow, the pressure distribution near separation is determined primarily by a Reynolds number-dependent interaction (free interaction) and only secondarily by the inviscid flow distribution.

The characteristics of laminar separated flow are different from those of transitional or turbulent flow. This is illustrated by the experimental study of Chapman et al. (1957); see Figs. 1.15 and 1.16. At supersonic speeds, the separated flow regions in front and behind the step are strongly dependent on whether flow is laminar, transitional, or turbulent, although the free-stream Mach number remains the same. This is seen in Figs. 1.15 and 1.16 (subscript o refers to the condition just upstream of separation). For a forward-facing step, the pure laminar regime has a plateau region of nearly constant pressure throughout the separated region, and the separation pressure p_s and plateau pressure p_p, respectively, are of the order of 15 and 30 percent greater than the pressure just upstream of the separated region. There is a small region near the step shoulder where local pressures on the face are higher than the pressure in the other part of the separated region. This is so because a portion of the separated layer is brought to rest on the step face.

At separation, if the separated layer is thick, then the pressure rise is small. If it is thin, then the area over which the pressure increase occurs is confined to a small area near the shoulder.

In the transitional regime, the boundary layer is still laminar at separation. Thus the pressure to cause separation remains about the same as that for pure laminar separation, but transition causes a greater pressure rise before reattachment at the step. Pressure variation on the step amounts to the order of 0.1 (p_0) and is attributable to the sizable subsonic velocities that exist

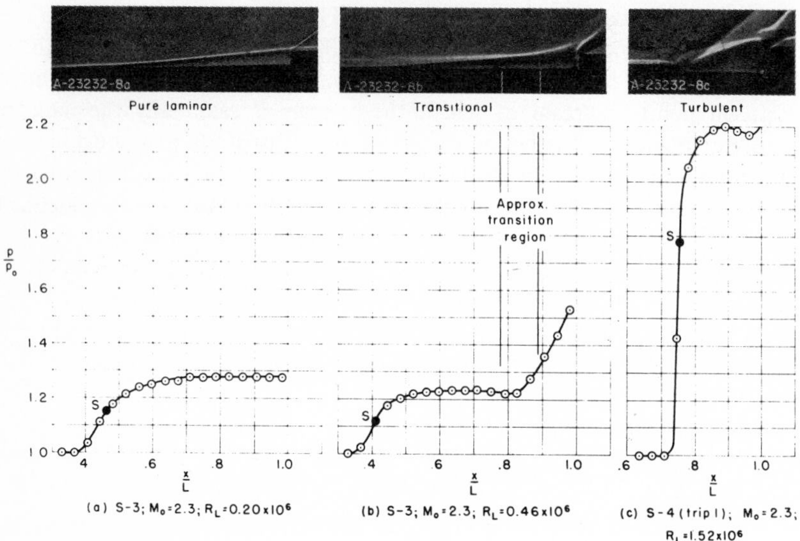

FIG. 1.15 The three regions for a step (M = 2.3 and R_L refers to Re_L) (*from Chapman et al., 1957*).

within the reverse flow region just upstream of the step. Since transition itself is fundamentally unsteady, flow in the region between transition and reattachment on the step is unsteady. In the turbulent region, the pressure rise to separation is much larger. There is no pressure plateau in the separated region because the eddying motion of the turbulent layer energizes the fluid. Pressure on the step face in turbulent flow is similar to that in transitional flow; the degree of steadiness is not as great for turbulent as for laminar flow, but it is larger than it would be for transitional flow. For a backward-facing step (Fig. 1.16), there is a region of constant pressure upstream of the reattachment for all three types of flow. But as would be expected, the magnitude of this pressure plateau is largest for turbulent flow, intermediate for transitional flow, and smallest for laminar flow (Chapman et al., 1957).

The transfer of heat to the wall within the separated region is high. Heat transfer has only a small effect on the pressure distribution, but it deepens the pressure gradient and consequently has a significant effect on the position of separation. The separation point may be shifted downstream by cooling the wall sufficiently. Thus the separated flow region can be reduced or completely eliminated, as evidenced by Lankford (1961). Conversely, by heating the wall it can be expected to increase the size of the separated flow region because of the separation that occurs upstream. Hence, the thermal treatment of the wall is also an effective technique for separation control. The behavior of the reattachment point is similar to the behavior of the separation point, but in a somewhat reversed sense.

At reattachment, where the flow velocity is reduced to zero and the shear layer becomes thin, the rate of heat transfer and the pressure rise are both large, even though the rate of heat transfer is low at the separation point. At hypersonic speeds, the pressure rises in the region of reattachment on a cone cavity by an amount which may reach about 50 percent of the value of the free stream, and the heat transfer rate is double the free-stream value (Chang, 1966; Nicoll, 1964). However, because of a decrease in the value of the recovery factor (Chapman, 1956), the heat transfer rate can be reduced by injecting gas into the separated region in laminar flow. This technique is practical because the amount of gas required is small.

Another method is to inject a high velocity jet through the trailing edge downstream of separation along the rear dividing streamline. The upstream separation may then be prevented because additional energy is supplied to the shear layer (Yuan et al., 1966).

Now consider wake flow. For the wake with a long trail of separated flow region, the pressure behind the base (the base pressure) is lower than the free-stream pressure because the momentum diffuses across the shear layer. Since this lowered pressure acting over the base gives a force component coincident with that of drag, the base pressure is closely connected with the total drag. The base drag for a body of revolution may reach 30 percent of the total drag, and the base drag for a missile may amount to 1000 or 2000 lb if the total drag reaches several thousand pounds. In the case of a thin streamlined body, the base drag is large compared to the relatively small

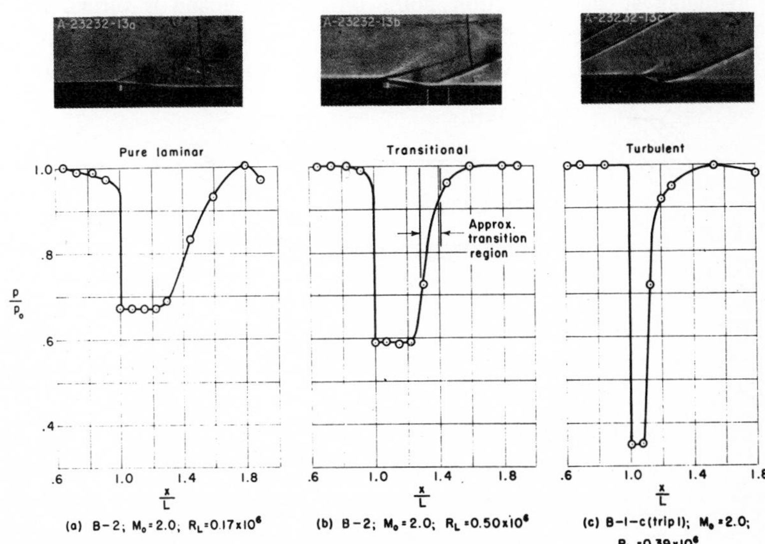

FIG. 1.16 The three regions for a backward facing step ($M = 2.0$ and R_L refers to Re_L) *(from Chapman et al, 1957)*.

friction drag. The wake may be very hot due to the compression shock that causes dissociation. Because the trailing wake is long, it then becomes a significant source of electronic observation if it contains electrons and other radiating species. For example, a meteor trail may reach a length of 15 miles at an altitude of about 100 miles or a body flying at 40,000 ft/sec may have a temperature of 6500°K at the outer part of the wake at a distance of 50–100 diameters downstream from the body (Lees & Hromas, 1962).

Downstream from a blunt trailing edge, the base flow at subsonic speeds is dominated by the periodic formation and shedding of eddies into the wake.

A splitter plate placed downstream of a blunt body such as a circular cylinder is useful to reduce drag because base pressure is increased, as shown in Fig. 1.17, reducing the shedding frequency. Although a short splitter plate with a length of one diameter changes the shedding frequency slightly and does not inhibit the vortex formation, it is possible to inhibit the periodic vortex formation and substantially decrease the drag by providing a longer length, equalling five diameters (Roshko, 1954).

The violent periodic behavior evident at subsonic speeds disappears at supersonic speeds, and a rather steady flow pattern is established. A hot "outer wake" is formed behind the blunt body when gas passes through the normal portion of the bow shock and is compressed and heated in the shock layer (Fig. 1.18).

The "inner wake" is formed by the turbulence that originates in the flow region with the highest velocity gradient and in the coalescence of the free shear layer shed from the body surface when the boundary layer leaves the surface. The zero velocity line, on which the tangential flow velocity is zero, originates at the separation point on the body and is turned at the neck because the pressure rises due to the flow deflection. Near the neck, the turbulence is confined to a narrow region around the wake axis, but it spreads

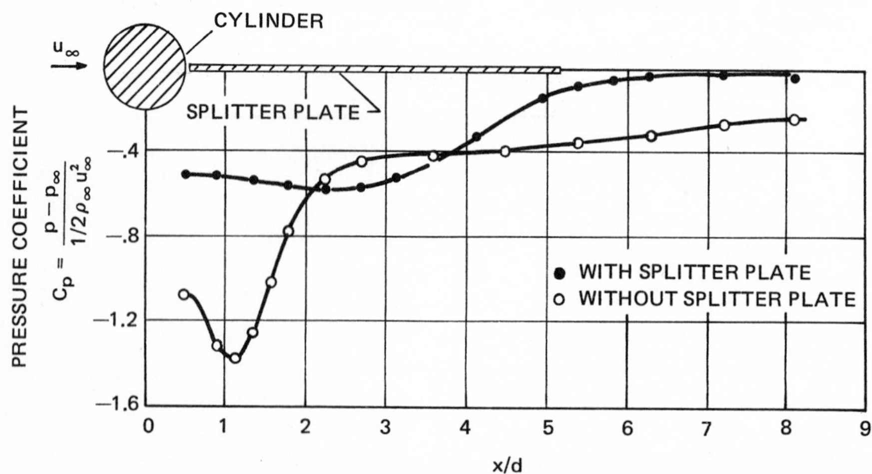

FIG. 1.17 Pressure on wake center line with and without splatter plate; $Re_d = (u_\infty d)/\nu =$ 14,500 (*from Roshko, 1954*).

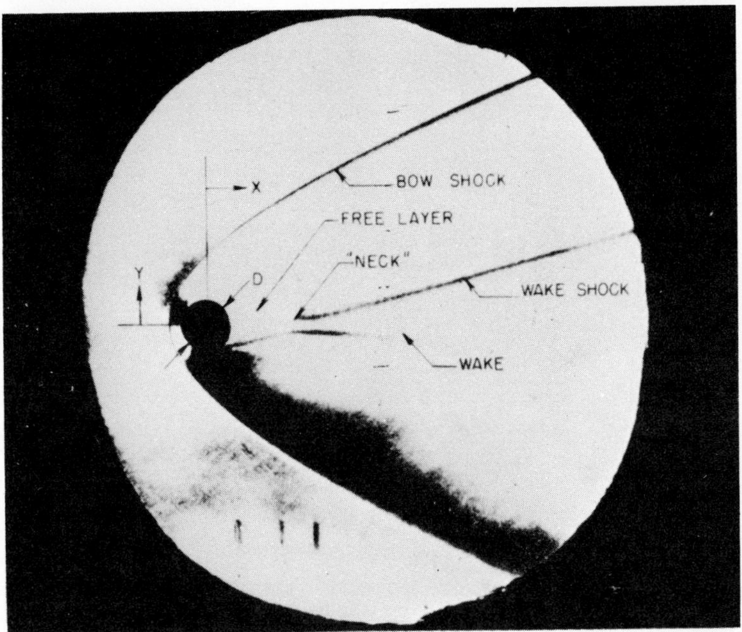

FIG. 1.18 Flow past a cylinder at Mach 5.8 (*courtesy of U.S. Naval Ordnance Laboratory*).

out rapidly by feeding on the surrounding gas. All of the streamlines originally in the "outer wake" are engulfed downstream by the turbulent wake.

For slender bodies (see Fig. 1.19), an oblique shock is wrapped close around the body, and the wake is "cold" behind the body.

In contrast to the "hot wake" which is primarily associated with the bow shock wave, the "cold wake" may be considered entirely due to the viscous effects associated with the boundary layer and base mixing. Thus the base drag can be reduced by bleeding gas into the wake region behind the projectile. The base drag can be reduced more effectively, (about two-thirds) by burning a gas, e.g., hydrogen, in the turbulent wake or by adding the heat of combustion. The combustion possibly influences the boundary layer thickness (Baker et al., 1951).

In order to achieve favorable aerodynamic performance, the shape of the shock may be changed, or a shock may be created artificially, thereby provoking the separation. In this case, the separation is not necessarily entirely detrimental. In fact, this technique could be regarded as provoking flow separation for a gainful engineering application. For example, a thin spike can be placed in front of a blunt axially symmetric nose exposed to the supersonic flow. The detached shock upstream of the blunt nose with strong compression may then be changed into a weaker oblique shock attached to the spike, but with a provoked separation flow region of conical shape (Fig. 1.20).

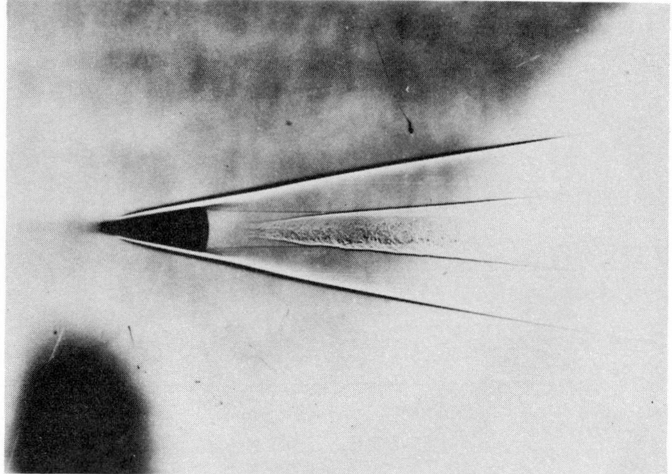

FIG. 1.19 Wake behind a slender body at hypersonic speeds (*courtesy of U.S. Naval Ordnance Laboratory*).

Separation may be caused by the pressure rise downstream from a spike (Fig. 1.21). Because of the inclination of the shock, the base pressure may become lower, and, consequently, the drag of the body decreases. However, this type of separated flow is unsteady, and oscillation may occur because of the nonequilibrium condition of the mass flow. Its occurrence depends on the configuration of the blunt body. For example, oscillation is observed for a flat-nosed body but not for a hemispherical nose.

A high rise in static pressure and a resulting large side force can be obtained by artificially provoking separation. For example, the injection of a

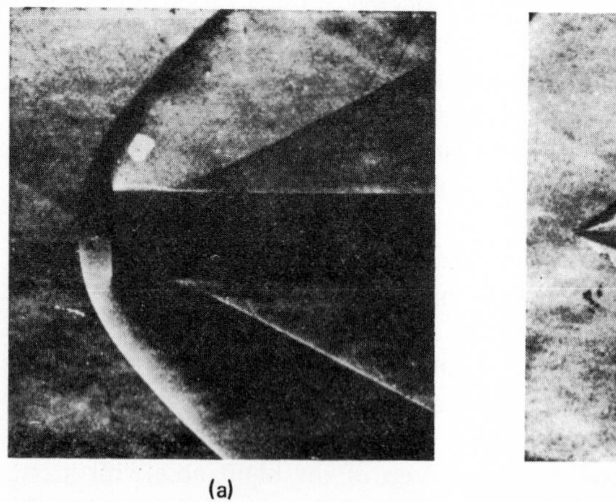

(a)

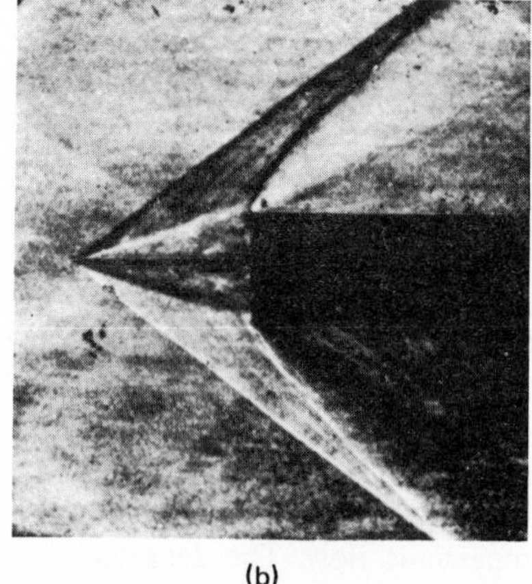

(b)

FIG. 1.20 Bow shock wave and separated flow in front of a blunt body (*from Mair, 1952*).

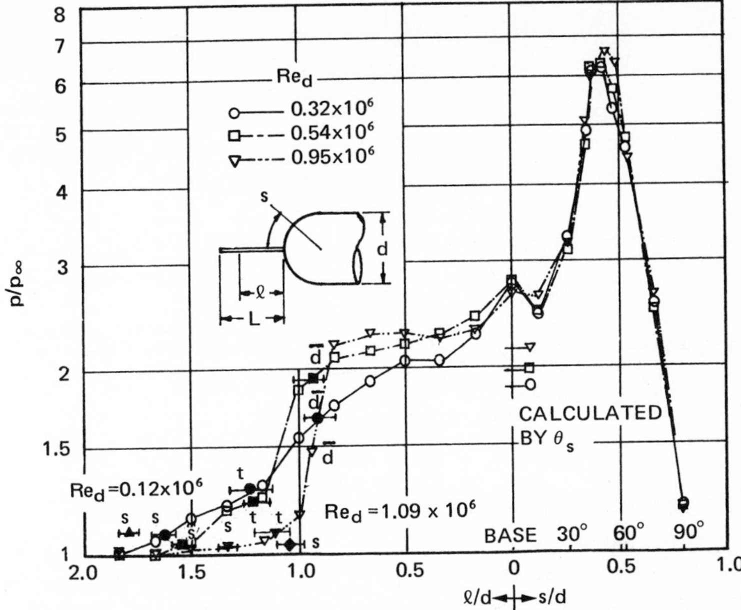

FIG. 1.21 Pressure distribution along a spike and the body surface for different Reynolds numbers ($L/d = 2$. Shaded symbols with *t, s,* and *d* denote transition, separation, and deflection points of the separated boundary layer, respectively; $M_\infty = 3.3$.) (*from Hahn, 1966*).

liquid into the hot supersonic gas stream may form a pseudobody with two components (Fig. 1.22). One component is formed by the vaporized liquid drop and the decomposition of the vapors downstream of the injection port. The other component is formed by the separation of the boundary layer upstream of the injection port. The separation is caused by the interaction of the boundary layer with the shock wave upstream of the vapor zone.

The effective configuration has the shape of a spiked blunt body. Because the pressure rises rapidly downstream from the shock, a large side force acts on the wall and, consequently, a clockwise moment occurs. Thus, fluid injection can be utilized to control thrust vector.

An analytical investigation of various injectants (hydrogen peroxide [H_2O_2], strontium perchlorate [$Sr(ClO_4)_2 + H_2O$], nitrogen tetroxide [N_2O_4], Freon 114-B_2 [$CBrF_2 - CBrF_2$], and water [H_2O], etc.) shows that the performance of the injectant is strongly dependent on the amount of energy in the injectant zone. Because the small amount of energy contained in the hot, free-stream gas entrained within the injectant drop zone is insufficient to evaporate all the liquid, the exothermic decomposition of the injectant becomes important. This importance may be illustrated by noting that the released energy of the decomposed hydrogen peroxide is several times larger than the energy of the entrained gas (LeCount et al., 1965).

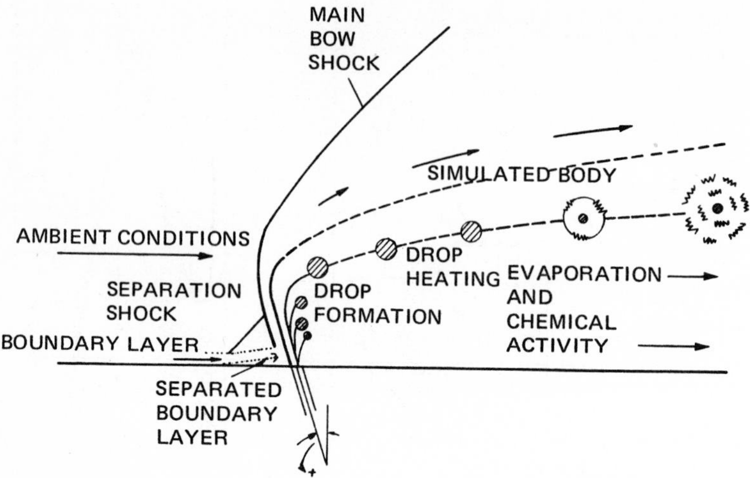

FIG. 1.22 Fluid injection from a side wall (*from LeCount et al., 1965*).

By provoking separation, a pseudobody shape may be formed by a free streamline flap that is useful for practical applications. In supersonic flight for example, a thin and sharp-edged airfoil is suitable for good cruising performance. However, a thicker airfoil is desirable at the low speeds necessary for takeoff and landing. With a thin airfoil at low speeds, lift is small, drag is large, controllability is poor, and buffeting occurs because of flow separation at the leading edge. Hence, as shown in Fig. 1.23, it is

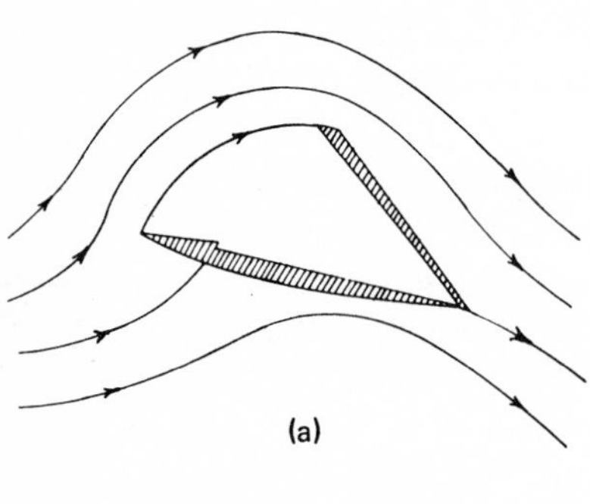

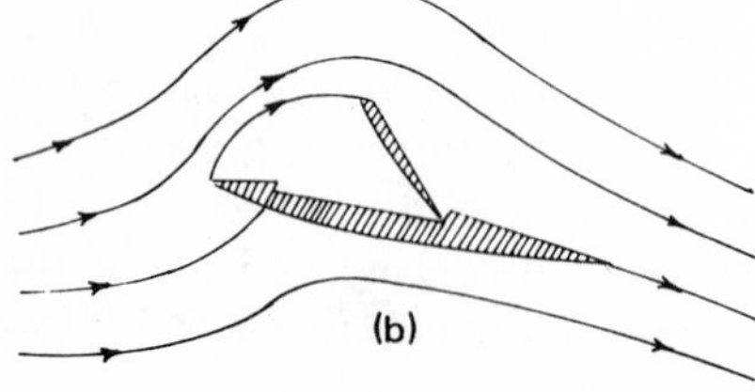

FIG. 1.23 Possible forms of a free-stream flap (*from Hurley, 1961*).

FIG. 1.24 Attached flow over an airfoil (*from Prandtl & Tietjens, 1934*).

advantageous to effectively transform a thin airfoil into a thicker one by provoking leading-edge separation and causing reattachment at the leading edge of the flap.

It is interesting to note that high polymer additives that greatly reduce skin friction may affect flow noise in the separated flow. Binder and Ishino (1967) report that the level of flow noise can be reduced below that of tap water by adding Polyox and Guar Gum in the marginally separated region of a diffuser. If there is no separation, the sound pressure level is above that of tap water. They attribute this phenomenon to the reduction in power spectral density values.

In the past, a great deal of investigation has been done on stall of aircraft wings and on the passages of compressors and pumps. Such studies have contributed to the enrichment of knowledge about flow separation and have led to superior designs for a wide range of engineering hardware, including household devices.

The flow field around an airfoil is studied next by referring to Figs. 1.24 and 1.25. An airfoil is designed to have optimum characteristics in

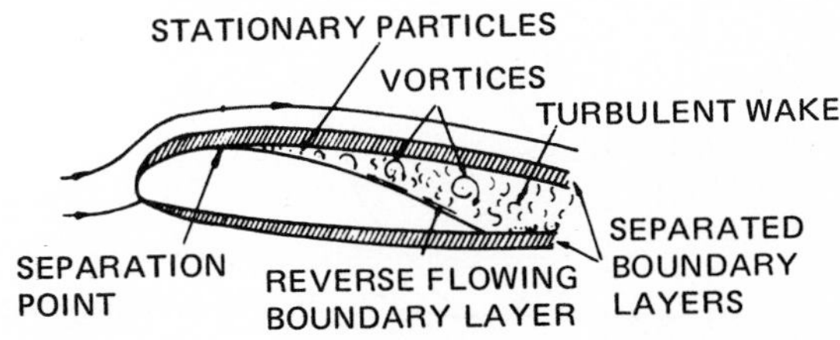

FIG. 1.25 Viscous effects in the flow around an airfoil at a positive angle of attack (boundary layer thickness is greatly enlarged) (*from Heinke, 1946*).

the operational range which usually requires attached flow over the airfoil surface, i.e., angles of attack that yield high lift and low drag. However, if the angle of attack increases sufficiently, then the flow over the upper surface separates. Vortices are then formed downstream and the stream pattern deviates from the optimum design condition, thus causing additional drag. Figure 1.26 shows the detrimental stall on an airfoil that suddenly losses lift but gains in drag. Such a stall is caused by the separation bubbles that are generated when the flow creates a circulating motion on the airfoil surface.

Additional details on the separation phenomenon over an airfoil at high angles of attack have recently been described by Hazen (1967). He classifies them roughly into the following four categories: separation at the trailing edge, thick-section bubble, short bubble, and long bubble. These are shown in Fig. 1.27 and described separately below. Some doubts exist concerning the details of the fundamental processes involved, and mixed flow types are not uncommon, particularly in the range of Reynolds number $2 \cdot 10^6$ to $10 \cdot 10^6$.

1. Trailing-edge separation. This type of flow separation is to be expected with turbulent flow. Thick (less than 15 percent) airfoils have a well-rounded suction peak and only a moderate adverse pressure gradient which covers the rear portion of the airfoil. For such airfoils, the flow separation is near the trailing edge, and the separation point moves upstream with increasing angle of attack. The thickness of the turbulent boundary layer may be reduced and some improvement in $C_{L\,\max}$ achieved by increasing the Reynolds number.
2. Thick-section separation bubble. With laminar flow at low Reynolds numbers, the flow separates on the forward portion of a thick airfoil. This type of separation is often followed by reattachment of the turbulent boundary layer downstream; it may reseparate near the trailing edge. If the Reynolds number is increased, the extent of the forward separation is reduced because the separation point shifts downstream until transition occurs upstream of the laminar separation point, thus reverting to the trailing edge type of separation.

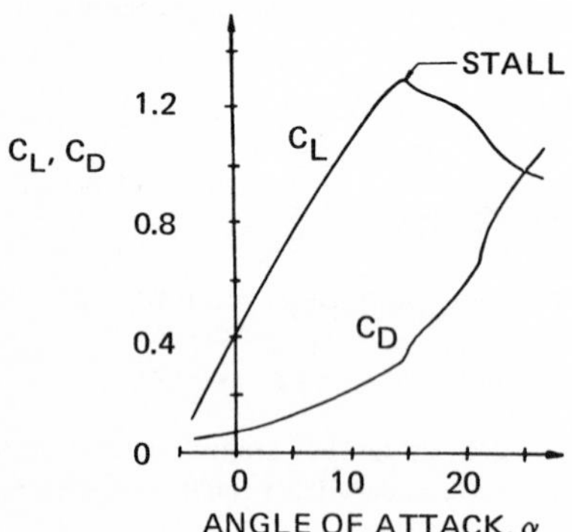

FIG. 1.26 Lift and drag of Clark Y airfoil (*from Weick & Bamber, 1932*).

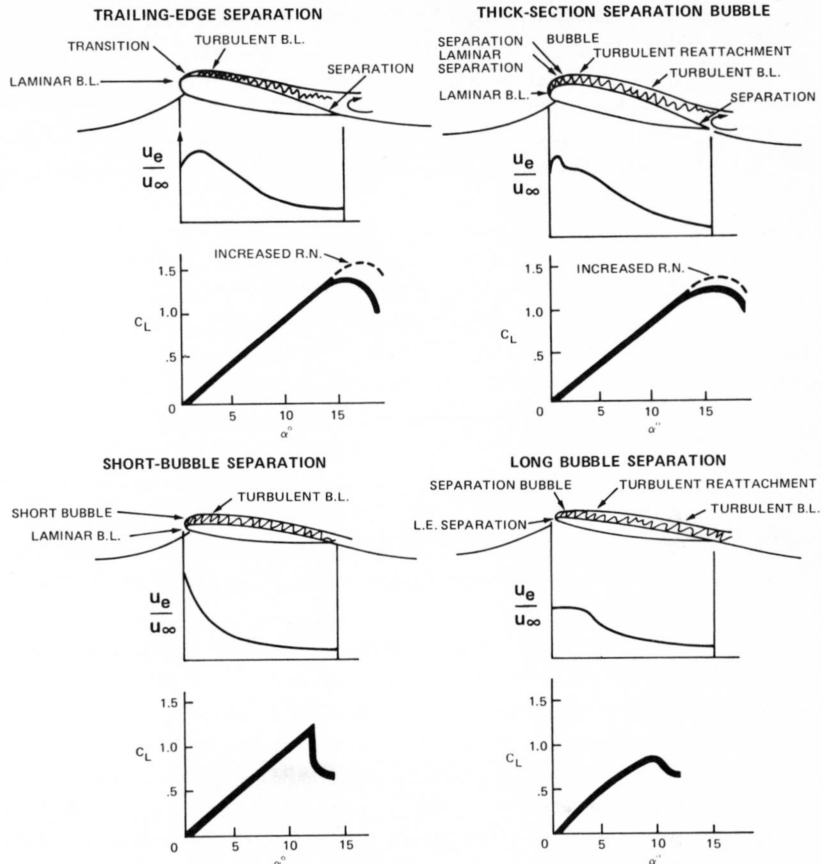

FIG. 1.27 Characteristics of two-dimensional airfoil separation (*from Hazen, 1967*).

3. Short-bubble separation. The size of the short bubble is of the order of 0.5–1 percent of chord length. On the thinner airfoils, where a suction peak of a sharper nature occurs close to the leading edge, a laminar separation starts. With an increase in angle of attack, the separation point moves up to the leading edge. The separated layer becomes turbulent and at medium or high Reynolds numbers, it reattaches to form a short bubble. With increasing Reynolds number, this bubble tends to contract until it suddenly bursts and causes an abrupt stall. At very high Reynolds numbers, a trailing-edge stall may form instead of the short bubble type. The very short bubble then acts merely as a transition-fixing device.
4. Long-bubble separation. Although this kind of bubble is termed "long," it is only approximately 2 percent of chord length. On very thin airfoils whose thickness is less than 6 percent, the laminar flow that separates at the leading edge at low Reynolds numbers reattaches following transition

to turbulent flow and forms a long bubble. With increasing angle of attack, this bubble extends to the trailing edge. Although this long-bubble separation causes a gentle stall in contrast to the abrupt stall of the short bubble, it nevertheless exerts a substantial influence on drag, lift, and pitching moment over a considerable range of angles of attack. Close to stall, a short-bubble type separation may occur when the Reynolds number becomes sufficiently large.

Because of the nature of the regimes of flow separation just mentioned and because of the influence of Reynolds number on them, it is readily understandable that experimental studies of small-scale airfoil models at high angles of attack are troublesome and unreliable. Hazen (1967) also studied the flow separation on three-dimensional wings. He found that current methods for predicting the subsonic characteristics of three-dimensional wings are inadequate. However, the maximum lift of a three-dimensional wing can be reasonably predicted from two-dimensional data for an unswept wing in a clean condition or with full span, high-lift devices.

1. Unswept wing. Although unswept wing is three-dimensional, the local wing sections have essentially the same chordwise loading distributions as in the two-dimensional case. The onset of flow separation may be reasonably deduced at the appropriate Reynolds number and spanwise C_L distribution. However, because of the wing-tip edge separation, the traditional theory does not yield satisfactory results for small aspect ratio wing. This is attributed to the strong vertical vortex sheets which stream from the inclined tip edges at high angles of attack. As shown in Fig. 1.28, these vortex sheets may be considered flexible end plates that induce extra lift near the tips (particularly toward the rear of the wing) with associated nonlinear lift curves and nose down moments. Although a recent analysis is in good agreement with experimental data on nonlinear lift, the applicability of such predictions is doubtful for conditions close to $C_{L\max}$.
2. Swept wing. The flow over a swept wing is three-dimensional. Thus even at the same Reynolds number, the separated flow characteristics of the local sections of a swept wing differ from those of two-dimensional sections of the same geometry because of the change in the pressure distribution. The root sections tend toward trailing edge stall whereas those at the tip exhibit short- or long-bubble separation. The tip stall can be counteracted by introducing positive camber. Since the three-dimensional viscous effects are complex and play an increasingly dominant role with increased angle of attack, the problem of swept wings is difficult to solve. Some "natural boundary-layer control" may be achieved at the root sections. There is a natural spanwise drift of the boundary layer from the root to the tip, particularly toward the leading edge, although a considerably thickened turbulent boundary layer is to be expected at the tip.

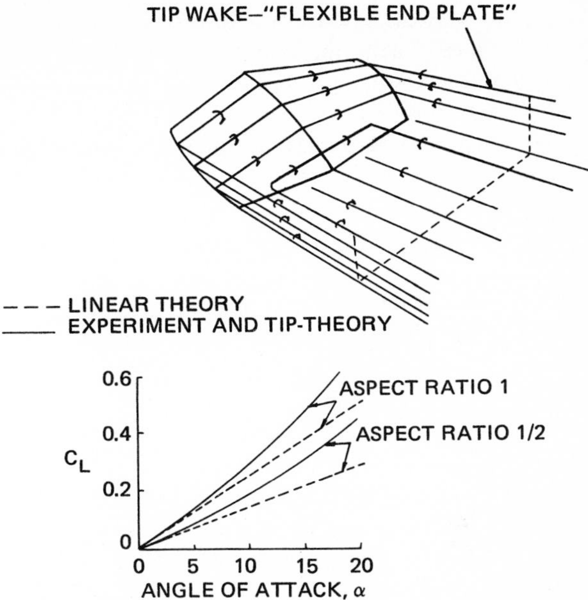

FIG. 1.28 Tip vortex sheet effect on low aspect ratio unswept wings (*from Hazen, 1967*).

3. Slender wing. Primary separation occurs along the whole length of effectively sharp swept leading edges. A pair of coiled vortex sheets starts to develop immediately from the wing apex (Fig. 1.29). As the angle of attack increases, these spiral vortices move inboard and upward. However, essentially the same flow regime is maintained up to high angles of attack,

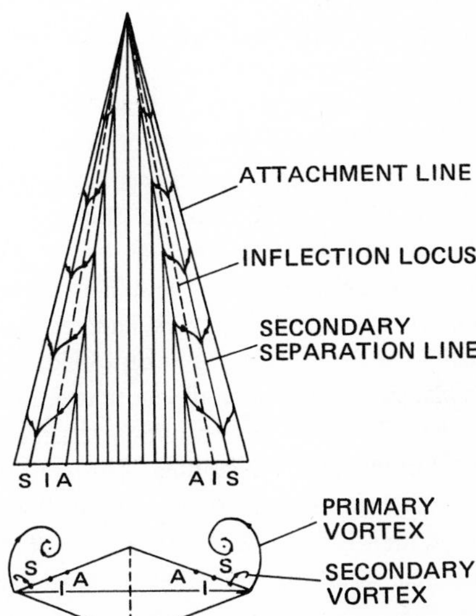

FIG. 1.29 Flow over the upper surface of a slender wing at incidence (*from Hazen, 1967*).

and the vortices extend well downstream of the trailing edge. Underneath these spiral sheets, the resultant velocities are large and create an area of suction on the upper surfaces of the wing. Thus, the generated lift increases more quickly with angle of attack than do the classical linear variations. The Reynolds number effects are less important for slender wings than for swept wings because the position of the primary separation is fixed. But the scale consideration should not be ignored, particularly for practical configurations with fuselages and excresences.

The flow separation at the leading and trailing edges of a wing may be controlled by powered boundary-layer control methods such as suction or blowing of fluid. The amount of fluid actually required to overcome separation is considerably more than predicted by theory. Since it is often difficult to provide fluid in these large quantities, both mechanical and powered systems should be considered.

Above the critical angle of attack at which the stall occurs, other devices (e.g., slats and flaps) can be used to augment the critical angle and maintain attached flow over the upper surface of the wing.

At stall conditions, the wing is unable to produce any additional lift. The stall is not reversible or nonunique. In other words, at a high angle of attack in the stall regime, decreasing the angle of attack does not restore the relation that existed for lift and drag prior to stall. This is indicated (Fig. 1.30) by the higher lift of path DAB when the angle of attack is increasing compared to the lower lift of path BCD when the angle of attack is decreasing (Thwaites, 1960). It seems that this phenomenon can be fully understood only after a detailed study of the interaction of separated and attached flows. Several examples of separation control in aviation practice are now described.

Stall is linked with the so-called "stall-spin" which accounts for the major portion of fatal crashes of light airplanes. At present, it appears that an intentional spin produces no tactical gain. Spin studies and stall studies were pursued simultaneously because the loss of rolling stability is closely related to stall. Analysis of gust statistics disclosed that there is a limiting speed below which the incremental change of angle of attack produced by a severe gust can result in a stall with no pitch. Commercial and military planes should not be intentionally operated at near stalling speed. It is known that the best action to take following a stall is to move the control column forward. This reduces the angle of attack and reclaims the unstalled or orderly flow regime. Stall can presently be controlled sufficiently to achieve stall-free light airplanes. However, stall continues to be a problem for swept-wing aircraft because it is difficult to obtain satisfactory stall characteristics with reasonable lift for takeoff and landing. The main difficulty is caused by the sweepback. The circulation of air about the forward section causes an up-flow and, consequently, the angle of attack of the aft section is increased. Therefore, the rear sections are first to reach the angle for maximum lift and stall.

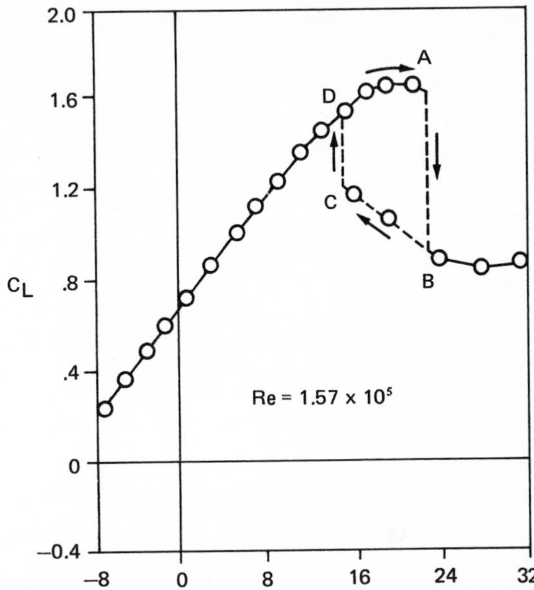

FIG. 1.30 Experimental lift curve for a NACA 103 airfoil (arrows indicate the sequence in which the incidence was varied. The curve was obtained by *Jacobs (1931)* and is given by *Thwaites, 1960*).

Because the tips of a swept wing are rear as well as outboard, the local stall affects both longitudinal and rolling stability (Soulé, 1967).

Shock stall can be more disastrous than the loss of total lift. For example, if shock waves form in front of the wing and decrease the wing lift, a comparable decrease in lift does not necessarily occur on the tail. The result can be a sudden terrific jolt and possibly an unexpected and dizzying dive (von Karman & Edson, 1967).

The phenomenon of stall in compressors and pumps is one of the most important and one of the least understood subjects in fluid machinery. Stalls that occur within the passages are closely coupled throughout the generally small interstage volume inside the machine. The unsteady behavior of the separated flow, the boundary layer, and the acoustic effects can provide interaction between passages. This complex stall phenomenon is governed by the upstream and downstream flow fields. The blade-passage characteristics may be better understood by reviewing the known features of flow separation over an airfoil and on a diffuser wall.

As a compressor or pump is throttled, one or more small stalled cells appear on the upper surface of the blade. The separation then spreads into the interblade region (the passage) which plays the role of a diffuser in the compressor (Fig. 1.31). This cell rotates relative to the passage (or the blade)

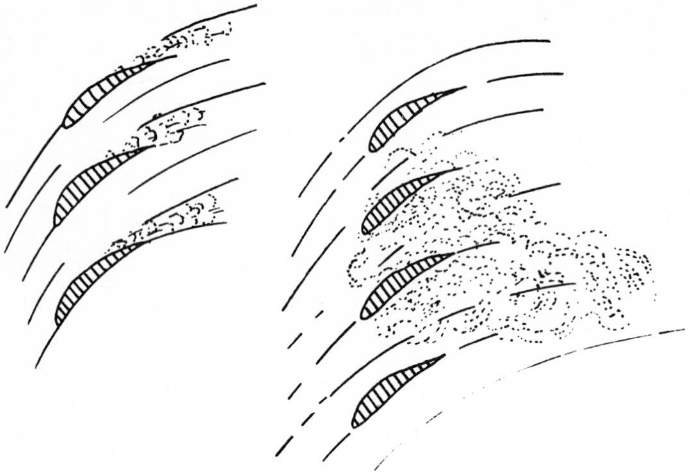

FIG. 1.31 Stall in a passage (*from Emmons, 1960*).

in a direction opposite to the rotor motion. Further throttling causes the successive formation of a number of cells–a dozen or more in some machines–and eventually the flow is unable to introduce more cells. Then when these small stalled cells merge into one large stall area, the flow separates from the leading edge and more or less fills the passage. The circulation around the blade that provides the lift is transferred (more or less) to a lump of fluid. This lump of fluid grows larger, affects the next blade, and begins to move out of the passage and downstream. By then, the preceding blade has begun to recover from the stall. As a blade recovers, it must rebuild its circulation and this requires rejection of a starting vortex (Emmons, 1960).

The control designer is primarily concerned with preventing the compressor from operating near those regions which are believed to be sensitive to stall. From the viewpoint of overall aircraft design, the engines, their compressors, and inlets must be designed to operate satisfactorily under all required conditions and without any stalls that adversely affect aircraft performance (Drabeck, 1960).

1.2 METHODS FOR PREDICTING SEPARATION OF STEADY FLOW

Control of flow separation depends on whether under the given conditions the flow separates around a body or within a flow passage. As previously mentioned, the position of separation is readily known for certain cases such as the trailing edge of a sharp corner where geometry of the body shape changes abruptly, e.g., with a rearward-facing step. The analytical predictions given in this section for the locations of separation deal with two cases of external flow. The first is laminar and turbulent flow along the body surface with continuous adverse pressure gradient at subsonic and supersonic speed, but without any jump conditions. The second is shock wave interaction with the boundary layer and resultant discontinuous adverse pressure gradients. For the first case of continuous adverse pressure gradient, it is necessary to determine the pressure distribution based on inviscid flow theory in order to predict the separation.

In a passage, unsteady pulsating separation and jet flow separation have to be considered. Thus, internal flow separation cannot be predicted by boundary-layer theory alone; the stability of the overall flow pattern must also be considered. However, no satisfactory method is available for the prediction of internal flow separation that involves unsteady pulsating flow and jet flow separation. Accordingly, it is often necessary to resort to steady boundary-layer theory.

An attempt is made here to present semiempirical methods for the determination of flow separation in passages with turbulent flow. Several methods are presented for predicting two-dimensional flow separation and a brief presentation of three-dimensional flow separation is then made. Only a few practically useful methods—all of which are approximate—are selected from among the available analytical and experimental methods of prediction. The reader is referred to the author's monograph (Chang, 1970) for a more detailed and extensive discussion of this complex topic.

1.2.1 Predictions for Two-dimensional Laminar Flow Separation

The minimum adverse pressure gradient is much smaller for laminar flow separation than for turbulent flow separation. Since the laminar boundary layer is more amenable to mathematical treatment, the position of laminar separation can be predicted with greater accuracy.

1.2.1.1 Prediction of steady, laminar, incompressible two-dimensional flow separation

As mentioned previously, the pressure distribution or the velocity distribution along the body surface must be known in order to compute the separation point. The separation point for laminar flow is generally independent of Reynolds number as long as the flow remains laminar. Thus, the separation point is geometrically fixed on the body surface. For example, the laminar incompressible flow separation on a circular cylinder is usually located at the same angular position regardless of its size and the free-stream velocity. This angular position is about 81 degrees, measured from the forward stagnation point. The prediction of the separation point is more accurate if the point of separation is located comparatively near the point of minimum pressure of potential flow.

The Pohlhausen method Pohlhausen (1921) developed an approximate solution to the boundary-layer equations by using the nondimensional form of velocity profile:

$$\frac{u}{u_e} = a\eta + b\eta^2 + c\eta^3 + d\eta^4$$

where $\eta = y/\delta$ and y and δ are distances measured perpendicularly from the wall and the thickness of boundary layer, respectively. A dimensionless parameter λ is defined by $\lambda = (\delta^2/\nu)(du_e/dx)$. The coefficients a, b, c, and d may be determined as a function of λ by using the boundary conditions at the wall and at the edge of boundary layer. When $(\partial u/\partial\eta)_{\eta=0} = 0$ is satisfied, the point of separation is given by the criterion of $\lambda = -12$. Thus computing δ and du_e/dx as a function of the streamwise distance x allows the point of separation to be determined at x location where $\lambda = -12$. As predicted by this simple method, the position of the separation point generally lies farther downstream than it is given by experimental data.

Method based on the Falkner-Skan equation One criterion of incompressible laminar flow separation (noted by Evans, 1968) uses the Falkner-Skan equation. The criterion of separation is given by $\beta = -0.198838$, where the pressure gradient parameter β is defined by $du_e/dx = Cu_e^{2(\beta-1)/\beta}$ and where C as well as β are constants.

The streamwise coordinate x is measured from the start of the boundary layer. At the position where $\beta = -0.198838$, the wall shear becomes zero because the dimensionless wall shear expressed by

$$f''_{\text{wall}} = \left[\frac{d}{d\eta}\left(\frac{u}{u_e}\right)\right]_{\text{wall}}$$

vanishes. The dimensionless independent coordinate η in a direction perpendicular to wall is defined by

$$\eta = y\left(\frac{1}{\nu\beta}\frac{du_e}{dx}\right)^{1/2}$$

where y is the physical distance perpendicular to the wall and is measured from the wall.

The Shvets method The Shvets method (Shvets, 1949) is developed from a second-order approximation for the solution of the boundary layer equations. The transformed boundary layer equation with the pressure gradient is

$$\nu\frac{\partial^2 u}{\partial y^2} = u\frac{\partial u}{\partial x} - \frac{\partial u}{\partial y}\int_0^y\left(\frac{\partial u}{\partial x}\right)dy - u_e\frac{du_e}{dx}$$

The boundary conditions are: at $y = 0$, $u = 0$ and $y = \delta$, $u = u_e$. The solution of this equation to a second-order approximation gives

$$\nu \frac{u}{u_e} = \frac{\delta^2}{24}\frac{du_e}{dx}(\eta^4 - 12\eta^2 + 11\eta) - \frac{u_e}{24}\frac{d\delta}{dx}\delta(\eta^4 - \eta) + \eta$$

where $\eta = y/\delta$.

At the point of separation where $\partial u/\partial\eta|_{\eta=0} = 0$, it reduces to

$$\nu\left(\frac{3}{8}\frac{du_e}{dx}\delta^2 + \frac{u_e}{8}\frac{d\delta}{dx}\delta\right) = 1$$

By assuming that the boundary layer thickness is zero at $x = 0$ and solving for δ^2,

$$\frac{1}{\nu}\delta^2 = \left(\frac{16}{u_e^6}\right)\int_0^x u_e^5\,dx$$

Evaluating the shear stress at the wall,

$$\tau_{\text{wall}} = \frac{l}{\sqrt{u}}\frac{\partial u}{\partial y}\bigg|_{\text{wall}} = \frac{u_e^4}{3}\left(\int_0^x u_e^5\,dx\right)^{-1/2}\left[1 + \frac{4(du_e/dx)}{u_e^6}\int_0^x u_e^5\,dx\right]$$

Since at separation $\partial u/\partial y|_{\text{wall}} = 0$, the criterion of separation is

$$\frac{4(du_e/dx)}{u_e^6}\int_0^x u_e^5\,dx = -1$$

The point of laminar separation is thus determined by evaluating the potential flow velocity relations only at the streamwise downstream distance x, although the separation of the flow is governed by the viscous flow phenomenon.

The Stratford method Stratford (1957) solved the outer and inner boundary layer equations. In the outer layer, the effects of a pressure gradient (in the absence of viscosity) can be considered separately from the effects of viscous forces (in the absence of pressure gradient). In the inner sublayer, the pressure gradient balances the viscous forces on the velocity profile. Therefore, a transition across the layer causes the pressure forces to be balanced by viscous forces at the wall and by inertia forces at the outer edge of the boundary layer.

Stratford originally obtained a mathematical expression for the condition of separation based on the Blasius solution that covered all possible values of the separation distance. Curle and Skan (1957) modified this Stratford prediction of the laminar separation by comparing it with the rigorous solution of Görtler (1957). The modified criterion of separation is

$$\left[\bar{x}^2 C_p \left(\frac{dC_p}{d\bar{x}}\right)^2\right]_s = 0.0104$$

$$\bar{x} = x_{\text{eq}} + x_s$$

The symbol x_{eq} refers to the equivalent length upstream of the maximum potential velocity given by

$$x_{\text{eq}} = \int_0^x \left(\frac{u_e}{u_{e_{\max}}}\right)^{8.17} dx$$

and x_s refers to the distance measured from the maximum potential velocity point to laminar separation. The pressure coefficient C_p is given by

$$C_p = \frac{p - p_{\min}}{\frac{1}{2}\rho u_{e_{\max}}^2}$$

This method is relatively simple to apply since, like the Shvet method, it is based on the properties of inviscid flow and does not involve the characteristics of viscous flow.

The method of Thwaites and Curle and Skan If in addition to the inviscid external velocity distribution the momentum thickness distribution is known, then the Thwaites method (1949), as modified by Curle and Skan (1957), is a useful means whereby the viscous and inviscid characteristics can be used to predict laminar flow separation. In a new approach Thwaites used the momentum thickness of the boundary layer as the principal dependent variable and established a relationship between

$$\left.\frac{\partial u}{\partial y}\right|_{y=0} \quad \text{and} \quad \left.\frac{\partial^2 u}{\partial y^2}\right|_{y=0}$$

to solve the boundary layer differential equations. With

$$\left.\frac{\partial u}{\partial y}\right|_{y=0} = \frac{u_e l}{\theta}$$

$$\left.\frac{\partial^2 u}{\partial y^2}\right|_{y=0} = \frac{u_e m}{\theta^2}$$

it is possible to form a relationship between l and m by examining the various known solutions of the boundary layer equations, such as those by Pohlhausen (1921), Falkner and Skan (1930), Howarth (1938), Hartree (1939), Blasius (1908), Iglisch (1944), and Schlichting and Bussmann (1943).

Integration of the momentum integral equation

$$\frac{d\theta}{dx} = -(H+2)\frac{du_e}{dx}\frac{\theta}{u_e} + \frac{\nu}{u_e^2}\left(\frac{\partial u}{\partial y}\right)\bigg|_{y=0}$$

leads to the Thwaites original solution

$$\theta^2 = 0.45\, u_e^{-6}\nu \int_0^x u_e^5\, dx$$

with

$$m = -\frac{\theta^2}{\nu}\frac{du_e}{dx} \quad \text{and} \quad \frac{\partial u}{\partial y}\bigg|_{y=0} = \frac{u_e l(m)}{\theta}$$

The separation criterion is found to be $m = 0.082$, $l(m) = l(0.082) = 0$, and $H(0.082) = 3.7$. The modified criterion by Curle and Skan (1957) gives $(\theta^2/\nu)(du_e/dx) = -0.09$.

Evaluation of the various prediction methods To compare the predictions given by the different methods, results were computed for a group of similar symmetrical struts of different sizes and at various angles of attack. The results obtained by the methods of Shvets, Stratford, and that of Thwaites as modified by Curle and Skan were then compared with experimental data (Chang & Dunham, 1961). Although the scope of the comparison was limited, a certain trend of prediction by the various analyses was evident.

The geometrical configuration of the symmetrical strut was a thin, two-dimensional biconvex given by $Y = \pm b[1-(X/L)^2]$. The coordinate system is shown in Fig. 1.32. The following potential flow velocity distribution (Fig. 1.33) around the biconvex strut placed at angle of attack α was used to compute the separation point:

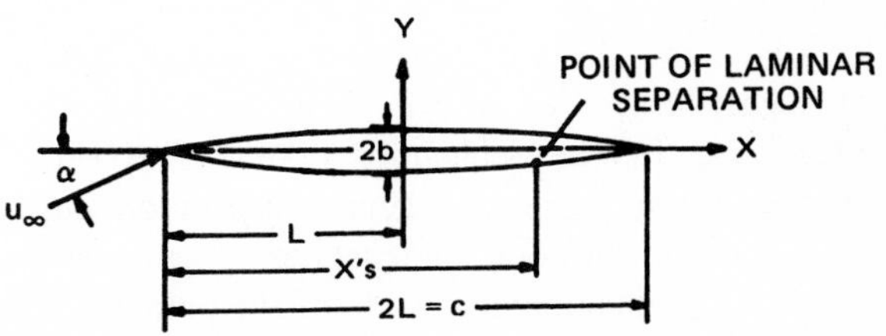

FIG. 1.32 Symmetrical strut (*from Chang & Dunham, 1961*).

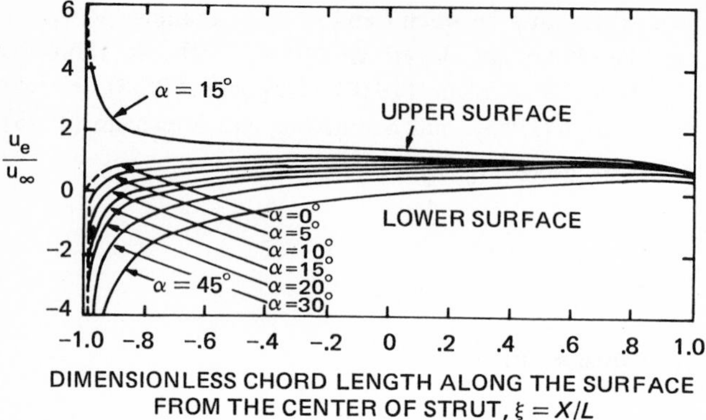

FIG. 1.33 Potential velocity distribution on a symmetrical strut (*from Chang & Dunham, 1961*).

$$\frac{u_e}{u_\infty} = 1 + \left(\frac{2}{\pi}\right)\beta\,[2 - \xi \ln F(\xi)]$$
$$+ \beta^2 \left\{\frac{3}{\pi^2}\,[2 - \xi \ln F(\xi)]^2 - \frac{1}{\pi^2} \ln^2 F(\xi) - (1 - \xi)^2\right\} \pm \alpha F(\xi)^{-1/2}$$
$$\mp \frac{\alpha}{\pi} \beta F(\xi)^{-1/2}\,[(1 + 2\xi) \ln F(\xi) - 4] - \frac{1}{2}\alpha^2$$

where $\xi = X/L$, $\beta = b/L$, and $F(\xi) = (1 + \xi)/(1 - \xi)$.

In this equation, the upper sign indicates the potential flow distribution on the upper surface and the lower sign that on the lower surface. At small positive angles of attack, the point of separation moves toward the leading edge of the upper surface and the flow over the upper surface becomes turbulent. But the flow remains laminar on the lower surface and measurements of the separation point were carried out by the reliable dust method. The details of this experimental technique are described later in Chapter 2.

The experimental and analytical results for the laminar separation points are given in Table 1.1 and Fig. 1.34 for a 9.42 percent thickness ratio.

These numerical values indicate that the Shvets method gave results that were closest to the experimentally determined separation point at zero angle of attack and that the agreement with this point improved for all the analyses at higher angles of attack. The discrepancy in these analytical results is explainable by the Schlichting (1968) analysis of the position of separation. As mentioned previously, laminar separation can be calculated only approximately when the point of separation is situated comparatively far behind the point of the minimum pressure of the potential flow. At zero angle of attack, the position of this minimum pressure on the thin biconvex strut is located at the midchord. As the angle of attack increases, the minimum pressure position

TABLE 1.1 Analytical and experimental determination of laminar separation points on the lower surface of a biconvex strut

Angle of attack (deg)	Distance from the leading edge to the point of laminar separation per chord length (X'_s/c)			
	Analysis			Experiment
	Thwaites method	Shvets method	Stratford method	Dust method
0	0.7037	0.7770	0.8346	0.7810
5	0.7910	0.9070	0.8750	0.7790
10	0.8400	0.9225	0.9050	0.8700
15	0.9300	0.9450	0.9425	0.8850
20	0.9510	0.9550	0.9550	0.9310
30	0.9540	0.9590	0.9580	0.9570
45	0.9730	0.9730	0.9540	0.9710

Note: Chord length 9.6 in., thickness ratio 9.42%. The experimental data are from Chang and Dunham (1961).

and the laminar separation point both shift toward the trailing edge. However, the position of minimum pressure is displaced more than the separation point.

Hence, at lower angles of attack, the distance between the positions of the minimum pressure and separation point is comparatively larger than at higher angles of attack. Thus, the analytical results can be expected to yield only the

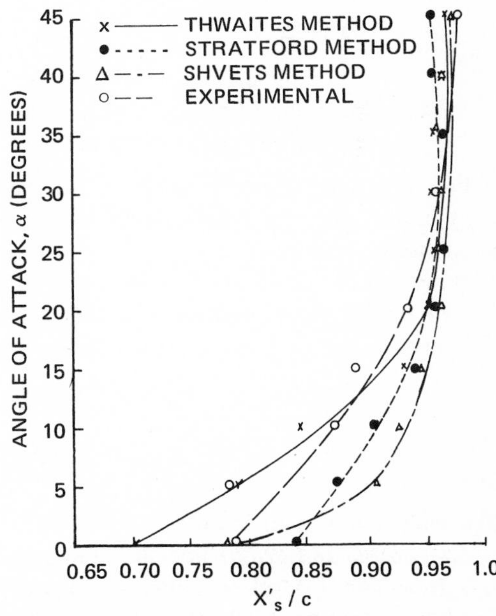

FIG. 1.34 Laminar separation point on the lower surface of a symmetric strut as a function of angle of attack (*from Chang & Dunham, 1961*).

approximate location of the laminar separation at smaller angles of attack.

Furthermore, values for the pressure coefficient at separation

$$C_{ps} = \frac{p_s - p_{\min}}{\frac{1}{2}\rho u_{e_{\max}}^2} = 1 - \left(\frac{u_{e_s}}{u_{e_{\max}}}\right)^2$$

were obtained by various approximate methods. These are given in Table 1.2 and compared with the accurate values obtained by numerical integration of the boundary layer equations.

The methods of Tani (1949) and Truckenbrodt (1952a, 1952b), which are based on the kinetic energy integral equation, are presented later in this chapter and in Chapter 5, respectively. As seen from Tables 1.2 and 1.3 (given later in the chapter), any of the methods noted may be used to predict the separation point for two-dimensional, incompressible, laminar flow. Among these methods, the predictions of Thwaites, Stratford, Timman, and Tani are satisfactory. Selection of the proper method for a particular application depends on such factors as the accuracy required, the ease of the computation, etc. For example, since the Stratford and Shvets methods are based on potential flow and do not involve boundary-layer behavior, they have the advantage of simplicity. But if information is available on boundary layer growth, then other methods (such as that of Thwaites) may also be used for a relatively easy and accurate analytical determination of the separation point.

1.2.1.2 Prediction of steady laminar compressible two-dimensional flow separation

For the same pressure gradient, an earlier separation is expected for compressible than for incompressible flows because compressibility increases the pressure gradient.

Howarth (1948) showed that the momentum equation of compressible flow derived by his transformation contains the term $u_e(du_e/dx)$, which may be evaluated for the Bernoulli equation and multiplied by a positive factor

$$\left[1 + \frac{\gamma - 1}{2}\frac{u_e^2}{a_e^2}\right]$$

This term results from compressibility.

The Loftin and Wilson method Loftin and Wilson (1953) extended the von Doenhoff (1938) simplified solution for the incompressible laminar separation to compressible flows by using a set of the Stewartson transformations (1949) which express the compressible laminar boundary layer equations

TABLE 1.2 Pressure coefficients C_{ps} at laminar separation of two-dimensional incompressible flow

Value given by approximation methods†	Case number*						
	1	2	3	4	5	6	7
A	–	0.305	0.192	0.118	0.067	–	–
B	0.072	0.230	0.140	0.080	0.043	–	–
C	0.049	0.204	0.119	0.067	0.035	–	–
D	0.063	0.217	0.133	0.074	0.040	0.041	0.046
E	0.069	0.231	0.138	0.080	0.047	0.047	0.051
F	–	0.225	0.142	0.088	0.057	–	–
G	–	0.215	–	–	–	–	–
H	–	0.236	0.141	0.080	0.044	0.066	–
I	0.069	0.194	–	–	–	–	–
J	0.067	0.194	0.135	0.071	0.051	–	–
K	0.074	0.204	0.125	0.078	0.050	0.048	–
L	0.088	0.227	0.141	0.088	0.055	0.062	0.068
Accurate value‡	0.082	0.226	0.142	0.089	0.056	0.056	0.062

Source: Rosenhead (1963).

*The following mainstream flows are classified by Cases 1–7:

Case	*Mainstream flow*
1	Hartree-Schubauer, ellipse Hartree (1939)
2	$u_e = u_0 [1 - (x/l)]$, Howarth (1938a)
3	$u_e = u_0 [1 - (x/l)^2]$, Tani (1949)
4	$u_e = u_0 [1 - (x/l)^4]$, Tani (1949)
5	$u_e = u_0 [1 - (x/l)^8]$, Tani (1949)
6	$u_e = (3\sqrt{3}/2)u_0 [(x/l) - (x/l)^3]$, Curle (1958)
7	$u_e = u_0 \sin(x/l)$, Terrill (1960)

Here u_0 and l are appropriate velocity and length and x refers to the distance along the surface.

†The methods indicated above as A–L are:

Symbol	*Identification*
A	Pohlhausen (1921)
B	Timman (1949)
C	Walz (1941)
D	Thwaites (1949)
E	Curle and Skan (1957) (Thwaites modification)
F	Tani (1954)
G	Truckenbrodt (1952a) (1952b)
H	Loitsianski (1949)
I	von Karman and Millikan (1934)
J	von Doenhoff (1938)
K	Stratford (1957) $[\bar{x}^2 C_p (dC_p/d\bar{x})^2]_s = 0.0076$ (1957)
L	Curle and Skan (Stratford modification) (1957) and Curle (1960b) $[\bar{x}^2 C_p (dC_p/d\bar{x})^2]_s = 0.0104$

‡The accurate value included for purposes of comparison with the results of the approximation methods was obtained by numerical integration of the boundary layer equations.

in terms of an equivalent incompressible laminar flow. The von Doenhoff criterion is given by the potential velocity decrement of incompressible flow before laminar separation, i.e., $\Delta u_e/u_{e\,\max}$ given in function F, defined by

$$F = \frac{x_t}{u_{e\,\max}} \frac{du_e}{dx}$$

as shown in Fig. 1.35. The quantity x_t is defined by ($x_t = x_{\mathrm{eq}} + x'$) where x_{eq} is the equivalent length mentioned in the discussion of the Stratford Method in Sec. 1.2.1.1, and x' is the distance (as shown by an example in Fig. 1.36) between the point of the maximum potential velocity and the initial point of the straight line approximation of the potential velocity variation along x downstream of the maximum velocity. The point indicated by p in Fig. 1.36 is the location near the point of maximum potential velocity where the measurement of the boundary layer is available.

For Pr = 1, the Stewartson transformations are

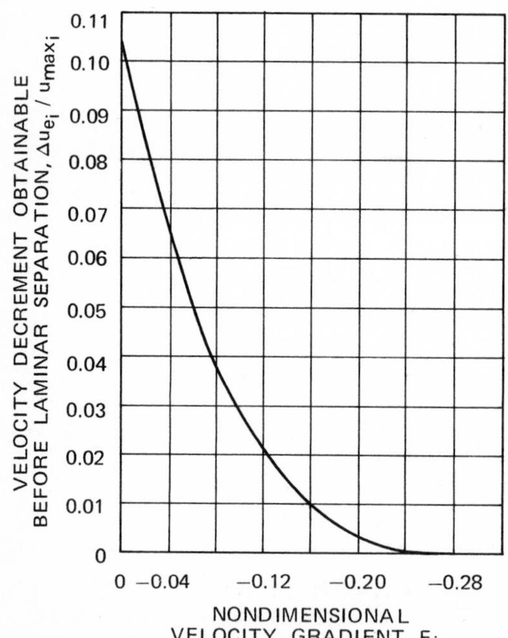

FIG. 1.35 Incompressible velocity decrement $\Delta u_{ei}/u_{\max i}$ obtainable before occurrence of laminar separation expressed as a function of the velocity gradient F_i (*from Loftin & Wilson, 1953; also von Doenhoff, 1938*).

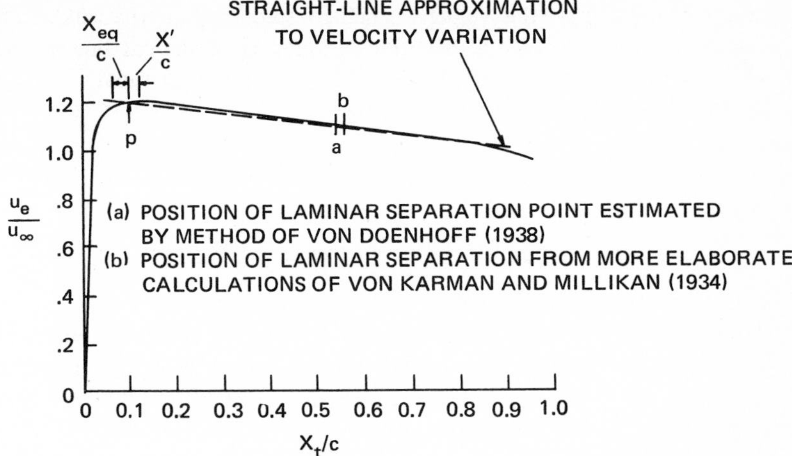

FIG. 1.36 Two methods for estimating the position of laminar separation point on NACA 0012 airfoil (von Karman-Millikan data are from *von Doenhoff, 1938*).

$$y_i = \frac{a_e}{a_{\max}\sqrt{\nu_{\max}}} \int_0^{y_c} \frac{\rho}{\rho_{\max}}\, dy_c$$

$$x_i = \int_0^{x_c} \left(\frac{a_e}{a_{\max}}\right)^{(3\gamma-1)/(\gamma-1)} dx_c$$

and

$$u_{e_i} = \frac{a_{\max}}{a_e} u_{e_c}$$

Here subscripts *i, c,* and max refer to incompressible, compressible, and maximum potential velocity, respectively.

By using the energy equation and the equivalent length in addition to these transformations, the criterion for laminar compressible flow separation is given as

$$F_i = \left(\frac{x_{\text{eq}}}{u_{\max}}\right)_i \left(\frac{du_e}{dx}\right)_i$$

$$= \left(1 + \frac{\gamma - 1}{2} M^2_{\max}\right) \left[\frac{x_{\text{eq}_i}}{u_{\max\, c}} \left(\frac{du_e}{dx}\right)_c\right]$$

For a given velocity distribution at a given Mach number, the laminar compressible flow separation is determined by the following procedure:

1. Evaluate graphically the velocity gradient $(du_e/dx)_c$ in the compressible plane. The given velocity distribution is faired by a straight line in such a way that the adverse velocity gradient begins at a distinct point.
2. Calculate the equivalent length by using

$$x_{eq_i} = \int_0^{x_{max_c}} \left(\frac{u_{e_c}}{u_{max_c}}\right)^{8.17} dx_c$$

3. Determine the factor of $1 + (\gamma - 1)/2\ M_{max}^2$ from Fig. 1.37.
4. Determine the nondimensional velocity gradient in the equivalent incompressible plane by using the equation for F_i which was given previously. From Fig. 1.35, find the velocity decrement which corresponds to laminar separation in the incompressible plane.

This approximate method of prediction rapidly yields results. It has been shown that for a linearly decreasing velocity distribution from the leading edge, the position of laminar compressible flow separation predicted by this method is in substantial agreement with results obtained by using the analyses of Stewartson (1949) and Howarth (1948).

The Morduchow and Grape method Morduchow and Grape (1955) extended the Morduchow and Clarke method (1952) to include heat transfer. Limitations of their method include uniform wall temperature, Prandtl number of unity, and linear variation of the viscosity coefficient with temperature. This method is based on the integral approach. The prediction of separation is entirely self-contained and does not rely on known exact solutions for either incompressible or compressible flow. Furthermore, this method is relatively simple to apply and provides a flexibility that permits a

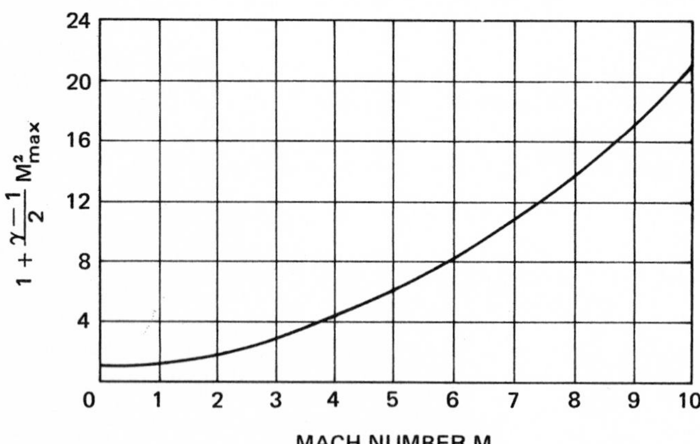

FIG. 1.37 The quantity $1 + (\gamma - 1)/2\ M_{max}^2$ as a function of Mach number (*from Loftin & Wilson, 1953*).

larger class of problems to be treated. This is possible because of the introduction of two parameters which lead to doubly infinite families of characteristics instead of to the singly infinite family of characteristics yielded by the conventional Pohlhausen method.

By introducing a variable t, defined by $y = \int_0^t (T/T_e)\, dt$, the boundary layer thickness in the $x - t$ plane is given by $\delta_t = \int_0^\delta (T_e/T)\, dy$. In order to satisfy the boundary conditions $u = v = 0$ at $t = 0$ and criterion of separation $(\partial^4 u/\partial t^4) = 0$, the following seventh-degree velocity and stagnation enthalpy profiles are introduced:

$$\frac{u}{u_e} = \left(\frac{7\tau}{4} - \frac{21\tau^5}{4} + 7\tau^6 - \frac{5}{2}\tau^7\right) + a_2\left(-\frac{1}{2}\tau + \tau^2 - \frac{5}{2}\tau^5 + 3\tau^6 - \tau^7\right)$$
$$+ a_3\left(-\frac{1}{5}\tau + \tau^3 - 3\tau^5 + \frac{16}{5}\tau^6 - \tau^7\right) \quad (1.1a)$$

$$\frac{H}{H_e} = G_1 + (1 - G_1)(35\tau^4 - 84\tau^5 + 70\tau^6 - 20\tau^7)$$
$$+ b_1(\tau - 20\tau^4 + 45\tau^5 - 36\tau^6 + 10\tau^7) \quad (1.1b)$$

where $\tau = t/\delta_t$

$H =$ stagnation enthalpy given by $H = u^2/2 + c_p T$

$G_1 = H_w/H_e$

$a_2 = -C/2(T_e/T_\infty)^{(2-\gamma/\gamma-1)}[(du_e/d\xi)/u_\infty]\, G_1 [1 + (\gamma - 1)/2\, M_e^2]\lambda$

$\xi = x/L$ (where L is a characteristic length)

$C = (\mu/\mu_\infty)/(T/T_\infty)$

$a_3 = (a_2/b_1)/3G_1$

$\lambda = \mathrm{Re}_\infty(\delta_t/L)^2$

Here b_1 is a coefficient of τ in Eq. (1.1b) and generally $\approx 2(1 - G_1)$ except for extreme cooling where G_1 is close to zero.

At separation

$$a_{2_s} = \frac{3.5 G_1}{G_1 + (2/15) b_1}$$

$$\lambda_{\mathrm{sep}} = -105C \frac{(T_e/T_\infty)^{-(2-\gamma/\gamma-1)}}{[d(u_e/u_\infty)]/d\xi\{1 + [(\gamma - 1)/2]M_e^2\}} \frac{1}{11G_1 + 4} \quad (1.2)$$

On the other hand, the momentum integral equation is reduced to a function of $\overline{F}_1 = f(a_2, b_1, G_1)$ by introducing the seventh degree velocity profile (Eq. [1.1a]) with the assumption that a_2 and b_1 are constants. Then, from this transformed momentum integral equation in a function of $\overline{F}$, the separation criterion is given by

$$\lambda_s = \frac{7C}{2\bar{F}_{1s}} \frac{\int_0^{\xi} (u_e/u_\infty)^{(2/\bar{F}_{1s})\phi_{1s}-1}(T_e/T_\infty)^{\{[(2\gamma-1)/(\gamma-1)]-(\phi_{1s}/\bar{F}_{1s})\}}\, d\xi}{(u_e/u_\infty)^{(2/\bar{F}_{1s})\phi_{1s}}(T_e/T_\infty)^{\{[(\gamma+1)/(\gamma-1)]-(\phi_{1s}/\bar{F}_{1s})\}}} \tag{1.3}$$

where
$$\bar{F}_{1s} = 0.1159 + 0.002525\, a_{2s} - 0.001454\, a_{2s}^2 - 0.0000572\, (b_1 a_{2s}/G_1)^2 - 0.000574\, (b_1 a_{2s}^2/G_1) + 0.000887\, (b_1 a_{2s}/G_1)$$
$$\phi_{1s} = 0.25\, G_1 + 0.0437 + 0.0738\, b_1 + 0.0348\, a_{2s} - 0.00291\, a_{2s}^2 + 0.00773\, (b_1 a_{2s}/G_1) - 0.001147\, (b_1 a_{2s}^2/G_1) - 0.0001145\, (b_1 a_{2s}/G_1)^2$$

Subscript s refers to separation. $\bar{F}_{1s}$ and ϕ_{1s} are functions of G_1 as shown in Fig. 1.38.

For a given adverse pressure gradient, the separation point for any given M_∞ and uniform temperature ratio of G_1 can now be located at the position at which the right-hand side of Eq. (1.2) and (1.3) are equal. Therefore, in order to determine the point of intersection of two curves drawn from Eq. (1.2) and (1.3), λ is plotted as a function of ϕ in the anticipated vicinity of

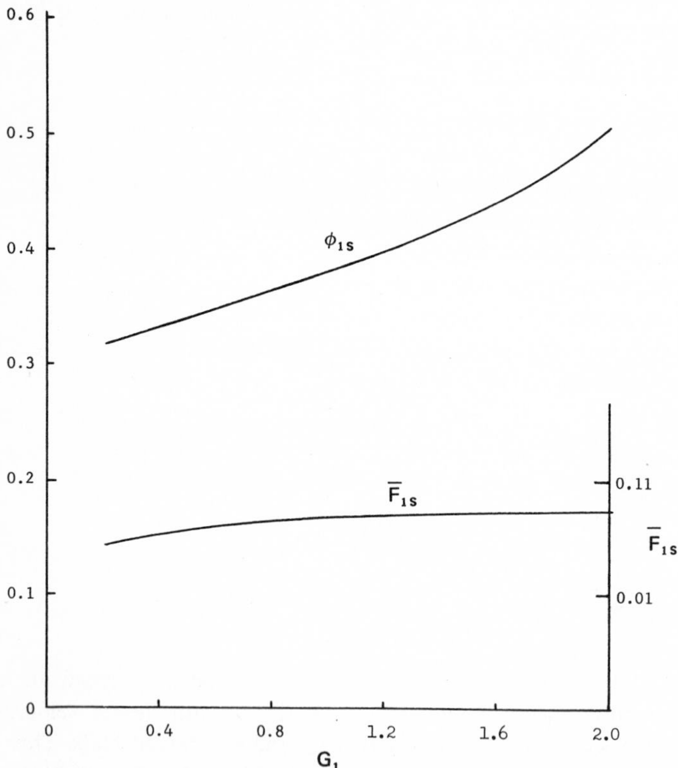

FIG. 1.38 $\bar{F}_{1s}$ and ϕ_{1s} as functions of G_1 (*from Morduchow & Grape, 1955*).

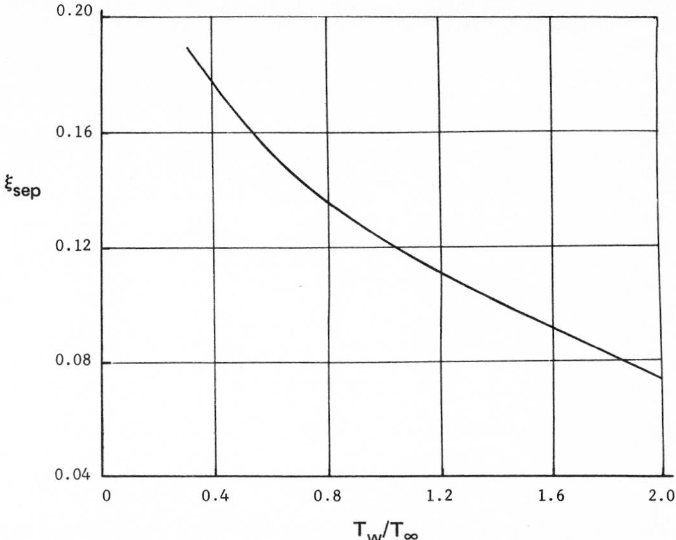

FIG. 1.39 Separation point as a function of wall temperature in incompressible flow ($u_e/u_\infty = 1 - \xi$ and $M_\infty = 0$) *(from Morduchow & Grape, 1955)*

separation by using the aforementioned equations. Since the point of separation is independent of C, it may be taken as unity. The effect of wall temperature on separation is felt from the momentum exchange or the pressure gradient and G_1. For a fixed potential velocity distribution and M_∞, Eq. (1.2) indicates that the λ_{sep} increases as G_1 becomes smaller. That is, when the wall is cooled, both G_1 and λ_{sep} are affected and the separation point moves downstream. This delays the separation because the influence of the pressure gradient is reduced.

Consider the case of a linearly decreasing velocity along a surface $u_e/u_\infty = 1 - \xi$. If the numerical example based on this formula is applied to a simple type of adverse pressure gradient, it may serve to illustrate the effect of heat transfer on laminar separation. For compressible flow, by taking $M_\infty = 0$ and $T_e/T_w = 1$ and by equating Eq. (1.2) and (1.3),

$$\xi_{sep} = 1 - \left(1 + \frac{60\phi_{1s}}{11G_1 + 4}\right)^{(-\bar{F}_{1s}/2\phi_{1s})}$$

In Fig. 1.39, ξ_{sep} is plotted as a function of T_w/T_∞ in a range of $0.3 < T_w/T_\infty < 2.0$ to indicate the delay of separation due to wall cooling.

Now consider compressible flow with $M_\infty \leqslant 5.31$, a fixed value of $T_w/T_\infty = 2$, and $0.3 \leqslant G_1 \leqslant 2$. For $M_\infty = 0$, the wall is heated at $G_1 = 2$, but for $M_\infty = 5.31$, the wall is cooled at $G_1 = 0.3$.

By using the relation of

$$\frac{T_e}{T_\infty} = 1 + \frac{\gamma - 1}{2} M_\infty^2 \left[1 - \left(\frac{u_e}{u_\infty}\right)^2\right]$$

and

$$M_e^2 = \left(\frac{u_e}{u_\infty}\right)^2 M_\infty^2 \left(\frac{T_e}{T_\infty}\right)^{-1}$$

we obtain

$$\lambda_{sep} = -105C \frac{(T_e/T_\infty)^{-0.5}}{[4 + 11(T_w/T_\infty) + 0.8M_\infty^2]\,[d(u_e/u_\infty)]/d\xi}$$

When λ_{sep} is evaluated for a given M_∞ as a function of ξ, the position of separation is determined by a value of ξ at which $\xi = \xi_{sep}$. This value is obtained numerically integrating Eq. (1.3) or by using Fig. 1.39. For the simple velocity distribution given by

$$\frac{u_e}{u_\infty} = 1 - \xi \quad \text{with} \quad \frac{T_w}{T_\infty} = 2$$

ξ_{sep} is plotted in Fig. 1.40 as a function of M_∞ and is compared to the case of zero heat transfer. Note the favorable effect of M_∞ for the case of zero heat transfer.

The accuracy of the separation prediction by the Morduchow and Grape method (which involves the isothermal wall) may not be determined at present for zero heat transfer, $Pr = 1$, $\gamma = 1.4$, and $\mu \sim T$.

However, by using the form

$$\lambda_s(\xi) = \frac{32.8 \int_0^\xi (u_e/u_\infty)^{6.13} (T_e/T_\infty)^{0.937} d\xi}{G(u_e/u_\infty)^{7.13} (T_e/T_\infty)^{2.44}}$$

where

$$G = \left(\frac{\rho_b}{\rho_\infty}\right)\left(\frac{T_b}{T_\infty}\right)^{-1/(\gamma-1)}$$

(subscript b refers to the condition immediately behind the shock wave) or

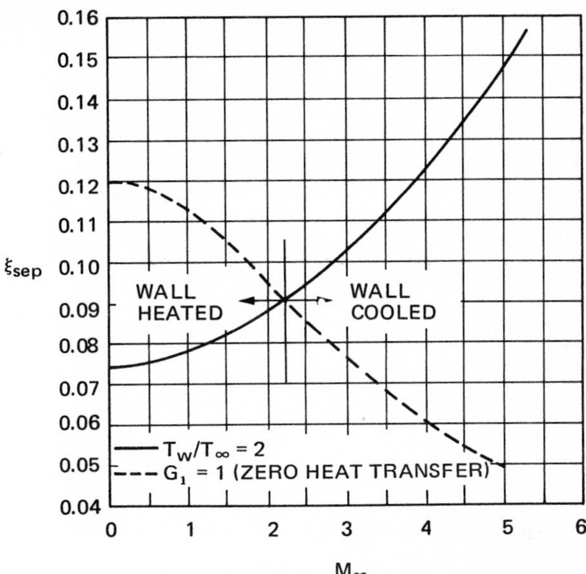

FIG. 1.40 Separation point as a function of Mach number ($u_e/u_\infty = 1 - \xi$) *(from Morduchow & Grape, 1955).*

$$\lambda_s(\xi) = -\frac{7}{G}\left(\frac{u_\infty}{du_e/d\xi}\right)\frac{(T_e/T_\infty)^{-1/2}}{1 + [(\gamma - 1)/2]M_\infty^2} \tag{1.4}$$

The Morduchow and Clarke (1952) prediction compares favorably with other available data for particular velocity distributions $u_e/u_\infty = 1 - \xi^n$ ($n > 0$) as shown in Tables 1.3 and 1.4 (Morduchow & Reyle, 1962).

1.2.1.3 Prediction of laminar flow separation caused by the interaction of shock waves with the boundary layer

Although the physical phenomenon of shock wave interaction with the boundary layer is reasonably well understood, as yet there is no completely satisfactory analysis method for predicting the position of separation and the critical pressure rise required to cause the separation. Determination of the critical pressure rise is simplified by introducing a convenient definition of

TABLE 1.3 Separation point for $u_e/u_\infty = 1 - \xi$, $\mu \sim T$, Pr = 1, and zero heat transfer

M_∞	0	1	3	6	10
ξ_s [Eq. (1.4)]	0.122	0.113	0.077	0.042	0.023
ξ_s (Stewartson, 1949)	0.120	0.110	0.077	0.044	0.024
ξ_s (Wrage, 1960)	0.125	0.112	0.072	0.037	0.021

TABLE 1.4 Separation point for $u_e/u_\infty = 1 - \xi^n$, $M_\infty = 0$

n	1	2	3	4	5	6
ξ_s [Eq. (1.4)]	0.122	0.268	0.375	0.452	0.510	0.625
ξ_s (Tani, 1949)	0.120	0.271	–	0.462	–	0.641
ξ_s (Görtler & Witting, 1957)	0.125	0.290	0.409	0.485	0.552	–

free interaction for fully developed separated flow. If it is assumed that the details in the separated region are unaffected by downstream disturbance, then the interaction of the separated flow with the shock wave can be defined as free interaction. In this case, the compression is directly responsible for the thickening boundary layer, and its ultimate separation is generated by the outward deflection of the external flow caused by the thickening itself. The separation characteristics are determined by flow conditions upstream from the interaction. For the laminar free interaction, the critical pressure rise required for separation is given by the Gadd (1957) analysis as

$$C_{p_s} = \frac{1.13}{\sqrt{\beta_0}\,\mathrm{Re}^{1/2}}$$

where $\beta_0 = \sqrt{M_0{}^2 - 1}$ and M_0 is a Mach number upstream from the interaction. On the other hand, Hakkinen and associates (1959) formulated the critical pressure rise for separation as

$$C_{p_s} = \sqrt{\frac{2c_{f_0}}{\beta_0}}$$

where c_{f_0} is the skin friction coefficient immediately upstream from the interaction. The validity of the above equation is confirmed by the theories of Crocco and Lees (1952); see also Bray et al. (1960) and Curle (1960a). These two formulas predict C_{p_s} not only for flow over a flat plate but also for compression corners and steps. Thus the problem of the shock wave interaction with the boundary layer is reduced to that of an abrupt rise of pressure, regardless of surface configuration. This is confirmed by Erdos and Pallone (1961); see Fig. 1.41.

The flow pattern shown in Fig. 1.42 was observed by Hakkinen et al. (1959) for flow over a flat plate with an inclined incident shock. Downstream from the point of the shock impingement, the boundary layer is thinner because the external streamline and the body surface converge. Therefore, a significant pressure over a rise can be supported over a small area of separation and without causing the main flow to separate.

In the vicinity of the shock, an almost constant pressure gradient on the surface is maintained by the external shock configuration. However, the

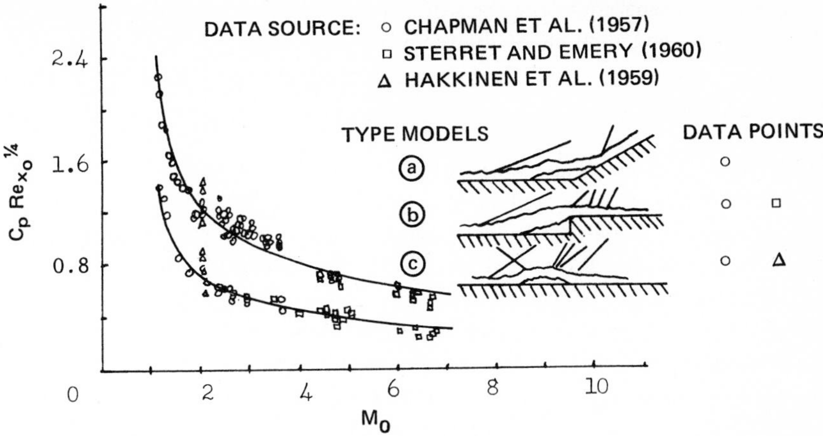

FIG. 1.41 Separation point and plateau pressure for insulated laminar flow (*from Erdos & Pallone, 1961*).

self-induced gradient would be negative because the boundary layer edge becomes convex. The local effect of the shock system cannot extend very far from the impingement point since (as confirmed by the experiment) the pressure jump is progressively cancelled by the expansion when the surface is approached. Thus the pressure rise is due primarily both to the self-induced

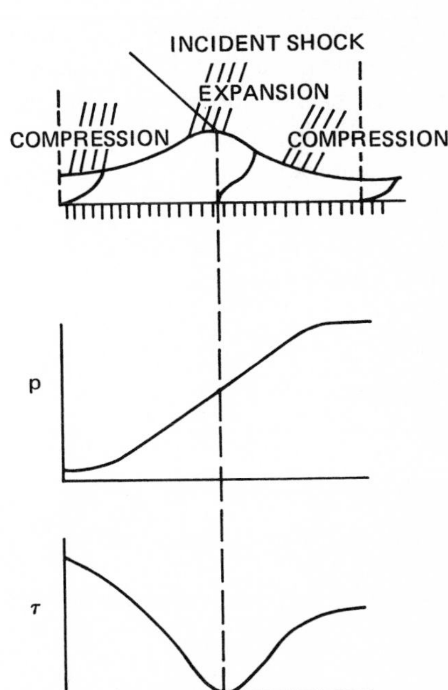

FIG. 1.42 Flow pattern at incident separation (*from Hakkinen et al., 1959*).

effect and to separation upstream of the shock. If symmetry is assumed for the thickening and thinning of the boundary layer, then

$$C_{p_{\text{incipient}}} = 2C_{p_s} \tag{1.5}$$

This is confirmed by the experiments of Hakkinen et al. (1959).

For a curved surface and laminar flow, Eq. (1.5) may be corrected by taking account of the particular feature of the curved wall. Greber (1959) used a simple model with a weak shock (Fig. 1.43) to study this case.

The pressure rise associated with the incident shock and its inviscid flow deflection is that appropriate for the shock deflection angle 2ϕ, the actual flow deflection effected by the shock and the expansion at A. The pressure rise imposed on the boundary layer is less than this value by the amount of flow deflection equal to the change of surface slope θ. Since this is the pressure rise imposed on the boundary layer that would correspond to $C_{p_{\text{incipient}}}$ of Eq. (1.5) for a curved surface, the shock strength is greater than the case given by Eq. (1.5). Greber (1959) showed that the shock strength necessary to cause incipient flow separation increases with the curvature on a convex surface, although the magnitude of the pressure rise at the surface decreases with curvature.

Erdos and Pallone (1962) formulated the separation pressure coefficient C_{p_s} and plateau pressure coefficient C_{pp} for the physical model of free interaction shown in Fig. 1.44 under the hypothesis that a universal function f_3 of constant value exists for laminar and turbulent flow. The validity of this hypothesis is verified by examining two values of f_3 which have special physical significance, namely, the value at the point of separation and the value at the beginning of the constant pressure plateau.

For laminar flow as well as for turbulent flow, C_{p_s} and C_{pp} were evaluated by

$$C_{p_s} = \mathrm{Re}_{x_0}{}^{-(1/2N)}\, \xi^{0.5} f_3\left(\frac{x - x_0}{l_i}\right) g\left(\mathrm{M}_0, \frac{T_{w_0}}{T_{e_0}}\right)$$

$$C_{pp} = \mathrm{Re}_{x_0}{}^{-(1/2N)}\, \xi^{0.5} f_3(1) g\left(\mathrm{M}_0, \frac{T_{w_0}}{T_{e_0}}\right)$$

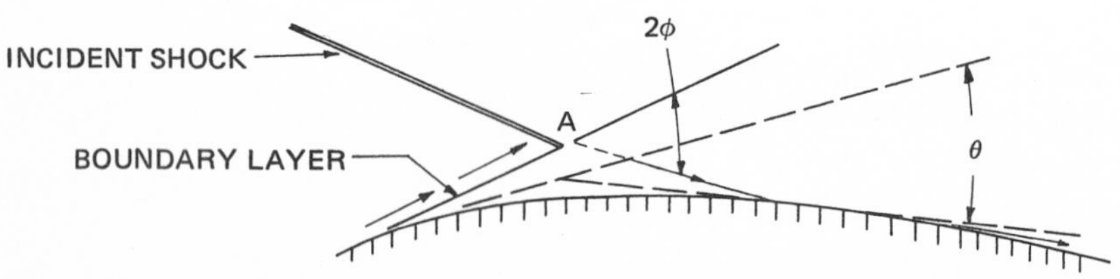

FIG. 1.43 Flow deflection angles on a convex surface (*from Pearcey, 1961; also Greber, 1959*).

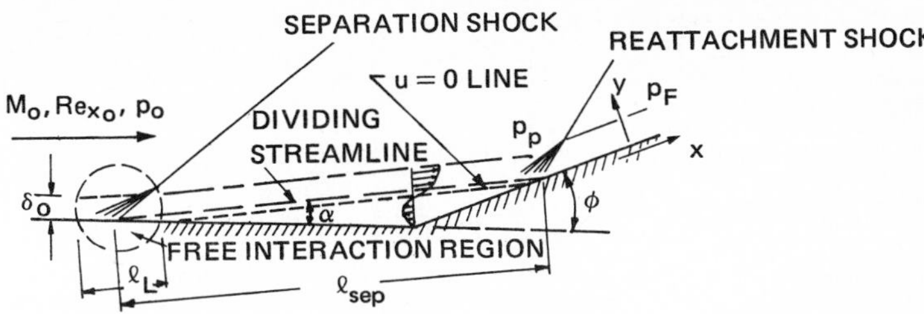

FIG. 1.44 Physical model for free interaction (*from Erdos & Pallone, 1962*).

For laminar flow,

$$f_3\left(\frac{x_s - x_0}{l_i}\right) = 0.81$$

$$f_3(1) = 1.47$$

$$N = 2$$

ξ is a correction factor defined by $\xi = C_{pp\,\text{exact}}/C_{pp\,\text{linear}}$. The value of $C_{pp\,\text{exact}}$ may be computed from shock tables such as NACA TR 1135 for an angle which is given by $C_{pp\,\text{linear}}$. The quantity ξ may be approximated by unity. Furthermore,

$$g\left(\mathrm{M}_0, \frac{T_{w_0}}{T_{e_0}}\right) = \left[\frac{2c_{f_0}\,\mathrm{Re}_{x_0}{}^{1/N}}{\beta_0}\right]^{-1/2}$$

The quantities x, l_i, M, and T are surface length, length of shock interaction region, Mach number at the edge of shear layer, and temperature, respectively. Subscripts 0, e, and w respectively refer to undisturbed flow just upstream from the shock interaction, at the edge of the shear layer, and at the wall. The separation point and the plateau pressure for insulated laminar flow for three different models are shown in Fig. 1.45. For laminar flow with heat transfer, the predicted separation point and plateau pressure are shown in Fig. 1.45 and compared to those of insulated flow.

1.2.2 Prediction of Two-dimensional, Turbulent Flow Separation

In practice, turbulent flow is more common than laminar flow, especially at subsonic speeds. However, because the mechanics of turbulent flow are not well understood, the solution for turbulent flow is not amenable to analytical treatment and requires empirical data. The prediction of separation for external turbulent flow may be based on the boundary layer concept. In

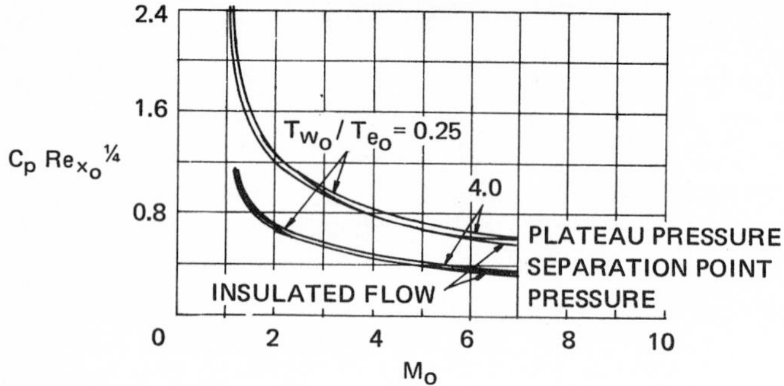

FIG. 1.45 Separation point and plateau pressure with heat transfer (*from Erdos & Pallone, 1962*).

general, such a prediction is less exact and more cumbersome than those for laminar flow. But boundary layer flow theory alone is insufficient for internal turbulent flow, e.g., flow through a diffuser. Stability and transient phenomena must also be considered in order to solve the problem of separation. At present, no applicable method can completely predict the behavior of diffuser flow separation.

1.2.2.1 Prediction of external turbulent incompressible, two-dimensional flow separation

In the past, the problem of incompressible turbulent flow separation on a circular cylinder has been investigated analytically and experimentally. Since turbulent flow is better able than laminar flow to overcome the pressure rise and friction, it separates further downstream. For turbulent flow, the separation occurs at about 110 deg (measured from the forward stagnation) compared to about 81 deg for laminar incompressible flow.

Extrapolation of the Ludwieg-Tillmann formula The following semiempirical formula, developed by Ludwieg and Tillmann (1949), may be used to predict the skin friction in incompressible turbulent flow with an arbitrary pressure gradient:

$$c_f = 0.246 \cdot 10^{-0.678H}\, \mathrm{Re}_\theta^{-0.268}$$

$$= 0.246 \cdot e^{-1.561H}\, \mathrm{Re}_\theta^{-0.268}$$

However, since this equation does not yield $c_f = 0$ at the separation point located at a position of vanishing shear stress, Maskell (1951) suggested an extrapolation skin friction curve to the point where $c_f = 0$. A reasonably accurate prediction of separation is possible inasmuch as the Ludwieg-Tillmann curve shows that c_f falls rapidly as the point of separation approaches and that it decreases monotonically toward zero. Such extrapolation causes error and for the best results, the starting point of the extrapolation should be taken on the skin friction curve where the errors caused by the extrapolation are small.

The von Doenhoff and Tetervin procedure For moderate adverse pressure gradients, such as on an airfoil surface, von Doenhoff and Tetervin (1943) developed a relatively simple analytical procedure for the determination of turbulent flow separation based on both the value of H (the shape parameter of the boundary layer) and the rapidity with which the value of H increases. The following two simultaneous equations are used to evaluate the values of H and θ along the body surface:

$$\frac{d\theta}{dx} + \frac{H+2}{2}\left(\frac{\theta}{q}\right)\frac{dq}{dx} = \frac{\tau_w}{2q}$$

$$\theta \frac{dH}{dx} = e^{4.680(H-2.975)}\left[\frac{\theta}{q}\frac{dq}{dx}\frac{2q}{\tau_w} - 2.035(H-1.286)\right]$$

$$\frac{2q}{\tau_w} = [5.890 \log_{10}(4.075\ \mathrm{Re}_\theta)]^2$$

Figures 1.46 and 1.47 are given to facilitate the numerical computations. The first equation for the momentum shows the rate at which momentum thickness changes along the surface. The second and third equations are relations needed to complete the description of the two quantities H and τ in the first equation. Also, x represents the distance along the surface from the starting point of turbulent flow. This method is relatively easy to apply since the criterion of separation is assumed to depend only on the value of H. If H reaches a value of 1.8 to 2.6 and H rapidly increases in a short distance, then turbulent flow separation is expected. The solution of the differential equation will supply the value of H at any particular point on the surface of the body, thus indicating whether or not separation will occur at that point. Step-by-step integration of the said equations presupposes certain well-defined starting values of speed, boundary layer momentum thickness θ, and H. The initial values of variables θ and H are substituted in the first two equations.

The above procedure yields the values of $d\theta/dH$ and dH/dx at the initial point. Next, suitable increments $\Delta\theta$ and ΔH are added to the initial values of θ and H, and the new, increased values of θ and H are substituted in the first two equations. The process is repeated until the cumulative values of the increments in x cover a sufficient range of integration. If $\Delta(dH/dx)$ is defined as the change in dH/dx between two successive values of x, then the product of $\Delta(dH/dx) \cdot \Delta x$ furnishes a criterion for measuring the maximum error in any one step of the integration. It is evident that the maximum error may be reduced by choosing smaller increments of Δx. The boundary layer thickness is not particularly sensitive to the initial value of H, if $\theta/q\ dq/dx\ 2q/\tau_w$ is very small or positive. Also, the position of separation is only slightly dependent on the initial value of H.

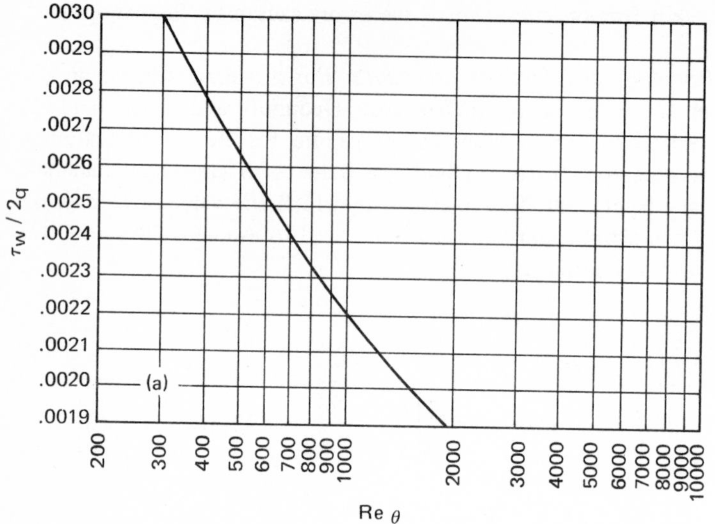

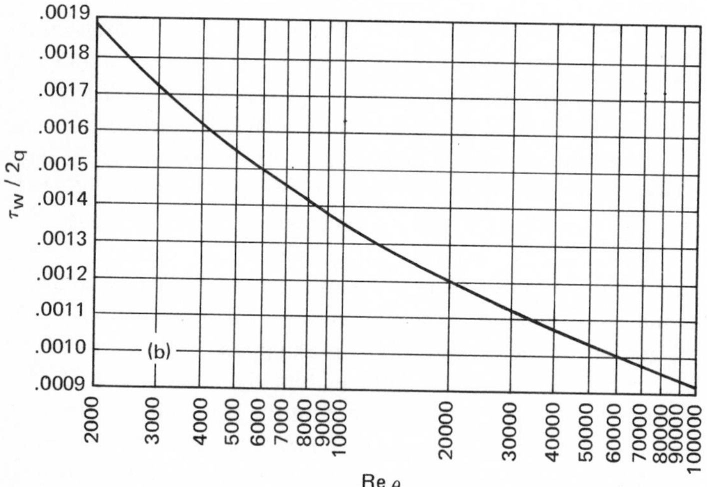

FIG. 1.46 Variation of $\tau_w/2q$ with Re_θ *(from von Doenhoff & Tetervin, 1943).*

In their numerical computations, Chang and Strandhagen (1961) used the von Doenhoff and Tetervin method to determine whether or not incompressible turbulent flow separates on a biconvex strut at zero angle of attack. The geometrical configuration and the potential velocity distribution of this biconvex strut were shown in Figs. 1.32 and 1.33, respectively, and the corresponding formulas were also presented previously (see *Evaluation of the various prediction methods* in Sec. 1.2.1.1 of this chapter). An initial value of $H = 1.4$ was used for turbulent flow and numerical computations were made for two different transition points at two different speeds. For comparison purposes, a point near the trailing edge at $X/L = 0.97$ was used.

Case I: The transition is assumed to occur at the point of maximum potential velocity.

1. For $u_\infty = 9.2$ ft/sec and a corresponding initial $\theta = 0.0022$ ft, the values of θ and H at $X/L = 0.97$ are $\theta = 3.5 \cdot 10^{-3}$ ft and $H = 1.56$.

   The value of $u_\infty = 9.2$ ft/sec corresponds to the so-called "limiting speed," which is defined as the speed which causes the transition to occur

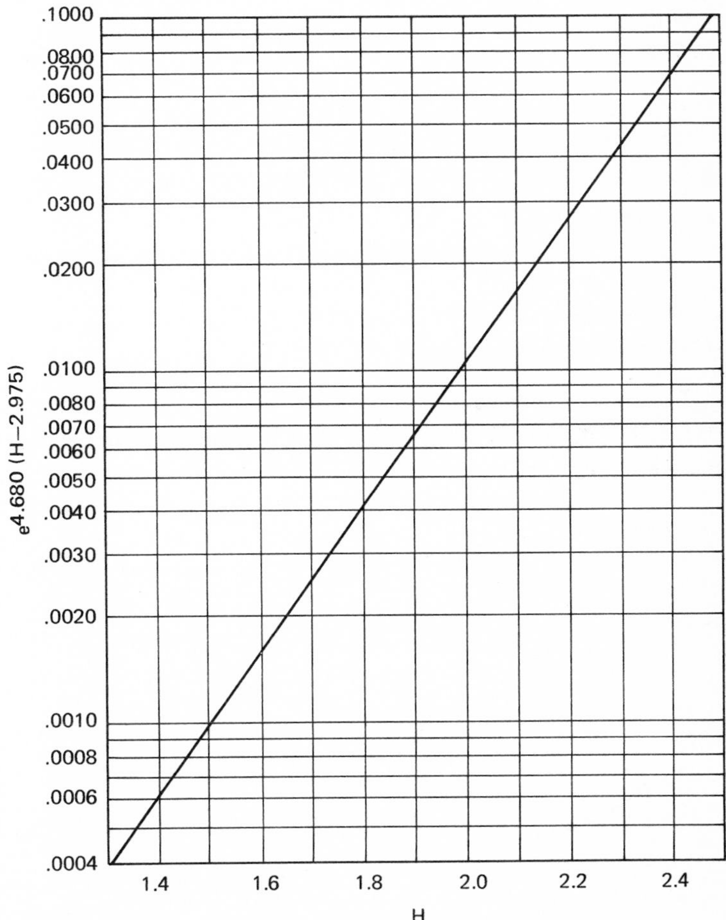

FIG. 1.47 Variation of $e^{4.680(H-2.975)}$ with H *(from von Doenhoff & Tetervin, 1943).*

at the position of the maximum velocity of potential flow. This speed is also near the lower extreme of a practical model speed in a water tunnel.

2. For $u_\infty = 20$ ft/sec and the corresponding initial $\theta = 0.00146$ ft, the numerical results plotted in Fig. 1.48 for this speed at $X/L = 0.97$ are $\theta = 2.99 \cdot 10^{-3}$ ft and $H = 1.58$.

The value of $u_\infty = 20$ ft/sec is close to the upper extreme of a practical model speed in a water tunnel.

A comparison of the numerical values of u_∞, θ, and H indicates that θ decreases but H remains almost constant with increasing u_∞. At speeds larger than 20 ft/sec, θ becomes even smaller than $2.99 \cdot 10^{-3}$ ft and H will remain smaller than 1.8, which may be considered as the lower limit for separation. Therefore, if transition is assumed at the point of maximum speed, the turbulent flow around this strut does not separate at speeds larger than the limiting speed.

Case II: Transition is assumed to take place at the leading edge.

1. For $u_\infty = 9.2$ ft/sec and the corresponding initial $\theta = 0.001$ ft, the values of θ and H at $X/L = 0.97$ are as follows: $\theta = 3.15 \cdot 10^{-3}$ ft and $H = 1.23$.
2. For $u_\infty = 20$ ft/sec and initial value of $\theta = 0.0001$ ft, the results are $\theta = 2.65 \cdot 10^{-3}$ ft and $H = 1.26$ at $X/L = 0.97$.

Again comparison of the numerical values of u_∞, θ, and H indicates that as u_∞ increases, θ decreases but H increases only slightly and remains below 1.8. Therefore, if the flow is turbulent at the leading edge, there will be no separation at speeds larger than the limiting speed.

The Garner criterion Garner (1944) combined the two existing methods (Howarth, 1938b, and von Doenhoff & Tetervin, 1943) and expressed the skin friction by a power law and the following two parameters:

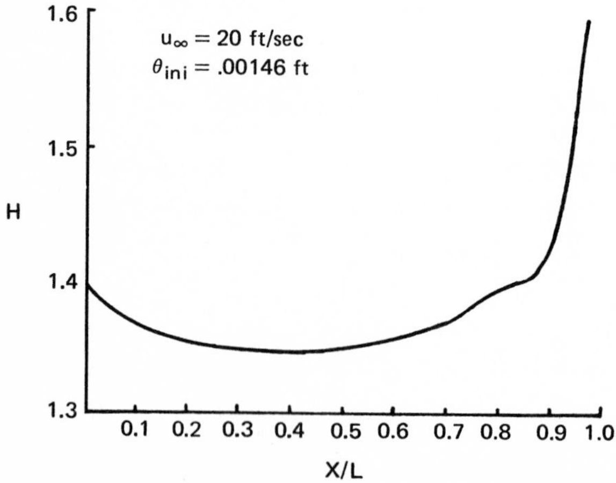

FIG. 1.48 Variation of H along biconvex strut surface (*from Chang & Strandhagen, 1961*).

$$\Theta = \theta \cdot \mathrm{Re}_\theta^{1/n} \quad \text{and} \quad \Gamma = \frac{\Theta}{u_e}\frac{du_e}{dx}$$

where n may be taken as 6 and x is the distance along the wall.

Garner carefully considered the effect of transition on turbulent separation and derived the following momentum and empirical equations:

$$\frac{d\Theta}{dx} = \frac{7}{6}\left[\zeta - \Gamma\left(H + \frac{13}{7}\right)\right] \tag{1.6}$$

$$\Theta\frac{dH}{dx} = e^{5(H-1.4)}\left[-\Gamma - 0.0135(H - 1.4)\right] \tag{1.7}$$

where $\zeta = (\tau_w/\rho u_e^2)\,\mathrm{Re}_\theta^{1/n}$

Separation is predicted to occur when $H = 2.6$. The relationship of Γ with respect to $\Theta(dH/dx)$ for $H = 1.5$, 1.55, 1.6, 1.7, 1.8, 1.9 is presented in Fig. 1.49 as a straight-line approximation of the empirical points shown.

For a given velocity distribution, the procedures of calculation are as follows:

1. Determine the development of the laminar boundary layer before transition.
2. Decide at which point on the surface transition takes place, if at all.
3. Compute the value of Θ immediately after transition. (The value of θ at the position of transition is determined from Step 1, and θ is continuous through transition.)

4*a*. If transition occurs at or downstream from the position of maximum velocity, assume that $dH/dx = 0$ there, i.e., that $H = 1.4 - (\Gamma/0.0135)$ immediately after transition.

4*b*. If transition is upstream from the point of maximum velocity, assume that H is constant between the positions of transition and the maximum velocity. The value of Θ at the position of the maximum velocity is given by

$$u_{e_{\max}}^k\Theta = [u_e^k\Theta]_{\mathrm{tr}} + 0.007623\int u_e^k\,dx$$

where the subscript tr denotes values at the assumed position of transition and the integral is taken from the position of transition to the point of the maximum velocity and $k = \frac{7}{6}H + \frac{13}{6}$.

5. Finally, determine the point of separation by integrating Eqs. (1.6) and (1.7).

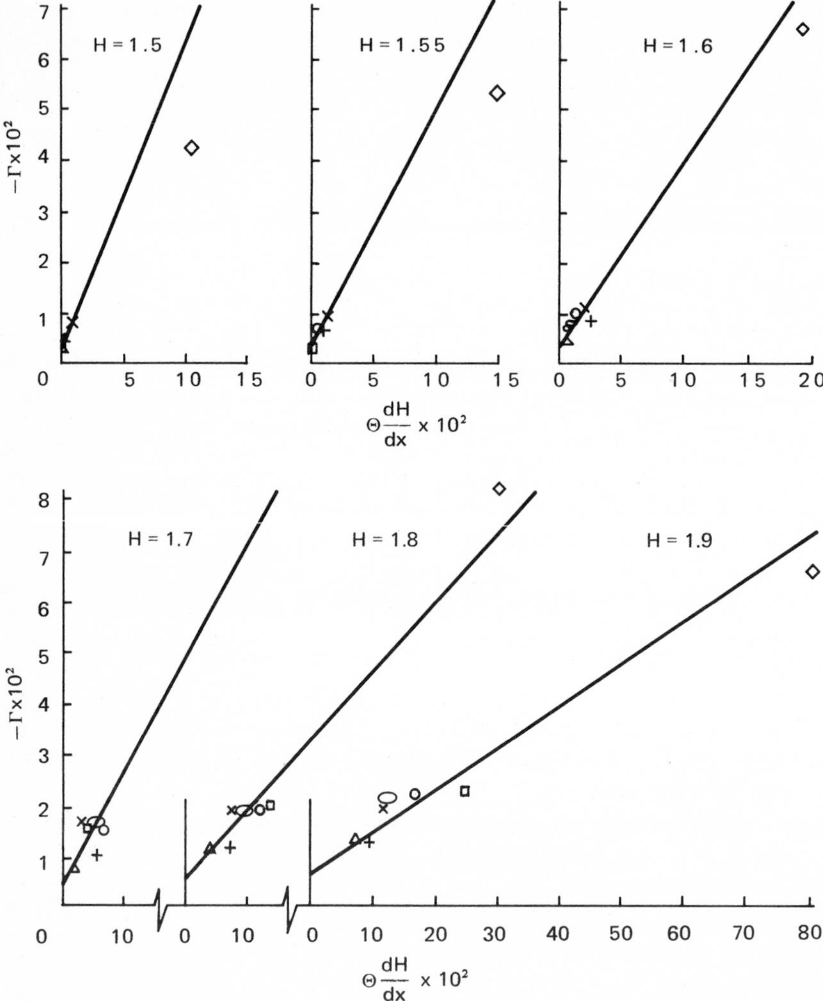

FIG. 1.49 Variation of $\Theta(dH/dx)$ with Γ for various values of H *(from Garner, 1944)*.

The Garner method is quicker to calculate than the von Doenhoff-Tetervin method and yields a relatively reliable prediction. This is confirmed by the good agreement between computed results and experimental data for an airfoil at angles of attack.

The magnitude of L as criterion By using H and $\overline{\overline{H}}$, the ratio of energy thickness with respect to momentum thickness, the separation criterion of incompressible turbulent flow is given by the nondimensional parameter L defined by

$$L = \int_{\bar{H}=\bar{H}_0}^{\bar{H}} \frac{d\bar{H}}{(H-1)\bar{H}}$$

where $\bar{H} = \delta^{**}/\theta$ and δ^{**} is energy thickness evaluated by

$$\delta^{**} = \int_0^\infty \left(\frac{u}{u_e}\right)\left[1 - \left(\frac{u}{u_e}\right)^2\right] dy$$

Separation is predicted at the position where $L = -0.13$ to -0.18, which corresponds to $H = 1.8$ to 2.4 (Schlichting, 1968).

1.2.2.2 Prediction of separation of incompressible two-dimensional turbulent flow through passages

Flow through a passage, or internal flow, is constrained by the surrounding wall; thus there may be differences between external and internal flow phenomena (for example, the different behavior of growth of stall). Kline (1959) indicated that the growth of stall in external flow is quite different from the growth in internal flow where there is a high ratio of length of diffuser wall to the two-dimensional width of the diffuser throat. Because the reason for the difference is not well understood, the reliable solution for external flow may not be used to determine the accuracy of predictions of turbulent internal flow separation. Moreover, since transitory flow is dominant in certain stages of internal flow, both stability considerations and transient elements should be considered in the formulation of separation. In practice, the flow through a subsonic diffuser is turbulent. Fox and Kline (1961) observed the following three different regimes of flow for both straight-wall and curved-wall diffusers.

1. A regime of well-behaved and unseparated flow.
2. A regime of large transitory stall in which the separation varies in size and intensity with time.
3. A regime of fully developed stall in which the flow is relatively steady and flows along the pressure wall of the diffuser. That portion of the diffuser adjacent to the suction surface is filled with a large eddying recirculation flow.

At present, no reliable theoretical prediction of internal separation is available which can be applied for all regimes of internal flow. However, since empirical data are available for diffusers, these may be used for the determination of separation. (The geometry of a curved diffuser is sketched in Fig. 1.50.)

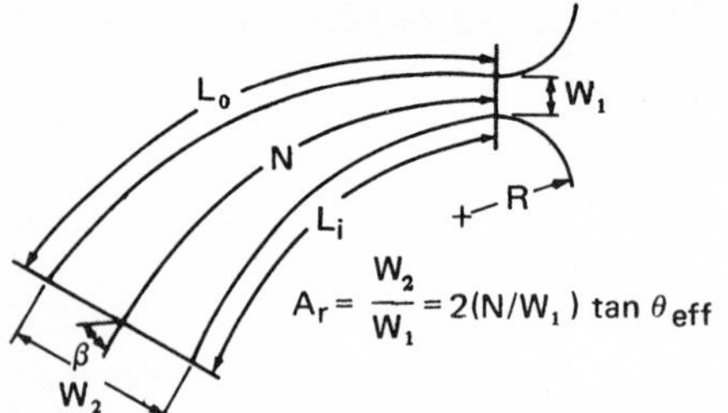

FIG. 1.50 Curved diffuser geometry (*from Fox & Kline, 1961*).

The information for a straight-wall diffuser may be obtained for the condition of $\beta = 0$. The quantity $2\theta_{eff}$ is the effective total divergence angle for curved diffusers computed from

$$\text{area ratio} = 1 + 2\frac{N}{W_1}\tan\theta_{eff}$$

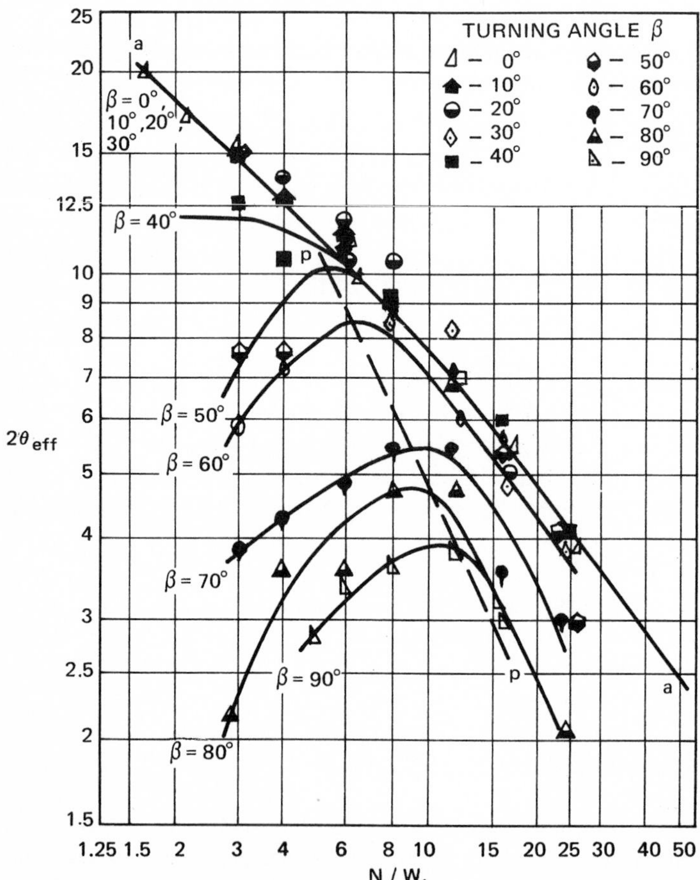

FIG. 1.51 Location of first appreciable stall (line a–a) as a function of turning angle β in curved diffusers with circular-arc centerline and linear area distribution (*from Fox & Kline, 1961*).

The locations of the first appreciable stall and the locations of transitions of fully developed stall are given in Figs. 1.51 and 1.52 as a function of turning angle β.

As shown in Fig. 1.51, the location of first appreciable stall is not significantly affected if β is less than or equal to 30 deg. At $\beta = 40$ deg, $2\theta_{eff}$ is slightly lower for small values of N/W_1, and for 50 deg $\leqslant \beta \leqslant$ 90 deg, all the lines of first appreciable stall show a peak, as indicated by the dotted line p–p. This line represents maximum diffusion for minimum N/W_1 in a passage without appreciable stall for a given value of β. Along the line p–p, the area ratio varies between 1.8 and 1.9. Because the area ratio on a plot of $2\theta_{eff}$ versus N/W_1 increases with N/W_1, a slightly greater diffusion without appreciable stall may be obtained by taking large values of N/W_1 and a corresponding lower $2\theta_{eff}$ for the turning angle β required. The drop in line a–a with decreasing values in the left portion of line p–p indicates a distinctly different characteristic of the curved diffuser geometry compared with the straight-wall diffuser. This phenomenon is considered attributable to increased values of the adverse pressure gradient on the inner wall downstream from the throat. At the same area ratio, N/W_1 and 2θ, this adverse pressure gradient is stronger on the curved diffuser than in a corresponding straight-wall diffuser. The location of the transition to a fully developed stall (line b–b) is shown in Fig. 1.52 as $2\theta_{eff}$, it decreases monotonically with increase in turning angle. The fully developed stall was found for all turning angles β with only one

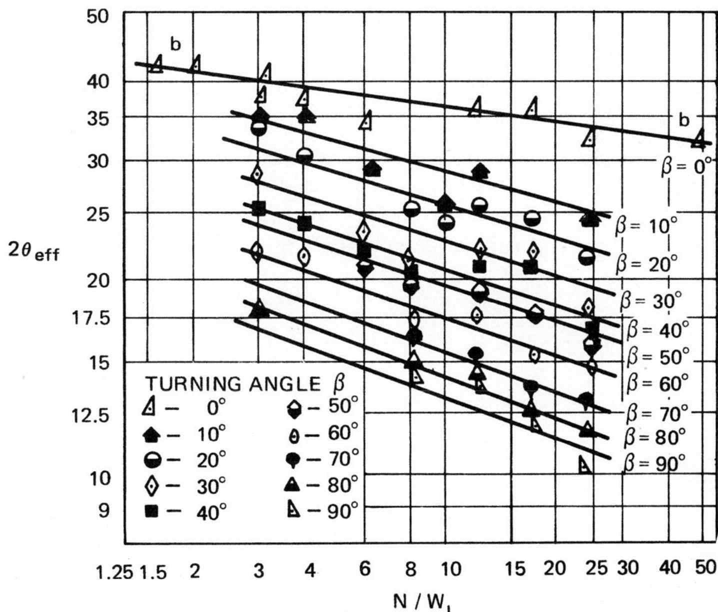

FIG. 1.52 Location of transition of fully developed stall (line b–b) as a function of turning angle β in curved diffusers with circular-arc centerline and linear area distribution (*from Fox & Kline, 1961*).

exception at $\beta = 10$ deg. Experimental investigations on incompressible diffuser flows are documented as early as 1898. Several references on the subject are listed in the monograph by Chang (1970) and elsewhere.

1.2.2.3 Prediction of compressible turbulent two-dimensional flow separation

Reshotko and Tucker (1957) developed an approximate method for the calculation of a compressible turbulent boundary layer with arbitrary pressure gradient and heat transfer. They employed the momentum integral and the moment of momentum equations simplified by the Stewartson transformation. The Ludwig-Tillmann formula for incompressible flow is transformed to that for compressible flow as follows:

$$c_f = \frac{\tau_w}{\frac{1}{2}\rho_e u_e^2} = 0.246 \cdot e^{-1.561 H_i} \left(\frac{M_0 a_0 \theta_{tr}}{\nu_0}\right)^{-0.268}$$

$$\left(\frac{T_e}{T_{\text{ref}}}\right)^{0.732} \left(\frac{T_e}{T_0}\right)^{0.268}$$

where subscript 0 indicates free-stream stagnation and subscripts *tr*, *i*, and ref respectively refer to transformed, incompressible, and reference conditions. The position of separation is determined by using the condition $c_f = 0$ and extrapolating the Ludwig-Tillman formula. The accuracy of the prediction of separation by the extrapolation depends on the degree of rapidity with which skin friction falls in a small increment of downstream distance and the speculative, but proper, qualitative assumptions which are made (Reshotko & Tucker, 1957). The separation is expected to occur where H_i equals 2.25 to 2.28, and the initial value of H_i may be assumed as 1.286. The following relations are useful for the computation of c_f:

$$\theta_{tr}^{1.2155} = \frac{0.01173 (T_e/T_{\text{ref}})^{0.732}\, x}{2.2155 (a_0/\nu_0)^{0.2155} M_e^{0.2155} (T_0/T_e)^{3.268}}$$

T_e/T_{ref} may be computed from

$$\frac{h_{\text{ref}}}{h_e} = \frac{1}{2}\left(\frac{h_w}{h_e} + 1\right) + 0.22\left(\frac{h_{aw}}{h_e} - 1\right)$$

where h is enthalpy and subscripts w and aw respectively refer to wall and adiabatic wall.

The separation point may also be determined by a numerical evaluation H_i. The point of separation is located at $H_i = 2.2$ to 2.28 and the following differential equations are used. These are applicable for axisymmetric as well as two-dimensional flow (Torrillo & Savage, 1964).

$$\frac{2}{H_i(H_i+1)^2(H_i-1)}\frac{dH_i}{dx}=\frac{1}{U_e}\frac{dU_e}{dx}\left[1-\left(1-\frac{h_w}{h_{se}}\right)\left(\frac{2H_i}{H_i+1}-\frac{H_i+1}{H_i+3}\right)\right]$$
$$-0.193(0.246)\frac{\rho^*}{\rho_e}\left(\frac{\mu^*}{\mu_e}\right)^{0.268}\left(\frac{\rho_e u_e\theta}{\mu_e}\right)^{-0.268}\frac{1}{\theta}$$
$$e^{-1.56H_i}\frac{(H_i+1.25)(H_i-1{:}3)}{(H_i-1)(2.45-H_i)} \qquad (1.8)$$

where $U = ua_0/a_e$ and the momentum thickness θ is evaluated by

$$\theta^{1.2155}=\frac{\rho_{e_0}^{\;1.2155}\left(\dfrac{\mu_{e_0}}{h_{e_0}^{\;0.5}}\right)^B h_{e_0}^{\;1/2}u_{e_0}^{\;0.2155}\,\theta_0^{\;1.2155}\left(\dfrac{R_0}{L}\right)^{1.2155j}}{\rho_e^{\;1.2155}\left(\dfrac{u_e}{h_e^{\;0.5}}\right)^B h_e^{\;0.5}\,u_e^{\;0.2155}\left(\dfrac{R}{L}\right)^{1.2155j}}$$

$$+\frac{0.01173\mu_{se}^{\;-0.0525}L\left[\displaystyle\int_{(x/L)_0}^{x/L}\rho^*(\mu^*)^{0.268}\left(\frac{R}{L}\right)^{1.2155j}\left(\frac{u_e}{h_e^{\;0.5}}\right)^B h_e^{\;0.5}\,d\left(\frac{x}{L}\right)\right]}{\rho_e^{\;1.2155}\left(\dfrac{u_e}{h_e^{\;0.5}}\right)^B h_e^{\;0.5}\,u_e^{\;0.2155}\left(\dfrac{R}{L}\right)^{1.2155j}}$$

Here subscript 0 refers to the initial condition, R is the radius of the axially symmetric body, and $B = 2.62 + 1.58\ (h_w/h_{se})$. For two-dimensional flow, $j = 1$. Superscript * refers to the reference condition of Eckert (1955). Note that if H_i exceeds 2.20, the provisional term, i.e., the second term on the right-hand side of Eq. (1.8) becomes very large and approaches infinity when H_i approaches a value of 2.45. This is certainly not realistic. Hence when Eq. (1.8) is used for the prediction or separation, the value of H_i should not exceed 2.28 (Torrillo & Savage, 1964).

Over the range of $1.00 \leqslant H_i \leqslant 3.00$, a good approximation is given by

$$1-\left(1-\frac{h_w}{h_{se}}\right)\left(\frac{2H_i}{H_i+1}-\frac{H_i+1}{H_i+3}\right)\cong a-bH_i$$

$$a = 1-0.40\left(1-\frac{h_w}{h_{se}}\right)$$

$$b = 0.16\left(1-\frac{h_w}{h_{se}}\right)$$

where subscript s refers to stagnation.

As shown in Fig. 1.53, Torrillo and Savage (1964) give the pressure rise for separation as a function of the initial Mach number.

1.2.2.4 Numerical solutions

A reliable, widely applicable, and economical method of solving the boundary layer equations has recently been developed by Patankar and Spalding (1967). They used a digital computer to eliminate the mathematical

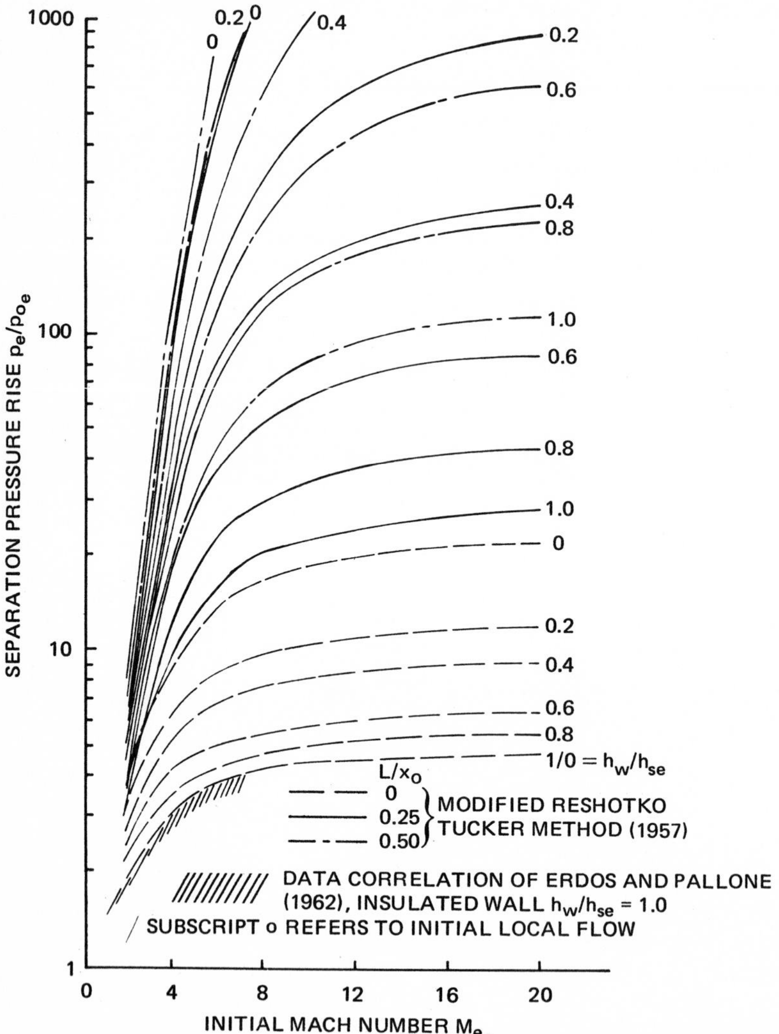

FIG. 1.53 Effect of Mach number and stagnation enthalpy on the pressure rise for separation ($\gamma = 1.4$) (*from Torrillo & Savage, 1964*).

difficulty caused by the absence of an adequate means for solving the differential equations. Written in Fortran IV language, the program is directly useful for incompressible two-dimensional and axially symmetric turbulent boundary layer with the time-averaged values of variables. The program is designed so that the partial differential equation for velocity u will always be solved; any desired number of conservation equations and other dependent variables can also be provided. Thus, although the monograph does not specially refer to the prediction of separation, no difficulty is anticipated in computing the location of separation.

Cebeci and Smith (1969) were successful in predicting the separation point of a two-dimensional incompressible turbulent boundary layer by using finite-difference solutions and the eddy-viscosity concept. They consider the boundary layer as made up of inner and outer regions and use a separate expression for eddy viscosity in each region. In the inner region, the eddy viscosity is based on the Prandtl mixing length theory; in the outer region, however, a constant eddy viscosity modified by an intermittency factor is applied. The boundary-layer equation is solved by an implicit five-point finite-difference method. The momentum equation has been transformed and linearized. Thus, this method is applicable not only for turbulent flow but also for the laminar boundary layer, which may prevail upstream of turbulent flow. The solution begins at the leading edge or at the stagnation point and proceeds downstream. Calculations can be started in the middle of a flow, however, provided that upstream information such as initial velocity profile is known and used as an input. The separation input is predicted accurately at the point where

$$f''_w = \left.\frac{\partial(u/u_e)}{\partial y}\right|_w$$

becomes negative or at a point where the local skin-friction coefficient shows an increase in the presence of an adverse pressure gradient.

Reyhner (1969) presented a numerical method for predicting the two-dimensional compressible turbulent boundary-layer separation close to the actual separation position. The solution of the finite-difference equations adopted differs from that of the partial differential equation by at least several percent. Problems in the Reyhner method are apparently due to the large errors in the difference quotients in the region of severe gradients near the wall and to the fact that the eddy viscosity concept is not very accurate for specifying the Reynolds stresses. Thus, the program is written so that the eddy viscosity can easily be changed and so that the Reynolds stresses can be specified directly if equations for them become available. Furthermore, a serious problem may arise because the normal pressure gradient is neglected for high Mach number flows with severe pressure gradients and because of truncation errors. These problems reflect the middle stage of development of the Reyhner (1969) numerical method.

1.2.2.5 Prediction of turbulent flow separation caused by the interaction of shock waves with the boundary layer

Prediction of the pressure rise by Erdos and Pallone For turbulent flow caused by shock wave-boundary layer interaction, Erdos and Pallone (1962) formulated the separation pressure coefficient C_{ps} and the plateau pressure coefficient C_{pp} based on the free interaction model shown in Fig. 1.44. The procedure is similar to that for laminar flow as discussed previously in Sec. 2.1.3. That is,

$$C_{ps} = \mathrm{Re}_{x_0}{}^{-(1/2N)}\xi^{1/2} f_3\left(\frac{x - x_0}{l_i}\right) g\left(M_0, \frac{T_{w_0}}{T_{e_0}}\right)$$

$$C_{pp} = \mathrm{Re}_{x_0}{}^{-(1/2N)}\xi^{1/2} f_3(1) g\left(M_0, \frac{T_{w_0}}{T_{e_0}}\right)$$

where M_0 is the Mach number of undisturbed flow, just upstream of the shock interaction and turbulent flow $f_3(x_s - x_0/l_i) = 4.22$, $f_3(1) = 6.00$ and $N = 5$. The separation point and plateau pressure for insulated flow and for flow with heat transfer are shown in Figs. 1.54 and 1.55.

The Kuehn measurement Kuehn (1959) measured the pressure rise and the flow deflection angle of turbulent flow for incipient separation due to incipient shocks as a function of Reynolds number. See Figs. 1.56 and 1.57.

DATA SOURCE

o . . . CHAPMAN ET AL. (1957)
Δ . . . STERRETT AND EMERY (1960)
□ . . . DROUGGE (1953)

TYPE OF MODELS (SEE FIG. 1.41)

(a) COMPRESSION CORNER . . . . o □
(b) FORWARD-FACING STEP . . . o Δ
(c) INCIDENT SHOCK . . . . o

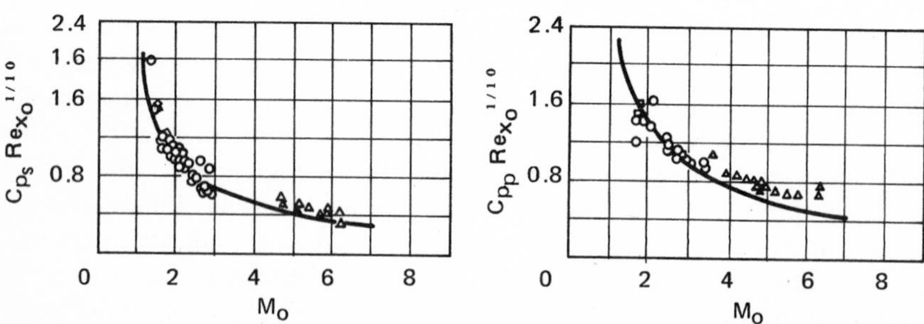

FIG. 1.54 Separation point and plateau pressure for insulated turbulent flow *(from Erdos & Pallone, 1962)*.

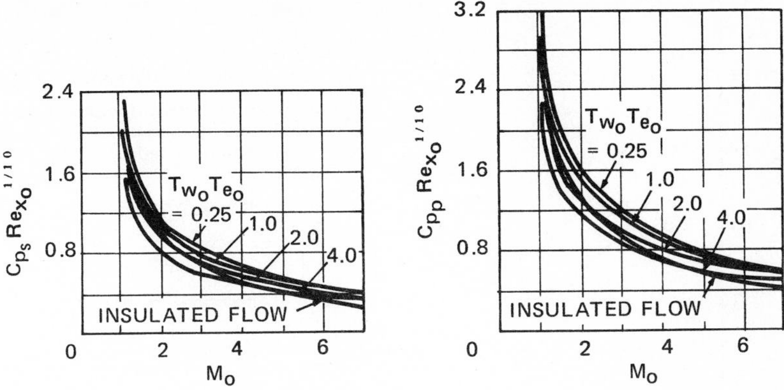

FIG. 1.55 Separation point and plateau pressure for turbulent flow with heat transfer (*from Erdos & Pallone, 1962*).

If Re_{δ_0} is held constant, the pressure rise required for separation and the flow deflection angle become larger with increasing Mach number. Subscript o refers to the value upstream of interaction. If the Mach number remains the same, these values become larger in the region where Re_δ is small.

1.2.3 Steady Flow Separation on Bodies of Revolution and Three-dimensional Solid Configurations

The problem of three-dimensional flow separation is undeveloped and difficult to solve, and solutions are known for only a few particular cases.

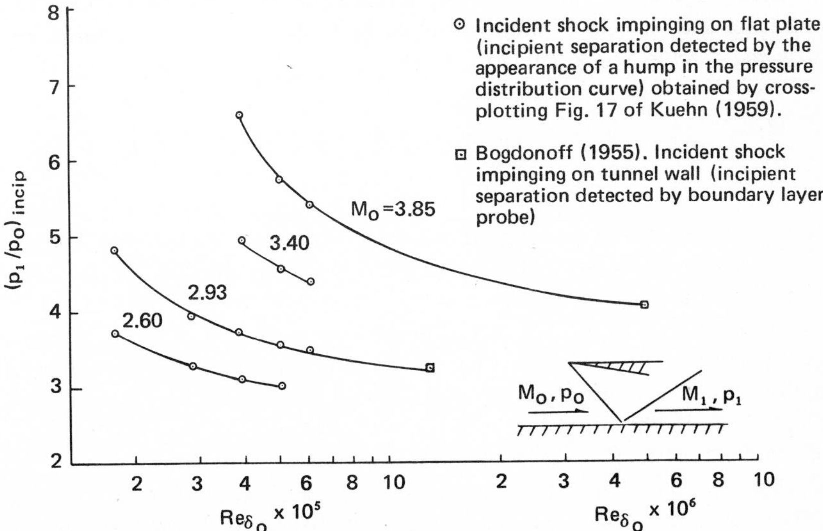

FIG. 1.56 Effect of Reynolds number on the pressure rise for turbulent separation incident shocks in turbulent flow (*from Kuehn, 1959*).

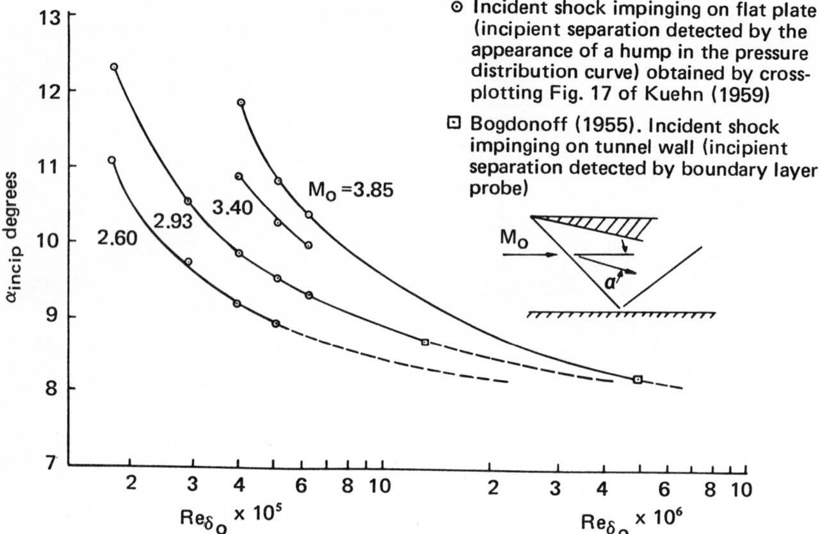

FIG. 1.57 Effect of Reynolds number on the flow deflection angle for incipient separation for incident shocks in turbulent flow (*from Kuehn, 1959*).

Although limited, some information on three-dimensional flow separation may be obtained from the Chang monograph (1970). The boundary layer in axisymmetrical flow about bodies of revolution is very similar to that for two-dimensional flow. This is true particularly because the vorticity source strength per unit is the same; thus vorticity production and diffusion are unaltered (Lighthill, 1963). Hence, a criterion for two-dimensional flow separation such as the Thwaites formula (see *The method of Thwaites and Curle and Skan* in Sec. 1.2.1.1) for laminar flow

$$\frac{\theta^2}{\nu}\frac{du_e}{dx} = -0.090$$

is often used for axial symmetric flow. But the applicability of the two-dimensional criterion becomes questionable if the thickness of the body of revolution increases.

If consideration is given to account for axial symmetry, then the Mangler (1948) transformation may be used for both incompressible and compressible laminar flows. For example, if the Görtler solution for two-dimensional laminar flow is transformed by the Mangler procedure to axisymmetric flow, then axisymmetric flow separation may be predicted on the basis of the two-dimensional criterion (Görtler, 1957, and Saljimilor, 1960). It is known from classical hydrodynamics that for a sphere submerged in subsonic laminar flow, the point of separation is located at about 83.5 deg, measured from the forward stagnation point, and approximately 2 deg downstream compared to about 81.5 deg for a cylinder. But the total drag of a sphere is much less than that of the cylinder because of the smaller difference in the static pressure distribution between inviscid and viscous flow around the sphere.

The results of three-dimensional boundary layer investigations show that the boundary layer is thin when the flow accelerates and that it is thick when the flow decelerates (in a manner similar to the two-dimensional case). However, the variation of thickness due to a pressure gradient is smaller than in the two-dimensional case. For three-dimensional flow, the low momentum fluid within the boundary layer does not flow against the adverse pressure gradient in the direction of the main flow but rather in a sidewise direction in which the pressure gradient is more favorable. As a result, the three-dimensional boundary layer is more capable of overcoming an adverse pressure gradient, and, consequently, may not separate. In contrast, a two-dimensional boundary layer would separate under the same adverse pressure gradient condition.

Three approximate analyses have been developed for predicting the separation of three-dimensional incompressible laminar flow: Eichelbrenner and Oudart (1955), Zaat (1956), and the Cooke (1959) modification of the Zaat method. All these methods are restricted to small cross flow, are based on the momentum equation, and use a streamwise coordinate system. However, although they use a one-parameter family for the streamwise flow, they differ in the choice of velocity profiles and that involves an additional parameter of the cross flow. Since at separation, the cross flow is not small, these approximate methods can yield only a rough determination of the position of separation. However, if the surface streamline (the limiting streamline) turns abruptly enough, then these approximate methods will still predict the separation accurately. After comparing the results obtained by these three methods, Cooke (1959) claimed that his method is best.

For flow over a delta wing, a separation takes place at the position where the angle between the surface streamline and the ray of the wing vanishes. The predicted positions of separation on a conically cambered thin wing of sweep angle 72 deg are shown in Fig. 1.58. In the figure η is defined by ($\eta = (z/x)$ tan 72 deg). Here x is taken along the centerline of the wing and z is perpendicular to x in the plane of the wing.

The angle between the leading edge and the centerline of the wing is 18 deg. At separation, the angle β between the potential streamline and surface streamline is about 16 deg. According to the three methods previously mentioned, the separation takes place at $\eta = 0.625$, 0.645, and 0.690. These values correspond to the occurrence of separation along the ray, making the centerline respectively 11.5, 11.8, and 12.7 deg. Separation is predicted latest by the Cooke method and earliest by the method of Eichelbrenner and Oudart.

Cooke (1965) applied his analysis to prediction of the incompressible laminar separation on an inclined cone surface. He used the polar coordinate

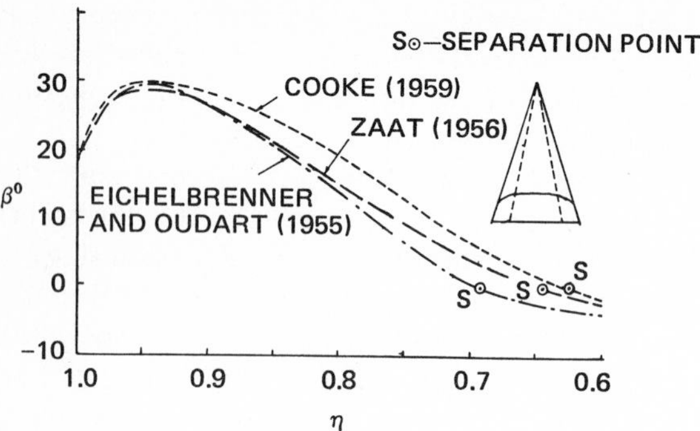

FIG. 1.58 Angle between limiting streamlines and rays for conically cambered delta wing (*according to Brebner (1957)*; with $n = 2$, $\eta = 0.6$, $C_L = 0.1$, sweepback = 72 deg; *from Cooke, 1959*).

r, the distance from the apex, and θ, the angle between any generator and a fixed generator measured in the plane into which the cone can be developed. By assuming that the external flow is conical, the transformed equations of continuity and momentum can be solved numerically by the Crank-Nicholson process (this is described by Hall, 1965, and others) with reasonable economy of machine time. Separation occurs at the position where angle β between the surface streamlines and the generators become zero; the occurrence of the

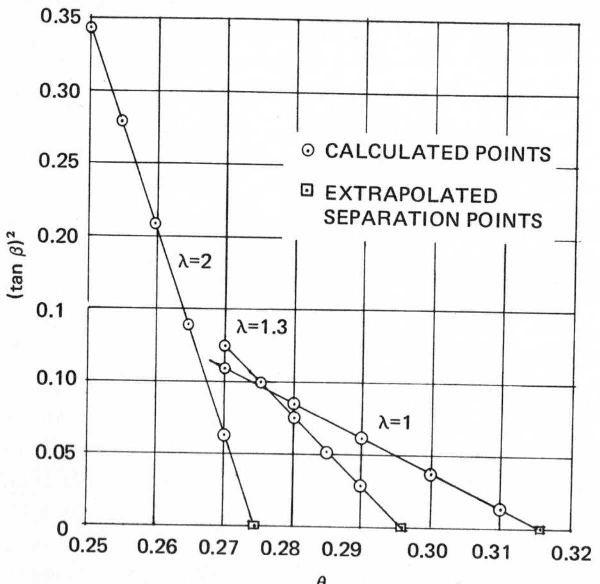

FIG. 1.59 Determination of the separation point on a delta wing $\lambda = 1, 1.3, 2$ (*from Cooke, 1965*).

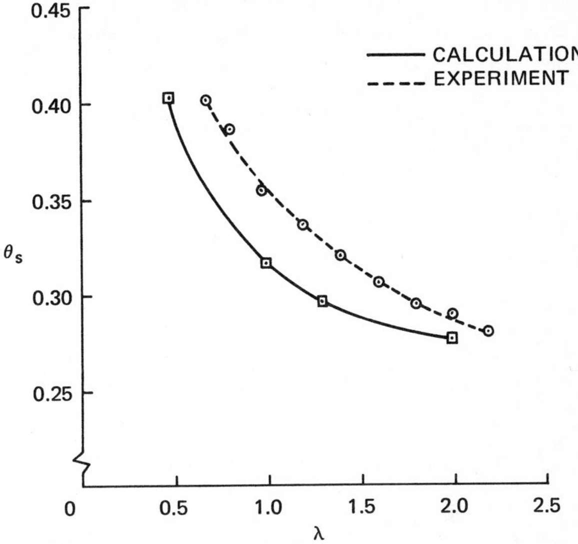

FIG. 1.60 Positions of separation on a delta wing (*from Cooke, 1965*).

separation may be judged by the value of λ, defined as $\lambda = (\alpha/\theta_c)$, where α is the angle of attack and θ_c is the cone half angle. If $\lambda < 0.5$, then separation does not occur at all. If $\lambda = 0.5$, the flow separates almost exactly at the leeward generator of the cone. Since the type of singularity encountered at separation is known (Brown, 1964), the separation position can be located fairly accurately by simply extrapolating the analytical solution as shown in Figs. 1.59 and 1.60. Note from Fig. 1.60 that the exact computation predicts separation earlier than do the experimental data of Rainbird et al. (1963). Commenting on this discrepancy, Cooke (1965) says that in this analysis, the external pressure is based on slender body theory (which excludes the vortex sheets associated with the separation) instead of the actual pressure distribution which may yield a better agreement between analysis and experiment.

Cooke (1966) recently extended his analysis for incompressible flow on an inclined cone to that of compressible laminar flow. The resulting equations are quite similar to those for incompressible flow, but the effect of compressibility requires an extra equation and other small changes. Table 1.5 shows the results of numerical calculations for the separation position of a cone of semiangle θ_c of 7.5 deg for Mach numbers of 3 and 6, two wall conditions (zero heat transfer and cooled), and two values of $\lambda = (\alpha/\sin \theta_c)$.

It can be seen from Table 1.5 that cooling slightly delays separation and that an increase in Mach number does not significantly affect this delay. The latter finding is somewhat unexpected inasmuch as separation occurs earlier at the higher Mach number for continual adverse pressure gradient with zero heat

TABLE 1.5 Position of laminar separation on a circular cone at incidence for two Mach numbers, two wall conditions, and two values of λ

M_∞	λ	Wall conditions*	θ_{sep} †
3	1	ZHT	0.326
3	1	C	0.336
3	2	ZHT	0.269
6	1	ZHT	0.353
6	1	C	0.361
6	2	ZHT	0.269

Source: Cooke (1966).
*Zero heat transfer (ZHT) and cooled (C) wall conditions.
†θ_{sep} is the angle between generators of the cone when developed into a plane at which separation occurs.

transfer. The explanation may be that an effective increase in the pressure gradient results at the higher Mach number, that this increases the crossflow, and that the need to counter the large crossflow counteracts the tendency for earlier separation at the higher Mach number.

NOMENCLATURE

a	Coefficient
B	Index of power
b	Coefficient. Also one-half thickness of a strut
C	$(\mu/\mu_\infty)/(T/T_\infty)$, one-half thickness. Also constant
c	Coefficient. Also chord
F	$F=(1+\xi)/(1-\xi)$. Also $F=(x_t/u_e)(du_e/dx)$
F_i	$F_i=[x_{eq}/(u_{max})_i]\,(du_e/dx)_i$
G_1	$G_1=H_w/H_e$
H	Shape parameter of the boundary layer, stagnation enthalpy
$\bar{H}$	$\bar{H}=\delta^{**}/\theta$
h	Enthalpy
j	$j=0$ two-dimensional flow $j=1$ axial-symmetric flow
k	$k=(7/6)H+13/6$
L	One-half of chord length, characteristic length. Also $\int_{\bar{H}_0}^{\bar{H}} d\bar{H}/[(H-1)\bar{H}]$
l	$\partial u/\partial y\|_{y=0}=(u_e/\theta)l$
m	$\partial^2 u/\partial_y{}^2\|_{y=0}=(u_e/\theta^2)m$
N	Index of power
q	Component of the velocity in the direction perpendicular to the separation line. Also dynamic pressure.

U $U = u(a_0/a_e)$
X Coordinate of biconvex strut
x_t $x_t = x_{eq} + x'$
x_{eq} Equivalent length
x' Distance between the point of the maximum potential velocity and the initial point of the straight line of potential velocity versus x which approximates the potential velocity distribution downstream from the maximum velocity
Y Coordinate of binconvex strut
y $y = \int_0^t (T/T_e)dt$
α_0 Defined by $\tan \alpha_a = \lim_{y \to 0} \tau_x/\tau_y$
β $\beta = b/L$, turning angle also angle between potential stream line and surface stream line
Γ $\Gamma = (\theta/u_e)(du_e/dx)$
δ_t $\delta_t = \int_0^\delta (T_e/T)dy$
δ^{**} Energy thickness
ζ $\zeta = (\tau_w/\rho u_e^2)\, \mathrm{Re}_\theta^{1/n}$
η $\eta = y/\delta$ also $\eta = (y/x) \tan 72°$
Θ $\Theta = \mathrm{Re}_\theta^{1/n}$
θ_c Cone half angle
θ_{eff} Effective divergence angle
λ $\lambda = (\delta^2/\nu)(du_e/dx)$ $\lambda = \mathrm{Re}_\infty(\delta_t/L)^2$ also $\lambda = \alpha/\theta_c$
ξ $\xi = X/L$, $\xi = C_{pp\,\mathrm{exact}}/C_{pp\,\mathrm{linear}}$ also $u_e/u_\infty = 1 - \xi$
τ $\tau = t/\delta_t$

Subscripts

aw Adiabatic wall
c Compressible
i Incompressible
0 Beginning of shock interaction, upstream of shock; also free-stream stagnation
p Plateau
ref Reference
s Separation; also stagnation
tr Transformed

REFERENCES

Baker, W. T. et al. (1951). "Reduction of drag of a projectile in a supersonic stream by the combustion of hydrogen in the turbulent wake," CM-673, Applied Physics Laboratory, The Johns Hopkins University, Silver Spring, Maryland.

Binder, R. C. and B. M. Ishino (1967). "Effect of high polymer additives on diffuser separation flow noise," AD648 088, Mechanical Engineering Department, University of Southern California, Los Angeles.

Blasius, H. (1908). "Grenzschichte in Flüssigkeiten mit kleiner Reibung," *Z. Math*, Vol. 56, pp. 1–37. (Translated as "The boundary layer in fluids with little friction," NASA TM 1256).

Bogdonoff, S. M. (1955). "Some experimental studies of the separation of supersonic turbulent boundary layers," presented at the Heat Transfer and Fluid Mechanics Institute, University of California at Los Angeles, June 23–25.

Bray, K. N. C. et al. (1960). "Some calculations by the Crocco-Lees and other methods of interaction between shock waves and laminar boundary layers, including effects of heat transfer and suction," ARC Report 21834.

Brebner, G. G. (1957). "Some simple conical camber shapes to produce low lift dependent drag on a slender delta wing," ARC C., p. 428.

Brown, S. N. (1964). "Singularities associated with separating boundary layers," *Phil. Trans., Roy. Soc. (London)* Ser. A., Vol. 257, No. 1084, p. 409.

Cebeci, T. and A. M. O. Smith (1969). "A finite-difference solution of the incompressible turbulent boundary-layer equations by an eddy-viscosity concept," Proceedings, Computation of Turbulent Boundary-Layers–1968, AFOSR–Stanford Conference, Vol. 1, Department of Mechanical Engineering, Stanford University, pp. 346–355.

Chang, P. K. (1966). "The reattachment of laminar cavity flow with heat transfer at hypersonic speeds," AFOSR Technical Report 66-1035.

Chang, P. K. (1970). *Separation of flow*, Pergamon Press, London.

Chang, P. K. and W. H. Dunham (1961). "Laminar separation of flow around symmetrical strut at various angles of attack," David Taylor Model Basin Report 1365.

Chang, P. K. and A. G. Strandhagen (1961). "The viscosity correction of symmetrical model lines for wave-resistance analysis," *Schiffstechnik*, Vol. 40, No. 8, pp. 28–34.

Chapman, D. R. (1956). "A theoretical analysis of heat transfer in regions of separated flow," NACA TN 3792.

Chapman, D. R. et al. (1957). "Investigation of separated flows in supersonic and subsonic streams with emphasis on the effect of transition," NACA Report 1356. (Replaces NACA TN 3869.)

Cooke, J. C. (1959). "Approximate calculation of three-dimensional laminar boundary layers," R.A.E. TN Aero2658. (Also ARC R&M 3201.)

Cooke, J. C. (1965). "The laminar boundary layer on an inclined cone," R.A.E. Technical Report 65178.

Cooke, J. C. (1966). "Supersonic laminar boundary layers on cones," R.A.E. Technical Report 66347.

Crocco, L. and L. Lees (1952). "A mixing theory for the interaction between dissipative flows and nearly isentropic streams," *J. Aeronaut. Sci.*, Vol. 19, No. 10, pp. 649–676.

Curle, N. (1958). "Accurate solutions of the laminar boundary layer equations for flows having a stagnation point and separation," ARC R&M 3164.

Curle, N. (1960a). "The effects of heat transfer on laminar boundary layer separation in supersonic flow," ARC Report 21986.

Curle, N. (1960b). "The estimation of laminar skin friction, including effects of distribution suction," *Aeronaut. Quart.*, Vol. 11, pp. 1–21.

Curle, N. and S. W. Skan (1957). "Approximate methods for predicting separation properties of laminar boundary layers," *Aeronaut. Quart.*, Vol. VIII, pp. 257–268.

Drabeck, S. (1960). "Stall control," presented at the Symposium on Compressor Stall, Surge and System Response, ASME, Houston, Texas, March 6–9.

Drougge, G. (1953). "An experimental investigation of the influence of strong adverse pressure gradients on the turbulent boundary layer at supersonic speeds," Aeronaut. Res. Inst. of Sweden, Report 46.

Eckert, E. R. G. (1955). "Engineering relations for friction and heat transfer to surfaces in high velocity flow," *J. Aeronaut. Sci.*, Vol. 22, No. 8, pp. 585–586.

Eichelbrenner, E. A. and A. Oudart (1955). "Méchode de calcul de la couche limite tri-dimensionelle. Application a un corps fusile incliné sur le vent," O.N.E.R.A. Publication 75.

Emmons, H. W. (1960). "Introduction," paper presented at the Symposium on Compressor Stall, Surge and System Response, Houston, Texas, March 6–9.

Erdos, J. and A. Pallone (1961). "Shock-boundary layer interaction and flow separation," AVCO/RAD TR-61-23.

Erdos, J. and A. Pallone (1962). "Shock-boundary layer interaction and flow separation," proceedings of the 1962 Heat Transfer and Fluid Mechanics Institute held at University of Washington, Stanford University Press, pp. 239–254.

Evans, H. L. (1968). *Laminar boundary-layer theory*, Addison-Wesley, Reading, Mass.

Falkner, V. M. and S. W. Skan (1930). "Some approximate solutions of boundary layer equations," ARC R&M 1314.

Fox, R. W. and S. J. Kline (1961). "Flow regimes in curved diffusers," Paper 61-WA-191, ASME Winter Annual Meeting, New York, N. Y.

Gadd, G. E. (1957). "A theoretical investigation of laminar separation in supersonic flow," *J. Aeronaut. Sci.*, Vol. 24, No. 10, pp. 759–771.

Garner, H. C. (1944). "The development of turbulent boundary layers," ARC R&M 2133.

Görtler, H. (1957). "A new series for the calculation of steady laminar boundary layer flows," *J. Math. Mech.*, Vol. 6, pp. 1–66.

Görtler, H. and H. Witting (1958). "Einige laminare Grenzschichtströmungen berechnet mittels einer neuen Reihen-Methode," *Z. Angew. Math. Phys.*, Vol. IXb, Fasc. 5/6.

Greber, I. (1959). "Interaction of oblique shock-waves with laminar boundary layers," Fluid Dynamics Research Group, Massachusetts Institute of Technology Technical Report 59-2.

Hahn, M. (1966). "Pressure distribution and mass injection effects in the transitional separated flow over a spiked body at supersonic speed," *J. Fluid Mech.*, Vol. 24, Part 2, pp. 209–223.

Hakkinen, R. J. et al. (1959). "The interaction of an oblique shock wave with a laminar boundary layer," NASA Memo 2-18-59W.

Hall, M. G. (1965). "A numerical method for solving the equations for vortex core," RAE Technical Report 65106.

Hartree, D. R. (1939). "A solution of the boundary layer equation for Schubauer's observed pressure distribution for an elliptic cylinder," ARC R&M 3966.

Hazen, D. C. (1967). "The rebirth of subsonic aerodynamics," *Astronaut. and Aeronaut.*, Vol. 5, No. 11, pp. 24–39.

Heinke, P. E. (1946). *Elementary applied aerodynamics*, Prentice-Hall, New York.

Howarth, H. (1938a). "On the solution of laminar boundary equations," *Proc. Roy. Soc. (London)*, Ser. A, Vol. 164, pp. 547–579.

Howarth, H. (1938b). "The theoretical determination of the lift coefficient for a thin elliptic cylinder," ARC R&M 1838. Also *Proc. Roy. Soc. (London)*, Ser. A. Vol. 149, pp. 558–586.

Howarth, H. (1948). "Concerning the effect of compressibility on laminar boundary layers and their separation," *Proc. Roy. Soc. (London)*, Ser. A. Vol. 194, pp. 16–42.

Hurley, D. G. (1961). "The use of boundary layer control to establish free stream-line flows," in *Boundary Layer and Flow Control*, Vol. I, G. V. Lachman, ed., Pergamon Press, New York.

Iglisch, R. (1944). "Exakte Berechnung der laminaren Grenzschicht an der längsangeströmten ebenen Platte mit homogener Absaugung," *Schr. d. Dt. Akad. d. Luftfahrt.*, Vol. 8B, pp. 1–51. (Translated as "Exact calculation of laminar boundary layer in longitudinal flow over a plate with homogeneous suction," NACA TM 1205.)

Jacobs, E. M. (1931). "The aerodynamic characteristics of eight very thick airfoils from tests in the variable density wind tunnel," NACA Report 391.

Kline, S. J. (1959). "On the nature of stall," *J. Basic Eng.*, Ser. D, Vol. 81, No. 3, pp. 305–320.

Kuehn, D. M. (1959). "Experimental investigation of the pressure rise required for the incipient separation of turbulent boundary layers in two-dimensional supersonic flow," NASA Memo 1-21-59A.

Lankford, J. L. (1961). "The effect of heat transfer on separation of laminar flow over axisymmetric compression surfaces: preliminary results at Mach number 6.78," U. S. Naval Ordnance Laboratory, Report 7402.

LeCount, R. et al. (1965). "An analysis of the physical and chemical aspects of liquid injection," LMSC-A-24-65-1, Lockheed Missiles and Space Company, Sunnyvale, California.

Lees, L. and L. Hromas (1962). "Turbulent diffusion in the wake of a blunt-nosed body at hypersonic speeds," Paper 62-71, IAS 30th Annual Meeting, New York, Jan. 22–24.

Lighthill, M. J. (1963). "Introduction to boundary layer theory," in *Laminar boundary layers,* L. Rosenhead, ed., Fluid Motion Memoirs, Oxford, at the Clarendon Press.

Loftin, L. K. and H. B. Wilson (1953). "A rapid method for estimating the separation point of a compressible laminar boundary layer," NACA TN 2892.

Loitsianski, L. G. (1949). "Approximate method of integration of laminar boundary layer in incompressible fluid," NACA TM 1293. (Translated from Russian article in *Prikl. Mat. i Mekh.*, Vol. 13.)

Ludwieg, H. and W. Tillmann (1949). "Untersuchung über die Wandspannung turbulenter Reibungsschichten," *Ing. Arch.*, No. 17, pp. 288–299. (Translated as "Investigations of the wall-shearing stress in turbulent boundary layers," NACA TM 1285.)

Mager, A. (1964). "Three-dimensional laminar boundary layers," in *Theory of laminar flows, High Speed Aerodynamics and Jet Propulsion,* Vol. IV, F. K. Moore, ed., Princeton University Press.

Mair, W. A. (1952). "Experiments on separation of boundary layers on probes in front of blunt nosed bodies in a supersonic air stream," *Philosophical Mag.*, 7th Series, Vol. 43.

Mangler, W. (1948). "Zusammenhang zwischen ebenen und rotationssymmetrischen Grenzschichten in kompressibler Flüssigkeiten," *Z. Angew., Math. Mech.*, Vol. 28, pp. 97–103.

Maskell, E. C. (1951). "Approximate calculation of the turbulent boundary layer in two-dimensional incompressible flow," RAE Report 2443.

Maskell, E. C. (1955). "Flow separation in three dimensions," RAE Report Aero 2565.

Morduchow, M. and J. H. Clarke (1952). "Method for calculation of compressible laminar boundary-layer characteristics in axial pressure gradient with zero heat transfer," NACA TN 2984.

Morduchow, M. and R. G. Grape (1955). "Separation stability and other properties of compressible laminar boundary layer with pressure gradient and heat transfer," NACA TN 3296.

Morduchow, M. and S. P. Reyle (1962). "On calculation of the laminar separation point, and results for certain flows," *J. Aero/Space Sci.*, Vol. 29, No. 8, pp. 996–997.

Nicoll, K. M. (1964). "A study of laminar hypersonic cavity flows," *AIAA J.*, Vol. 2, No. 9 (September), pp. 1535–1541.

Oswatitsch, K. (1958). "Die Ablösungsbedingung von Grenzschichten," Symposium on Boundary Layer Research, International Union of Theoretical and Applied Mechanics, Görtler, ed., Springer-Verlag, Berlin.

Patankar, S. V. and D. B. Spalding (1967). *Heat and mass transfer in boundary layers*, Morgan-Grampian, London.

Pearcey, H. H. (1961). "Shock-induced separation and its prevention," in *Boundary-layer and flow control*, Vol. 2, G. V. Lachmann, ed., Pergamon Press, New York.

Pohlhausen, K. (1921). "Zur näherungsweisen Integration der Differentialgleichung der laminaren Reibungsschicht," *Z. Angew., Math. Mech.*, Vol. 1, pp. 252–268. (Translated as "The approximate integration of the differential equation for the laminar boundary layer," by R. C. Anderson, University of Florida, Gainesville, 1965, AD 645 784.)

Prandtl, L. and O. G. Tietjens (1934). *Applied hydro- and aero-mechanics,* McGraw-Hill, New York. (Also published by Dover Publications, New York, 1957.)

Rainbird, W. J. et al. (1963). "A water tunnel investigation of the flow separation about circular cones at incidence," National Research Council (Canada) Aero Report LR-385.

Reshotko, E., and M. Tucker (1957). "Approximate calculation of the compressible turbulent boundary layer with heat transfer and arbitrary pressure gradient," NACA TN 4154.

Reyhner, T. A. (1969). "Finite-difference solution of the compressible turbulent boundary-layer equations," Proceedings, Computation of Turbulent Boundary-Layers–1968, AFOS–Stanford Conference, Vol. 1, Department of Mechanical Engineering, Stanford University, pp. 375–381.

Rosenhead, L. (ed.) (1963). *Laminar boundary layer, fluid motion memoirs,* Oxford, at the Clarendon Press. p. 331.

Roshko, A. (1954). "On the drag and shedding frequency of two-dimensional bluff bodies," NACA TN 3169.

Saljimilor, V. (1960). "Übertragung der Görtlerschen Reihe auf die Berechnung von Grenzschichten an Rotationskörpern," *Dt. Versuchsanstalt f. Luftfahrt, Bericht,* No. 133, October.

Schlichting, H. (1968). *Boundary layer theory*, Sixth Edition, McGraw-Hill, New York.

Schlichting, H. and K. Bussmann (1943). "Exakte Lösungen für die laminare Grenzschicht mit Absaugung und Ausblasen," *Schr. d. Dt. Akad. d. Luftfahrtf.* Vol. 7B, pp. 25–69.

Shvets, M. E. (1949). "Method of successive approximations for the solution of certain problems in aerodynamics," NACA TM 1286. (Translated from a Russian Article in *Prikl. Mat. i Mek.*, Vol. 13, No. 3.)

Soulé, H. (1967). "Stalls and Spins," *Astronaut. & Aeronaut.*, Vol. 5., No. 4 (April).

Sterrett, J. R. and J. C. Emery (1960). "Extension of boundary layer separation criteria to a Mach number of 6.5 by utilizing flat plates and forward facing steps," NASA TN D-618.

Stewartson, K. (1949). "Correlated incompressible and compressible boundary layers," *Proc. Roy. Soc. (London)*, Ser. A, Vol. 200, pp. 84–100.

Stratford, B. S. (1957). "Flow in the laminar boundary layer near separation," ARC Technical Report R&M 3002.

Tani, I. (1949). "On the solution of the laminar boundary layer equations," *J. Phys. Soc. Japan,* Vol. 4, pp. 149–154.

Tani, I. (1954). "On the approximate solution of the laminar boundary-layer equations," *J. Aero Sci.*, Vol. 21, pp. 487–495 and 504.

Terrill, R. M. (1960). "Laminar boundary layer flow near separation with and without suction," *Phil. Trans., Roy. Soc. (London)*, Ser. A, Vol. 253, pp. 55–100.

Thwaites, B. (1949). "Approximate calculation of the laminar boundary layer," *Aeronaut. Quart.*, Vol. 1, pp. 255–280.

Thwaites, B. (1960). *Incompressible aerodynamics*, Oxford at the Clarendon Press.

Timman, R. A. (1949). "One-parameter method for the calculation of laminar boundary layers," Rep. Trans. Nat. Luchtvlab., Amsterdam, Vol. 15, F29-45. (Also available as Rep. Nat. Luchtvlab., Amsterdam, F35.)

Torrillo, D. T. and S. B. Savage (1964). "A brief review of compressible laminar and turbulent boundary layer separation," RAC 2325, Republic Aviation Corporation (AD 809020).

Truckenbrodt, E. (1952a). Ein Quadraturverfahren zur Berechnung der laminaren and turbulenten Reibungsschichten bei ebener und rotationssymmetrischer Strömung, *Ing. Arch.*, Vol. 20, pp. 211–228.

Truckenbrodt, E. (1952b). "An approximate method for the calculation of the laminar and turbulent boundary layer by simple quadrature for two-dimensional and axially symmetric flow," *J. Aeronaut. Sci.*, Vol. 19, No. 6, pp. 428–429.

von Doenhoff, A. E. (1938). "A method of rapidly estimating the position of the laminar separation point," NACA TN 671.

von Doenhoff, A. E. and N. Tetervin (1943). "Determination of general relations for the behavior of turbulent boundary layers," NACA Report 772.

von Karman, T. and L. Edson (1967). *The wind and beyond, pioneer in aviation and pathfinder in space,* Little, Brown & Company, Boston.

von Karman, T. and C. B. Millikan (1934). "On the theory of laminar boundary layers involving separation," NACA Report 504.

Walz, A. (1941). "Ein neuer Ansatz für das Geschwindigkeitsprofil der laminaren Reibungsschicht," Ber. Lilienthal-Ges. Luftfahrtf. No. 141.

Weick, F. E. and M. J. Bamber (1932). "Wind tunnel tests of a Clark Y wing with a narrow auxiliary airfoil in different positions," NACA Report 428.

Wrage, E. (1960). "Entwicklung und Anwendung einer allgemeinen Reihenmethode zur Berechnung laminarer, kompressibler Grenzschichten," Bericht No. 134 (Nov.), Deutsche Versuchsanstalt für Luftfahrt.

Yuan, S. W. et al. (1966). "Investigation of circulation control airfoils by means of jets," USA AVLABS Technical Report 66-72. (Prepared by Department of Aerospace Engineering, University of Texas, Austin, Texas.)

Zaat, J. A. (1956). "A simplified method for the calculation of three-dimensional laminar boundary layers," Rep. Nat. Luchtvlab, Amsterdam, F184.

SUPPLEMENTARY REFERENCES

Ackerberg, R. C. (1970). "Boundary-layer separation at a free streamline: Part 1, Two-dimensional flow," *J. Fluid Mech.*, Vol. 44, p. 211.

Ackerberg, R. C. (1971). "Appendix to boundary-layer separation at a free streamline: Part 2." *J. Fluid Mech.*, Vol. 46, p. 727.

Ackerberg, R. C. (1971). "Boundary-layer separation at a free-streamline: Part 2, Numerical results." *J. Fluid Mech.*, Vol. 46, p. 727.

Ackerberg, R. C. (1973). "Boundary-layer separation at a free streamline: Part 3, Axisymmetric flow and the flow downstream of separation." *J. Fluid Mech.*, Vol. 59, p. 645.

Addy, A. L. et al. (1973). "A study of flow separation in the base region and its effects during powered flight." AGARD Aerodyn. Drag, p. 15.

Albers, J. A. (1971). "Two-dimensional potential flow and boundary layer analysis of the airfoil of a STOL wing propulsion system," Ph.D. thesis, Michigan State University, East Lansing.

Amick, J. L. and T. Masoud (1973). "Reattachment of a separated boundary layer to a convex surface," *AIAA J.*, Vol. 11, No. 8, p. 1426.

Appels, C. (1973). "Turbulent boundary layer separation," Tech. Note 90, von Karman Institute for Fluid Dynamics, March 12, Rhode Saint Genese, Belgium.

Appels, C. (1974). "Incipient separation of a compressible turbulent boundary layer," Tech. Note 99, von Karman Institute for Fluid Dynamics, Rhode Saint Genese, Belgium.

Badri Narayanan M. A. et al. (1974). "Similarities in pressure distribution in separated flow behind backward facing steps." *Aeronaut. Quart.*, Vol. XXV, Pt. 4, pp. 305–312.

Ball, K. O. W. (1971). "A summary of the factors influencing the extent of separation of a laminar boundary layer due to a compression corner at moderately hypersonic speeds," ARL-71-0065, Aerospace Research Lab., U. S. Air Force Systems Command, Wright-Patterson AFB, Ohio.

Batson, J. L. and R. H. Sforzini (1970). "Swirling flow through a nozzle," *J. Spacecraft and Rockets*, Vol. 7, p. 159.

Berger, E. (1964). "Unterdrückung der laminaren Wirbelströmung und des Turbulenzeinsutzes der Kármánschen Wirbelstrasse im Nachlauf eines schwingenden Zylinders bei kleinen Reynoldszahlen," *Jahrb. 1964 der W.G.L.R.*, pp. 164–172.

Berger, S. A. (1971). *Laminar Wake*, Am. Elsevier, New York.

Betham, J. P. (1972). "An experimental study of turbulent separating and reattaching flows at a high Mach number. *J. Fluid Mech.*, Vol. 52, p. 425.

Bevilaqua, P. M. and P. S. Lykoudis (1971). "Mechanism of entrainment in turbulent wakes," *AIAA J.*, Vol. 9, No. 8, p. 1657.

Boggess, A. L. (1972). "An investigation of the unsteady flow associated with plume induced flow separation," NASA-CR-112218, BER-149-02, Alabama University.

Bordner, G. L. and R. T. Davis (1971). "Compressible three-dimensional laminar boundary layers on cones at incidence to shear and axisymmetric wake flows," ARL-71-0262, Aerospace Research Lab., U. S. Air Force Systems Command, Wright-Patterson AFB, Ohio.

Briley, W. R. (1971). "A numerical study of laminar separation bubble using the Navier-Stokes equations," *J. Fluid Mech.*, Vol. 47, p. 713.

Brodkey, R. S. (1975). *Turbulence in mixing operations: Theory and application to mixing and reaction*, Academic, New York.

Brown, C. E. (1973). "Aerodynamics of wake vortices," *AIAA J.*, Vol. 11, No. 4, p. 531.

Butter, D. J. et al. (1971). "A numerical method for calculating the trailing vortex system behind a swept wing at low speed," *Aero. J. Roy. Aero. Soc.*, Vol. 75, pp. 564–568.

Cebeci, T. et al. (1972). "Calculation of separation points in incompressible turbulent flows," *J. Aircraft*, Vol. 9, p. 618.

Chang, P. K. (1966). "The reattachment of laminar cavity flow with heat transfer at hypersonic speed," Tech. Report, The Catholic University of America, Washington, D.C.

Chang, P. K. (1971). "A numerical solution for the incompressible turbulent wake downstream of a body of revolution," TN 9197, U. S. Naval Ordnance Lab., White Oak, Maryland.

Chang, P. K. (1972). "Discussion to the paper: Experiments on flow about a yawed circular cylinder: Trans. ASME," *J. Basic Eng.*, Vol. 94-D-4, p. 776.

Chang, P. K., W. J. Kelnhofer, and S. K. Min (1974). "Analysis for flow field around buildings," The Catholic University of America, Washington, D.C.

Chang, P. K., S. K. Oh, and R. A. Smith (1972). "Inviscid hypersonic flow around a cone at angles of attack," *Zeitschrift f. Flugwissenschaften*, Vol. 20, No. 4, pp. 141–146.

Chang, P. K. and Y. H. Oh (1968). "Axially symmetric incompressible turbulent wake downstream of a single body," *J. Hydronaut.*, Vol. 2, No. 4, p. 223.

Chang, P. K. and Y. H. Oh (1972). "Analysis of the incompressible rotationally symmetric turbulent wake downstream of screw propeller," Tech. Report, The Catholic University of America, Washington, D.C.

Chang, P. K. and Y. H. Oh (1973). "Reply by authors to R. H. Cramer," *J. Hydronaut.*, Vol. 7, No. 2, p. 96.

Chang, P. K. et al. (1967). "The investigation of the aerodynamic and heat transfer characteristics of slender hypersonic vehicles over a wide range of angles of attack," Tech. Report No. 1, The Catholic University of America, Washington, D.C.

Chang, P. K. et al. (1968). "Analysis of laminar flow and heat transfer on a hypersonic cone at high angle of attack," Tech. Report No. 2, The Catholic University of America, Washington, D.C.

Chang, P. K. et al. (1969). "Analysis of inviscid flow field in shock layer, separated flow in leeward side and three-dimensional attached flow," Tech. Report No. 3, The Catholic University of America, Washington, D.C.

Chang, P. K. et al. (1969). "Analytical investigation of laminar flow field and heat transfer on leeward side of a sharp nosed hypersonic cone at large angle of attack," presented at 8th Navy Symposium on Aeroballistics, May 6–8, Corona, Calif.

Chang, P. K. et al. (1969). "Numerical analysis of inviscid flow in shock layer of sharp nosed hypersonic cone at angles of attack," Tech. Rept., The Catholic University of America, Washington, D.C.

Chang, P. K. et al. (1971). "The investigation for the aerodynamic and heat transfer characteristics of slender hypersonic cones over a wide range of angles of attack," AD-72-5606, Tech. Rept. No. 4, The Catholic University of America, Washington, D.C.

Chieng, C. C. et al. (1974). "Investigation of the turbulent properties of the wake behind self-propelled, axisymmetric bodies," Aerospace and Ocean Engineering Department, Virginia Polytechnic Institute and State University, Blacksburg, Virginia.

Chigier, N. A. and V. R. Corsiglia (1972), "Wind-tunnel studies of wing wake turbulence," *J. Aircraft,* Vol. 9, p. 820.

Chou, F. K. and V. A. Sandborn (1973). "Prediction of the turbulent boundary layer separation," AD-766845, CER-73-74FKC-VAS3, Fluid Dynamic and Diffusion Lab., Colorado State Univ., Fort Collins. THEMIS-CER-TR-22, avail: NTIS CSCL 20/4.

Coder, D. W. (1971). "Location of separation on a circular cylinder in crossflow as a function of Reynolds number," AD-733983, NSRDC-3647, Naval Ship Research and Development Center, Bethesda, Md.

Coe, P. L. (1972). "Stationary vortices behind a flat plate normal to the freestream in incompressible flow," *AIAA J.*, Vol. 10, No. 12, p. 1701.

Coleman, G. T. and J. L. Stollery (1972). "Heat transfer in hypersonic turbulent separated flow," IC-Aero-72-05, Imperial Coll. of Science and Technology, London.

Coleman, G. T. and J. L. Stollery (1974). "Incipient separation of axially symmetric hypersonic turbulent boundary layers," *AIAA J.*, Vol. 12, No. 1, pp. 119–120.

Colenhour, J. L. and B. W. Farquhar (1971). "Inlet vortex," *J. Aircraft*, Vol. 8, p. 39.

Crabbe, R. S. (1965). "Flow separation about elliptic cones at incidence," Aeronautical Rept. LR-436, National Research Council of Canada, Ottawa.

Dadone, L. U. and T. Fukushima (1974). "Investigation of rotor blade element airloads for a teetering rotor in the blade stall regime," NASA-CR-137534, D210-10792-1.

Danberg, J. E. and K. S. Fansler, (1974). "An investigation of the Moore-Rott-Sears criterion for laminar boundary layer separation," Tech. Rept. No. 172, University of Delaware, Newark.

Danberg, J. E. and K. S. Fansler (1974). "Additional two-dimensional wake and jet-like flows," Report 1727, U.S.A. Ballistic Research Laboratories, Aberdeen Proving Ground, Maryland.

DePaul, M. V. (1972). "Airfoil stall prediction in incompressible flow," Office National d'Etudes et de Recherches Aerosatiales, Paris, in French. (In AGARD, Fluid Dyn. of Aircraft Stalling.)

Diamant, B. and P. K. Chang (1970). "Hypersonic aerodynamic characteristics of sharp, slender, right-circular cones at angles of attack including the effects of separated flow region," AIAA Paper, 70-9797, presented at AIAA Guidance, Control and Flight Mechanics Conference, Aug. 17–19, Santa Barbara, Calif.

Diamant, B. and P. K. Chang (1971). "Hypersonic characteristics of sharp cones at angles of attack," *J. Spacecraft and Rockets*, Vol. 8, p. 1105.

Ericsson, L. E. and J. Peter Reding (1971). "Dynamic stall simulation problems," *J. Aircraft,* Vol. 8, p. 579.

Ericsson, L. E. et al. (1971). "Analytic difficulties in predicting dynamic effects of separated flow," *J. Spacecraft and Rockets*, Vol. 8, p. 872.

Fannelop, T. K. and G. D. Wadman (1968). "Displacement interaction and flow separation on cones at incidence to a hypersonic stream," AGARD-CP-30, pp. (21-1)-(21-20).

Fidler, J. E. (1974). "Approximate method for estimating wake vortex strength," *AIAA J.*, Vol. 12, No. 5, pp. 630–635.

Finson, M. L. (1973). "Hypersonic wake aerodynamics at high Reynolds numbers," *AIAA J.*, Vol. 11, No. 8, p. 1137.

Fluid Physics of Hypersonic Wakes (1967). Vol. 1 (May), Vol. 2 (May), AGARD Conference Proceedings, No. 19.

Fox, J. and H. Rungaldier (1972). "Electron density measurements in projectile wakes," *AIAA J.*, Vol. 10, No. 6, p. 790.

Geissler, W. (1974). "Three-dimensional laminar boundary layer over a body of revolution at incidence and with separation," *AIAA J.*, Vol. 12, No. 12, pp. 1743–1745.

Gerhart, P. M. (1974). "On prediction of separated boundary layers with pressure distribution specified," *AIAA J.*, Vol. 12, No. 9, pp. 1278–1279.

Gersten, K. and R. Löhr (1962). "Studies on lift increase of wings with simultaneous blowing on the flap of the trailing edge and on the profile nose," (in German) paper presented at the Fifth European Aviation Congress, Venice, Italy, September 12–15. (English translation, Boeing Scientific Research Laboratories, Translation Request No. 171, 1963.)

Ginoux, J. J. (1965). "Effect of Mach number on streamwise vortices in laminar reattaching flows," T.N. 26, von Karman Institute for Fluid Dynamics, Rhode-Saint-Genese, Belgium, July.

Ginoux, J. J. (1965). "Laminar separation in supersonic and hypersonic flows," Contract AF EOAR Final Report, von Karman Institute for Fluid Dynamics, Rhode-Saint-Genese, Belgium, October.

Ginoux, J. J. (1965). "Investigation of flow separation over ramps at $M_\infty = 3$," AEDC-TR-65-273, von Karman Gas Dynamics Facility, Arnold Engineering Development Center, December.

Ginoux, J. J. (1966). "Laminar separation in supersonic and hypersonic flows," Contract AF EOAR 66-6, Final Report, von Karman Institute for Fluid Dynamics, Rhode-Saint-Genese, Belgium, September.

Ginoux, J. J. (1967). "Laminar separation in hypersonic flows," Grant AF EOAR 67-09, Final Scientific Report, von Karman Institute for Fluid Dynamics, Rhode-Saint-Genese, Belgium, November.

Ginoux, J. J. (1969). "Supersonic separated flows over wedges and flares with emphasis on a method of detecting transition," ARL 69-0009, Aerospace Research Laboratories, Wright-Patterson Air Force Base, Ohio, January.

Goldberg, T. J. (1973). "Three-dimensional separation for interaction of shock waves with turbulent boundary layers," *AIAA J.*, Vol. 11, No. 11, p. 1573.

Granger, R. A. (1974). "Observations of surge in a forced vortex flow," *J. Ship Research*, Vol. 18, No. 1, pp. 12–15.

Granville, P. S. (1975). "Modified law of the wake for turbulent shear flows," Report 4639, Naval Ship Research and Development Center, Carderock, Maryland.

Green, M. and J. Rom (1970). "Measurements of heat transfer rates behind axissymmetric backward facing steps in the shock tube: Part 1," AD-735745, TAE-115, ARL-71-0305-Pt.1, Technion-Israel Inst. of Tech., Haifa.

Guffroy, D. et al. (1968). "Etude theorique et experimentale de la couche limite autour d'un cone circulaire place en incidence dans un courant hypersonique," AGARD-CP-30, pp. (20-1)-(20-18).

Hartzuiker, J. P. (1972). "An experimental investigation of the effect of the approaching boundary layer on the separated flow behind a downstream facing step," NLR-TR-70033-U, National Aerospace Lab., Amsterdam (Netherlands).

Hasinger, S. H. (1971). "Limit of centrifugal separation in free vortex," *AIAA J.*, Vol. 9, No. 4, p. 644.

Hodgson, J. W. (1970). "Heat transfer in separated laminar hypersonic flow," *AIAA J.*, Vol. 8, No. 12, p. 2291.

Hoffman, J. D. and H. R. Velkoff (1971). "Vortex flow over helicopter rotor tips," *J. Aircraft*, Vol. 8, p. 739.

Holden, M. S. (1971). "Establishment time of laminar separated flows," *AIAA J.*, Vol. 9, No. 11, p. 2296.

Holzapfel, C. (1971). "Ueber elektrische Verluste im MHD-generator," Angewandte Magneto-Hydrodynamik, Heft 7 (Electrical losses in the MHD generator), Kernforschungsanlage Juelich, W. Germany, Jul-742-TP.

Horstman, C. C. and F. K. Owen (1974). "New diagnostic technique for the study of turbulent boundary-layer separation," *AIAA J.*, Vol. 12, No. 10, pp. 1436–1438.

Horton, H. P. (1974). "Separating laminar boundary layers with prescribed wall shear," *AIAA J.*, Vol. 12, No. 12, pp. 1772–1774.

Inger, G. R. and B. Dutt (1973). "Analytical study of subsonic laminar boundary layer separation and reattachment including viscous-inviscid interaction," AD-767602, VPI-E-73-20, Virginia Polytechnic Inst. and State Univ., Blacksburg. (Dept. of Aerospace Engineering, AFOSR-73-1686 TR. Avail: NTIS CSCL 20/4.)

Inouye, M. et al. (1972). "Turbulent-wake calculations with an eddy-viscosity model," *AIAA J.*, Vol. 10, No. 2, p. 216.

Johnson, G. M. (1971). "An empirical model of the motion of turbulent rings," *AIAA J.*, Vol. 9, No. 4, p. 763.

Kawamura, R. and T. Nakajima (1974). "Hypersonic flow around blunt body at angle of attack," Rept. No. 515, Institute of Space and Aeronautical Science, University of Tokyo, 39(12).

Kayser, L. D. and J. E. Danberg (1974). "Experimental study of separation from the base of a cone at supersonic speeds," *AIAA J.*, Vol. 12, No. 11, pp. 1607–1609.

Kirkpatrick, A. et al. (1966). "Progress in electrostatic probe measurements in a hypersonic turbulent wake," Canadian Armament Research and Development Establishment, CARDE.N. 1733/66, Project D46-95-51-10.

Klemp, J. B. and A. Acrivos (1972). "High Reynolds number steady separated flow past a wedge of negative angle," *J. Fluid Mech.*, Vol. 56, p. 577.

Korkegi, R. H. (1973). "A simple correlation for incipient turbulent boundary-layer separation due to a skewed shock wave," *AIAA J.*, Vol. 11, No. 11, p. 1578.

Kraemer, K. (1962). "Die Potentialströmung mit Totwasser an einer geknickten Wand," Aerodynamische Versuchsanstalt 62 A 02, Göttingen.

Kraemer, K. (1964). "Die Druckverteilung am Keil inkompressibler Strömung, ein Beitrag zum Totwasserproblem," Mitteilungen aus dem Max-Planck-Institut für Strömungsforschung und der Aerodynamischen Versuchsanstalt, Nr. 30, Göttingen.

Kubin, J. S. (1973). "An analysis of steady asymmetric vortex shedding from a missile at high angles of attack," M.S. thesis, AD-774390, GAM/AE/73A-13, U. S. Air Force Inst. of Tech., Wright-Patterson AFB, Ohio.

Kuethe, A. M. (1972). "Effect of streamwise vortices on wake properties associated with sound generation," *J. Aircraft*, Vol. 9, p. 715.

Kuhn, G. D. and J. N. Nielsen (1974). "Prediction of turbulent separated boundary layers," *AIAA J.*, Vol. 12, No. 7, pp. 881–882.

Law, C. H. (1974). "Supersonic, turbulent boundary layer separation," *AIAA J.*, Vol. 12, No. 6, p. 794.

Leal, L. G. (1973). "Steady separated flow in a linearly decelerated free stream," *J. Fluid Mech.*, Vol. 59, p. 513.

Lien, H. and J. Eckerman (1966). "Interferometric analysis of density fluctuations in hypersonic turbulent wakes," *AIAA J.*, Vol. 4, No. 11, pp. 1988–1994.

Mabey, D. G. (1972). "Analysis and correlation of data on pressure fluctuations in separated flow," *J. Aircraft*, Vol. 9, p. 642.

McErlean, D. P. and C. E. G. Przirembel (1970). "The turbulent near-wake of an axisymmetric body at subsonic speeds," AFOSR-70-0449-TR, RU-TR-132-MAE-F, Rutgers–The State University, New Brunswick, New Jersey.

Madorskii, E. Z. (1973). "Generalized dependences for the parameters at the flow separation boundary in compressor blades," Foreign Technology Div., U. S. Air Force Systems Command, Wright-Patterson AFB, Ohio. (Transl. into English from *Energomashinostroenie (USSR)*, Vol. 2, pp. 39–40.)

Mager, A. (1971). "Incompressible viscous swirling flow through a nozzle," *AIAA J.*, Vol. 9, No. 4, p. 649.

Mager, A. (1974). "Steady, incompressible; swirling jets and wakes," *AIAA J.*, Vol. 12, No. 11, pp. 1540–1547.

Mason, W. H. and J. F. Marchman III (1973). "Far-field structure of aircraft wake turbulence," *J. Aircraft*, Vol. 10, p. 86.

Matthews, R. D. and J. J. Ginoux (1974). "Correlation of peak heating in the reattachment region of separated flows," *AIAA J.*, Vol. 12, No. 3, pp. 397–399.

Mays, R. A. (1971). "Inlet dynamics and compressor surge," *J. Aircraft*, Vol. 8, p. 219.

Mehta, U. B. and Z. Lavan (1975). "Starting vortex, separation bubbles and stall: A numerical study of laminar unsteady flow around an airfoil," *J. Fluid Mech.*, Vol. 67, Pt. 2, pp. 227–256.

Miles, J. W. (1971). "Upstream boundary-layer separation in stratifield flow," *J. Fluid Mech.*, Vol. 48, p. 791.

Miller, E. H. and D. Migdal (1970). "Separation and stability studies of a convergent-divergent nozzle," *J. Aircraft*, Vol. 7, p. 159.

Moss, W. D. (1972). "Flow separation at the upstream edge of a square-edged broadcrested weir," *J. Fluid Mech.*, Vol. 52, p. 307.

Mukerjee, T. et al. (1971). "A note on turbulent shear-layer reattachment downstream of a backward-facing step in confined supersonic two-dimensional flow," *J. Fluid Mech.*, Vol. 46, p. 293.

Narashima, R. and A. Prabhu (1972). "Equilibrium and relaxation in turbulent wakes," *J. Fluid Mech.*, Vol. 54, p. 1.

Norton, D. J. et al. (1973). "Surface temperature effect on subsonic stall," *J. Spacecraft and Rockets*, Vol. 10, No. 9, pp. 581–587.

Ohrenberger, J. T. and E. Baum (1972). "A theoretical model of the near wake of a slender body in supersonic flow," *AIAA J.*, Vol. 10, No. 9, p. 1165.

Page, R. H. (1971). "Separated flows," AD-733729, RU-TR-138-MAE-F, AFOSR-71-2325 TR, Rutgers University, New Brunswick, N. J.

Peake, D. J. et al. (1972). "Three-dimensional flow separations on aircraft and missiles," *AIAA J.*, Vol. 10, No. 5, p. 567.

Peregrine, D. H. (1971). "A ship's waves and its wake," *J. Fluid Mech.*, Vol. 49, p. 353.

Perrier, P. and J. J. Deviers (1972). "Calculs tri dimensionnels d'hypersustentation (Three-dimensional calculation of hypersustentation)," Soc. Nat'l. Ind. Aerospatiale, in French. Presented at 9th Conf. on Appl. Aerodyn., Nov. 8–10, 1972 Paris, and Nov. 9, 1972 in Saint-Cyr-l'Ecole, France.

Peters, A. R. and G. A. Phelps (1974). "The effect of inlet velocity profile shape on flow separation in a confined two-dimensional channel," *Trans. ASME, J. Fluids Eng.*, Vol. 96, No. 3, Ser. 1, pp. 305–306.

Petersohn, E. G. M. (1957). "Berücksichtigung der Ablösung bei Widerstandsbestimmungen durch Modellversuche," *Schiffstechnik,* Vol. 4, No. 20, pp. 71–74. ("Consideration of flow separation in determining resistance by means of model test," translation 278 by E. N. Labouvie, June 1959, David Taylor Model Basin.)

Pick, G. S. and S. E. Dawson (1975). "Multi-channel pressure telemetry system for base pressure measurements on small models," Report 4698, Naval Ship Research and Development Center, Carderock, Maryland.

Pick, G. S. et al. (1975). "Aerodynamic characteristics of a 10° sharp cone at hypersonic speeds and high angles of attack," Report 4692, Naval Ship Research and Development Center, Carderock, Maryland.

Pirri, A. N. (1972). "Decay of boundary-layer turbulence in near wake of a slender body," *AIAA J.*, Vol. 10, No. 5, p. 657.

Ranger, K. B. (1972). "An extension of the MHD Jeffery-Hamel flow," AD-753149, MRC-TSR-1280, University of Wisconsin, Mathematics Research Center, Madison.

Rao, D. M. and A. H. Whitehead, Jr. (1972). "Lee-side vortices on delta wings at hypersonic speeds," *AIAA J.*, Vol. 10, No. 11, p. 1458.

Riddhagni, P. R. et al. (1971). "Measurements in the turbulent wake of a sphere," *AIAA J.*, Vol. 9, No. 7, p. 1433.

Roache, P. J. and T. J. Mueller (1970). "Numerical solutions of laminar separated flows," *AIAA J.*, Vol. 8, No. 3, p. 530.

Roberts, M. L. (1970). "Transitional flow separation upstream of a compression corner," *J. Spacecraft and Rockets*, Vol. 7, p. 1113.

Sarpkaya, T. (1971). "Vortex breakdown in swirling conical flows," *AIAA J.*, Vol. 9, No. 9, p. 1792.

Sato, H. (1970). "An experimental study of non-linear interaction of velocity fluctuations in the transition region of a two-dimensional wake," *J. Fluid Mech.*, Vol. 42, p. 289.

Seath, D. D. (1971). "Equilibrium vortex positions," *J. Spacecraft and Rockets*, Vol. 8, p. 72.

Sedney, R. (1973). "A survey of the effects of small protuberances on boundary-layer flows." *AIAA J.*, Vol. 11, No. 6, p. 782.

Sfeir. A. A. (1973). "On the integral properties of separated laminar boundary layers," *J. Fluid Mech.*, Vol. 60, p. 97.

Sforzini, R. H. and J. E. Essing (1970). "Swirling flow through multiple nozzles," *J. Spacecraft and Rockets*, Vol. 7, p. 1366.

Shang, J. J. et al. (1971). "A rearward facing step in a hypersonic stream," AD-725066, ARL-71-0030, DOD-Element-61102F, DOD-Subelement-68-1307, U. S. Air Force Systems Command, Wright-Patterson AFB, Ohio.

Sirieix, M. (1973). "Drag and separation," N 74-14709 06-01. Office National d'Etude et de Recherches Aerospatiales, Paris, in French. In AGARD, Aerodyn. Drag, p. 23.

Smith, C. R. and S. J. Kline (1974). "An experimental investigation of the transitory stall regime in two-dimensional diffusers," *Trans. ASME, J. Fluids Engineering,* Ser. 1, Vol. 96, No. 1, pp. 11–15.

Smith, R. A. and P. K. Chang (1970). "Analysis of the laminar compressible three-dimensional boundary layer with heat transfer or a cone at high angle of attack," *AIAA J.*, Vol. 8, No. 11, p. 1921.

Snel, H. (1974). "A method for the calculation of the flow field induced by a jet exhausting perpendicularly into a cross flow," AGARD-CP-143, pp. (18-1)–(18-16).

Spaid, F. W. (1972). "Cooled supersonic turbulent boundary separated by a forward facing step," *AIAA J.*, Vol. 10, No. 8, p. 1117.

Spalding, D. B. (1972). "Transfer of momentum and heat in turbulent separated flows," Imperial Coll. of Science and Technology, London. Dept. of Mechanical Engineering. Presented at the 4th All-Union Conf. on Heat and Mass Transfer, Minsk, EF/TN/A/49.

Stetson, K. F. (1971). "Experimental results of laminar boundary layer separation on a slender cone at angle of attack at $M_\infty = 14.2$," ARL-71-0127, Aerospace Research Lab., U. S. Air Force Systems Command, Wright-Patterson AFB, Ohio.

Stetson, K. F. (1972). "Boundary-layer separation on slender cones at angle of attack," *AIAA J.*, Vol. 10, No. 5, p. 642.

Stewartson, K. (1970). "Is the singularity at separation removable?" *J. Fluid Mech.*, Vol. 44, p. 347.

Stratford, B. S. (1973). "The prevention of separation and flow reversal in the corners of compressor blade cascades," *Aeron. J. Roy. Aero. Soc.*, Vol. 77, pp. 249–256.

Tanner, M. (1964). "Zur Bestimmung des Totwasserwiderstandes mit Anwendung auf Totwasser hinter Keilen," Mitteilungen aus dem Max-Planck-Institut für Strömungsforschung und der Aerodynamischen Versuchsanstalt, Herausgegeben von W. Tollmien und H. Schlichting, Nr. 31, Gottingen.

Tanner, M. (1965). "Theorie der Totwasserströmungen um angestellte Keile," Deutsche Luft- und Raumfahrt, Forschungsbericht 65-14 (DLR FB 65-14), March.

Tanner, M. (1965). "Einflusz der Reynoldszahl und der Grenzschichtdicke auf den Totwasserdruck bei der Umströmung von Keilen," Deutsche Luft- und Raumfahrt, Forschungsbericht 65-18 (DLR FB 65-18), May.

Tanner, M. (1966). "Über die Strömungsvorgänge beim periodischen Totwasser," Deutsche Luft- und Raumfahrt, Forschugsbericht 66-71 (DLR FB 66-71).

Tanner, M. (1967). "Ein Verfahren zur Berechnung des Totwasserdruckes und Widerstandes von stumpfen Körpern bei inkompressibler, nichtperiodischer Totwasserströmung," Mitteilungen aus dem Max-Planck-Institut für Strömungsforschung und der Aerodynamischen Versuchsanstalt, Herausgegeben von W. Tolmien und H. Schlichting, Nr. 39, Göttingen.

Tanner, M. (1968). "Druckverteilungsmessungen an einem Kreiszylinder im kritischen Reynoldszahlberich," Deutsche Luft- und Raumfahrt, Forschungsbericht 68-08 (DLR FB 68-08).

Tanner, M. (1970). "Druckverteilungsmessungen an Keilen bei kompressibler Strömung," *Z. Flugwiss.*, Vol. 18, No. 6, pp. 202–208.

Telionis, D. P. and D. Th. Tsahalis (1973). "Numerical investigation of unsteady boundary-layer separation," *Physics of Fluids*, Vol. 16, No. 7, pp. 968–973.

Telionis, D. P. and D. Th. Tsahalis (1973). "Unsteady laminar separation over cylinders started impulsively from rest," presented at 24th International Astronautical Congress, Oct. 7–13, 1973, Baku, USSR, Session: Fluid Mechanic Aspects of Space Flight.

Telionis, J. P. and D. Th. Tsahalis (1974). "Response of separation to impulsive changes of outer flow," *AIAA J.*, Vol. 12, No. 5, pp. 614–619.

Teyssandier, R. G. (1973). "Internal separated flows expansions, nozzles and orifices," Ph.D. thesis, Rhode Island University, Kingston.

Tritton, D. J. (1971). "A note on vortex streets behind circular cylinders at low Reynolds numbers," *J. Fluid Mech.*, Vol. 45, p. 203.

Tsahalis, D. Th. and D. P. Telionis (1974). "Oscillating laminar boundary layers and unsteady separation," *AIAA J.*, Vol. 12, No. 11, pp. 1469–1476.

Wang, K. C. (1972). "Separation patterns of boundary layer over an inclined body of revolution," *AIAA J.*, Vol. 10, No. 8, pp. 1044–1050.

Wang, K. C. (1973–74). "Three-dimensional laminar boundary layer over body of revolution at incidence," AFOSR Scientific Report, AFOSR-TR-73-1045 Part VI, AFOSR-TR-73-1265 Part VII, and AFOSR-TR-74-14C Part VIII. (Also MML-TR-73-02C Part VI, MML-TR-73-08C, and MML-TR-74-14C, Martin Marietta Lab., Baltimore, Md.)

Wang, K. C. (1974). "Boundary layer over a blunt body at high incidence with an open-type of separation," *Proc. R. Soc. (London)*, Ser. A., Vol. 340, pp. 33–55.

Wang, K. C. (1974). "Laminar boundary layer near the symmetry plane of a prolate spheroid," *AIAA J.*, Vol. 12, No. 7, pp. 949–958.

Wehrmann, O. (1957). "Hitzdrahtmessungen in einer aufgespaltenen Kármánschen Wirbelstrasse," Forschungsberichte des Wirtschafts–und Verkehrsministeriums Nordsheim–

(Continued on p. 504)

chapter 2
EXPERIMENTAL TECHNIQUES APPLIED TO FLOW SEPARATION

INTRODUCTORY REMARKS

The interpretation of flow phenomena associated with separation is not always straightforward and requires both skill and experience. Therefore, it is necessary to examine in detail the various experimental techniques available in order to make the right choice for the given conditions of the experiment as well as for the particular problem of flow separation that is to be studied. Concern for experimental techniques is not particularly evident in the literature. It is hoped that special attention will be given them in the future so that the physical phenomena of flow separation may be understood better and determined more accurately.

Among the various human organs of sense, the eyes are the most effective; they account for over one-half of all information received by all these organs. Thus the actual observance of flow may eliminate some mental block and lead to an understanding of an obscure flow phenomenon. Therefore flow visualization based solely on observation and needing no tedious data reduction is a very effective and satisfactory technique for obtaining qualitative information. However, considerable experience is needed to interpret the flow pictures correctly.

Flow visualization techniques are often grouped in two categories; those applicable to wind tunnels and those applicable to liquid channels. This chapter employs a more direct classification and groups the methods as optical and nonoptical. Optical methods are especially useful for compressible flow. Although all real flows are compressible to some extent, optical methods are most advantageous at high Mach numbers. Thus, optical methods are an almost indispensible tool for high-speed gas flow observation. The most commonly used techniques are based on density changes, but the most recently developed method is based on waves.

The flow visualization methods that are applicable to water tunnels (various tracer techniques and the tuft method) are rather limited because they are useful only for low speeds. In contrast, a wind tunnel is useful for all other methods of flow visualization, not only for the tuft and dust methods (which come under tracer technique) but also for rarefied gas, high speed, compressibility, and heat transfer effects.

Nevertheless, for the visualization of incompressible flow, a liquid tunnel (especially a water tunnel) is preferred over a low-speed wind tunnel, e.g., one that uses smoke techniques, because of the following advantages (Clutter et al., 1959):

1. The building cost, the time, and the power required for operation are less.
2. A large range of Reynolds numbers may be employed. (Very large Reynolds numbers are obtainable by heating the water, and low ones are obtainable using glycerin.)
3. Although the turbulence of water at normal operation is nearly zero, the level of turbulence may be raised by moving a grid in front of the model.
4. A water channel allows the effect of blowing, suction, and internal flow to be easily and cheaply simulated.

The simplest method that can produce reliable results at a reasonable expenditure of money, effort, and time is obviously the most desirable technique from the practical point of view. However, since no single method can provide the required information on quality and quantity, a preliminary test by a flow visualization technique is recommended in order to observe the overall flow pattern and to determine whether or not separation takes place. If the flow does separate and precise quantitative information is needed on the position of separation and the characteristics of the separated flow, then the physical methods that measure the pertinent characteristics are very desirable. These are described later in Sec. 2.3.

Parameters that are observable by direct visualization techniques include pathlines or particle paths (the path of the fluid particles), streakline or filament line (all fluid particles that pass through a certain position of space), and–in a limited case–streamlines (instantaneous picture of the flow field).

Pankhurst and Holder (1952) described these three observable parameters as follows. A *pathline* is the path followed by a particular particle of the fluid, for example, as sketched in Fig. 2.1. If a photographic record of the motion of the particle is taken at some time with an exposure δt, then its trace δs indicates the particle path during the time δt. Thus $\delta s/\delta t$ is the mean velocity. The motion of the particle may be visualized by introducing into the flow marking elements such as solids or liquids, puffs of smoke, etc.

The *streakline* is defined as the line that joins the positions of particles at some particular time, and it generally does not coincide with the pathline. Suppose that at a fixed point A in the fluid (see Fig. 2.2), a continuous

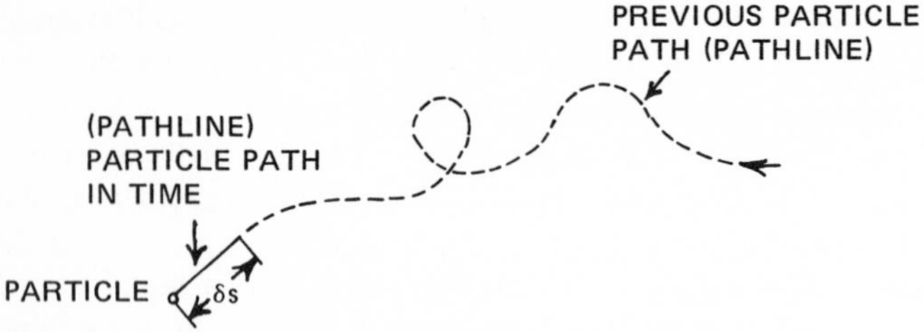

FIG. 2.1 A particle path (also called *pathline; from Pankhurst & Holder, 1952*).

stream, designated by the particular paths of Particles 1 and 2, is introduced and the motions observed. Particle 1 reaches point C at the time of exposure, whereas Particle 2 arrives at point B. However, the streakline can be made visible by introducing a continuous stream of a tracer at some fixed point into the flow. An instantaneous photograph will then indicate the streakline of the particles of fluid which previously passed through point A.

Stream filaments can also be made visible by the spark technique, wherein the thermal capacity of the electrodes between which the spark is discharged produces a continuous filament of heated air between the spark discharge.

A *streamline* is defined as a line that gives the velocity direction of the fluid at each point along the line. Thus, the instantaneous position of a streamline is a line with a normal velocity component of zero. In other words, a streamline must at any instant be tangential everywhere to the direction of particle paths. The form of the streamline for this particular case is shown in Fig. 2.2. The streamlines cannot be visualized directly for unsteady flow, but their positions in steady motion can be deduced from a photographic record of the paths of a number of particles taken over a time interval. Figure 2.3 indicates the pathlines of a series of particles introduced originally at Point A. During a short time interval, the streamline has the form shown in Fig. 2.2 or Fig. 2.3.

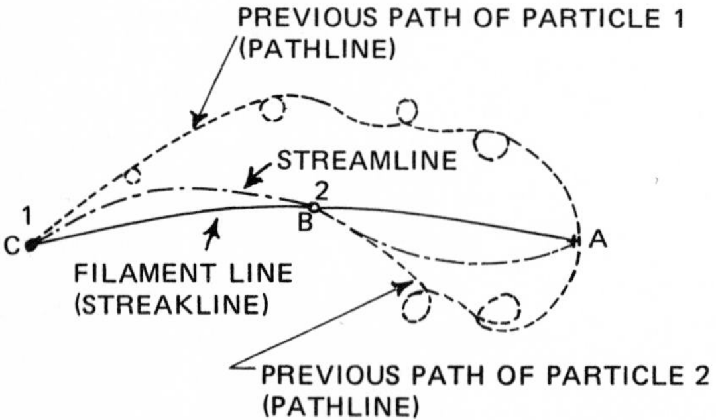

FIG. 2.2 Particle paths, filament line and streamline (also called *pathline, streakline,* and *streamline,* respectively; *from Pankhurst & Holder, 1952*).

The relationship between pathlines, streaklines, and streamlines is illustrated in Fig. 2.4. A pathline (the track actually followed by a particle of fluid) and a streakline that joins the instantaneous positions of a succession of particles are clearly distinguishable in the figure. In order to form the streaklines, particles are issued from one point source. The change in the form of streaklines with time in unsteady flow is also shown. For steady motion, however, streaklines, streamlines, and pathlines are identical to each other.

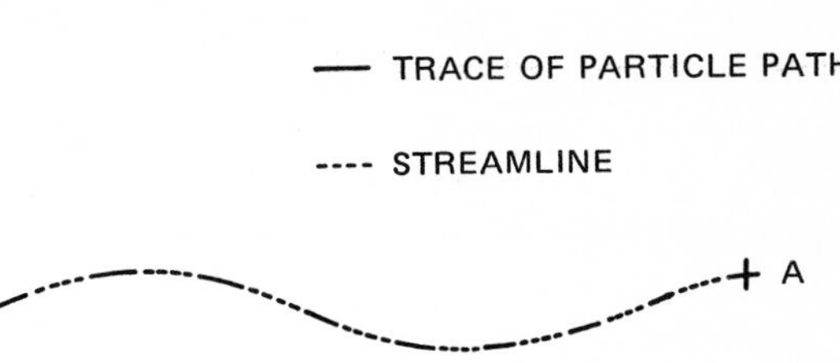

FIG. 2.3 Streamline and traces of particle paths (particle path is also called *pathline; from Pankhurst & Holder, 1952*).

It should be pointed out that a flow pattern differs according to whether the observer is at rest relative to the body (such as in wind tunnel practice) or at rest relative to the undisturbed fluid far upstream.

It was pointed out in Chapter 1 that the criterion of flow separation is determined by the surface streamline—the streamline nearest to the wall surface. Therefore, the observable streaklines of particle paths may serve to investigate separation in steady flow. Furthermore, steady flow is more

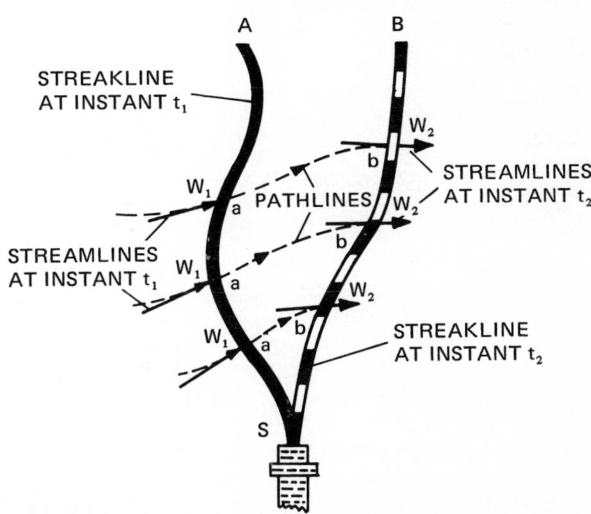

FIG. 2.4 Streamline, streakline, and pathline (*from Vallentine, 1959*).

common, and most analytical and experimental investigations are made for steady cases. Hence, an attempt is made to present the various experimental techniques of flow visualization that are useful in determining separation rather than to categorize them according to observable streaklines and pathlines.

When one of the nonoptical flow visualization techniques is selected, attention should be given to the undesirable aspects of contamination of the working fluid and tunnel when using the bubble technique, smoke technique, dye injection, dust method, oil vapor (oil mist) method, and fluorescent-oil flow method. Among the visualization techniques that contaminate the working fluid within the water tunnel, tracer techniques are widely used. Tracer techniques comprise various methods, including gas bubbles, smokes, dyes, etc.

Recording of the flow phenomena, which can be visualized by photographs, may constitute a major problem. Since the film is more sensitive to different wavelengths of light than is the human eye, lighting that is good for visualization may not be good for photographing or vice/versa. Maltby (1962) and Maltby and Keating (1960) give a table of details for photographing various oil flow (see Sec. 2.1.7) patterns. Friberg (1965) reported that separation on a cone at an angle of attack using the bubble technique was clearly visible by eye, but he had many problems in obtaining good photographs.

In order to photograph flow separation, an additional special technique is required for unsteady eddy motion. Since the eddy is moving with time, it is

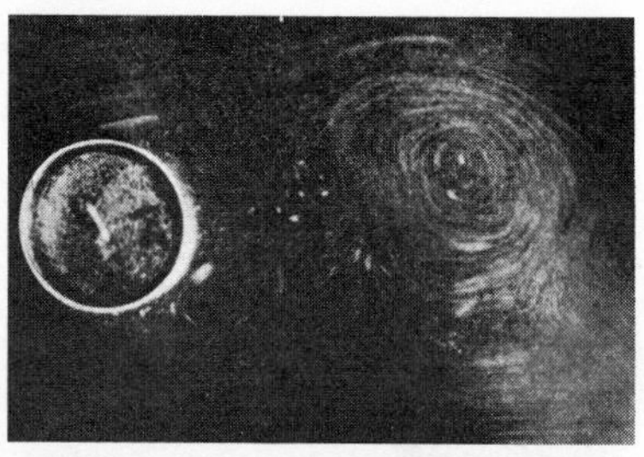

(a)

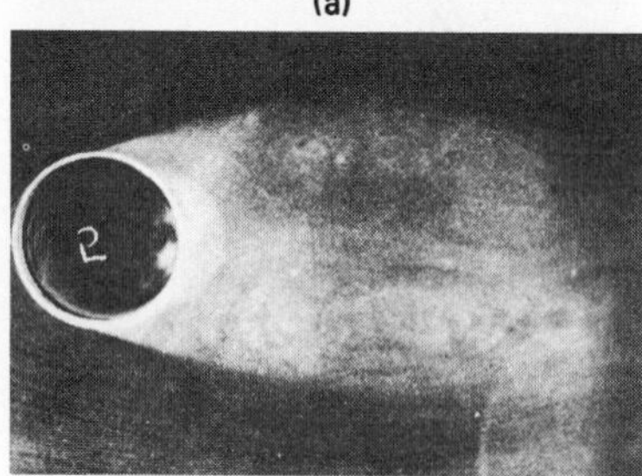

(b)

FIG. 2.5 Near wake of circular cylinder: (*a*) exposure time much less than vortex shedding period; (*b*) exposure time much greater than vortex shedding period (*from Roshko, 1967; photos by F. Hama*).

necessary to shorten the exposure time to obtain a photograph of instantaneous phenomena. For example, consider subsonic separated flow downstream of circular cylinder. Photographs taken in a time span that is short compared to the vortex shedding period (see Fig. 2.5*a*) show the vortex pattern extending far downstream. But when the exposure time is longer than the vortex shedding period (see Fig. 2.5*b*), then a closure point of wake in the flow at the separated region is indicated several cylinder diameters downstream (Roshko, 1967).

In general, a series of photographs of an unsteady flow phenomenon such as flow separation will give more information because it enables a comparison of various instantaneous photographs. However, if a single instantaneous photograph does not show the time average of a slightly unsteady phenomenon, then a longer exposure time is preferable. For example, it is advantageous to use a longer exposure time in order to observe the shape of a region of separated flow between a separated turbulent boundary layer and a wall. The images of the boundaries of the region may be confused by transient phenomena, such as eddies, in an instantaneous photograph.

2.1 NONOPTICAL FLOW VISUALIZATION TECHNIQUES

Various nonoptical methods are now presented, with emphasis on their effectiveness for visualizing flow separation.

2.1.1 The Use of Bubbles

The bubble technique is the most suitable and inexpensive method for visualizing two-dimensional, as well as three-dimensional, flow in a water tunnel. Geller (1955) developed a direct method for obtaining the velocity profile of the boundary layer by marking a line element of fluid perpendicular to the surface over which the boundary layer is to be studied. The working fluid for this technique is necessarily water because of the needed chemical reaction. A fine platinum wire is mounted perpendicular to the surface over which the velocity profile is to be obtained. When a voltage pulse is applied between the wire and another electrode in contact with the water, electrolysis forms a row of small bubbles along the wire, thus marking the necessary fluid element. As it is swept downstream, the row of bubbles is distorted into the shape of the velocity profile.

Clutter et al. (1959) used a fine wire as a cathode to generate hydrogen bubbles by electrolysis of water. The streamers produced by very small bubbles may be generated in any area of interest by moving the cathode. Because of the electrical nature of the generation, they may be pulsed or timed. Hydrogen bubbles are generally preferred over oxygen bubbles since the electrolysis of water volumetrically forms twice as much hydrogen as oxygen.

This technique requires quite simple equipment, power supply, controls, etc. A variable-voltage supply of 10 to 250 V is advisable since the required voltage depends on electrolyte concentration, distance between electrodes, and geometry. An electrolyte such as 0–15g/liter of sodium sulfate (Na_2So_4) is needed in soft water, and it is desirable to have a variable resistor to tune the circuit. The choice of material for the bubble-generating wire is not critical provided it is resistant to corrosion and not fragile. Further details of the method may be found in Schraub et al. (1965).

The bubbles thus generated dissolve rapidly since hydrogen is soluble in water. Schraub et al. (1965) estimate the half-mile as 3 sec, and even less in turbulent flow.

The position of the onset of flow separation is found by noting the most upstream location to which bubbles are carried in the reversed flow separated region where the behavior of the flow can be studied by observing the motion of the bubbles. Since the bubble size is very small and because the magnitude of the velocity is small, it is possible to determine the boundary of the separated flow region by observing the opposite direction of bubble motion outside and inside the region. The separation line on a slender body at an angle of attack may be seen as a definite ray of bubbles, although the ray is faint. Figure 2.6 shows a velocity profile obtained from the bubbles produced by a 60-cycle voltage.

FIG. 2.6 Velocity profiles obtained from the bubbles produced by a 60-cycle voltage at the exit of the three-wall duct (90 V 120 pulses/sec, free-stream velocity: $\frac{1}{2}$ ft/sec; *from Clutter et al., 1959*).

Another advantage of the bubble technique is that hydrogen bubbles do not lose their identities in wakes or in turbulent flows as do the filaments of dyes because of diffusion. Unlike dyes, moreover, bubbles do not require special ducting. It is possible to observe the flow very close to a boundary surface and in the separated region. Furthermore, the effects of main stream disturbance can be photographed for detailed examination (Clayton & Massey, 1967).

One drawback of this technique is the tendency of the bubbles to rise because of buoyancy. The rate of rise depends on the size of bubbles. This, in turn, depends on the current between the anode and cathode and on the flow velocity. However, such a rise should give no serious distortion to the streamline for the speeds at which channels usually operate (speeds of 0.5 to 2 ft/sec are usual although velocities up to 10 ft/sec may be used) because the rise due to buoyancy for the optimum streamlines is about 0.05 to 0.06 in/sec. The buoyancy problem is more serious for a channel where the water is stationary than for a water tunnel, and it is least serious for a water tunnel with vertical test sections (Clutter et al., 1959). If there is considerable streamline curvature, the bubbles do not follow the direction of water flow. Extensive investigations of the accuracy of the water bubble technique has been carried out by Lukasiek and Grosch (1959) and by Roos and Willmarth (1969). Roberson (1955) also investigated the accuracy of the bubble method for flows of different density.

Clutter and Smith (1961) accidentally discovered an interesting method that involved a loop of wire (this can be easily produced by pulling the wire through a gear couple). Application of the same voltage on all parts of the loop formed hydrogen pearls along the entire wire. These investigators found that the hydrogen pearls wander to the loop and separate.

Air bubbles may be introduced directly into the liquid through a probe, as if often done with dye (see Sec. 2.1.3).

Similar to the hydrogen bubble technique, there is an electrolytic process which generates a dye by using potassium-iodide starch solution as the electrolyte. (Iodine is liberated by the electrolysis of a potassium-iodide solution.) This method is based on the fact that a starch suspension gives an intense blue color when iodine is present and a colorless solution when iodine is not present. The iodine-starch technique is best for studying the wakes of bodies or regions of separated flow where the velocity is slow enough for the iodine to have time to produce intense blue filaments. This method is particularly suitable for small tanks where the amount of chemicals required is small and where the milkiness of a starch solution does not obscure the flow patterns. This milkiness constitutes the drawback of this technique because good photographic contrast is not simple; in fact, it is sometimes impossible to obtain (Clutter et al., 1959).

2.1.2 The Use of Smoke

Smokes consist of suspensions of small solid or liquid particles in transparent gas. Flow is observed by the scattering and reflection of light by these particles.

Although the smoke technique is the most widely used method for observing flow patterns, in especially designed wind tunnels, its usefulness for flow separation studies has been limited because it only works at low Reynolds numbers. Recently, however, V. P. Goddard (University of Notre Dame) has successfully developed a smoke tunnel for supersonic flow.

The flow pattern of the entire flow around a model, the behavior of the separated flow involving the vortices, and the positions of flow separation are made by carefully observing the clearly visible smoke lines that emanate from single or multiple tubes. The position of separation is observed at the location where the smoke line nearest to the model leaves the surface. Downstream of laminar separation, the filament continues to be well defined for some distance, but careful examination shows that wisps of smoke are swept against the stream in the separated flow region between the body surface and the main filament (Pankhurst & Holder, 1952). For example, on the surface of a bluff-tailed wing, the point of turbulent separation is observed at the position where the diffused smoke band leaves the surface (Pankhurst & Holder, 1952). The behavior of the separated flow is indicated by the unstable motion of the smoke lines or smoke lumps. The flow field that involves separation may also be observed by the flow paths of different smoke lines, the variation of the width between adjacent smoke lines, and the speed of smoke particles. In order to avoid disturbance, smoke may be introduced into shear layer in a region of strongly favorable pressure gradient. In general, a slit about $\frac{3}{16}$ in. long in the spanwise direction and $\frac{3}{64}$ in. wide is most suitable for the introduction of smoke in a low-speed tunnel.

Since the integrity of the smoke stream is very important, diffusion of the smoke must be avoided by the design of the wind tunnel and by selecting the proper substance for the smoke. Furthermore, the smoke lines should be clearly visible; for example, sufficient lighting should be available from high intensity lamps and a dark background (such as black or black velvet walls) should be provided. Reflections from certain parts of the model surface can cause lighting difficulty in three-dimensional tunnels. Since a dull black surface reflects any diffused light, it is almost dark. If light is reflected in only one direction, a polished surface may reflect no light back to the camera. Thus, a highly polished surface is generally better than a dull black surface. The smoke should be light so that the smoke filament is not appreciably influenced by gravity. Furthermore, the inertia of smoke should be low so that transient and unsteady phenomena can be observed.

The observation of smoke lines is instantaneous; thus the experiment must be recorded photographically, preferably by a high-speed motion-picture camera.

The choice of a suitable smoke is not easy. It must not only preserve the integrity of the smoke lines in the wind tunnel around the model but it also must be dense, white, and free from toxic reaction, corrosion, clogging, and condensing.

Figure 2.7 shows the flow visualization around a sphere in a smoke tunnel. Such pictures may show unknown flow phenomena or clarify the doubtful aspects of flow that involve separation.

In the smoke or gas filament technique, local separation with downstream reattachment is indicated by a region of stagnant flow close to the surface, as shown by the dark area in Fig. 2.8 (Pankhurst & Holder, 1952). As shown in this photograph, gas is introduced from various holes in the model. Where the smoke streams aft from the hole, the flow is attached. The reverse flow in the separated region carries the gas forward, against the airstream.

Smoke may be produced by burning any substance that produces smoke, e.g., straw, certain chemicals, signal flares, or even a cigarette. Many chemical reactions will also produce a smoke that is satisfactory for flow visualization. Other chemicals which have the characteristic of smoke may be introduced into the flow, and their other properties may be of use in certain instances.

Smoke visualization techniques produced by burning an organic substance have been developed by Marey (1900 and 1901), by Lippisch (1939), and by Preston and Sweeting (1943). Farren (1932), Clark (1933), Tanner (1931), and Hazen (1955) were successful in using chemical substances such as titanium and tin tetrachloride to visualize the flow. Although these substances are in the form of a liquid at normal temperatures, under the influence of humidity in an airstream, they produce a thick, white smoke from the metallic oxygen by excluding the hydrogen chloride. Because of corrosive effects, this method is recommended for experiments only in wind tunnels whose walls are made of wood. A very thick smoke can also be produced by mixing hydrochloric acid with ammonia and blowing this white smoke into the airflow through single or multiple tubes. This method is practical in that there is no danger of congestion in the narrow (about 1-mm diameter) tube.

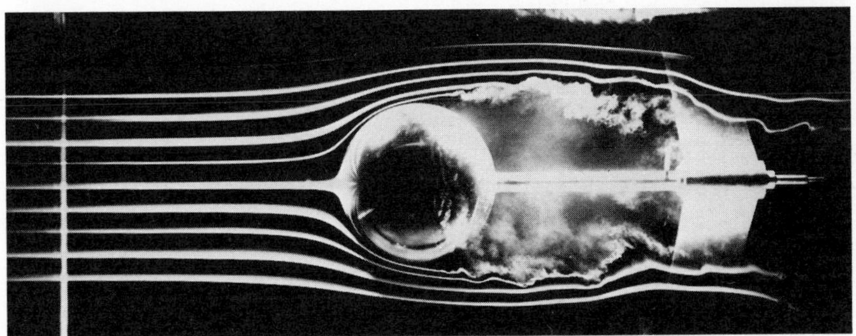

FIG. 2.7 Subsonic flow around a sphere in a smoke tunnel (*from Goddard, 1962*).

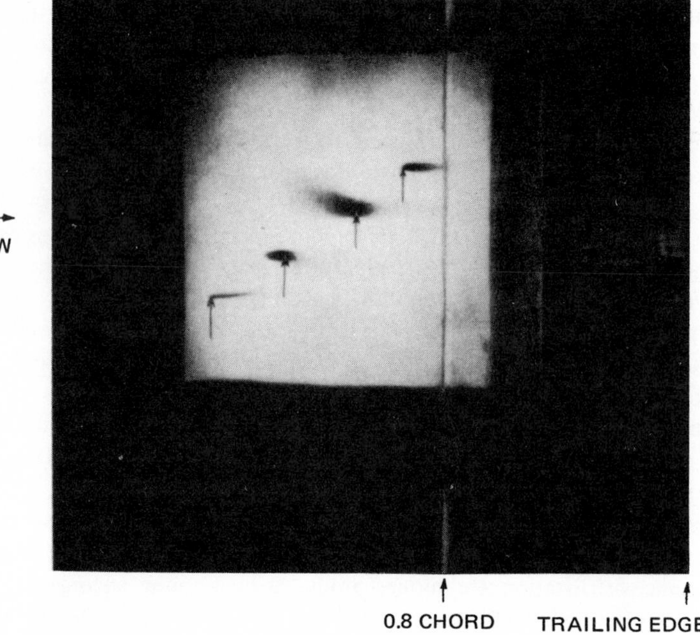

FIG. 2.8 Laminar separation followed by transition and reattachment (gas filament technique); arrows indicate position of holes at which gas was introduced (*from Pankhurst & Holder, 1952*).

The oil smoke produced by the cracking technique (i.e., dripping oil onto a hot surface) is superior to straw smoke. The cost of an oil smoke generator system is less, and its construction and maintenance are far simpler than a comparable smoke generator system (Goddard, 1962). Figure 2.9 is a sketch of the oil smoke generator developed at the University of Notre Dame and later improved by setting the entire unit an angle of about 60 deg.

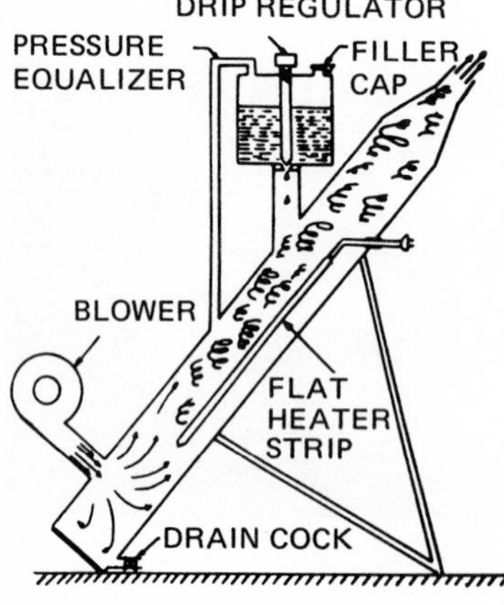

FIG. 2.9 Oil smoke generator (*from Goddard, 1962*).

Introduction of smoke before the tunnel passage contraction helps to damp out the fine turbulence introduced and results in a superior turbulence free flow. It also reduces the flow velocity at the screen to an extremely low value. Thus, differences between the stream and the smoke velocities at the point of introduction and the energy loss at the screen are both minimized. To maintain the integrity of the smoke streamline, i.e., to avoid diffusion of the smoke, a large contraction ratio of about 12:1 should be used. Extensive use can also be made of antiturbulence screening by introducing the smoke filament at the upstream side of the screens.

Encouraged by Brown's successful design of a smoke tunnel, Goddard (1962) built a supersonic smoke tunnel by further raising the contraction ratio to the order of 100:1. This tunnel is suitable for straw smoke (grain straw is best), oil smoke (kerosene is recommended), and nitrous oxide (N_2O is a practical source; standard gas tanks are available) as supersonic streamline tracers. The magnitude of free-stream velocity reaches Mach 1.38 and photographs are taken by schlieren optics. As developed by Goddard et al. (1959) and extended by Mueller et al. (1969) at the University of Notre Dame, the technique enables the streamline and the shock and expansion wave pattern that occur in supersonic flow to be indicated simultaneously on a single negative. This is shown in Fig. 2.10.

There are two ways to obtain this combination type of photograph. One method (used for Fig. 2.10) employs a standard schlieren system with a horizontal light source and knife edge. In place of smoke, nitrous oxide (N_2O) is employed to obtain a stream tube whose thermodynamic properties are quite close to that of air yet sufficiently different optically so that the streamlines along with the shock pattern are visible by using ordinary schlieren techniques.

Another method is to modify the standard schlieren system, described later in Sec. 2.2.2, to a so-called opaque-stop system. This involves replacing the slit source by a circular source and the knife edge by a piece of glass with a $\frac{1}{16}$-in. diameter opaque spot known as a circular opaque stop. The result is that the undeflected light hits the opaque stop and produces the black background (unexposed on photographic film) necessary for simultaneous schlieren and smoke photography. This type of oil smoke technique is used to indicate the streamlines (Mueller et al. 1969).

One of the drawbacks to using smoke is that particles may settle on the surface of the model. These deposits may trip the boundary layer and destroy the true flow pattern. This problem is nonexistent in some cases, e.g., for oil smoke, but it becomes very important in others, e.g., straw smoke.

A note of caution is in order concerning the use of the chemicals mentioned above. Since they may be poisonous, adequate ventilation is necessary, and extra care must be used when generating a poisonous smoke. In some cases, the model or the tunnel or both may corrode.

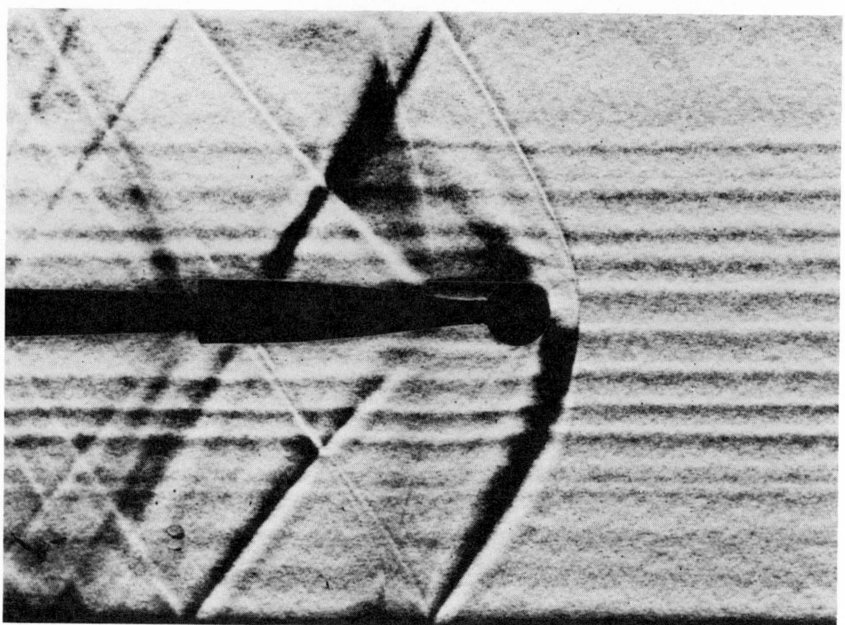

FIG. 2.10 Streamline and shock pattern in supersonic flow around a sphere using nitrous oxide (*from Goddard, 1962*).

One very interesting use of the smoke technique has been to determine the pattern of the exhaust from the funnels of ships. Figure 2.11 demonstrates the similarity of the smoke flow for a full-size aircraft carrier and its small-scale model.

Sovran (1959) introduced smoke formed by the fog from very small droplets of kerosene into an airstream through small tubes from the trailing edge of an airfoil-shaped reservoir. He then visualized the air flow and qualitatively measured the rotating stall phenomena in an axial flow compressor and in a two-dimensional cascade. Rotating stall is indicated by the periodic occurrence of regions in which reverse flow components exist; it can be detected if the passing frequency is less than about five per second. Although it is not always possible to judge whether the stall has started in a rotor or stator, it is possible to trace the origin of the stall to either the curb or the casing end of the blading. Table 2.1 compares various streamline tracers.

Several other methods such as those of Preston and Sweeting (1943) and Bergh (1958), etc., have been described in the book by Pankhurst and Holder (1952), and in papers by Wuest (1963) and by Maltby (1962).

(a)

(b)

FIG. 2.11 Similarity of smoke flow from (*a*) a full-scale ship and (*b*) a model (*courtesy of the U.S. Naval Air Engineering Center, Philadelphia, Pennsylvania*).

TABLE 2.1 Comparison of streamline tracers

	Smoke		Gas
	Straw	Oil	Nitrous oxide
Initial cost of generator	Highest	–	None
Maintenance of generator (cost & labor)	Highest	–	None
Operating cost	–	Lowest	Highest
Streamline distinction	Good	Good	Good but smoke better
Vortex trail distinction	Good	Good	–
Directly photographable	Yes	Yes	No
Direct photo + modified schlieren	Yes	Yes	–
Schlieren only	Occasionally	–	Yes
Adaptability to an existing facility	Yes, if indraft	Yes, if indraft	Yes
Safety	Avoid excessive inhalation	Avoid excessive inhalation	Anesthetic, provide adequate ventilation

Source: Goddard (1962).

Sanders and Thompson (1967) dipped a wire in an oil and quickly heated it to generate a filament of smoke. They then used this technique for quantitative velocity measurements by taking a series of photographs.

The vapor screen technique is related to that of smoke and is often used in a high-speed wind tunnel. Condensation is caused in the working section by deliberately using humid air. The vapor is then illuminated—as in the smoke technique—to visualize such flow phenomena as the formation of free vortices. A detailed description of the application of the vapor screen method is given by McGregor (1961) and by Maltby (1962).

The importance of lighting for photographic recording has already been mentioned. It is again emphasized that care must always be taken to ensure that the best illumination is available for photographs that record smoke flow visualizations. Furthermore, as Maltby (1962) remarks, since it is not unusual for smoke patterns which are invisible in general lighting to reveal important details of the flow when properly lit, the best available quality of lighting in the wind tunnel is also needed for successful smoke flow visualization.

2.1.3 The Injection of Dye

Dye is injected into the ambient liquid through holes in the model, and flow phenomena in the neighborhood of the wall are observed by the formation of dye filament. The dye can also be injected from a probe which may be placed at different locations in order to investigate the flow field about the model. Precautionary steps should be taken to keep the dye velocity from the tube equal to that of the uncolored ambient fluid that passes over the tube. Parallel dye filament may be produced by injecting the dye through a rake.

In order to produce excellent pictures, it is important to select a dye that absorbs light effectively. The intensively colored dark red chamber filter is well suited. Pinakryptol red or green are also often used. If a large quantity is required, then inexpensive eosin A (salt-free) or green PLX are used. These dyes are dissolved in a small quantity of alcohol or water and then diluted with water (about 1 : 100). The exit velocity of the dye can be controlled by the altitude position of the reservoir which is connected with the probe by the tube. A control can also be carried out by using pressurized air with an adjustable throttle.

At the Offices National Études de Recherches Aérospatiales (ONERA), Werlé (1953, 1957, 1958) successfully used white and colored skim milk to visualize a wide range of flow phenomena, and he obtained beautiful pictures and films. The phenomenon of turbulence spread can also be measured by using skim milk as done by Nikuradse (1930), for example, for the turbulent secondary flow in a noncircular pipe. By simply mixing a detergent in the water tank and generating air bubbles, the ONERA vertical water tunnel can also be used to investigate the characteristics of flow separation, vortices, wake, etc., by the bubble motion.

Other dyes such as nigrosine black, acid chrome blue, and methylene blue N. F. have been used by Hama (1960). These yield black, red, and blue solutions, respectively. The black solution is best suited for black-and-white pictures, and this dye also has the least diffusion for single-streak observations. However, the black-solution method is not recommended for two-dimensional objects since it rapidly makes the entire water dark and opaque. The red solution is used for plan-view observation by injecting it two-dimensionally through a narrow slit. The blue solution is used occasionally to distinguish flow patterns from red ones which have a relatively faster diffusion.

To avoid buoyancy effects, the dye should have approximately the same specific gravity as the tunnel fluid. This can be achieved by adding other liquids of suitable specific gravity to the dye.

Figure 2.12 shows the surface streamlines or limiting streamlines obtained for an elliptic cone by injecting dye through the surface of the model. In this three-dimensional flow, the limiting streamlines coalesced from both sides to form the envelope for separation as discussed in Chap. 1.

2.1.4 Tracers on a Water Surface

When an incompressible flow field can be assumed as two-dimensional, a very simple and classic method is available to observe only the motion on the water surface of a tank or a channel. A visualization of the whole flow field

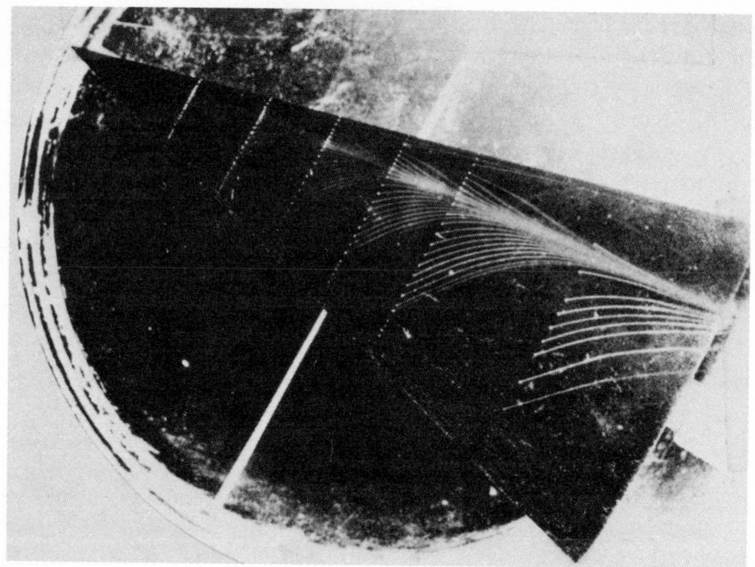

FIG. 2.12 Limiting streamlines on a 1.5 : 1 elliptic cone at 30-degree incidence (*from Rainbird et al., 1966*).

can be obtained by sprinkling a powdered tracer, such as aluminum flakes, lycopodium powder, mica chips, or plastic or glass beads, on the surface. Separation that involves vortex formation in a channel can be photographed by using the aluminum powder flow visualization technique (Prandtl & Tietjens, 1934). Chapter 1 shows examples of the flow field and of separation visualized by means of aluminum powder. This substance is widely known for its effectiveness in flow visualization. Because of its low specific weight, the aluminum powder swims over the water and its movements correspond to the fluid motion. Because the powder is reflective, the motion of the fluid is easily observed–the water should be colored in order to increase the contrast; for example, a dark red may be used for darkroom light. For a given illumination time, the length of the pathline is the measure of the flow velocity.

It is important to ensure that the water surface is clean. This can be checked by blowing the aluminum-covered surface perpendicularly so that a hole is created. If the hole closes after blowing, then the water surface is unclean and must be renewed. The wetting of a towed surface can be prevented by adding paraffine to the body surface.

Lycopodium powder is another frequently used solid material and one that enables visualization of the finest detail of the fluid motion. The spores of lycopodium clavatum are smaller and more regular in size (range of 28-50 microns) than those of commercially available aluminum powder (range of 20–400 microns). These spores resemble a tetrahedron, except that the base plane is like one-half a sphere and the three other faces are slightly convex. Lycopodium powder is better suited for photography than is aluminum powder (Frey & Vasuki, 1966).

The disadvantage of this method is the capillary wave which occurs at a velocity of 23-24 cm/sec. Note that the capillary wave can be caused by the undisturbed free-stream velocity at a much lower velocity than 23-24 cm/sec. For subsonic flow around a circular cylinder, this magnitude of local velocity corresponds to a free-stream velocity of 11-12 cm/sec upstream of a circular cylinder. Theoretically, the magnitude of the maximum velocity on the cylinder surface will be twice that of the free-stream velocity.

2.1.5 The Use of Tufts

The tuft method is popular because of its rapidity and simplicity; it needs no special lighting arrangement and does not contaminate the tunnel. Tufts are also used for free-flight studies and in high-speed tunnels. One end of a thread is fixed onto the body or onto a grid in the flow field so that the flow visualization may be made close to the surface or in the separated region. Because reverse flow occurs and vortices are formed in a separated flow region, the free end of the tuft points in a reverse direction to the main flow. The separated regions and the core of the vortices can be located by tracing the positions of such tuft motions. Observations show that the regions of separation are marked by more or less violent and random fluctuations of the tuft, depending on the severity of the separation. The exact location of the separation point cannot be easily determined by this method since a tuft can cover a broad region of flow. Tufts are most effective for observing regions of stall inception. Figure 2.13 shows an example of motion as visualized by a

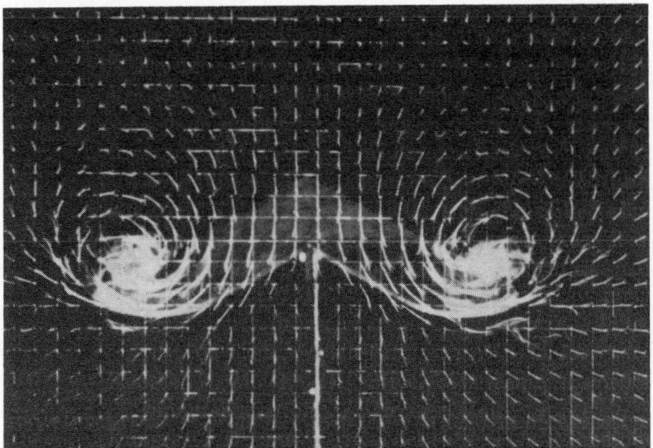

FIG. 2.13 Tuft-grid picture of flow behind a swept wing, looking upstream (angle of sweep 60 deg, angle of incidence 30 deg; *from Bradshaw, 1964*).

tuft mounted on a grid downstream of a swept wing (Bird, 1952, and Bradshaw, 1964).

This method is applicable for both gas and liquid flows. It can be used not only for purely external flow but also for other types of flow such as cascade flow in fluid machinery. Encouraged by Smith and Murphy's (1955) successful application, Saylor (1960) observed stall phenomena by the tuft method in a study of the blow angle on a four-stage, low-speed compressor.

Various types of threads have been used for tufts—silk, nylon, knitting wool, and others. Threads made of carbon have been used to withstand the high temperatures present at supersonic and hypersonic speeds. Any thread used must be flexible, have no preferred direction, and have a sufficiently small density so that the thread will easily align itself with the flow. In most instances, a $\frac{3}{4}$-in. length is satisfactory.

The tufts may be viewed through a simple low-power telescope with a cross-hair reticule which rotates in the lens to align the crosshair with the tuft. The flow angle is then read from the eyepiece protractor with an error of 1 deg. The tufts can usually be photographed during tests, but it is not always practical to record the angle data by photography because of the fluctuation of tufts in turbulent flow. If the exposure time is short, it may be possible to photograph the tufts in one of the extreme positions of the oscillation, but if the exposure time is long, a blur may result.

2.1.6 Radiating Grid

Schindel and Chamberlain (1967) used the glow from an electrically heated wire grid to "see" the position of separation or the location of the vortex core. In this method, a grid of continuous fine conducting wire on a supporting frame is heated until it glows. This grid then can be placed at various points in the flow field. Since the glow of the wires depends on the heat transfer, it will diminish in areas of high velocity. The radiating grid has been used only at very low (about 2 mph) wind speeds.

2.1.7 Oil Film and Oil Drop Techniques

The surface streamline or flow direction on the surface may be visualized by coating a surface with oil drops, oil film, or other chemical substance. Since a separation line is generally an envelope of surface streamline, the separation line is detected by this technique, as shown in Figs. 2.14 and 2.15.

The oil film is basically different from, for example, a thin film of a colored liquid observed under natural light because the oil film cannot be made more perceptible by increasing the intensity of the lighting. The special physical and optical properties of oil permit flow to be observed at two or more conditions during a single test, provided that sufficient time is allowed

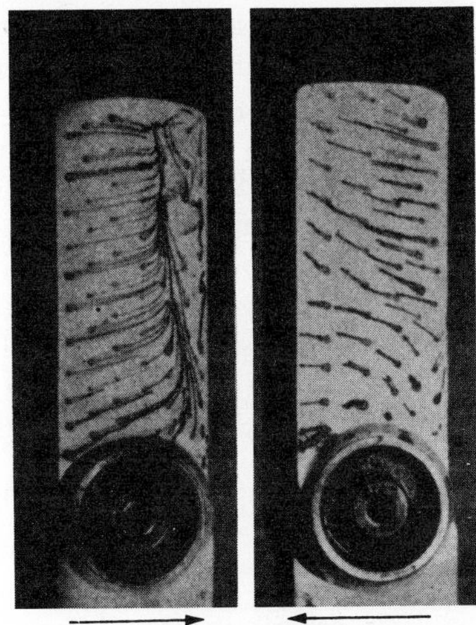

FIG. 2.14 Oil-drop indication of the separation line on a propeller VWS 18 rotating at 5 RPS with no forward velocity (*from Gutsche, 1940*).

for each preceding pattern to be obliterated and the next succeeding one to be established.

In most cases, a fluorescent oil is more visible than a nonfluorescent one. The fluorescent oil film technique was first applied by Loving and Katzoff (1959) at Langley Research Center, and separation was detected by means of

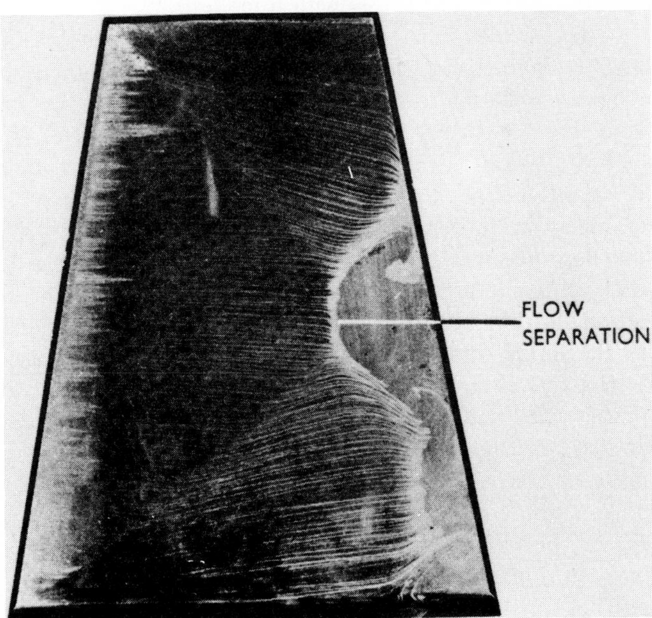

FIG. 2.15 Example of an oil flow pattern showing separation (*from Maltby, 1962*).

separation bubbles and transition of gas flow under different fluorescent illumination. This technique is also applicable for observing hydrodynamic flow behavior, e.g., the operation of a hydrofoil under water.

The materials required for wind tunnel tests are usually readily available. There is no need to select them for precision or to adjust them for different test conditions such as change of angles of attack and Mach number.

The fluorescent oil film method consists simply of smearing a model with petroleum-base lubricating oil (such as kerosene or light oil) that contains a soluble fluorescent additive and observing the oil film in the dark under ultraviolet light. During a test the action of liquid flow sweeps the oil along the surface and, to some extent, causes it to evaporate so that the oil film soon develops a pattern indicative of the surface shear intensity and direction. The thickness of the oil layer is affected by the shearing action of the boundary layer, and their relative thicknesses are determined by use of ultraviolet light. These oils are among the most brilliantly fluorescent of the available organic liquids. If the oil film is too thin to be seen under visible light, it is easily detectable under ultraviolet light.

Although a typical metal model requires no special preparation before the oil is applied, wooden models or nonmetal areas may present a problem because these absorb oil and are fluorescent. Therefore, it is first necessary to coat such models with some light-colored nonfluorescing material and allow the coating to dry thoroughly before applying the oil. Since oil film produces no appreciable effect on the aerodynamic forces, these flow-visualization tests can be made simultaneously with force tests.

Other liquids such as methyl anthranilate and ethyl anthranilate yield brightly fluorescent and acceptable wetting characteristics. However, their range of applicability is limited because they are highly volatile compared to kerosene.

For a hydrofoil, SAE 30W lubricating oil is applied to the model before it is submerged in the water. On the wing or on a hydrofoil surface, for example, the forward position is laminar; a separation bubble then forms and the flow becomes turbulent downstream of the bubble. Under illumination, these different types of flow are distinguishable by the oil film in laminar regions, by the heavy accumulation of oil in the separation-bubble region, and by the almost complete absence of oil over the rear turbulent flow region.

Hopkins et al. (1960) utilized fluorescent oil to study vortex formation on models with forward-facing steps, rear-facing steps, wires, and discrete surface particles or unswept flat surface with sharp leading edges. Their study was made at Mach 3.03, zero angle of attack, and a Reynolds number of 10.7 million per foot. These flow studies confirmed the Görtler (1940) theory that spacing of vortices on an unswept wing model is smaller at higher Reynolds number. For this study, a mixture of yellow fluorescent powder and SAE 40W oil (1:40) was brushed on the model, and after the flow pattern was established the model was photographed under ultraviolet light. A time exposure of 5 sec at f/16 was used to give satisfactory contrast on a panchromatic film with an ASA rating of about 100. Approximately 12 coats of a saturated solution of napthalene and petroleum ether were sprayed on the model. After the petroleum ether had evaporated and before the model was installed in the wind tunnel, the napthalene was rubbed smooth with a piece of paper to reduce the possibility that slight roughness would cause transition.

A temperature-sensitive (thermographic) phosphor which can be made up into a paint may also be successfully used. When painted on a surface and observed under ultraviolet light, the fluorescence intensity of this material is sensitive to small changes in temperature. Such a phosphor paint can be used to detect the difference between laminar and turbulent flow since the recovery temperatures are different for the two types of flow.

A different technique involves placing drops of oil on the surface of a body. Since the oil filament will flow in the direction of the surface shear stress, the flow direction can be visualized. Hence, the separation can be detected by determining the region of reverse flow as indicated by the oil drop streak whose direction differs from that in an attached flow region (Meyer, 1966).

Gutsche (1940) used this method to investigate the separation of a three-dimensional flow due to rotation of a propeller involving the centrifugal force and the Coriolis force. He placed drops or streaks of linseed oil, varnish, turkey red oil, or lamp black on the propeller blade and demonstrated the spanwise separation line of the three-dimensional flow which is the envelope of the surface streamline. This is clearly shown in the photograph in Fig. 2.14 of the suction side of the propeller. A dye is sometimes mixed with the oil to give greater color contrast with the model.

The surface oil flow techniques described here indicate the local direction of the flow at the surface. Questions have arisen concerning the effect of gravity and pressure gradients on the surface pattern and the possibility that the flowing oil will interfere with the boundary layer. Maltby (1962) cites a mathematical analysis of this problem by Squire. These sources and a paper by Maltby and Keating (1960) contain much helpful information on the selection of an oil and the interpretation of oil flow patterns; see Fig. 2.15.

2.1.8 The Dust Method

The dust method (sometimes known as the "oil and talc" method) is used to determine the most upstream location of the onset of flow separation and the entire separated flow region. Since the flow reverses its direction, in the separated flow region, dispersal of a dust downstream of the trailing edge of a model enables the entire separated region to be traced. Talcum (either white or colored) is the most frequently used dust. When the model surface is covered with a thin oil film, the definite dusted area is marked permanently by the dust particles which attach to the oil film. Thus, the most upstream position of the separation can be determined simply and accurately. Since the dusted model may be stored indefinitely, the investigator can evaluate the position of separation at his convenience, and photographic recording is not absolutely necessary.

Figure 2.16 shows the method of dispersing talcum powder introduced by W. Spangenberg of the National Bureau of Standards. It has been used by Smith and Murphy (1955) and by Chang and Dunham (1961). The model is prepared for testing by applying a thin coat of SAE 30 lubricating oil to the surface of the model. Dust particles dispensed under air pressure are carried forward by the reverse flow in the separated region to the farthest position of separation from the trailing edge and then away to the free stream. Thus, it may not be possible to detect the change of separation position caused by the fluctuation of flow by observing the farthest upstream dust-covered boundary.

The region of the separated flow which is permanently recorded by the talcum powder may be observed more clearly by painting the model a contrasting color or by using colored powder. Figure 2.17 shows the white separated region recorded on the black surface of a strut.

One drawback is that the dust method contaminates a wind tunnel and cleaning is required following a test. Cleaning is comparatively simple, however, if the tunnel wall is placed sidewise and the dust blown out into the ambient air.

Murphy (1954 and 1955) used the dust method to detect axisymmetric turbulent flow. (See also Sec. 2.4.1.) He found that the density of the dust deposit increased rapidly from zero to a maximum value in about 0.10 in. This may indicate a possible fluctuation of the separation position or the effect of inertia of the dust particles. To rule out this latter possibility, Murphy suggests that separation occurs at the point where the deposited powder has the maximum density. He also found that the dust method in this axisymmetric turbulent flow repeatedly gave the separation position within ±0.05 in. on a 5.9-in.-long model.

The use of a soap solution to locate areas of separated flow is very similar to the dust technique. The solution is fed through tubes leading to the surface holes in the model used for pressure plotting. The soap solution will form

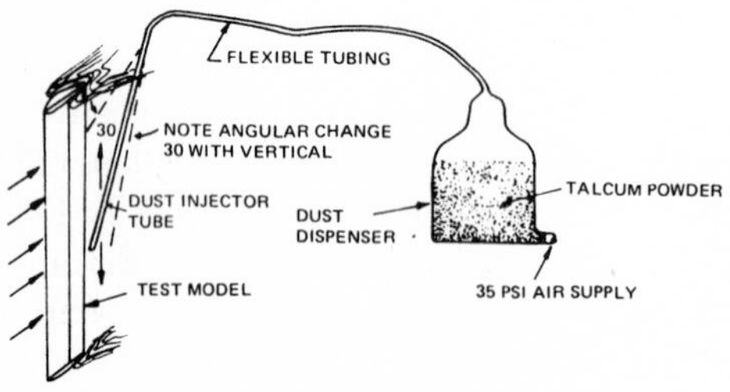

FIG. 2.16 Method of introducing dust into the separated region of flow (*from Chang & Dunham, 1961*).

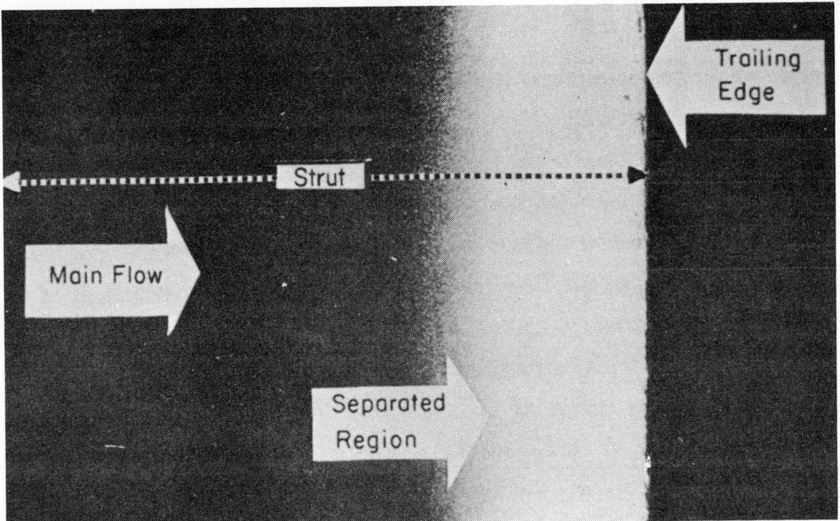

FIG. 2.17 Laminar separation of flow using the dust technique (a part of strut is shown at left; *from Chang & Dunham, 1961*).

bubbles at the model surface. These bubbles will be blown away in attached flow but they will congregate and mark the separated area (Maltby, 1962, and Maltby & Keating, 1960).

2.1.9 Evaporation Techniques (China Clay, Etc.)

Since the heat transfer in separated flow is different from that in attached flow, it is sometimes possible to distinguish these regions by coating the body surface with an evaporating material. However, evaporation of the surface indicator may not always provide pronounced contrast between laminar, separated, and attached flow regions. Although the rate of evaporation from the surface is generally greater downstream of turbulent separation than with an attached laminar layer, it is often less than with an attached turbulent boundary layer. A special lighting arrangement is often required to improve the contrast between separated and attached flow regions. If the laminar separation is followed by turbulent reattachment, a surface indicator such as china clay (kaolin) may show the position of reattachment (Pankhurst & Holder, 1952).

The china clay technique utilizes the different drying times in the various flow regions and the consequent local changes in the color of the deposit. The china clay in the form of a suspension is first applied onto the surface to be examined. After this liquid dries, a white coating remains over the surface and can be smoothed out by a fine abrasive paper. Before a wind-tunnel test, the surface is treated with a developer (ethyl-salicylate) so that it becomes transparent. During the test, the white deposit emerges as the developer dries. The surface may be repeatedly treated with the developer, and thus the china clay method enables a series of experiments to be carried out by such surface treatments. Schindel and Chamberlain (1967) used "Parepectolin," an anti-diarrhetic medicine (which is more than 90-percent kaolin) in a water-soluble base as a china clay. They also used a mixture of turpentine and kerosene as the developer. Since kerosene evaporated much slower than turpentine, the proportions of the mixture determined the time scale of the process.

In the past, the china clay method has been widely used to detect transition since the developer dries faster in the turbulent flow region and the white deposit can be seen in the turbulent region long before it appears in the laminar region. This technique has also been studied by Pankhurst and Holder (1952), Richards and Burstall (1945), and Gazely (1950).

The position of separation is made visible by the concentration of the evaporating liquid on the separation line. Thus, with the china clay technique, a separation line will appear black against a white model. Careful and continuous observation is required for three-dimensional models with several separation lines since the secondary separation line may disappear before the primary separation line fully forms (Schindel & Chamberlain, 1967).

Murphy and Phinney (1951) developed this method further by exposing the traced china clay suspension directly to the airflow to permit not only the transition point but also the streamline near the wall to be detected. Morkovin, et al. (1952) report that a reflected shock wave at supersonic speed may be visualized by this method.

Gray (1946) and Örnberg (1954) employed a method wherein the surface is coated with a light volatile oil. This oil dries quickly in the turbulent flow region, and the laminar flow region is made clearly visible by powdering with a chalk dust.

Muesmann (1959) used a method whereby a saturated solution of napthalene in gasoline was traced by a spray gun. Since the evaporation of gasoline causes a white deposit and the napthalene evaporates faster in turbulent flow, that region is distinguishable from the laminar flow region. This technique is similar to that of china clay.

Stalker (1956) was successful in using the china-film technique not only to detect the transition but also to visualize the shear stress line on the wall.

Lambourne and Pusey (1959) and Kraemer (1961) coated a black-lacquered surface with a solution of 19 percent titanium white powder, 4 percent other powder, 56 percent petroleum, and 21 percent water. During exposure to an airstream, a colored strand was formed in the direction of local wall shear stress. Örnberg (1954) successfully used a diluted white oil with petroleum; Bazzochi (1956) traced suspended lampblack in petroleum at a small velocity.

2.1.10 Ammonia-azo Trace Technique

An ammonia-azo trace technique may be used to detect separation bubbles. Velkoff et al. (1969) located the laminar separation bubble region on a helicopter rotor blade by injecting ammonia through orifices on the surface of the rotor. Figure 2.18 shows a trace left on the blue line paper attached to the blade. This technique is especially useful on a rotor since the density of air and ammonia are essentially equal. In Fig. 2.18, the laminar separation bubbles lies in the open, severely discontinuous space between L_1 and L_2. The forward discontinuity and the outward moving trace indicate separation; the aft portion of the discontinuity marks indicate reattachment. Thus, the separation bubble region was detected by observing the discontinuous space.

The chemistry of the reaction involves the joining of a diazonium salt and a naphtha-coupling component to form a diazo or azo dye. The colors of the reactants and products are yellow and blue-jack, respectively. Ammonia provides the necessary basic pH for this reaction to proceed. The ammonia-azo process can be used by placing chemically coated blue line paper (such as the paper used for blue prints.)

Velkoff et al. (1969) obtained the trace shown in Fig. 2.18 by using the ammonia feed apparatus and rotor stand sketched in Fig. 2.19. The ammonia vapor and air mixture was directed to the transfer ring at the base of the rotor stand drive shaft. It passed up through the hollow drive shaft and then out a lateral port at the top of the shaft. After flowing into the hollow blade spar, the ammonia vapor was mixed with air and injected through small orifices located on the blade surface. The blue line paper was pricked at each

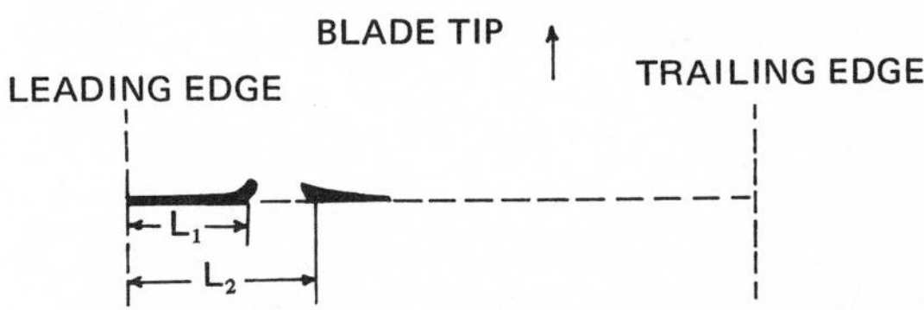

FIG. 2.18 Ammonia trace showing discontinuity for the 4-inch chord blade (shown for the chord blade at 72 percent, $L_1 = 13.5$ percent, $L_2 = 30$-percent chord, 400 rpm, blade pitch angle $\theta = 10$ deg; *from Velkoff et al., 1969*).

orifice location to allow the ammonia to enter into the boundary layer. The ammonia mixed quickly with the air in the region of the orifice and was carried along with the local air flow. As it diffused to the surface, a permanent trace was left on the paper.

The streamwise velocity near the surface is very small in the region immediately upstream of separation, but the crosswise velocity there is not necessarily small. Thus, the fluid particles in this adverse pressure gradient

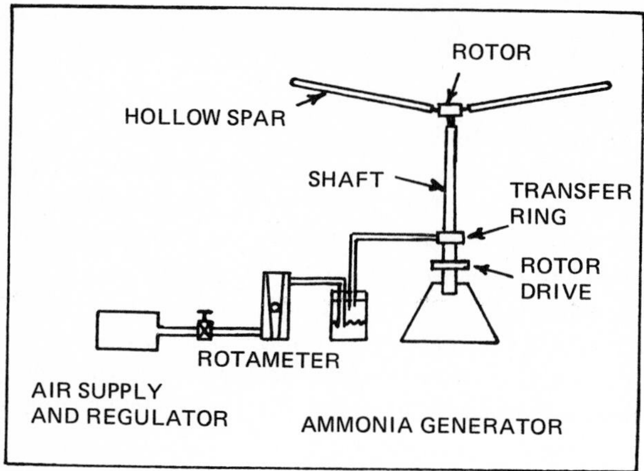

FIG. 2.19 Ammonia feed apparatus and rotor stand *(from Velkoff et al., 1969)*.

appear to be more strongly subject to both the outward spanwise pressure gradient and the centrifugal force because of blade surface curvature and Coriolis force. Therefore, the inner layers of the boundary layer tend to move outward under the action of these forces and ammonia diffuses to the surface, leaving the forward portion of the discontinuity.

A significant portion of ammonia in the upper layers of the boundary layer or in the potential flow above the boundary layer is carried by the main stream over the separation bubble in a chordwise direction. With reattachment of the flow, a portion of the ammonia returns to the surface. It may be noticed that at reattachment the shape of the aft portion of the discontinuity is similar to the forward portion of separation.

Because the chordwise velocity near the surface in the reattachment region is similar to the case of separation and is very small, the same pattern of outward movement of the ammonia particles observed for separation is also marked on the paper in the reattachment region.

Earlier, Schubauer (1939) used a similar technique to investigate the details of separation (the reaction of litmus to ammonia or gas). He applied a concentrated water solution of litmus to the surface (when dry, the surface turned red from the fumes of hydrochloric acid) and released a small amount of ammonia gas through a hypodermic needle in the area to be investigated. The flow direction was definitely indicated when the litmus turned blue even in the low speeds occurring at separation.

2.1.11 Electric Spark Technique

As mentioned previously, the spark technique is useful for visualizing streaklines. This technique was introduced by Townend (1937) for the determination of the velocity profile in the boundary layer. Saheki (1950) improved the method by increasing the conductivity of the ionized spark path, thus enabling quantitative measurements of subsonic as well as supersonic flow characteristics.

The first discharge between tip or wire produces a spark path. When the same electrode is exposed to a high voltage impulse whose impulse sequence is shorter than the deionization time, then the plasma filament becomes luminescent periodically with the impulse sequence. After a certain number of ignition impulses, the spark segment shifts downstream, and a new plasma filament must be ignited. In its present form, the technique is essentially that developed by Bömelburg in cooperation with Herzog and Weske (1959), as well as an electronics engineer at the University of Maryland.

It is possible to produce spark lengths up to 50 cm. Wuest (1963) has reported application of this method up to Mach 18. The method has also been applied to the three-dimensional flow within the blade passages of an axial flow compressor rotor wheel.

This technique is particularly well suited for the investigation of steady or nonsteady instantaneous three-dimensional flow phenomena at high subsonic, transonic, or supersonic speeds. However, it is not adaptable for low air speed; displacement patterns taken at low spark frequency are subject to errors because of the systematic deviation of the spark path. Like hot wire anemometry, this technique is useful for measuring both steady and nonsteady flow phenomena that occur within time intervals of the order of a microsecond, but it does not lend itself to the study of low-intensity, fine-scale turbulence.

2.1.12 Electrochemiluminescence Technique

The separation position can be visualized by a darker glow in the flow field as indicated in an electrochemiluminescence technique reported by Howland et al. (1966). When a potential is applied between two electrodes immersed in a flowing chemiluminescent solution, then a blue glow appears at the surface of the anode. Therefore, it is necessary to make the anode in the shape of the model about which the glow is to be studied. Only the platinum or platinum-plated anodes yield a good glow. The cathode can be of any shape, but the surface areas of the cathode and anode must be kept equal in order to prevent bubbling of the solution. The cathode may be made of a conducting material other than platinum, for instance, aluminum. For best results, the cathode should be located in the flow close to the anode.

The glow is generated within a flow wavelength of the anode, i.e., at a distance smaller than the thickness of the boundary layer. Thus, the technique of electrochemiluminescence is particularly useful for studying those phenomena that occur at the wall, such as separation. Because the flow covers the entire surface, this technique is also used to study local effects. The intensity of the glow depends on the mass transfer of the active electrolyte to the surface of the anode.

TABLE 2.2 Composition of chemiluminescent solution

Substance	Amount*	Remarks
H_2O	–	Solvent, should be distilled
KCl	$1N$	Supporting electrolyte
KOH	$0.01N$	Adjusts pH; Luminol exhibits chemiluminescence in alkaline solution only
H_2O_2	$8.5 \times 10^{-4}N$	Active electrolyte; concentration strongly affects light intensity
Luminol†	$4.4 \times 10^{-3}N$ (150 mg/l)	Chemiluminescent substance, soluble in alkaline solutions only
EDTA‡	Very small	Ties up traces of iron and copper; only very small amount should be added

Source: Howland et al. (1966).
*The symbol N refers to normal in the chemical sense.
†Eastman Kodak Co. EDTA = (ethylenedinitrite) tetraacetic acid.
‡Eastman Kodak Co. Luminol–5-amino-2, 3-dihydrol-1, 4-phtalazinedione.

FIG. 2.20 Indication of flow (left to right) past a circular cylinder (dark line shows position of separation, Reynolds number = 6000, based on cylinder diameter of 0.315 in.; *from Howland et al., 1966*).

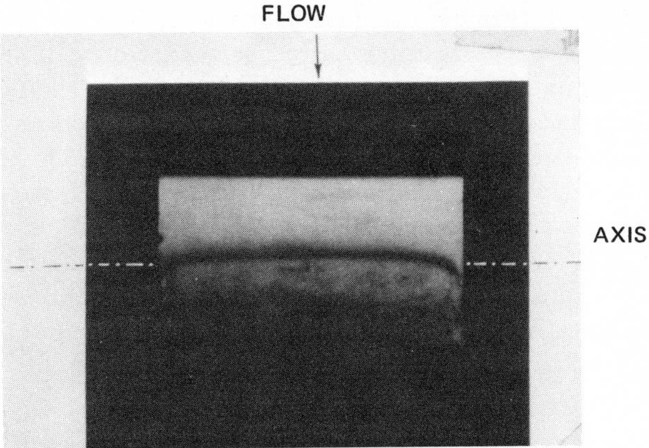

FIG. 2.21 Indication of flow (top to bottom) past a circular cylinder (dark line shows position of separation, Reynolds number = 130,000, based on cylinder diameter of 1.5 in.; *from Howland et al., 1966*).

The composition of the chemiluminescent solution and the applied voltage are the two major variables that affect the glow. The composition of a chemiluminescent solution is given in Table 2.2 (Howland et al., 1966).

All of the chemicals, including the water, should be free of impurities since traces of iron or copper will catalyze the chemiluminescent reaction and result in an objectionable bulk glow of the solution. Small changes in these specified concentrations alter the intensity and contrast of the glow but have little effect on the appearance of the glow pattern.

Howland et al. obtained a satisfactory glow by using methanol, acetone, and mixtures of water and glycerine as solvents. The flow visualization may be made by using a small rotating flow chamber or a large flow down-type water tunnel. The Howland study employed a rotating flow chamber and separation was visualized by the darker lines on a 0.315-in.-diameter cylinder immersed in the solution.

The separation shown in Fig. 2.20 is for a Reynolds number of 6000; flow is from left to right, and the view is normal to both the cylinder axis and the flow direction. Figure 2.21 shows the separation line observed on a cylinder at higher Reynolds number (130,000 based on a cylinder diameter of 1.5 in.). This test was performed in a water tunnel. The separation line at the end of the cylinder was affected by the wall of the test section.

2.1.13 Chemical Indicators

Hauser and Dewey (1942) developed a chemical technique wherein the region of two-dimensional flow separation is determined by streaming double refraction of dilute colloidal bentonite suspension. The suspension flows through a channel containing the model to be investigated. The portion of the channel to be studied is placed between crossed plane polarizers or circular polarizers and is illuminated with transmitted white or monochromatic light. With the suspension present, a flow pattern appears around the model because of the double refraction induced in the suspension by the presence of velocity gradient. When monochromatic light is used, this pattern is composed of black fringes and regions of varying light intensities. When white light is used, the pattern is composed of fringes of various colors. These fringes represent different amounts of double refraction in the fluid which, in turn, represent different values of the velocity gradient. This is one of several chemical techniques developed by Hauser and Dewey (1942).

2.1.14 Outdoor Flow Visualization

During the winter time, it is simple to achieve an outdoor flow visualization on a wing. If a large model wing is hinged on a pole (like a wind vane on the top of a mountain), then the surrounding fog forms a layer of frost crystals on the surface of the model. These indicate the direction of the flow and, more particularly, the distinct pattern of the vortex motion at the wing tip.

The most useful application, however, has been the observation of flow around helicopter rotors on outdoor test stands when the weather is right. On a humid day, the water vapor in the low pressure region at the center of vortices forms a visible vapor trail of the vortex core. This vapor trail persists for as much as a full rotor revolution, i.e., about one-third of a second. Sikorsky used this method to study blade-vortex interference (Jenney et al., 1968). High-speed photographs may be taken for analysis either with a camera mounted on and rotating with the rotor head or from a distance with a telescopic lens.

2.1.15 Visualization of Low Density Flows

The phenomena of flow separation at low density are scarcely known. At very low density of air in the region of free molecule flow, the boundary layer effect diminishes. Thus, flow separation as defined on the basis of continuous flow is not a problem in this flow regime.

In rarefied gas fields outside the limits of applicability of standard techniques, flow visualization may be accomplished by two approaches, attenuation and luminescence (Hurlbut, 1960). The first employs the attenuation of the radiant energy of particle beams in some density-dependent fashion as determined by the gas within the flow field. Among various possibilities, the following have received widest attention: attenuation of soft X-rays, the absorption by ozone of 2537 Å radiation (mercury resonance line), the use of the strong absorption band of oxygen in the region of 1400 to 1500 angstroms, the attenuation of electron beams largely by scattering, and similar attenuation of fast atomic beams.

Flow visualization by the absorption of X-rays was first developed by Dimeff et al. (1952) at Ames Research Laboratories and also by E. Winkler (1954). X-rays of 6 Å wavelength were collimated at Ames to a 0.5-mm-diameter pencil and used to scan the test section of a 10- by 14-in. supersonic wind tunnel at gas densities of $2\text{-}\times 10^{-5}$-g/cm^3 (approximately 1500 microns of mercury). Although, in principle, the mass absorption coefficient can be increased by increasing the wavelength of the incident radiation, intensity imposes a serious limitation.

It appears that one of the most promising techniques for density measurements is the observation of the attenuation of electron beams. Measurements of the density variation in a flow field may be made either by direct measurement of the electron current or by the observation of related effects, such as ion recombination luminescence along the beam path. The electron beam does not visualize the entire flow field at one time, nor can it be made to do so readily because of the peculiar limitation on the aperture of the detector. An atomic beam attenuation system would be similar to the electron beam system in many respects.

Luminescence has a wide range of applicability for flow visualization as an alternative to beam attenuation measurements. Flows are visualized by the luminescence of the working fluid itself, and the condition of luminescence is created by chemical effects following electrical discharge. Since there are many chemiluminescent flows in gases other than atmospheric constituents that might be used as working fluids, this discussion of useful glow mechanisms is not comprehensive.

Excitation is accomplished by means of a radio frequency electrodeless discharge within the pyrex pipe. The time of discharge is controlled, and glows are recorded by conventional photographic techniques. The afterglow technique is used simply to visualize the entire flow field, but it is doubtful whether this technique can be applied for quantitative evaluation of flow phenomena beyond the observation of the salient structures of the flow (Hurlbut, 1960).

2.2 OPTICAL METHODS

Special knowledge of optics, photography, and electronics is needed for an elaboration of the optical methods of flow visualization. However, an attempt is made here to present various optical methods in a manner that is meaningful to readers who do not have such background knowledge. Greater detail and extensive bibliographies are available from several sources: Tapia (1965) 538 references, Holder and North (1963) 269 references, and Holder et al. (1956) over 300 references.

If light passes horizontally through a gas current rising above some heat source, brighter and darker regions can be seen either directly or on a screen. Such changes in illumination are caused by variations in the optical density of gas which corresponds to the variation of the mass density. The density differences cause deflections of the incident rays and a consequent wavering of the illumination in the plane of observation. The main disadvantage of optical methods is that all the density changes in the light path are integrated. Thus, for example, serious errors may occur at the ends of airfoils. Optical methods are therefore usually restricted to two-dimensional flows (sometimes for axisymmetric flow) unless special provisions are made for three-dimensional flow. The effect of vibration and temperature fluctuations are difficult to minimize.

Because they are optical, these methods do not interfere with the object under observation. Normal motion of gases is not impeded, as is the case when pitot tubes or yaw heads are inserted in the gas stream. Since these probes may distort the experimental data by setting up shock waves, it is readily seen that the optical method is particularly valuable at high gas velocities. The best known optical methods are schlieren, shadowgraph techniques, and interferometry. Light passing through one part of the field is retarded differently from that passing through another part, where the density of the medium is different. Thus, two effects can be distinguished: (1) a refraction of rays, i.e., the turning of the wavefront and (2) a relative phase shift of the different rays. The shadowgraph and the schlieren methods use the first effect, and the interferometer is based on the second effect. The separated region has a different appearance depending on which method is employed. This is illustrated in Fig. 2.22.

These three optical methods are used in both wind tunnels and shock tubes. The shock tube is most frequently used for research on fundamental problems of nonstationary fluid mechanics; therefore the shock tube may be called the test tube of nonstationary fluid mechanics. The validity of applying test results for nonstationary flow (which does not possess a fully developed thermal and velocity boundary layer) to stationary flow problems has not been established (Cords, 1967).

2.2.1 Shadowgraph

The first discovery of a shadowgraph system probably came from seeing heat waves rise from a heated object. The shadowgraph is sometimes called a direct-shadow-graph, and of the three optical methods, it requires the simplest apparatus. The shadowgraph may be explained as follows. When a parallel beam of light travels in the x-direction and passes through a region whose refraction index has a gradient in the y-direction, each ray will be deflected through an angle. This is given by $\theta = \int 1/n\,(\partial n/\partial y)dx$, where n is the refractive index (defined by $n = c_0/c$) and c_0 and c refer to the speed of light in vacuum and in the medium, respectively. The value of $(n-1)$ is

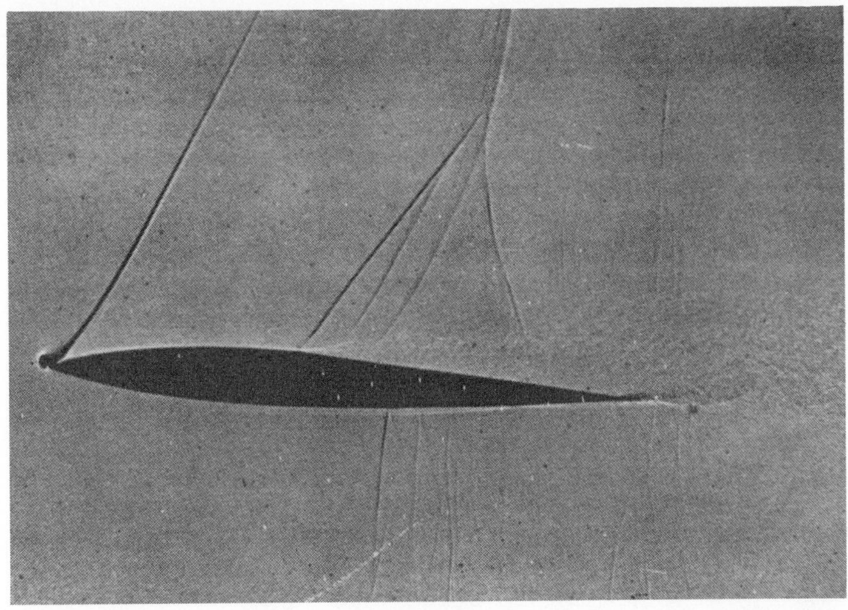

(a)

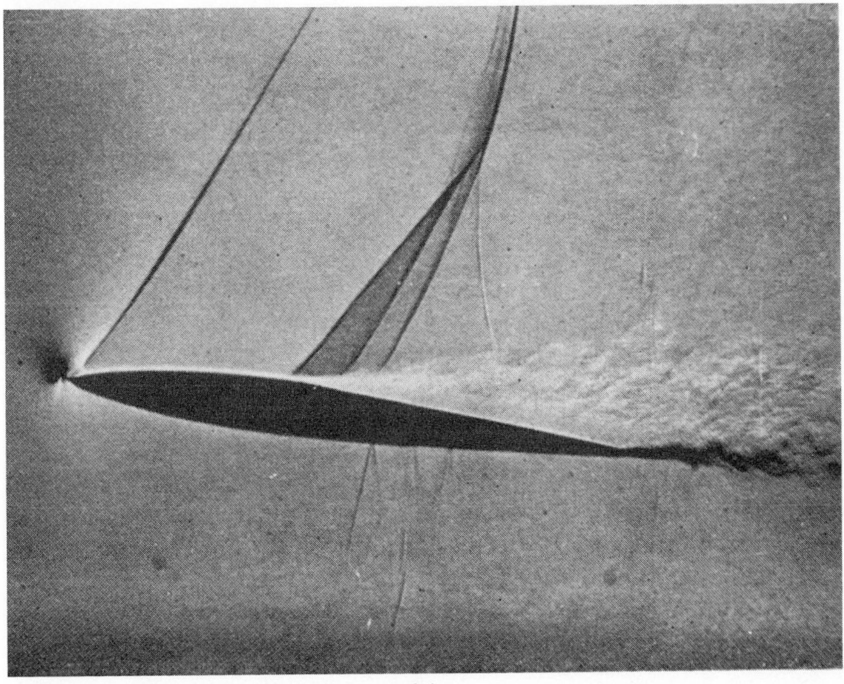

(b)

FIG. 2.22 Flow past a two-dimensional airfoil as depicted by various optical methods: (*a*) shadowgraph (*from Holder & North, 1963*); (*b*) Töpler schlieren (*from Holder & North, 1963*);

(c)

FIG. 2.22 *(continued)* Flow past a two-dimensional airfoil as depicted by various optical methods: (*c*) interferometry (*from Bradshaw, 1964*).

proportional to the density for a given gas; it is equal to 0.00029 for air at standard sea level density.

If this beam then falls on a screen, the illumination will be increased where the rays are converging, that is, where $\partial\theta/\partial y$ is negative or $\partial^2 n/\partial y^2$ is negative. The illumination is decreased where $\partial^2 n/\partial y^2$ is positive since the rays diverge.

For wind tunnel experiments, the parallel beam is produced by a point source and a converging lens or mirror. It is then passed through the test section parallel to the span of the model (Bradshaw, 1964). If the screen is placed at some position other than the focal plane of the test section, the effect of the ray deflections may be visible as indicated in Fig. 2.23. The best position for the screen is close to the test section (Liepman & Roshko, 1957).

As shown in Fig. 2.22*a*, a shock wave appears in a shadowgraph picture as a dark line followed by a light line that gives the general shape of the density profile through a shock wave.

Because the deflection of the light rays is a measure of the first derivative of density with respect to distance (i.e., the density gradient), the shadowgraph indicates only those regions of flow in which the second derivative of the density is not zero. The change of illumination is roughly proportional to the rate of change of density gradient. Since the optical density is very nearly proportional to the mass density (because the speed of light varies with the density of the medium through which it is passing), the shadowgraph method uses the extent of the over-illumination by a gas layer of specified thickness proportional to the second derivatives of the density of gas normal to the direction of observation. Because of this sensitivity, it is most suitable when there are rapid changes in the flow field. Thus, for example, this method is very useful for the visualization of wakes and shock waves in which there are large changes in the second derivative of the density. It is not suitable for observing regions of expansion or gradual compression where changes in the derivatives are small. If the gradient of the density is constant in the plane disturbance, there will be no change in the light intensity at the screen.

A point source of light is usually used. The light does not necessarily need to be exactly parallel when it enters the test section. This simplicity makes the shadowgraph technique less expensive compared to schlieren.

Separation of the boundary layer can be determined through a very interesting phenomenon. The presence of a boundary layer leads to a region of low illumination close to the surface, that is, to an extension of the shadowgraph image of the surface in a direction roughly perpendicular to the surface. When the boundary layer separates from the surface, then this effect is reduced so that the image of the apparent surface is displayed away from the flow after the occurrence of separation. This displacement of the boundary of the dark image may take place quite suddenly; thus it enables the position of separation to be determined.

Although the thickness of the shock waves is infinitesimal, in shadowgraph pictures the shock waves appear as two distinguishable bands because the rays of light are all refracted. The resulting change in the position of the rays at

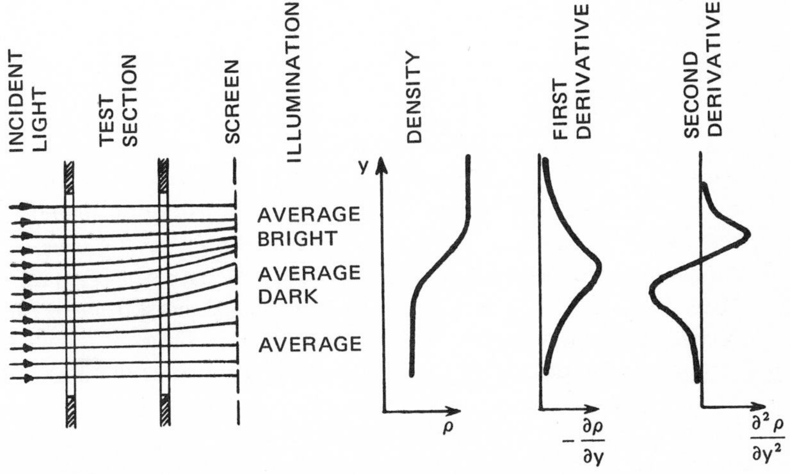

FIG. 2.23 The shadow effect (*from Liepmann & Roshko, 1957*).

the screen bears no relation to the thickness of the shock waves. Since the density is different for various flows, the shadowgraph also is used to determine whether the flow in the boundary layer is laminar or turbulent as well as to indicate the position along the surface at which transition from laminar to turbulent takes place.

Schmidt (1932) successfully utilized the shadowgraph method to measure heat transfer from a cylinder. When a beam of light is projected parallel to the axis of a cylinder and passes through the boundary layer, the rays of light are deflected through an angle which is proportional to the temperature gradient. A maximum deflection is caused by rays that graze the surface, and rays on the edge of the thermal boundary layer are not deflected at all. If the distance between screen and specimen is sufficiently large, then the thermal boundary layer remains dark, and an illuminated zone around the dark shadow may be observed because of the rays of light which are deflected out of the boundary layer. The outer edge of the illuminated zone is formed by rays which just graze the surface; consequently, their deflection is proportional to the density gradient at the surface, i.e., to the local coefficient of heat transfer. (This method is effective only when the thermal boundary layer is not too thick.) Such details of practical application as numerical evaluation of the temperature can be found in Boelter et al. (1965) and Jakob (1964).

2.2.2 Schlieren Methods

For convenience, the schlieren method is classified as conventional schlieren method (Töpler schlieren) which reproduces the flow field in black and white and as color schlieren which reproduces in multiple colors.

2.2.2.1 The Töpler schlieren

Töpler schlieren is normally used for aerodynamic investigations. The basic idea of the schlieren system is that part of the deflected light is intercepted by a knife edge before it reaches the viewing screen or photographic plate; thus the parts of the field which it has transversed appear darker (see Fig. 2.22*b*). Figure 2.24 is a sketch of the Töpler schlieren apparatus.

If a knife edge or graded filter is inserted at the focus so as to cut off a fraction of the image of the light source, then the intensity of illumination on a screen placed beyond the focus is reduced uniformly by the same fraction. The exceptions are regions illuminated by deflected rays which appear lighter or darker according to whether the ray is deflected away from or onto the knife edge or the more opaque side of the filter. The change of illumination is proportional to the density gradient of the field in the direction normal to the knife edge and not to the rate of change of density gradient as in the shadowgraph method. The light source is usually a slit parallel to the knife edge; thus the image at the focus is also a slit of finite height. The knife edge is either vertical or horizontal. The boundary layer can be observed clearly, by

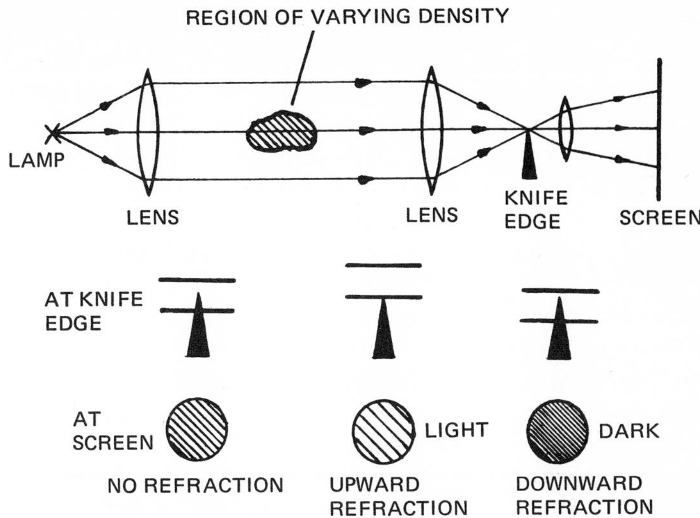

FIG. 2.24 Töpler schlieren apparatus (*from Bradshaw, 1964*).

the schlieren method, but this method is generally employed for qualitative observation rather than for quantitative evaluation of the flow phenomena.

The sensitivity of the system (percentage change in screen illumination for a given change in density gradient) is inversely proportional to the the height of the unobstructed part of the image at the knife edge. The sensitivity of the schlieren can be varied at will to suit requirements. The sensitivity can be increased to detect temperature difference as small as 10°F in an air stream. This would, for example, disclose the currents of heated air that rise from a person's fingers. On the other hand, the sensitivity can be reduced to record and show the presence of shock waves or other phenomena in the exhaust of a liquid-fueled rocket with a total temperature of more than 5000°F. The equipment needed is relatively simple and inexpensive, and the results are usually not too difficult to interpret.

Conventional schlieren methods measure the total angular deflection of a light ray in crossing the working section. Even when this measurement is accurate, however, there is still the problem of determining the density gradient from the measured deflection. If a ray of light were in a region of constant density gradient throughout its passage across the working section, then the determination would be simple. However, since the ray is continuously refracted, such a condition may not be realized, even in two-dimensional flow, if the flow contains strong density gradients. The method is further restricted for three-dimensional flow because of the complicated variation of density across the working section. It is therefore possible, for example, for two rays to experience the same total deflection even though they follow different paths and traverse regions of different density gradients. Nevertheless, it is possible to investigate three-dimensional flow phenomena by modifications that allow the use of such devices as steroscope, rangefinder, sharp focus, shock-wave plotting, and conical flows.

The quantitative evaluation of temperature distribution by measuring the light beam deflection using schlieren picture is reported by Oh and Didion (1966). The light beam deflection affected by temperature is evaluated by using Ronchi grid. In a twin mirror type schlieren system, the knife edge is replaced with the Ronchi grating which has black and clear stripes of equal unit width a_0 of the grid, as the slit width of the light source. This grid is positioned in such a way that the focus of the undisturbed light beam can just pass the free aperture. The deflected light beams which fall on the grid away from the focus of the undisturbed light beam at the distances $\pm a_0$, $\pm 3a_0$, and $\pm 5a_0$ perpendicular to the direction of the grid bar, are screened out by the black grating. Light falling at distances 0, $\pm 2a_0$, and $\pm 4a_0$ can pass through the clear aperture freely. These lines are called *Isophots*, and Fig. 2.25 shows the Isophot schlieren pictures around the circular cylinder taken with vertical and horizontal slits. From such pictures the deflection angle of light beam α is computed by

$$\alpha = \frac{Ka_0}{f}$$

where K is the number of fringe lines shifted measured in the Isophot schlieren picture, and f is the focal length of concave mirror at stations of the flow field.

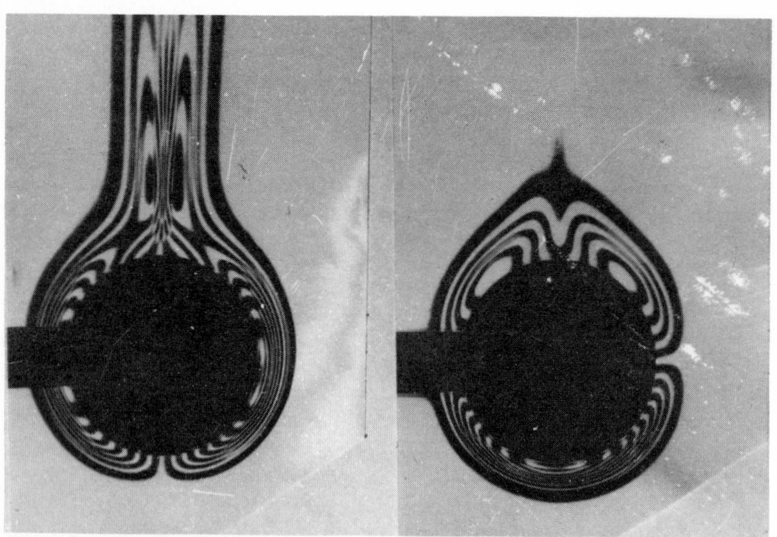

VERTICAL GRID AND SLIT HORIZONTAL GRID AND SLIT

FIG. 2.25 Heat convection from a circular cylinder (schlieren with Ronchi grating used to obtain the fringes; *from Oh & Didion, 1966*).

On the other hand, since

$$\frac{n_0 - 1}{n - 1} = \frac{T}{T_0}$$

$$\alpha_x = \frac{1}{n_0} \int_{z_1}^{z_2} \frac{\partial n}{\partial x} d_z$$

and

$$\alpha_y = \frac{1}{n_0} \int_{z_1}^{z_2} \frac{\partial n}{\partial y} dz$$

where n refers to index of refraction of light and subscript 0 refers to reference conditions such as ambient air.

By using these equations the temperature distribution in the flow field can be computed. For the free convection around the circular cylinder, Oh and Didion (1966) obtained a good agreement of reduced temperature distribution with the data measured by using other techniques, such as thermocouple.

Because temperature is closely related to density, a quantitative evaluation of density distribution may also be made by Isophot schlieren technique applying Ronchi grid.

Flow separation, forward positions of shock waves, and expansion and transition points can be observed with a fairly simple arrangement that does not require excessively high quality optical components. Details of the schlieren method are given by Töpler (1867), Schardin (1934 and 1941), Barnes and Bellinger (1945), Ladenburg (1954), and others.

Gawthrop et al. (1931) used the schlieren technique to show the emergence of the vortex ring from the mouth of a tube (Fig. 2.26). This vortex ring is formed by each successive plane shock wave as it emerges from the mouth of the tube and becomes a spherical wave. The pressure in the vortex was less than atmospheric; mathematical theory gives this as a condition of stability for such a ring. The changed optical density existing within the ring makes it possible to detect the vortex ring which is invisible to the naked eye. The vortex ring travels more slowly than the shock wave and is essentially the same as a smoke ring blown by a smoker.

Keto (1962) developed an improved schlieren technique using a number of mirrors, the so-called mirror method which involves the tight sealing of interchangeable fluidic elements to optical windows. This technique is suitable for the visualization of attached and separated flow through the narrow channel of a fluid amplifier; an exposure of no more than 1 μsec is required at a rate as high as 6000/sec. The resolution of flow details in passages 0.010 to 0.030 in. wide is possible with a maximum 8-in.-diameter field of view. This system is versatile enough to permit such variations as directional, nondirectional, phase contrast, color,

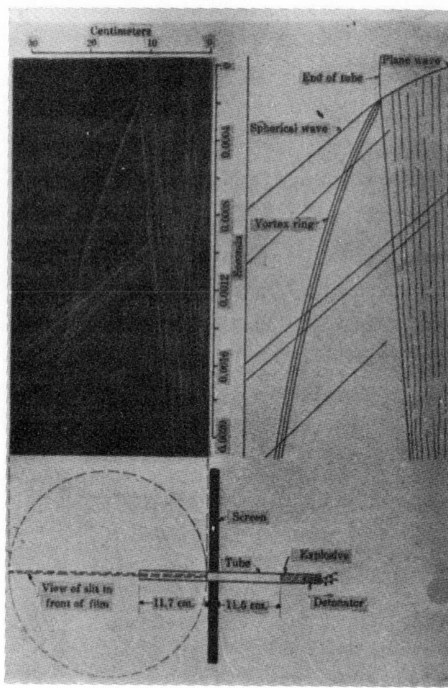

FIG. 2.26 Vortex ring at the end of cellophane tube using schlieren (*from Gawthrop et al., 1931*).

or monochromatic schlieren. Furthermore, it may be utilized as a polarizing or a diffraction grating-type interferometer. The sensitivity is restricted to 10^{-6} radian deflection. The behavior of a pure fluid element depends on flow in the small area in the immediate vicinity of the control jet; therefore, no leakage of fluid should occur. A tight sealing was originally achieved by strongly clamped transparent plastic windows. Since those windows introduced a strong schlieren effect, a special arrangement was provided to allow a leakage rate of less than 30×10^{-6} scfm at 30-psig pressure across a $\frac{1}{16}$-in. wall. As shown in Fig. 2.27, the photographs produced by the mirror method permit the flow separation and the shock at supersonic flow conditions to be observed. The same method can also be used to aid schlieren visualization of incompressible flow involving low sensitivity. If a volatile transparent liquid, such as trichlorethylene, is introduced onto the surface of the window, the flow pattern can be visualized within the errors caused by inertia effects. Because this technique works extremely well and requires little sensitivity of the schlieren system itself, color may be used to enhance the visualization.

Leonard and Keck (1962) showed that projectile wake flow visualization at low densities (say less than 1 percent of the atmosphere) may be made if the schlieren sensitivity is increased by resonance radiation. Sodium vapor and light near 5896 Å were used as the working media. After studying their own low density schlieren photographs, Leonard and Keck (1962) supported the hypothesis of Goldburg and Fay (1962) that the transition to unsteady motion in a compressible wake is the same vortex-shedding process that occurs in the incompressible case. Washburn and Goldburg (1963) utilized the enhanced index of refraction obtained with light operating near the resonant

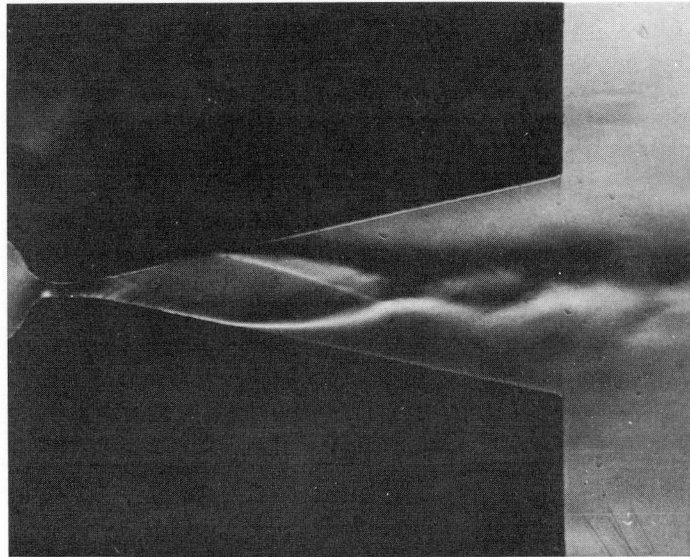

FIG. 2.27 Divergent nozzle showing flow separation using the mirror schlieren method (the throat is 0.030-inch wide; *from Keto, 1962*).

transition of sodium vapor and investigated the development of turbulence in the hypersonic wake by means of sodium schlieren pictures.

The sensitivity of schlieren is affected by the index of refraction of the working fluid. The refractivity may possibly be enhanced by using a gas with strong resonance in the ultraviolet or visible region. For low density, the optical system can be the usual schlieren off-axis paraboloidal mirror arrangement, modified by placing a narrow-band interference filter just in front of the knife edge. The wake photographs shown in Fig. 2.28 were taken with a wide-band filter at 2-percent atmospheric pressure. The horizontal wavelike striations in the first picture are several body lengths long and originate from a point near to the base region which corresponds to an area of small perturbations. The second picture shows the formation of the striations into a nearly ordered flow; the structure appears as a procession of vortex filaments with looplike shapes. In the third picture, the original perturbations have been greatly amplified. The pattern resembles vortex loops (stacked one behind the other) and exhibits the nearly periodic component of flow. The fourth picture gives an indication of the deterioration of the large-scale vortex filaments as they fill up with small-scale eddies. The last photographs indicate the continuation of the decay process toward fully developed turbulence.

Allan (1961) utilized the thermal changes caused in the boundary layer by a transfer of heat from a warm test object to cool water. In his successful application of schlieren at low speeds, Allan observed the boundary layer and the vortices in the wake of a live fish and obtained excellent photographs. In

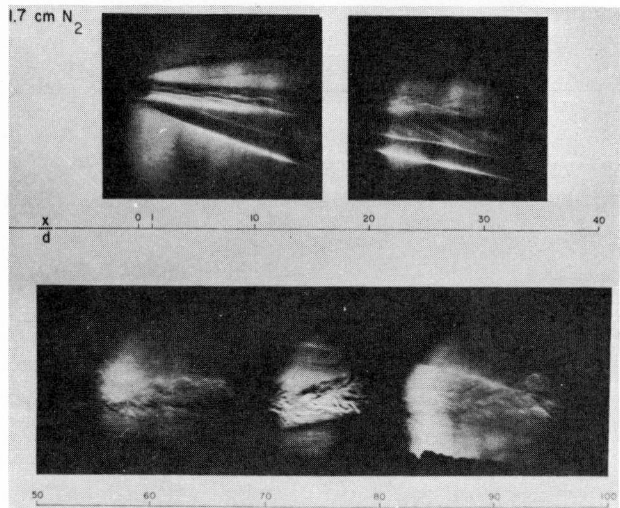

FIG. 2.28 Sodium schlieren composite of the wake of a 0.22-inch square at hypersonic speed in nitrogen (pressure = 1.7 cmHg, velocity = 14,000 fps; *from Washburn & Goldburg, 1963*).

order to record the boundary layer of the fish, he first raised its temperature by placing it into water that was warmer than the ambient temperature of the water in the test section of the flume. The transfer of heat from the fish changed the density of the water in the boundary layer. The vortical motion of the water shows more clearly in a still-water photograph since the vortical motion is essentially a circulation around a fixed point rather than a translation of the water movement. To a fixed camera, the vortical motion in the moving water appears to be a sinusoidal motion rather than a circulation phenomenon. The schlieren technique also shows large irregular static thermal lines in the water.

2.2.2.2 Color schlieren

Color schlieren photographs provide better information on flow phenomena than do black-and-white schlieren since they give the observers three factors to evaluate, i.e., hue, saturation, and brightness, rather than merely shades of gray. The eye can discriminate a larger number of different hues than it can detect shades of monochrome illumination and so the color system is more sensitive. Furthermore, a color schlieren system has a greater range for a given sensitivity level. Color is especially useful for boundary layer investigations and makes it easy to differentiate boundary layer phenomena (colored) from the boundaries of the model and the test section windows (black). However, the system poses a severe problem in diffraction and its sensitivity is limited. The sensitivity of a single-pass schlieren system is generally controlled by the size of the light source at the knife-edge plane.

The first color schlieren technique was developed by Holder and North (1952a and 1952b). They placed a constant-deviation dispersion prism between a white light source and the schlieren optics to produce a spectrum of the light source at the knife-edge plane. They then adjusted a slit in this plane to allow only a single color to pass onto the screen in the absence of disturbances in the test section. When a density gradient in the test section differed from that in the surrounding field, part of the spectrum was displaced, and light of a different color passed through the slit. Thus the image of the density gradient region was a different color than that of the surrounding image.

Holder introduced a second color technique which is easier to employ than the prism method. As described by North (1954), it consisted of simply arranging narrow strips of colored gelatin side by side and clamping them between glass plates. The filter was arranged at the schlieren knife-edge plane so that the image of the white light source was slightly greater than the width of one color strip. Thus as light was passed through one color strip, a small amount passed through both adjacent strips. A slight shift in the light source due to a different density gradient caused the corresponding screen image to be illuminated through an adjacent filter. Both Holder color techniques used a slit at the knife-edge plane, but in the multicolor filter system, each filter strip acted as a slit.

In the case of black-and-white schlieren, where a single knife-edge is used, diffraction produces a halo around the edge but does not significantly affect the apparent edge sharpness, even when the schlieren system is adjusted for higher sensitivities. For the existing color schlieren methods, however, the edge sharpness and the ability of the system to resolve small objects drop off rapidly when the slit is used. Compare the results for black-and-white versus color schlieren photographs in Fig. 2.29.

The color light source configuration used in a new dissection color schlieren method (Cords, 1967) is achieved by cutting rectangular holes in white graph paper and then photographing the paper over a black background. A high contrast negative of suitable reduction is obtained and color filter material is then secured over each clear rectangle. An image of the light source configuration is produced at the knife-edge plane. Two knife-edges are used to permit only the light from the desired rectangles to reach the camera image plane in the absence of disturbances in the test section. This dissection color schlieren method has two principal advantages: (*a*) the image of the light source configuration is affected in its entirety, and (*b*) the camera is focused on the schlieren test plane where light from all of the light source rectangles is integrated. The dissection method matches the black-and-white method in both image sharpness and sensitivity.

When color schlieren photography is used for tests in the ballistic range, a more efficient light-gathering optical system is needed than for wind-tunnel tests. Cords (1967) met this need at the U. S. Naval Ordnance Laboratory

(a)

(b)

FIG. 2.29 Separated flow over an axisymmetric body with a forward spike: (*a*) black-and-white schlieren; (*b*) color schlieren (Mach 2.5, spike diameter 0.25 in., after body diameter 1.0 in.; *courtesy of Professor R. H. Page, Rutgers University*).

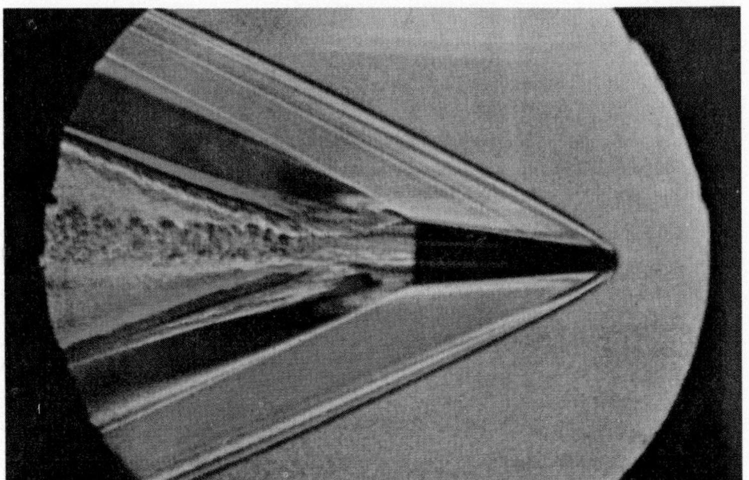

FIG. 2.30 Color schlieren of a ballistics model flying at Mach 2 in the NOL pressurized ballistics range (*from The Oak Leaf, NOL,* July 1967*).

(NOL)* by developing an improved focused shadowgraph recording device. It contains a spherical mirror which very efficiently uses the light from a self-contained spark source. The spark light, which lasts only three-fourths of one-millionth of a second (300 nanosec), passes twice through the air flow around the model, thus making the NOL system at least four times more sensitive than the standard parallel system with the same focal length mirror. The shadowgraph system is converted to color use by substituting for the camera aperture a color knife-edge consisting of a laminate of 0.015-in.-wide gelatin color-filter strips. The gelatin strips produce a fuzzy image by diffracting the light, but the NOL developed schlieren system produced much sharper photos useful for free-flight range tests, as can be seen in Fig. 2.30 for flight tests at Mach 2 in the NOL pressurized ballistic range.

Both black-and-white and color schlieren are used mainly for qualitative analysis of the flow field. Maddox and Binder (1970) recently developed a color schlieren system which enables quantitative analysis of the flow field. The system uses a diffraction grating to disperse white light into the spectrum and a multiple/slit arrangement at the knife edge. Conventional black and white schlieren systems are said to be easily converted. Details of the apparatus and the derivation of the equation for use in calculating the flow field are available in the original paper.

2.2.3 Interferometer

Because the principle of interference yields the density field directly, interferometry is the most suitable method for quantitative determination of

*Now Naval Surface Weapon Center.

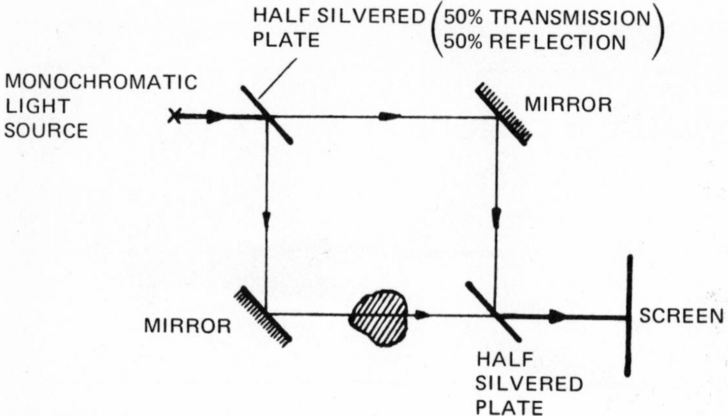

FIG. 2.31 Mach-Zehnder interferometer (*from Bradshaw, 1964*).

the density field. The most widely used system for wind-tunnel work is the Mach-Zehnder interferometer; see Fig. 2.31. In regions of large density gradient such as boundary layers, the refraction introduces a position error since the path of a ray through the test section is not the same as in the no-flow condition. Much care must be taken in adjusting the instrument, and the disadvantage is the difficulty in interpreting the results. An important consideration in the design of an interferometer is the great cost of mounting to isolate vibration.

The Mach-Zehnder interferometer uses a half-silvered plate to split a single beam from a light source into two beams, one of which bypasses the working section. The two beams are recombined at a second half-silvered plate before falling on a screen. The interferometer must be calibrated to enable absolute values of the density to be obtained. An excellent guide to the use of the interferometer is given by Holder et al. (1956) and by Ladenburg (1964). An example of an interferometer picture was shown in Fig. 2.22*c*.

The interferometer has been applied to the qualitative examination of hypersonic flows at low densities because it gives a higher sensitivity than possible with conventional schlieren methods (Oertel, 1960).

Eckert and Soehngen (1948) used an interferometer to visualize the temperature field around a horizontal circular cylinder heated in air (Fig. 2.32). In this figure the flow is laminar over the entire surface. The closer spacing of the interference fringes over the attached flow region on the lower portion of the cylinder indicates a steeper temperature gradient. Thus, the local unit-surface conductance in the attached flow region is larger than in the separated flow region over the top portion of the cylinder.

The Mach-Zehnder interferometer is also an ideal tool for the study of time-varying density fields in gases or liquids. In many cases, this interferometer gives indirect information on fields of temperature and velocity. A

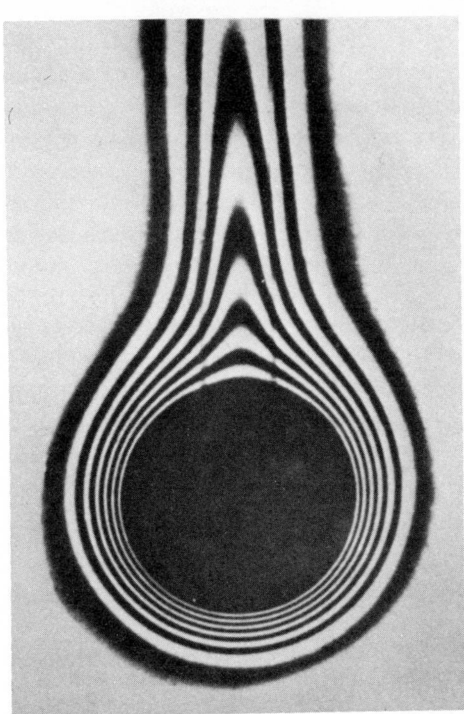

FIG. 2.32 Interferometer illustration of temperature field around a horizontal cylinder in laminar flow (*from Eckert & Soehnghen, 1948*).

spark from a high pressure mercury lamp with a duration of only a few microseconds will give sufficient illumination for a small-scale interference photograph on sensitive film. With proper adjustment, the resolution of the interference fringes is excellent and allows the interferogram to be greatly enlarged. A properly timed series of such sparks permits the study of processes which vary rapidly with time.

For the free convection phenomena, the interferogram reveals the growth of the boundary layer and permits the quantitative evaluation of the time variation of the shape of temperature profiles within the boundary layer (Eckert, 1960).

2.2.4 Holography

Holography, a new method of interferometry, is used to recreate a three-dimensional scene. This method consists of recording a portion of the stationary interference pattern produced by the passage of two coherent waves from a laser source through one another. The recorded pattern is called a hologram. Unlike a conventional camera, which records only the image of a focused subject, the hologram records the complex wave pattern scattered from a laser-illuminated object. This hologram can reconstruct the recorded waves faithfully in both amplitude and phase at a later time. Since the process allows the reconstructed waves to be observed in the same way as the waves scattered from the real object, the hologram provides a close substitute for the real object in an optical sense. Since the hologram may contain an immense amount of information about the subject, the method becomes extremely useful for recording transient or changing subjects.

The limitation on depth of field imposed by a conventional photographic system is bypassed by holography. In addition it permits recording small events of a larger distribution, such as spray, insect flights, cultures, and so on. Holograms have been made for bullets in flight, bubbles, cavitation, gas jets, aerodynamic wakes, chemical dissociation, cones, nozzle exhausts, etc.

This new method of interferometry has many significant advantages, including these pointed out by Heflinger et al. (1966) and Beamish et al. (1969) for application of the technique to fluid mechanics.

1. It does not require accurate alignment or precision optical elements.
2. It can easily measure small changes in either optical path or the position of complex subjects by using differential interferometry.
3. It can give a complete three-dimensional record of flow phenomena and thus enables postexposure focusing and examination from various directions.
4. It can provide holographic interferograms of corners and ducts (the more usual optical methods would require a light beam to be passed through a part of the model).

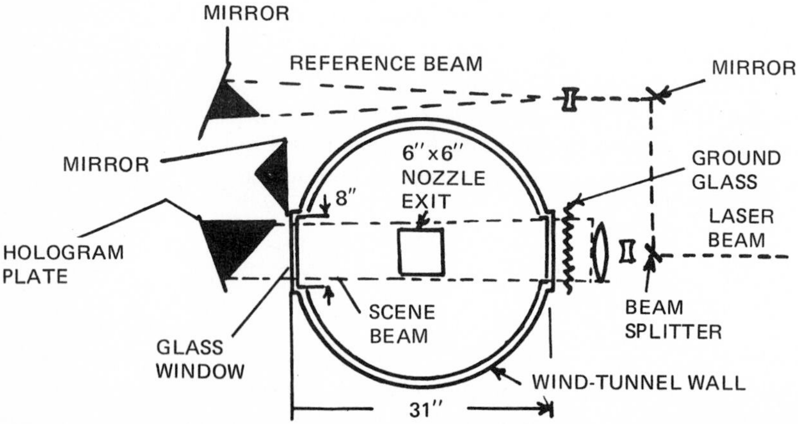

FIG. 2.33 Optical arrangement of holocamera around wind tunnel (*from Beamish et al., 1969*).

The arrangement shown in Fig. 2.33 is described by Beamish et al. (1969). The interference pattern recorded on the photographic hologram plate is formed by a reference beam and a scene beam. To make a hologram, the plate is first exposed without flow in the tunnel but at a static pressure approximately equal to that in the tunnel with flow. After the flow is established, the same plate is exposed again. After this (doubly exposed hologram) plate has been developed, it is illuminated with a replica of the reference beam to produce images of both exposures simultaneously. Any change in air density between the exposures causes the two reconstructed wavefronts to assume different shapes, resulting in an interference pattern from which density variation may be found. The interference pattern from one hologram may be viewed or photographed from any aspect, thus resulting in an interferogram from that perspective.

Distortion of the wavefront by optically poor elements or transparent model components does not affect the fringe pattern since the beams for both exposures follow the same optical path, except in the test section.

2.3 PHYSICAL MEASUREMENTS

The position of separation and the characteristics of separated flow may be determined by quantitative measurements, e.g., of the velocity profile, skin friction, and heat transfer in the neighborhood of separation and within the separated region. Detailed explanations of experimental methods may be found in the texts by Pope (1947), Pankhurst and Holder (1952), Howarth (1953), Ladenburg (1964), Rosenhead (1963), Ower and Pankhurst (1966), and others. These physical measurements are generally tedious, expensive, and require considerable amount of data. Furthermore, since the shear layer of attached flow is usually thin, the measuring devices must be accordingly small, and special techniques for their manufacture are required.

An adverse pressure gradient is a necessary condition for flow separation on a continuous body surface. Hence, the position of separation may be estimated for some cases by surveying the pressure distribution, for example, in the case of laminar flow over a surface where a sufficiently strong adverse pressure gradient prevails. In this case, laminar flow separation may be expected to occur downstream of the pressure minimum.

As separation is approached, the shear layer grows rapidly. Therefore, an estimate of separation point generally requires both a streamwise pressure survey and measurement of the shear layer growth.

2.3.1 Measurement of the Velocity Profile

The velocity profiles of shear flows are commonly measured with a pitot tube or a hot-wire anemometer. Separation on a two-dimensional surface occurs where the derivative normal to the surface of the tangential velocity is zero at the surface.

2.3.1.1 The pitot tube

The pitot tube is a useful instrument for determining local, but time averaged, velocity u from a single measurement of the difference between the static pressure p and total pressure p_t. For incompressible flow, the velocity u is determined by

$$\frac{1}{2}\rho u^2 = p_t - p \tag{2.1}$$

The static pressure p is often measured by a pressure orifice on the surface. Since the static pressure may be assumed to remain constant across the boundary layer, the value for p measured with a pressure orifice on the surface can be regarded as that of the inviscid flow at the outer edge of the shear layer. A standard pitot tube is essentially an open-ended tube with its mouth facing upstream. The shape of the mouth is generally circular, but a flattened mouth is often used in order to measure the desired flow properties close to a wall.

For most practical purposes, yaw and free-stream Reynolds number have negligible influence on pitot tube measurements. Near the wall, however, the effect of low Reynolds number (based on the internal radius of the tube or the internal height of flattened tubes and local flow velocity) may be considerable and require corrections; see pages 47–48 in Ower and Pankhurst (1966) and pages 619-622 in Rosenhead (1963). There are four basic corrections that need to be made for a pitot tube measurement:

1. Viscosity errors—only for very small tubes—as shown in Fig. 2.34, a pressure coefficient C_p different from one results. Thus Eq. (2.1) should read

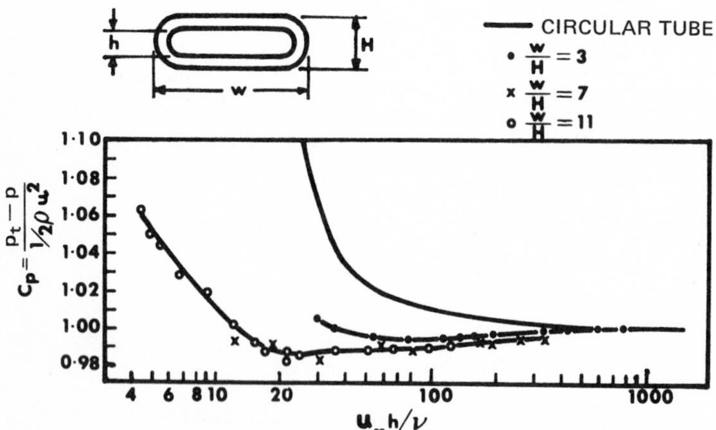

FIG. 2.34 Viscosity errors of small pitot tubes in incompressible flow (*from Ower & Pankhurst, 1966*).

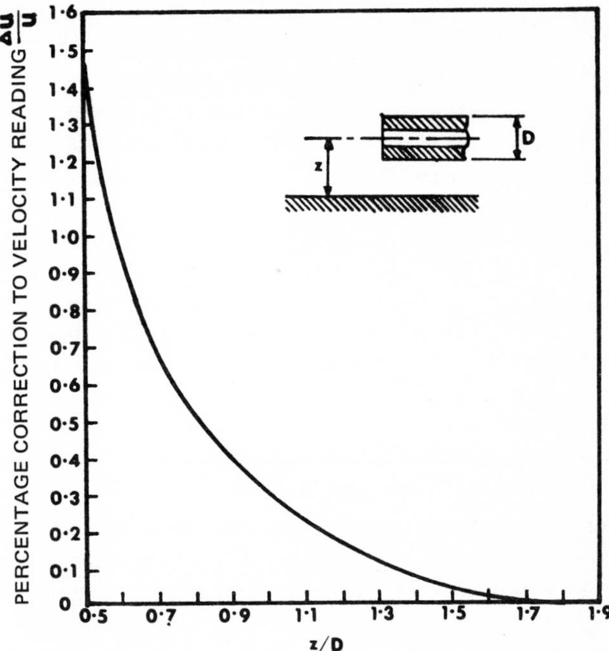

FIG. 2.35 Wall proximity correction for pitot tube (*from Ower & Pankhurst, 1966*).

$$p_t - p = C_P \left(\frac{1}{2}\right) \rho u^2 \tag{2.2}$$

2. Displacement effect–because of the transverse velocity gradient in the boundary layer, the effective center of the pitot tube is displaced toward the higher velocities, i.e., away from the wall. For a square-ended circular tube (Young & Maas, 1936)

$$\frac{\Delta z}{D} = 0.13 + \frac{0.08\,d}{D} \tag{2.3}$$

   where Δz is the displacement and d and D are the inner and outer diameters of the tube. Later MacMillan (1956) suggested that this equation may give high values. Also Livesey (1956) found no displacement effect for conical-nosed, sharp-edged probes. There appears to have been no investigation into the displacement effect of flattened pitot tubes.
3. Proximity effect–when a pitot tube is closer to the wall than twice the tube diameter, the measured velocity should be increased by the quantity given in Fig. 2.35.
4. Turbulence effect–in turbulent flow, pressures measured by a pitot tube include the effects of the fluctuating velocities. This correction is often

neglected, but it should be considered when measuring turbulent boundary layers.

This discussion has been concerned with incompressible flows. For information on the use of pitot tubes in compressible flow, the reader is referred to the studies of Ower and Pankhurst (1966) and an article by P. L. Chambre and S. A. Schaff in the book by Ladenburg (1966).

The so-called boundary layer "mouse" consists of a number of flat pitot tubes mounted at slightly varying heights. Just one setup is used to measure the velocity through the thickness of the boundary layer. However, the usefulness of this "mouse" for curved surfaces is very limited. Hence, this device may not be very useful for separation because adverse pressure gradients are usually associated with curved surfaces.

2.3.1.2 The hot-wire anemometer

The standard hot-wire anemometer consists essentially of a short length of very fine wire supported between two metal prongs and heated electrically. The hot wire is aligned normal to the airstream. The air velocity is deduced from the cooling produced by the airflow, either by recording the current needed to maintain a constant wire temperature or by measuring the potential difference across the wire with a constant current passing through it. Either method is satisfactory for studying separation air speeds less than 100 ft/sec, but above that level, the first method is more accurate.

A hot-wire anemometer has two main advantages over a pitot tube:

1. Its small size makes it possible to place the instrument very close to the wall.
2. Its extremely rapid response allows instantaneous velocity fluctuations, especially those of turbulent flow, to be measured.

However, in order to achieve accurate measurements close to a wall, a correction must be made for the transfer of heat between the hot wire and the surface—otherwise, the air velocity appears too high. The requirement that the length of wire must be short may make it necessary to amplify the voltage before their fluctuations can be measured. The standard type of the hot-wire anemometer is insensitive to the direction of the flow, but this shortcoming may be overcome by using two or more identical wires symmetrically disposed. Detailed descriptions of the use of hot-wire anemometers are available in Lowell (1950), Davies and Davis (1966), Davies et al. (1967), and Wieghardt and Kux (1967).

When high mechanical strength is required for a measurement, hot film may be used in place of hot wire. The film consists of a thin film of metal (platinum fused to the surface of an insulator, such as pyrex glass tubing). Hot-film probes are available in such shapes as wedges, cylinders, rings, etc. The hot-film apparatus was developed by S. C. Ling (1960) and is described in his paper.

Hot film may also be used in liquid flows (Ling and Hubbard, 1956). In addition, it can be calibrated to measure skin friction; see the papers by Bellhouse and Schultz (1964, 1968) and Sec. 2.3.3 of this chapter.

2.3.1.3 Interpretation of separation in a measured velocity profile

Schubauer (1939) measured the two-dimensional boundary layer flow on an elliptic cylinder with a major axis of 11.78 in. and a minor axis of 3.98 in. The major axis was parallel to the undisturbed subsonic air stream wherein the turbulence could be varied. Velocity profiles were measured with a standard type of hot-wire anemometer. The Schubauer investigation of the locations of transition, laminar, and turbulent separation and reattachment serve as an excellent example of the detailed interpretation required when one attempts to measure these phenomena. See his data for the mean velocity as presented in Figs. 2.36 and 2.37.

Figure 2.36 shows the measured velocity profiles at eight positions along the body surface. Figure 2.37 presents the contour diagram of the boundary layer; each curve represents a particular value of u/u_∞. Such simultaneous presentation of the velocity profile with a contour diagram is very useful in the interpretation of complex flow phenomena.

Note from Fig. 2.36 that about the only change for the profiles that correspond to $x/D = 0.251$ to 2.01 (where x is the distance from the leading edge and D is the minor axis of the ellipse) was caused by a thickening of the boundary layer. Further, the shapes of the profiles are indicative of a laminar boundary layer. At $x/D = 2.52$, the profile shows the beginning of separation.

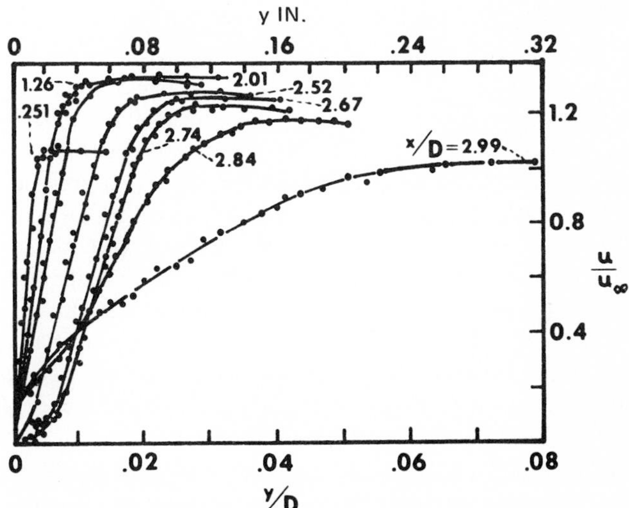

FIG. 2.36 Speed distribution in the boundary layer of an elliptic cylinder (air speed $u_\infty = 70$ fps; $Re_d = 139{,}000$; stream turbulence 0.85 percent; *from Schubauer, 1939*).

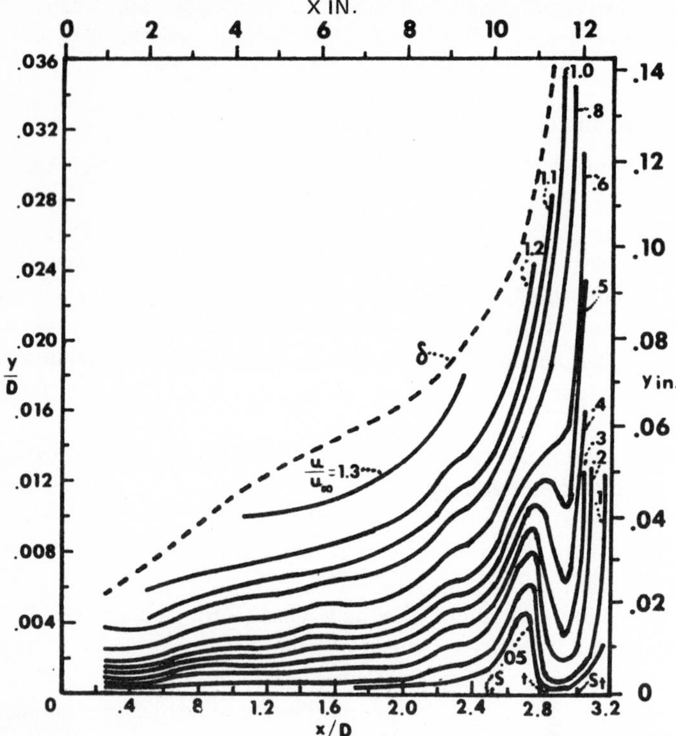

FIG. 2.37 Contours of equal speeds in boundary layer of elliptic cylinder (air speed $u_\infty = 70$ fps; $\text{Re}_d = 139{,}000$; stream turbulence 0.85 percent; (*from Schubauer, 1939*).

The very small initial slope of the profile from 2.67 to 2.74 indicates that separation has taken place. A marked change occurs between 2.74 and 2.84 and all evidence of separation disappears at 2.84. The profile at 2.99 is distinctly not that ordinarily ascribed to a laminar boundary layer. Although this profile characterizes the turbulent boundary layer with a strong adverse pressure gradient, it does not exhibit the characteristics found in a turbulent boundary layer on a flat plate (represented by the $\frac{1}{7}$ power law).

The contour diagram of Fig. 2.37 gives a broad picture of the boundary layer as a whole. The dotted curve indicates the boundary layer thickness δ, defined as the distance where $u/u_\infty = 0.995(u_e/u_\infty)$. The occurrence of separation followed by the reattachment of the boundary layer to the wall is evidenced by the hump in the contours from $x/D = 2.5$ to 2.9. Beyond the hump, the very rapid increase in slope of the contour curves indicates the approach of the separation point for the turbulent boundary layer.

Schubauer (1939) observed the occurrence of a weak and very incomplete transition at $x/D = 1.53$ by detecting the changes in the average speed near the surface. Thus, the boundary layer separated, not in a purely laminar way, but rather in the manner of a transition so incomplete that the layer continued to exhibit mostly laminar properties.

The hot-wire anemometer was insensitive to the direction of flow in the immediate vicinity of the separation point. Accordingly, Schubauer (1939) utilized kerosene and lamp black to determine the points S (laminar separation) and St (turbulent separation) at $x/D = 2.51$ and 3.03, respectively (see Fig. 2.37). Thus he found that the attached turbulent flow extended from $x/D = 2.81$ to 3.03.

By using the measured data at the laminar separation point ($x/D = 2.51$), Schubauer obtained a value of −11.8 for the Pohlhausen parameter $\lambda = \delta^2/\nu \, du_e/dx$. Since the criterion for laminar separation by this parameter is −12 (as discussed in Chap. 1), there was close agreement between the measured and the theoretical values.

2.3.2 Measurement of Skin Friction

Because $\tau_w = \mu(\partial u/\partial y)_{y=0}$, the position of separation on a two-dimensional body surface or axisymmetric body may be determined by measuring the point where the skin friction becomes zero. The skin friction can be measured by using a floating element, a Stanton tube (surface tube), or a Preston tube. Skin friction may also be determined by measuring heat transfer as described in Sec. 2.3.3.

2.3.2.1 The floating-element skin friction meter

A floating-element skin friction balance was first used by Dhawan (1953) in subsonic flow. Since that time, this device has been used in transonic flows (Hakkinen, 1955), in supersonic flow (Hopkins & Keener, 1966), and in flows with mass injection (Dershin et al., 1966). A floating-element system has also been designed for the direct measurement of skin friction in actual flight (Moore et al., 1966).

A small portion of the surface is arranged to form a floating element which is separated from the rest of the surface by a narrow gap. The element is attached to some form of balance housed within the model, and the tangential force on the element is obtained by direct measurement. The advantage of direct measurement makes it probably the most accurate means of measuring skin friction (O'Donnell & Westkaemper, 1965). On the other hand, the floating element is expensive, delicate, and involves electrical devices.

Details of the construction of this device may be found in the above references or in the paper by Westkaemper (1967). Since the gap may cause a discontinuous change in the variation of the shear stress, consideration must be given to the effect produced when the size of the gap is varied. This has been investigated by Smith and Walker (1959) for incompressible flow and by O'Donnell and Westkaemper (1965) for supersonic Mach numbers. The floating element has been applied successfully only to flat surfaces. Since separation generally occurs with adverse pressure on a curved surface, the use of the floating element to investigate separation is limited except in such cases as a straight wall diffuser.

A skin friction balance for measurement in flows with heat transfer and a pressure gradient has been built at NOL [now Naval Surface Weapon Center] (Bruno et al., 1969). This device is water-cooled with a 0.5-in^2 circular measuring element, and is used on the flat plate side of a boundary layer channel. A pressure gradient is imposed by adjusting the opposing wall.

2.3.2.2 The Stanton tube (surface tube)

The Stanton tube is essentially a small pitot tube with one of its walls by the surface itself; see Fig. 2.38. It was originally developed by Stanton et al. (1920).

For a typical tube, the upper lip is about 0.003 in. from the surface and ranges in the same order of magnitude as the laminar sublayer. Because of the low Reynolds number and severe velocity gradient involved in measurements by a Stanton tube, the instrument should be calibrated in known flow conditions in order to evaluate the local wall shear stress.

As an example, separation that involves shock interaction can be detected by the difference in pressure as measured by the Stanton tube and by an orifice on the wall. Gadd et al. (1954) indicate that near the separation point, the static pressure measured by a wall orifice rises steeply and the difference between Stanton tube pressure and static pressure decreases. Eventually, the static pressure slightly exceeds the tube pressure. Separation is considered to occur in the neighborhood of the point where the static and tube pressures are identical. Along the separated region, the measured static pressure remains slightly larger than that measured by the Stanton tube probably because of reverse flow. As reattachment is approached, this trend reverses. Thus, in the

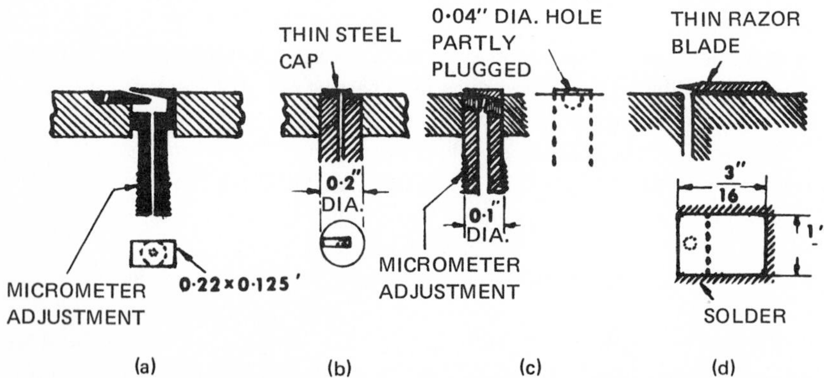

FIG. 2.38 Forms of surface tubes: (*a*) *from Stanton et al., 1920*; (*b*) *from Fage and Falkner, 1930*; (*c*) *from Fage and Sargent, 1947*; (*d*) *from Hool, 1956; taken from Rosenhead, 1963.*

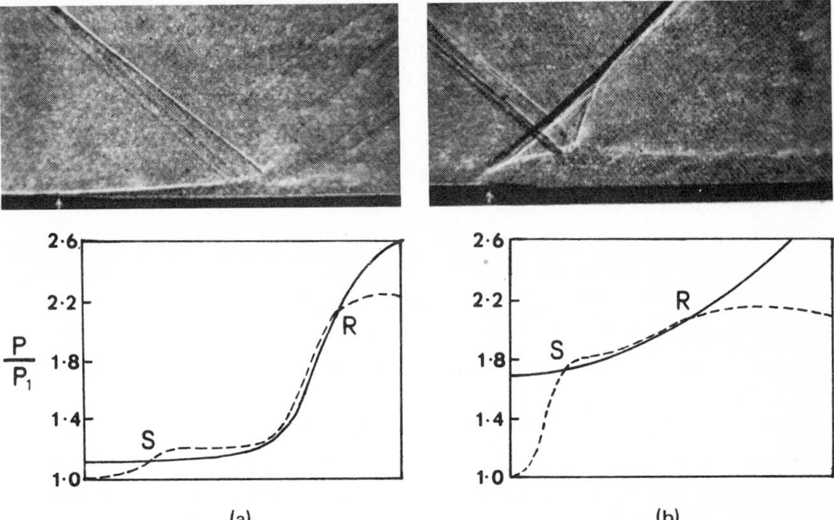

FIG. 2.39 Examples of the method used to detect separation with a surface tube: (*a*) laminar separation; (*b*) turbulent separation (*from Gadd et al., 1954*).

neighborhood of the point where this difference diminishes, reattachment may be considered to occur (similar to the case for separation). This may be seen in Fig. 2.39.

The Stanton tube may be used on any model where provision has been made for measuring the pressures to determine the skin friction. The use of a portion of a razor blade (Hool, 1956) converts the pressure holes into surface pitot tubes, as shown in Fig. 2.38.

Extensive calibration of the Stanton tube was carried out by Smith et al. (1962) on various body shapes and Lewis and Kubota (1966) at hypersonic speed. Hopkins and Keener (1966) indicate that the Preston tube (see next section) is superior to the Stanton tube for measuring local skin friction because the calibration of the latter is nonlinear and is greatly affected by the streamwise location of the blade leading edge relative to the static orifice. Nevertheless, as indicated above, separation is fairly clearly indicated by Stanton tube measurements without the need for any calibration.

2.3.2.3 Preston tube

Preston (1954) measured skin friction in turbulent flow by means of an ordinary circular flat-ended pitot tube in contact with the wall. His experiment showed that the following relationship exists between the wall shear stress τ_w and the reading of a pitot tube resting on the wall P:

$$\frac{(P-p)\,d^2}{\rho\nu^2} = F\left(\frac{\tau_w d^2}{\rho\nu^2}\right)$$

if $u = u^* f(y^*)$ and $y^* = yu^*/\nu$. In the above equation, d is the outside diameter of the pitot tube, p is static pressure on the surface, ρ is density, ν is kinematic viscosity, and u^* is friction velocity given by $u^* = (\tau_w/\rho)^{\frac{1}{2}}$. Preston found that this equation could be approximated by a straight line on log-log paper for large ranges of $(\tau_w d^2)/(\rho\nu^2)$. Later calibrations in incompressible flow by Smith and Walker (1959), Head and Rechenberg (1962), Patel (1965), and Ferriss (1965) suggest that the original Preston calibration may have been in error.

Calibration in turbulent supersonic flow was completed by Hopkins and Keener (1966) for adiabatic wall conditions and by Keener and Hopkins (1969) for hypersonic flow. The effects of heat transfer, compressibility, and favorable pressure gradient were reported by Yanta et al. (1969).

Because the relation of $u = u^* f(y^*)$ does not apply in the neighborhood of separation, the Preston tube has limited use for measurement of the separation point (see Sec. 2.4.1).

Allen (1970) recently attempted to use the Preston tube on a body of revolution at supersonic ($M_\infty = 2.5$ to 4.5) speeds. The agreement between various indirect methods was insufficient for him to assess the overall accuracy of the tube.

Murphy (1955) suggested that the Preston tube technique could be improved by using a flattened pitot tube resting on the surface and by providing a correction for the effective center of the probe and the boundary condition of $u = 0$ at $y = 0$ to evaluate the velocity gradient on the wall.

2.3.3 Measurement of Heat Transfer

The heat transfer characteristics of a flow change rather abruptly downstream of separation. Therefore the position of separation may be determined by measuring the rate of change of the heat transfer, e.g., by the use of a radiometer or a direct heating element.

The radiometer determines the location of boundary layer transition and flow separation by measuring the abrupt change in radiation level (the radiation levels are also different in the different flow regimes). The intensity of the radiant energy is measured either as total radiation or as radiation in a particular frequency of infrared radiation. A radiometer system consists of a sensitive detector, an optical system, and some means of measuring the output of the detector. The sensitive detector may be a thermocouple and a galvanometer or an ultrasensitive voltmeter used to measure the thermocouple voltage. A typical optical system could, for example, consist of a parabolic mirror aluminized on its front surface to form an image of the object or the selected area on the thermocouple.

Loving and Katzoff (1959) claim that a radiometer that measures the far infrared radiation from a surface is capable of detecting temperature differences of 0.04°F at room temperature. A Brown self-balancing recording potentiometer is used to obtain a record of the infrared radiation received by the instrument. The separated flow regime is indicated by a lower infrared radiation level than that of the attached flow region. However, the degree and sharpness of the temperature change at the edge of the separated flow region are not as pronounced as at the transition line between laminar and turbulent flow regions.

Not only can a radiometer be mounted in an open-throat channel, but it can also be adapted to flight studies. In the latter case, the instrument is mounted in the fuselage and arranged to survey the wing surface through a window that is transparent to infrared rays.

Because the rate of local heat transfer from a solid surface to the moving fluid is related to the local skin friction, the measurement of heat transfer can also be used for an indirect determination of the position of flow separation. This technique was used at subsonic speeds by Fage and Falkner (1931) for laminar flow and by Ludwieg (1949) for turbulent boundary layers. Liepmann and Skinner (1954) extended it to high-speed flow by using an ordinary hot wire cemented into a groove in the surface.

The heat transfer may also be measured (and thus separation, skin friction, and transition determined) by baking a thin film (of the type discussed in Sec. 2.3.1.2) on the surface of the body rather than a probe. Bellhouse and Schultz (1964) measured the current necessary to keep the resistance of the film constant at various positions on a circular cylinder. The separation points determined by Stanton tube measurements were close to the minima of their current versus angle (position) curve. Also voltage fluctuations were found near separation in laminar flow.

2.4 DETERMINATION OF SEPARATION; A COMPARISON OF VARIOUS TECHNIQUES

As stated in the introduction to this chapter, much experimental work needs to be done. The two experiments discussed in this section not only illustrate differences in various techniques but also serve to indicate the use of more than one experimental method for interpreting flow phenomena.

2.4.1 Separation of Axially Symmetric Turbulent Boundary Layers

Murphy (1954 and 1955) used the dust method (Sec. 2.1.8) and the china clay method (Sec. 2.1.9) to visualize separation. He made pitot tube and Preston tube measurements (Sec. 2.3.2.3) on a slender body of revolution (5-ft long, 5.9-in.-diameter model with blunt nose and tail) in axial incompressible flow. The flow was parallel to the axis of the body, and measurement indicated axisymmetric flow. The two visualization techniques (dust and china clay) gave good agreement on the position of separation.

The use of the Preston tube shear stress data required extrapolation to zero at separation. Murphy (1955) found that the stress rapidly decreased in the region of the adverse pressure gradient. The extrapolation gave fair agreement with the dust method. The latter gave a separation position about 0.3 in. upstream of the position obtained from shear stress measurements.

The detailed measurements confirmed some of the characteristics of boundary layers at separation as discussed in Chap. 1. All boundary layer characteristics, δ, δ^*, θ (boundary layer, displacement, and momentum thickness, respectively) and $H = \delta^*/\theta$ (shape parameter) grow at a rapid rate in the adverse pressure gradient. In fact, before separation occurred, the boundary layer thickness δ was found to exceed the body radius. The measured wall velocity gradient decreased as the separation approached (Murphy, 1955).

2.4.2 Flow Separation on an Inclined Cone

The separation of flow from an inclined cone is the simplest case of three-dimensional geometry. It has been studied in detail by Rainbird et al. (1963), Friberg (1965), and Schindel and Chamberlain (1967). At the National Aeronautical Establishment in Canada, Rainbird used the dye injection technique (Sec. 2.1.3) and reported unpublished oil film experiments (Sec. 2.1.7) at the College of Aeronautics, Cranfield. Similar oil film tests and hydrogen bubble visualizations (Sec. 2.1.1) were made by Friberg (1965) at the Massachusetts Institute of Technology. Schindel and Chamberlain (1967) followed up the Friberg experiments with china clay (Sec. 2.1.9), oil drop (Sec. 2.1.7), radiating grid (Sec. 2.1.6), tuft grid (Sec. 2.1.5), and smoke (Sec. 2.1.2).

The major differences in the angular position of separation derived by several of the methods indicate that they may be sensitive to different aspects of the phenomena of three-dimensional flow separation. The interpretation according to Friberg (1965) and Schindel and Chamberlain (1967) is given here with reference to Fig. 2.40. Such techniques as dye injection (from surface holes), china clay, oil film, or oil drop are responsive to the flow at the bottom of the boundary layer or to the limiting surface streamline. In this manner, they indicate separation when the surface streamline becomes a generatrix of the cone. On the other hand, hydrogen bubbles are off the surface of the model, probably at the outer edge of the boundary layer. Thus they follow a stagnation line of the cross flow at the outer edge of the boundary layer. Since the boundary layer is comparatively thick at separation, these techniques give different results.

It is interesting to note some results of a comparison of theory with the previously described experiments. Boundary-layer calculations to predict separation are in much better agreement with methods based on flow at the bottom rather than on the surface of the boundary layer (Friberg, 1965). But

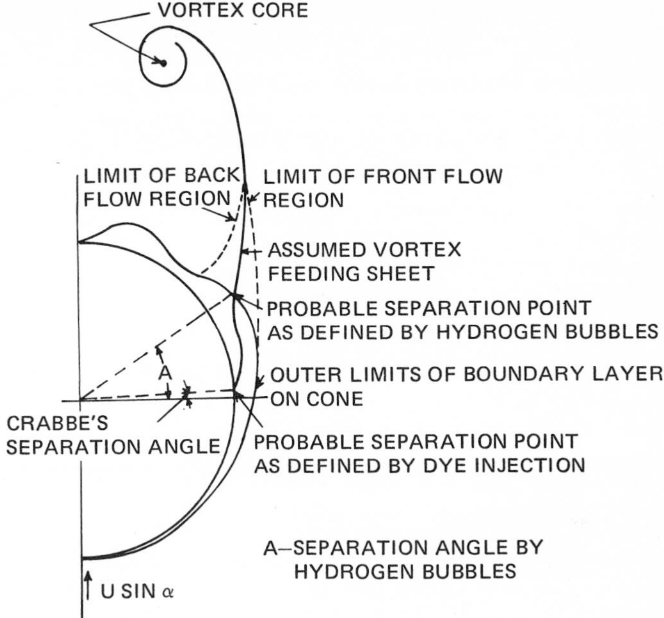

FIG. 2.40 Possible explanation for the different values of separation angle measured by dye injection and hydrogen bubbles (*from Friberg, 1965*).

the separation line as indicated by bubble tests give better results when used in an inviscid model of the flow which requires the separation position as an input. This is a further indication of the complexity of three-dimensional flows with separation and illustrates the problems of interpreting experiments and setting up analytical models.

REFERENCES

Allan, W. H. (1961). "Underwater flow visualization techniques," NAVWEPS Report 7778, NOTS TP 2759.

Allen, M. (1970). "Experimental Preston tube and law-of-the-wall study of turbulent skin friction on axisymmetric bodies at supersonic speeds," NASA TN D-5660.

Barnes, N. F. and S. L. Bellinger (1945). "Schlieren and shadowgraph equipment for air flow analysis," *J. Opt. Soc. Am.*, Vol. 35, p. 497.

Bazzochi, E. (1956). "Boundary layer flow-visualization tests in a low velocity wind tunnel," *(Ital) Aerotechnica*, Vol. 36, p. 315.

Beamish, J. K. et al. (1969). "Wind-tunnel diagnostics by holographic interferometry," *AIAA J.*, Vol. 7, No. 10 (October), pp. 2041–2043.

Bellhouse, B. J. and D. L. Schultz (1964). "Determination of skin friction, separation and transition with a thin film-heated element," ARC R&M 3445 (AD 807673).

Bellhouse, B. J. and D. L. Schultz (1968). "The measurement of fluctuating skin friction in air with heated thin-film gauges," *J. Fluid Mech.*, Vol. 32, Part 4, pp. 675–680.

Bergh, H. (1958). "A method for visualizing boundary layer flow," Symposium, Grenzschichtforschung, Freiberg, i. Br. Berlin-Göttingen-Heidelberg, Springer-Verlag, p. 173.

Bird, J. D. (1952). "Visualization of flow fields by use of a tuft grid technique," *J. Aeronaut. Sci.*, Vol. 19, No. 7 (July), pp. 481–485.

Boelter, L. M. K. et al. (1965). *Heat transfer notes*, McGraw-Hill Book Company, New York.

Bömelburg, H. J. et al. (1959). "The electric spark method for quantitative measurements in flowing gases," *Z. Flugwissenschaften*, Vol. 7, pp. 321–329. (Also AFOSR TN-59-273 [AD 230 263].)

Bradshaw, P. (1964). *Experimental fluid mechanics*, Pergamon Press, New York.

Bruno, J. R. et al. (1969). "Balance for measuring skin-friction in the presence of heat transfer," NOLTR 69-56.

Chang, P. K. and W. H. Dunham (1961). "Laminar separation of flow around symmetrical struts at various angles of attack," David Taylor Model Basin Report 1365.

Clark, K. W. (1933). "Methods of visualizing air flow with observations on several aerofoils in the wind tunnel," ARC R&M 1552.

Clayton, B. R. and B. S. Massey (1967). "Flow separation in an aerofoil cascade," *J. Roy. Aero. Soc.*, Vol. 71, No. 680, pp. 559–565.

Clutter, D. W. and A. M. O. Smith (1961). "Flow visualization by electrolysis of water," *Aerospace Eng.*, Vol. 20, No. 1, p. 24ff, January.

Clutter, D. W. et al. (1959). "Techniques of flow visualization using water as the working medium," Report ES29075, April, Douglas Aircraft Division, El Segundo, California.

Cords, P. H., Jr. (1967). "A high resolution, high sensitivity color schlieren method," Preprint of Society of Photo-Optical Instrumentation Engineers, paper presented at 12th Annual Technical Symposium, 7–11 August, Los Angeles, California.

Davies, P. O. A. L. and M. R. Davis (1966). "The hot wire anemometer," I.S.A.V. Report 155 (N67-39257), University of Southampton, England.

Davies, P. O. A. L. et al. (1967). "Operation of the constant resistance hot-wire anemometer," I.S.A.V. Report 189, University of Southampton, England.

Dershin, H. et al. (1966). "Direct measurement of compressible turbulent boundary layer skin friction on a porous flat plate with mass injection," General Dynamics/Pomona Report CR-332-761-001. (Also NASA CR-79095 [N66-39999].)

Dhawan, S. (1953). "Direct measurements of skin friction," NACA Report 1121 (supersedes NACA TN 2567, 1952).

Dimeff, J. et al. (1952). "X-ray instrumentation for density measurements in a supersonic flow field," NACA TN 2845.

Eckert, E. R. G. (1960). "The application of interferometer to time-varying flow conditions," Symposium on Flow Visualization, ASME Annual Meeting, 30 November, New York.

Eckert, E. R. G. and E. Soehnghen (1948). "Studies on heat transfer in laminar free convection with Zehnder-Mach interferometer," USAF Technical Report 5747.

Fage, A. and V. M. Falkner (1930). "An experimental determination of the intensity of friction on the surface of an aerofoil," *Proc. Roy. Soc. (London)*, Series A, Vol. 129, pp. 378–410. (Also ARC R&M 1315 [1930–31 Vol.].)

Fage, A. and V. M. Falkner (1931). "Further experiments on the flow round a circular cylinder," ARC R&M 1369 (1930–31 Vol.).

Fage, A. and R. F. Sargent (1947). "Shock-wave and boundary-layer phenomena near a flat surface," *Proc. Roy. Soc. (London)*, Series A, Vol. 135, pp. 656–677.

Farren, W. (1932). "Airflow with demonstrations on the screen by means of smoke," *J. Roy. Aeronaut. Soc.*, Vol. 36, p. 454.

Ferriss, D. H. (1965). "Preston tube measurements in turbulent boundary layers and fully developed pipe flow," ARC C.P. 831, AD-479412.

Frey, K. P. H. and N. C. Vasuki (1966). *Detached flow and control,* published by author at Box 584, Newark, Delaware, 19711.

Friberg, E. G. (1965). "Measurement of vortex separation, part II: Three-dimensional circular and elliptic bodies," MIT Aerophysics Laboratory Technical Report 115 (AD 628019).

Gadd, G. E. et al. (1954). "Interaction between shock waves and boundary layers," *Proc. Roy. Soc. (London)*, Ser. A., Vol. 226, pp. 227–253.

Gawthrop, D. B. et al. (1931). "The photography of waves and vortices produced by the discharge of an explosive," *J. Franklin Institute*, Vol. 211, pp. 67–86.

Gazely, C. (1950). "The use of the china-clay lacquer technique for detecting boundary layer transition," General Electric Report R49A0536.

Geller, E. W. (1955). "An electrochemical method of visualizing the boundary layer," *J. Aeronaut. Sci.*, Vol. 22, No. 12 (December), pp. 869–870.

Goddard, V. P. (1962). "Development of supersonic streamline visualization," Aero Eng. Dept., University of Notre Dame, March.

Goddard, V. P. et al. (1959). "Visual supersonic flow patterns by means of smoke line," *J. Aerospace Sci.*, Vol. 26, No. 11 (November), pp. 761–762.

Goldburg, A. and J. Fay (1962). "Vortex loops in the trail behind hypersonic pellets," AMP 75, Avco-Everett Research Laboratory.

Görtler, H. (1940). "Über eine dreidimensionale Instabilität laminarer Grenzschichten an konkaven Wänden," *Nachr. Ges. Wiss. Göttingen, Math.-phys. Kl.*, Vol. 1, pp. 1–26. (Translated as "On the three-dimensional instability of laminar boundary layers on concave walls," NACA TM 1375.)

Gray, W. E. (1946). "A simple visual method of recording boundary layer transition," RAE Tech. Note Aero. 1816.

Gutsche, F. (1940). "Versuche an umlaufenden Flügelschnitten mit abgerissener Strömung," *Jahrb. d. Schiffbautech. Gesellschaft,* Vol. 41.

Hakkinen, R. J. (1955). "Measurement of turbulent skin friction on a flat plate at transonic speeds," NACA TN 3486.

Hama, F. R. (1960). "The injection of dye for flow visualization," Symposium on Flow Visualization, ASME Annual Meeting, 30 November, New York.

Hauser, E. A. and D. R. Dewey (1942). "Visual studies of flow patterns," *J. Phys. Chem.,* Vol. 46, pp. 212–213.

Hazen, D. C. (1955). "Some results of the Princeton University smoke flow visualization program," Proceedings of 5th Anglo-American Aeronautical Conference, Roy. Aeronaut. Soc.

Head, M. R. and I. Rechenberg (1962). "The Preston tube as a means of measuring skin friction," *J. Fluid Mech.,* Vol. 14, Pt. 1, pp. 1–17.

Heflinger, L. O. et al. (1966). "Holographic interferometry," *J. Appl. Phys,* Vol. 37, No. 2 February, pp. 642–649.

Herzog, J. and J. R. Weske (1959). "The electric spark method for quantitative measurements in flowing gases," *Z. Flugwiss,* Vol. 7, No. 11, p. 322ff.

Holder, D. W. and R. J. North (1952a). "Colour in the wind tunnel," *The Aeroplane,* 4 January.

Holder, D. W. and R. J. North (1952b). "A schlieren apparatus giving an image in colour," *Nature*, Vol. 169, March, p. 466.

Holder, D. W. and R. J. North (1963). "Schlieren methods," Notes on Applied Science, No. 31, National Physical Laboratory.

Holder, D. W. et al. (1956). "Optical methods for examining the flow in high-speed wind tunnels," AGARDograph 23.

Hool, J. N. (1956). "Measurement of skin friction using surface tubes," *Aircraft Eng.,* Vol. 28, pp. 52–54.

Hopkins, E. J. and E. R. Keener (1966). "Study of surface pitots for measuring turbulent skin friction at supersonic Mach numbers–adiabatic wall," NASA TN-D 3478.

Hopkins, E. J. et al. (1960). "Photographic evidence of streamwise arrays of vortices in boundary layer flow," NASA TN-D 328.

Howarth, L. (editor) (1953). *Modern development in fluid dynamics, high speed flow,* Vol. II, Clarendon Press, Oxford.

Howland, B. et al. (1966). "Use of electrochemiluminescence in visualizing separated flows," *J. Fluid Mech.,* Vol. 24, Pt. 4, pp. 697-704.

Hurlbut, F. C. (1960). "Techniques of flow visualization applicable in low density fields," Symposium on Flow Visualization, ASME Annual Meeting, 30 November, New York.

Jakob, M. (1964). *Heat transfer,* John Wiley and Sons.

Jenney, D. S. et al. (1968). "A reassessment of rotor hovering performance prediction methods," *J. Amer. Helicop. Soc.,* Vol. 13, No. 2 (April), pp. 1-26.

Keener, E. R. and E. J. Hopkins (1969). "Use of Preston tubes for measuring hypersonic turbulent skin friction," NASA TN-D 5544.

Keto, J. R. (1962). "Flow visualization," Proceedings of the Fluid Amplification Symposium, Diamond Ordnance Fuze Laboratories, Vol. 1, pp. 109-123.

Kraemer, K. (1961). "Windkanal Untersuchungen an einem Deltaflügel bei mässigen Geschwindigkeitten," Report 61A, Aerodyn. Versuchsanstalt, Göttingen.

Ladenburg, R. W. (Ed.) (1954). *Physical measurements in gas dynamics and combustion,* Vol. IX, High Speed Aerodynamics and Jet Propulsion, Princeton University Press.

Lambourne, N. C. and P. S. Pusey (1959). "Some visual observations of the effects of sweep on the low speed flow over a sharp-edged plate at incidence," ARC R&M 3106.

Leonard, D. A. and J. C. Keck (1962). "Schlieren photograph of projectile wakes using resonance radiation," *Am. Rocket Soc. J.,* Vol. 32, No. 7, pp. 1112-1114, July.

Lewis, J. E. and T. Kubota (1966). "Stanton tube calibration in a laminar boundary layer at Mach 6," *AIAA J.,* Vol. 4, No. 12, pp. 2251-2252, April.

Liepmann, H. W. and A. Roshko (1957). "*Elements of gas dynamics,*" John Wiley and Sons.

Liepmann, H. W. and G. T. Skinner (1954). "Shearing-stress measurements by use of a heated element," NACA TN 3268.

Ling, S. C. (1960). "Heat transfer characterictics of hot-film sensing element used in flow measurement," *J. Basic Eng.,* Vol. 82, No. 3 (September), pp. 629-634.

Ling, S. C. and P. G. Hubbard (1956). "The hot film anemometer," *J. Aeronaut. Sci.,* Vol. 23, pp. 890-891.

Lippisch, A. (1939). "Results from the Deutsche Forschungsanstalt für Segelflug Smoke Tunnel," *J. Roy. Aeronaut. Soc.,* Vol. 43, pp. 653-672.

Livesey, J. L. (1956). "The behavior of transverse cylindrical and forward-facing total-head probes in transverse total-pressure gradient," *J. Aeronaut. Sci.,* Vol. 23, No. 10 (October), pp. 949-955.

Loving, D. L. and S. Katzoff (1959). "The fluorescent-oil film method and other techniques for boundary-layer flow visualization," NASA Memo 3-17-59.

Lowell, H. H. (1950). "Design and applications of hot wire anemometers for steady-state measurements at transonic and supersonic airspeeds," NACA TN 2117.

Ludwieg, H. (1949). "Ein Gerät zur Messung der Wandschubspannung turbulenter Reibungsschichten," *Ingr. Arch.,* Vol. 17, pp. 207-218. (Translated as "Instrument for measuring the wall shearing stress of turbulent boundary layers," NACA TM 1284).

Lukasiek, S. J. and C. E. Grosch (1959). "Velocity measurements in thin boundary layers," Stevens Institute of Technology, Davidson Laboratory, Tech. Memo 122.

MacMillan, F. A. (1956). "Experiments on Pitot Tubes in Shear Flow," ARC R&M 3028.

Maddox, A. R. and R. C. Binder (1970). "A new dimension in the schlieren technique: Flow field analysis using color," AIAA Paper 70-223, 8th Aerospace Sciences Meeting, 19-21 January, New York.

Maltby, R. L. (1962). "Flow visualization in wind tunnels using indicators," AGARDograph 70.

Maltby, R. L. and R. F. A. Keating (1960) "Flow visualization in low-speed wind tunnels," RAE (Bedford) Tech. Note AERO 2715.

Marey, M. (1900). "Des mouvements d l'air lorsqu'il recontre des surfaces de differentes formes," *Comptes Rendues,* Vol. 131, p. 160.

Marey, M. (1901). "Changement de direction et de vitesse d'un courant d'air," *Comptes Rendues,* Vol. 132, p. 1291.

McGregor, I. (1961). "The vapor screen method of flow visualization," *J. Fluid Mech.,* Vol. 11, Pt. 4, pp. 481–511, December.

Meyer, R. F. (1966). "A note on a technique of surface flow visualization," National Research Council of Canada, Aero Report LR-457, July.

Moore, J. W. et al. (1966). "The design of a skin friction meter for use in extreme environmental conditions," Research Laboratories for the Engineering Sciences, University of Virginia Report EME-4029-103B-66u (N66-39288).

Morkovin, M. V. et al. (1952). "Experiments on interaction of shock waves and cylindrical bodies at supersonic speeds," *J. Aeronaut. Sci.*, Vol. 19, No. 4 (April), pp. 237–248.

Mueller, T. J. et al. (1969). "Supersonic wake flow visualization," *AIAA J.,* Vol. 7, No. 11, pp. 2151–53, November. (Also AIAA Paper 69-346, AIAA Fourth Aerodynamic Testing Conference, Cincinnati, Ohio, April).

Muesmann, G. (1959). "Messungen and Grenzschichtbeobachtungen an affin verdickten Gebläseprofilen in Abhängigkeit von der Reynolds Zahl," *Flugwiss.*, Vol. 7, No. 9, p. 253.

Murphy, J. S. (1954). "The separation of axially symmetric turbulent boundary layers, part I, preliminary results on several bodies in incompressible flow," Douglas Aircraft Company Report ES 17513 (AD 432 666).

Murphy, J. S. (1955). "The separation of axially symmetric turbulent boundary layers, part II, detailed measurements in the boundary layers on several slender bodies in incompressible flow," Douglas Aircraft Company Report ES 17513 (AD 432491).

Murphy, J. S. and R. E. Phinney (1951). "Visualization of boundary layer flow," *J. Aeronaut. Sci.,* Vol. 18, No. 11 (November), pp. 771–772.

Nikuradse, J. (1930). "Untersuchungen über turbulente Strömugen in nicht kreisförmigen Rohren," *Ingr.-Arch.*, Vol. 1, p. 306.

North, R. J. (1954). "A colour schlieren system using multicolor filters of simple construction," National Physical Laboratory Aeronote/Aero/266, August.

Oertel, H. (1960). "High-speed photography of hypersonic phenomena by schlieren-interferometric methods," Proceedings of the Fifth International Congress on High-Speed Photography, Washington, D.C.

Oak Leaf, The (1967). The U. S. Naval Ordnance Laboratory, July.

O'Donnell, F. B. and J. C. Westkaemper (1965). "Measurements of errors caused by misalignment of floating element skin-friction balances," *AIAA J.*, Vol. 3, No. 1, pp. 163–165.

Oh, Y. H. and D. A. Didion (1966). "A quantitative schlieren-grid method for temperature measurement in a free convective field," Master's Thesis, Department of Mechanical Engineering, Catholic University of America.

Örnberg, T. (1954). "A note on the flow around delta wings," KTH Aero TN 38, Stockholm.

Ower, E. and R. C. Pankhurst (1966). *The measurement of air flow,* Pergamon Press, New York.

Pankhurst, R. C. and D. W. Holder (1952). *Wind tunnel techniques,* Pitman and Sons, London.

Patel, V. C. (1965). "Calibration of the Preston tube and limitations on its use in pressure gradients," *J. Fluid Mech.,* Vol. 23, Pt. I, pp. 185–208.

Pope, A. (1947). *Wind-tunnel testing,* John Wiley & Sons, Inc., New York.

Prandtl, L. and O. G. Tietjens (1934). *Applied hydro-and-aeromechanics,* McGraw-Hill, New York. (Also published by Dover Publications, New York, 1957).

Preston, J. H. (1954). "The determination of turbulent skin friction by means of Pitot tube," *J. Roy. Aeronaut. Soc.,* Vol. 58, pp. 109–121.

Preston, J. H. and N. E. Sweeting (1943). "An improved smoke generator for use in the visualization of airflow, particularly boundary layer flow at high Reynolds numbers," ARC R&M 2023 (ATI 22, 169).

Rainbird, W. J. et al. (1963). "A water tunnel investigation of the flow separation about circular cones at incidence," National Research Council of Canada, Aero Report LR-385.

Rainbird, W. J. et al. (1966). "Some examples of separation in three-dimensional flow," National Research Council of Canada. Reprint of article from DME/NAE Quarterly Bulletin, No. 1966(1).

Richards, E. J., and F. H. Burstall (1945). "The china-clay method for indicating transition," ARC R&M 2126.

Roberson, E. C. (1955). "The development of a flow visualization technique," National Gas Turbine Est. Report R 181.

Roos, F. W. and W. W. Willmarth (1969). "Hydrogen bubble flow visualization at low Reynolds numbers," *AIAA J.,* Vol. 7, No. 8, pp. 1635–1637, August.

Rosenhead, L., ed. (1963). *Laminar boundary layers, Fluid motion memoirs,* Chapter X, Clarendon Press, Oxford.

Roshko, A. (1967). "A review of concepts in separated flow," *Proc. Canadian Congress of Appl. Mech.,* Vol. 3, Laval University, 22–26 May.

Saheki, Y. (1950). "On the measurement of wind velocity distribution by the electric spark method," Hokkaido University, Faculty of Engineering Memo 8, 1947. (English Translation by Scient. Intelligence Defense Research Board, Canada.)

Sanders, C. J. and J. F. Thompson, Jr. (1967). "An evaluation of the smoke-wire technique for measuring velocities in air," U. S. AAVLABS TR 67-29 (AD652276).

Saylor, D. R. (1960). "Flow visualization using tufts in turbomachinery," Symposium on Flow Visualization, ASME Annual Meeting, 30 November, New York.

Schardin, H. (1934). "Das Töplersche Schlierenverfahren: Grundlagen für seine Anwendung und Quantitative Answertung," Forschg. Ing.-Wes. 5 (also available as DTMB Translation 156.)

Schardin, H. (1941). "Die Schlierenverfahren und ihre Anwendungen," *Ergeb. d. Exakt. Naturwissenschaften,* Vol. 20.

Schindel, L. H. and T. E. Chamberlain (1967). "Vortex separation on slender bodies of elliptic cross section," MIT Aerophysics Laboratory Tech. Report 138 (AD 662 689).

Schmidt, E. (1932). "Schlierenaufnahman der Temperaturfelde in der Nähe Wärmeabgeleiteten der Körper," *Forschg. Ing.-Wes.* Vol. 3, pp. 181–182.

Schraub, F. A. et al. (1965). "Use of hydrogen bubbles for quantitative determination of time-dependent velocity fields in low-speed water flows," *J. Basic Eng.,* Vol. 87, June, pp. 429– 444. (Also Paper 64-WA/FE-20, ASME Winter Annual Meeting, New York, 1964; and Report MD-10, Thermosciences Division, Dept. of Mechanical Eng., Stanford University, February 1964).

Schubauer, G. B. (1939). "Air flow in the boundary layer of an elliptic cylinder," NACA Report 652.

Smith, A. M. O. and J. S. Murphy (1955). "A dust method of locating the separation point," *J. Aeronaut. Sci.,* Vol. 22, No. 4, pp. 273–274.

Smith, D. W. and J. H. Walker (1959). "Skin friction measurements in incompressible flow," NASA TR R-26. (Supersedes NACA TN 4231).

Smith, K. G. et al. (1962). "The use of surface Pitot tubes as skin friction meters at supersonic speeds," RAE (Bedford) Report Aero 2665 (AD 285 977).

Sovran, G. (1959). "Pre-measured and visualized behavior of rotating stall in an axial-flow compressor and in a two-dimensional cascade," *J. Eng. for Power*, Ser. A, Vol. 81, No. 1., pp. 24–34, January.

Stalker, R. J. (1956). "A note on the china-film technique for boundary layer indication," *J. Roy. Aeronaut. Soc.,* Vol. 60, p. 543.

Stanton, T. E. et al. (1920). "On the conditions at the boundary of a fluid motion," *Proc. Roy. Soc. (London)*, Ser. A, Vol. 97, pp. 413–434.

Tanner, T. (1931). "Movement of smoke in the boundary layer of an aerofoil without and with slot," ARC R&M 1352.

Tapia, Elizabeth W. (1965). "Bibliography on high-speed photography 1960–1964," Eastman Kodak Company, Rochester, New York, Distributed at 7th International Congress on High-Speed Photography, Zürich, Switzerland.

Töpler, A. (1867). "Beobachtungen nach einer neuen optischen Methode," *Ann. d. Phys. Chem.*, Vol. 131, p. 33.

Townend, H. C. H. (1937). "Visual and photographic method of studying boundary layer flow," ARC R&M 1803.

Vallentine, H. R. (1959). *Applied hydrodynamics*, Butterworth Scientific Publications, Washington, D.C.

Velkoff, H. R. et al. (1969). "Boundary layer discontinuity on a helicopter rotor blade in hovering," Paper 69-197, AIAA/AHS VTOL Research, Design, and Operations Meeting, February, Georgia Inst. Technol., Atlanta.

Washburn, W. K. and A. Goldburg (1963). "Transition structure in the hypersonic sphere wake as shown by the sodium schlieren," AMP 111, AVCO Everett Research Laboratory.

Werlé, H. (1953). "Visualization en Tunnel Hydrodynamique," *La Rech. Aeronaut.*, Vol. 33, p. 3.

Werlé, H. (1957). "Le Tunnel hydrodynamique visualization et ses applications á l'aerodynamique," Communication á l'Assoc. Tech. Maritime et Aéronaut, session de juin.

Werlé, H. (1958). "Apercuser les possibilitiés du tunnel hydrodynamique á Visualization de l'ONERA," Offices National Etudes, Recherches Aerospatiales Tech Note 48.

Westkaemper, J. G. (1967). "The design and construction of floating element skin-friction balances for use at 50° to 150° F," NASA CR-66423 (N67-36703).

Wieghardt, K. and J. Kux (1967). "Experimental methods in wind tunnels and water tunnels, with special emphasis on the hot-wire anemometer," AGARD Report 558.

Winkler, E. M. (1954). "X-ray technique," in *Physical Measurements in Gas Dynamics and Combustion*, R. W. Ladenburg (ed.), Vol. IX (High Speed Aerodynamics and Jet Propulsion), Princeton University Press, Princeton, N.J.

Wuest, W. (1963). "Sichtbarmachung von Strömungen," Forschungsbericht, Aerodynamische Versuchsanstalt, Göttingen. (Also *Archiv Für Technische Messen*, Lieferung 328, Vol. 144, No. 2, pp. 99–102, and Lieferung 329, Vol. 144, No. 3, pp. 125–128.)

Yanta, W. J. et al. (1969). "An experimental investigation of the Preston probe including effects of heat transfer, compressibility and favorable pressure gradient," Paper 69-648, AIAA Fluid and Plasma Dynamics Conference, 16–18 June, San Francisco, California.

Young, A. D. and J. N. Maas (1936). "The behavior of a Pitot tube in a transverse total-pressure gradient," ARC R&M 1770.

SUPPLEMENTARY REFERENCES

Abbott, I. H. and A. Sherman (1938). "Flow observations with tufts and lampblack of the stalling of four typical airfoil sections in the NACA variable-density tunnel," NACA-TN-672.

Allen M. (1959). "Better way to trace liquid flow patterns," *Chemical Eng.*, Vol. 147, p. 150.

Allen, M. (1960). "Fluorescent particles for tracing liquid flow," *Chemical Eng.*, Vol. 136, p. 138.

Allen, M. and A. J. Yerman (1960). "Neutral density beads for flow visualization," Symposium on Flow Visualization, ASME, Annual Meeting, 30 November, New York.

Balint, F. (1953). "Techniques of flow visualization: A discussion of the funadmentals with examples of techniques used in research," *Aircraft Eng.*, Vol. 25, pp. 161–167.

Bradfield, W. S. and J. J. Sheppard (1959). "Microschlieren," *Aero-Space Eng.*, Vol. 18 pp. 37–40.

Brown, F. N. M. (1952). "An American method of photographing flow patterns," *Aircraft Eng.*, Vol. 24, pp. 164–169.

Brown, F. N. M. (1953). "A photographic technique for the measuration and evaluation of aerodynamic patterns," *Photographic Eng.*, Vol. 3, p. 164.

Bull, G. V. and C. B. Jeffery (1957). "Wake visualization studies in the aeroballistics range," Ballistics Res. Lab., U.S.A., Rept, 1005, Pt. 1, pp. 127–148.

Chang, P. K. (1970). *Separation of Flow*, Pergamon Press, New York.

Clark, K. W. (1933). "Methods of visualizing airflow with observations on several airfoils in the wind tunnel," ARC-R&M-1552.

Cox, A. P. (1959). "Measurements of the velocity at the vortex centre on an A.P.I. delta wing by means of smoke observations," ARC-CR-511.

Criminale, W. O., Jr., and R. W. Nowell (1965). "An extended use of the hydrogen bubble flow visualization method," *AIAA J.*, Vol. 3, No. 6 (June), p. 1203.

Erlich, E. (1974). "Methods of visualizing the leading edge separation bubble and analysis of the results," *La Rech. Aerospatiale,* ESTRO-TT-90, pp. 41–53.

Evans, R. A. (1954). "A new method of flow visualization for low density wind tunnels," Engineering Proj. Rept. HE-150-119, University of California.

Fruengel, F. (1960). "Bewegungsaufnahmen rascher Luftstroemungen und Stoszwellen durch hochfrequente Hochspannungsfunken," Jb WGL, p. 175.

Geller, E. W. (1954). "A method of studying vortex systems in the wake of circular cylinders," Research Note 2, Eng. Research Station, Mississippi State College.

Giesbrecht, H. et al. (1972). "Schlieren visualization of hypersonic entropy wake," *AIAA J.*, Vol. 10, No. 12, p. 1694.

Gough, M. N. and E. Johnson (1932). "Methods of visually determining the air flow around airplanes," NACA-TN-425.

Grün, A. E. et al. (1953). "Electron shadowgraphs and afterglow pictures of gas jets at low densities," *J. Appl. Phys.*, Vol. 24, p. 1527.

Hansen, A. G. et al. (1953). "A visualization study of secondary flow in cascades," NACA TN 2947.

Hauser, E. A. and D. R. Dewey, II (1942). "Visual studies of flow patterns," *J. Phys. Chem.*, Vol. 46, pp. 212–213.

Herzig, H. Z. et al. (1952). "A visualization study of secondary flows in cascades," NACA Rept. 1163. (Formerly NACA/TIB/3111 [NACA-RME-52F19] and NACA-TN-2947.)

Johnson, H. I. and R. G. Mungall (1954). "A preliminary flight investigation of an oil-flow technique for air-flow visualization," NACA/TIB/4472, NACA-RM-L54G14a.

LeManach, J. and E. Robert (1958). "Contribution of visualization on the study of low velocity flow in models of centrifugal compressors," MOA-TIL/T.-5068, *La Recherche Aéronautique*, Vol. 67, pp. 21–34.

Manoni, L. R. and J. D. Cadwell (1953). "An oil spray technique suitable for visual transonic flow observations," Res. Dept., United Aircraft Corp., U.S.A. Rept, R-14107-11.

Meijer Drees, J. and W. P. Hendal (1951). "The field of flow through a helicopter rotor, obtained from wind tunnel smoke tests," NLL Amsterdam, Rept. A-1205.

Melvill Jones, B. and J. A. G. Haslam (1932). "Airflow about stalled and spinning aeroplanes shown by cinematographic records of the movements of wool-tufts," ARC-R&M-1494.

Merzkirch, W. (1974). *Flow visualization*, Academic, New York.

Perry, C. C. (1955). "Visual flow analysis," *Product Engineering*, pp. 154–161.

Persoz, B. and G. Grenier (1957). "Enduits de visualisation à base de corps gras," *La Recherche Aéronautique*, Vol. 60, pp. 19–22.

Preston, J. H. (1946). "Visualization of boundary layer flows," ARC-R&M-2267, ARC-10-094.

Ragsdale, W. C. (1972). "Flow visualization workshop report," NOL-TR-72-94, U.S. Naval Ordnance Lab., White Oak, Md.

Salter, C. (1950). "Multiple jet white smoke generators," ARC-10296, 13004, ARC-R&M-2657.

Sellerio, A. (1953). "Impiego del ghiaccio secco nella visualizzazione delle corrent i d'aria," *L'Aerotehnica*, Vol. 33, pp. 398–400.

Sherman, P. M. (1957). "Visualization of low density flows by means of oxygen absorption of ultraviolet radiation," *J. Aero. Sci.*, Vol. 24, p. 93.

Simmons, L. F. G. and N. S. Dewey (1934). "Photographic records of flow in the boundary layer," ARC R&M 1335.

Smith, A. M. O. and D. W. Clutter (1960). "The electrochemical technique of flow visualization," Symposium on Flow Visualization, ASME Annual Meeting, 30 November, New York.

Smith, L. H., Jr. (1955). "Three-dimensional flow in axial-flow turbomachinery," Wright Air Development Center Report 55-348, Vol. 1.

Spring, W. C. and W. C. Ragsdale (1973). "An investigation of the use of hologrpahy in studying supersonic flow in inlet, NOL-TR-73-80, U.S. Naval Ordnance Lab., White Oak, Md.

Srinivasan, P. et al. (1972). "Decollement tridimensionnel sur une aile delta en ecoulement supersonique" [Three dimensional supersonic flow separation on a delta wing], Soc. Nat'l. Ind. Aerospatiale, presented at 9th Conf. on Appl. Aerodyn., 8–10 Nov. 1972, Paris, and 9 Nov. 1972, Saint-Cyr-l'Ecole, France.

Stanbrook, A. (1959). "Experimental observation of vortices in wing-body junctions," ARC-R&M-3114.

Symposium on Flow Visualization (1960). ASME Annual Meeting, 30 November, New York.

Taylor, M. K. (1950). "A Balsa-dust technique for airflow visualization and its application to flow through model helicopter rotors in static thrust," NACA-TR-2220.

Tietjens, O. (1931). "Beobachtung von Strömungsfermen," in *Handbuch der Experimental Physik*, W. Wien and F. Harms, eds., Springer Verlag, Leipzig.

Townend, H. C. H. (1932). "Hot wire and spark shadowgraphs of the airflow through an airscrew," R&M-1434.

Waltrup, P. J. and J. M. Cameron (1974). "Wall shear and boundary-layer measurements in shock separated flow," *AIAA J.*, Vol. 12, No. 6, pp. 878–880.

Witte, A. B. (1972). "Holographic interferometery of a submarine wake in stratfield flow," *J. Hydronautics*, Vol. 6, p. 114.

chapter 3
PRINCIPLES FOR THE CONTROL OF FLOW SEPARATION

INTRODUCTORY REMARKS

There are two important aspects to the control of flow separation: (*a*) When flow separation is detrimental, it should be eliminated or reduced, and (*b*) When flow separation is useful, then it should be exploited to the full extent possible.

Friction is generally considered undesirable because it causes a loss of energy; therefore, investigators have attempted to reduce the friction of flying or fast-moving vehicles–for example, by designing an optimal geometrical configuration for a high-performance aircraft. On the other hand, because large friction is needed for braking, designers attempt to utilize its characteristics fully when designing efficient brakes for airplanes, cars, etc.

These basic principles are straightforward; their practical application is rather diverse and depends on purpose, economy, state of the art, etc. For example, the shapes of a fighter plane and a glider are not the same, and brakes come in various designs (shoe or disc type), etc. Similarly, the details of a practical technique to be selected and applied to the control of flow separation may differ from one set of circumstances to another. Therefore, the principles of flow separation are described only in general form in this chapter. More specific features of the control techniques will be treated in connection with the detailed presentation of the various techniques in Chaps. 4, 5, and 6.

The principles may be formulated most effectively by observing optimum examples. Hence, an attempt is made to present two separate optimum control situations wherein separation is prevented (or reduced) on the one hand and deliberately provoked on the other. Limited cases of practical application for a wing and for a blunt body will be presented to facilitate an understanding of the principles of control and the consequences of their application.

3.1 PRINCIPLES OF SEPARATION CONTROL BY THE PREVENTION OR REDUCTION OF SEPARATION

The loss of momentum or energy due to flow separation is detrimental for an airfoil or a diffuser. Attempted solutions include prevention of the initial occurrence, early elimination, or some reduction. Under certain conditions of flow, e.g., a cascade, efficiency is higher with weak separation than when it is completely absent. In this case, the optimum control of flow would involve the reduction rather than the prevention or elimination of separation.

From the point of view of a control technique, however, prevention or reduction are essentially the same; they differ only in the degree of control required. Therefore, only the control technique of prevention of separation is presented.

3.1.1 Devices without Auxiliary Power

This section is concerned with devices such as vortex generators, fences, etc., that are mounted to a main body but are not operated by an external energy source. Control techniques, such as suction or blowing, that use mechanical devices operated by an external energy source, for example, a pump driven by a motor, are described in Sec. 3.1.2. The details of separation control devices that require no power to prevent separation may be found in Chap. 4.

As previously mentioned in Chap. 1, the separation of flow from a continuous surface is governed by two factors, adverse pressure gradient and viscosity. In order to remain attached to the surface, the stream must have sufficient energy to overcome the adverse pressure gradient, the viscous dissipation along the flow path, and the energy loss due to the change in momentum. This loss has a more pronounced effect in the neighborhood of the wall where momentum and energy are much less than in the outer part of the shear layer. If the loss is such that further advancement of the fluid is no longer possible, then the flow leaves the surface, i.e., the flow separates. In two-dimensional flow, the criterion of separation is formulated by zero velocity gradient at the wall

$$\left.\frac{\partial u}{\partial y}\right|_{y=0} = 0$$

or zero wall friction.

Therefore, if the momentum and the energy in the neighborhood of the wall are sufficient, then no separation occurs. Hence, the conceivable techniques for separation control are either (*a*) to design the body surface configuration in such a way that a sufficiently high energy level is maintained along the flow path (reaching to the final distance) in the neighborhood of the wall or (*b*) to augment the energy level by an auxiliary device placed at a suitable position along the flow path.

The problem then is how to secure a sufficiently high energy level of the fluid along the entire flow path in order to overcome the pressure rise and the viscous friction in the neighborhood of the wall. Thus, the optimum control criterion may be formulated by the energy level of the fluid in the neighborhood of the wall, which is equal to the work done by the fluid to overcome pressure rise and friction. Therefore, control is optimum when the flow conditions produced are at the brink of separation but still in a state of attached flow. In the past, this has been achieved by designing a proper two-dimensional wall configuration so that nearly zero friction prevails along the flow path. Such a critical condition is closely related to the minimum convective heat transfer because the latter depends directly on friction.

3.1.1.1 Proper design of the basic wetted surface configuration

If a certain continuous surface configuration causes separation for a given condition, then it may be asked whether separation can be prevented by changing this surface configuration.

Consider a smooth and continuous two-dimensional wall or axisymmetric body. The surface configuration along which fluid flows is related to the boundary layer growth over the surface. Hence, from the viewpoint of fluid mechanics, the problem of surface contouring for separation control is to so arrange the proper potential pressure distribution along the surface that the supply of energy to the inner boundary layer close to the surface equals or exceeds the amount of energy dissipated during the flow process. The streamwise pressure gradient may be made favorable or adverse by designing convex or concave surfaces. With a streamwise favorable pressure gradient, the energy level of the inner shear layer may be enhanced by using the great energy available just outside the viscous flow region and the thickness growth of the shear layer may be reduced. No separation occurs with a streamwise favorable pressure gradient. Thus, for example, leading-edge separation may be prevented by shaping the nose, such as by drooping. With a streamwise adverse pressure gradient, the outer inviscid fluid may not be able to supply the energy required to the inner fluid because the outer fluid itself is losing kinetic energy due to reduced velocity along the flow path. Furthermore, the shear layer grows rapidly along the wall and dissipates energy. Hence for a given length of flow path, proper arrangement of the region of favorable or adverse pressure gradient and the skillful selection of their magnitudes are essential to control the behavior of the boundary layer and to balance the gain and loss of energy in the neighborhood of the wall. The final shape to be selected from among those that have proven satisfactory from the viewpoint of fluid mechanics should be the one that is most simple to manufacture.

Another possible technique for preventing extended downstream separation is to provide an abrupt change of the geometrical configuration in a region of the flow path that is continuous elsewhere. Such an abrupt change causes separation locally; thus the possible choice of location to insert the abrupt geometrical change may be that region where separation would be expected on a continuous surface. An abrupt change of geometrical configuration, e.g., by means of a groove, generally forms a strong vortex within the groove. Thus, when vortices are trapped in a groove, the attached flow above them bridges over the groove. It passes through the outer part of the vortices and remains attached downstream of the groove. This technique of creating localized separation may eliminate an extended downstream trail of separated flow.

3.1.1.2 Transition control

By nature, turbulent flow has a higher level of momentum and energy in the neighborhood of the wall than does laminar flow. Thus, separation may be prevented by causing the transition of laminar to turbulent flow to occur upstream of what would be a laminar separation point. Transition is affected by Reynolds number, roughness of the surface, pressure gradient, surface curvature, heat transfer, compressibility, free-stream turbulence, etc. It appears, however, that control techniques that cause transition immediately downstream of the point of minimum pressure are simple and effective.

3.1.1.3 The increasing momentum of the surface fluid

The mixing of shear layer particles can be increased by employing an auxiliary device mounted to the main body. The mixing raises the turbulence level so that momentum and energy in the neighborhood of the wall are enhanced to prevent the separation that would otherwise occur. For example, vortex generators are used to transport energy into the shear layer from the outer flow. Furthermore, the shed vortices downstream of a vortex generator also bring higher kinetic energy into the more slowly moving surface fluid. Thus, vortex generators help to reenergize the fluid near the surface. One way to utilize vortices for securing a substantial interchange between high and low momentum air is to provide vortex paths close to the edge of the boundary layer. The vortices must be in a region where the transverse gradient of momentum is significant. Brown et al. (1968) prevented separation in a diffuser by using a wing-type vane with a height of 1.2 times the thickness of the boundary layer. An energizing effect may also be attained by machining a series of lateral grooves on a wall upstream of separation.

The energy level of the surface fluid may be increased by imposing a rotational motion within the shear layer. For example, rotational motion with a velocity component proportional to the radius can be produced by inserting radial deflector vanes into a pipe ahead of a diffuser. After leaving the deflectors, the fluid in the pipe rotates as if it were a solid body, i.e., the tangential velocity is zero at the axis but maximum at the walls.

A screen that diverts the flow toward the wall can be used to increase the velocity gradient at the wall, but the shear stress is also increased. The mechanics of flow through the screen is similar to that of natural turbulent flow. Thus a screen may be used to control separation if efficiency is not of prime importance. Note that a screen is also effective in causing reattachment of separated flow and in stabilizing the flow.

3.1.1.4 Splitting the region

Downstream of a solid body of finite dimensions, fluid particles separate from the solid wall and are oscillated by eddy motions within the separated region downstream of the trailing edge. The vortex street drifts downstream at a speed less than that of the free stream; the frequency of the fluctuations is nearly proportional to the speed of the free stream and inversely proportional to the width of the obstacle. At subsonic speeds, the periodic wake extends to several wake diameters.

These oscillations may be prevented by using a solid surface to split the oscillating region into two subregions. For example, the wake may be divided into two parts by placing a trailing plate immediately downstream of a blunt body. The effectiveness depends on the length of the plate. If it is, say, five diameters of the body, the plate may prevent regular oscillation completely. If it is short, it alters the frequency of oscillation because the periodic wake is influenced by the early stages of the wake rather than by the downstream flow. Another advantage is that the splitter plate reduces the drag of the body.

3.1.2 Auxiliary Powered Devices

Separation may be effectively prevented by using a powered device that eliminates viscosity, energizes the surface fluid, or properly regulates skin temperature. However, power is required to operate these devices; thus to determine feasibility, the energy gained by the effective control of separation must be compared to the power required by the device.

The application of powered devices to an aircraft wing also requires the designer to find the best compromise between conflicting aerodynamic requirements and to consider structural weight and the necessary stiffness of the structure as well as the space needed for a powered device. Optimization for minimum operating cost is involved with such parameters as wing area and span, thickness of chord, sweep angle, choice of cruising altitude, additional weight for the powered equipment, its power requirement, etc. This tradeoff does not lend itself to a general mathematical approach and must be performed on the basis of detailed design and cost studies.

3.1.2.1 Suction

Elimination of viscosity effect (suction) is tantamount to the elimination of the shear layer. This can be achieved effectively by suction of the boundary layer.

Chapter 1 outlined a classical experiment wherein boundary layer suction was used to prevent separation in a subsonic diffuser. Suction removes the decelerating flow particles in the neighborhood of the wall and so prevents flow separation. If the suction area is limited, then it is necessary to examine whether the newly formed boundary layer is capable of overcoming the adverse pressure gradient downstream of the suction area. Because the amount of fluid that must be removed by suction is small and because, in addition, suction delays transition, increases lift, reduces drag, and improves stability, etc., the practical application of this technique is widespread for both subsonic and supersonic flows.

Suction may be provided either continuously or discontinuously. For the prevention of separation, it appears more effective to provide the suction mainly in the region of adverse pressure gradient.

Suction is frequently used to improve aircraft performance. The equivalent "pump drag" can serve as a criterion for the gainful application of suction to an aircraft because this term allows correlation with the drag of conventional aircraft. For continuous suction operation, the sucked air mass must be discharged with an efflux speed equal to or larger than flight speed, and power is needed to drive a compressor for such discharge. If the fuel flow is assumed to be constant, then the power taken from the turbine to pump the boundary layer air causes a loss of thrust. This loss may be considered equivalent to a drag resulting from the pump power load. Thus with constant engine fuel flow, the equivalent pump drag may be defined as loss of engine thrust caused by the absorption of power from the engine jet gas flow. This equivalent pump drag is added to the wake and induced drags to formulate the total aircraft drag. Then this sum of all drag components is used to compare the lift-to-drag ratio with that of conventional aircraft.

The equivalent pump drag coefficient per unit span C_{D_p} depends on the characteristics of the aircraft engine and the compression; it is evaluated by

$$C_{D_p} = \eta_a \frac{P}{(1/2)\rho u_\infty^3}$$

where u_∞ is flight velocity, P is the power required by the suction compressors, and η_a is the efficiency of the propulsive system. A value of approximately 40 percent may be used for a typical straight turbojet engine at cruise setting. If the engine is throttled, the value of η_a increases.

3.1.2.2 Blowing

Additive energy may be supplied to the surface fluid by injecting a foreign fluid in the neighborhood of a wall; this technique is called *blowing*. If the fluid is injected parallel to the wall, the momentum of the shear layer is augmented whereas if it is injected normal to the wall, the mixing rate is increased. Both effects are advantageous. The former is termed *tangential blowing* and the latter *normal blowing*. Blowing is considered an effective means for separation control, but its side effects must be considered. Lift and position of transition are influenced by blowing. If the foreign fluid is injected normal to a supersonic main stream, then shock is formed upstream of injection. Because of the rise in pressure downstream of such a shock, flow separation is provoked there. This case will be presented later.

The problems of blowing can be solved mathematically in a manner similar to those of suction because the sign of velocity for blowing is opposed to that for suction.

The equivalent pump drag is also an important criterion of the effectiveness of the blowing technique to control separation.

The shear layer flow close to the wall may be energized by other techniques that are entirely different from blowing, e.g., by moving the wetted wall surface or using a rotating cylinder. The wetted wall is moved in the same direction as the main flow. Prandtl used a rotating cylinder to demonstrate this classical technique for preventing separation.

3.1.2.3 Combined suction and blowing

As mentioned previously, techniques for separation control by suction or by blowing are based on different principles. Suction eliminates the viscosity effect whereas blowing supplied energy to the surface fluids. But it may be possible to combine these two techniques into a single auxiliary device and reap the benefits of lower weight and less space. This involves placing the control areas so that the fluid removed by suction can conveniently serve as the fluid needed for blowing.

3.1.2.4 Heat transfer

Heat transfer can be a useful technique since it is possible to vary the streamwise pressure gradient by controlling surface temperature. If the wall surface is cooled, the streamwise gradient may be deepened and thus delay the separation and reduce the separated region.

If heat is added by burning gas in the wake, the thickness of the boundary layer is affected. Application of this technique to the wake region may reduce base drag.

3.2 PRINCIPLES OF FLOW CONTROL BY PROVOCATION OF SEPARATION

Flow separation can be provoked intentionally or it can occur as a byproduct of certain flow processes. Either way, the phenomenon can be controlled and/or exploited.

When flow separation is provoked and arrangements made for a proper reattachment downstream, a free separation streamline is formed between the points of separation and reattachment. These separating streamlines can be exploited to function as a pseudo-solid body. As mentioned in Chap. 1, a thick pseudo-airfoil is an example of such practical use. A thin solid airfoil is transformed into a thick airfoil by provoked separation. The upper part of a thin airfoil is split and lifted upward (like a forward-facing flap) in such a manner that the leading edge separation at the forward part of the airfoil reattaches aft of the leading edge on the upper part of the airfoil. The free separating streamline has thus provided an aerodynamically favorable substitute for the thin leading edge.

The pressure rise along a solid surface in the separated flow region may be utilized to obtain a side force. Shock waves may be produced or the shock shape may be changed in order to control flow. If these phenomena are the main reason for flow control, then the accompanying separation downstream of the shock is a by-product or an auxiliary control. This supplementary control may be utilized to improve the efficiency of overall performance by exploiting the characteristics of provoked separation to eliminate or reduce the unfavorable aspects of such separations.

3.2.1 Thick Pseudo-airfoil

The control technique is to provoke separation at the desired location and arrange for its subsequent reattachment at a desirable position. It is also necessary to maintain the separating streamline in an aerodynamically favorable shape. Two vortices can be formed inside the thick pseudo-airfoil behind the separating flow and this vortex motion may endanger the stability of the separating streamline. An auxiliary powered control device may then be needed in order to accomplish successful control.

3.2.2 Obstacle to the Main Stream

When an obstacle is placed on a solid surface or when a fluid jet is injected at a certain angle to the surface, then flow separates upstream of such an obstacle to the main stream. Downstream of such separation, control of the separated flow region may be relatively easy if eventual flow reattachment can be promoted. For example, the pressure rise in the separated region downstream of a jet issuing from a wall is often used to obtain an aerodynamic moment and thus adjust flight direction; this was mentioned in Chap. 1. There is the possibility of oscillating motion in the separated region, and separation generally causes the loss of energy. These undesirable aspects may be eliminated or reduced by control.

3.2.3 Change of Shock Shape

A curved and detached bow shock upstream of a blunt body submerged in supersonic flow may be changed to an oblique shock by placing a spike as described in Chap. 1. In this case, the flow downstream of the spike tip may separate because of the pressure rise and the existence of a solid surface on which a boundary layer forms. Depending on the length of the spike, the shape of the blunt body where the spike is mounted, and the flow conditions, the separated flow could reattach to the blunt body and cause a high rate of heat transfer. An oscillatory motion of separated flow and loss of energy due to separation may result and should be controlled.

3.3 PRACTICAL EXAMPLES OF SEPARATION CONTROL

Certain examples of practical application of these principles will serve to familiarize the reader with separation control. Accordingly some of the available control and the subsequent design techniques for prevention of separation (restricted to a wing) are presented, together with some examples of the use of a spike to provoke separation.

3.3.1 Separation Control by Preventing Separation

The performance of a wing depends on many different factors. A particular control technique may be favorable to one phenomenon and unfavorable to another. The prevention of flow separation is the most important control technique for a wing, but caution is needed in applying a particular technique because unjustifiable penalties may have to be paid under certain conditions.

Wallis (1965) conducted a wind-tunnel investigation of the wing model sketched in Figs. 3.1 and 3.2. The basic wing is a quarter-scale model of the outer wing of a Victor bomber, except that no aileron is fitted. The wing sections do not correspond exactly to any standard profile, but they are similar to the NACA "low drag" airfoils, with a small amount of camber which increases as the tip is approached.

Flow separation control for the Victor wing was considered for six different possible sources of separation:

1. Laminar separation bubble.
2. Flow separation at the tip extremity which causes tip vortex.

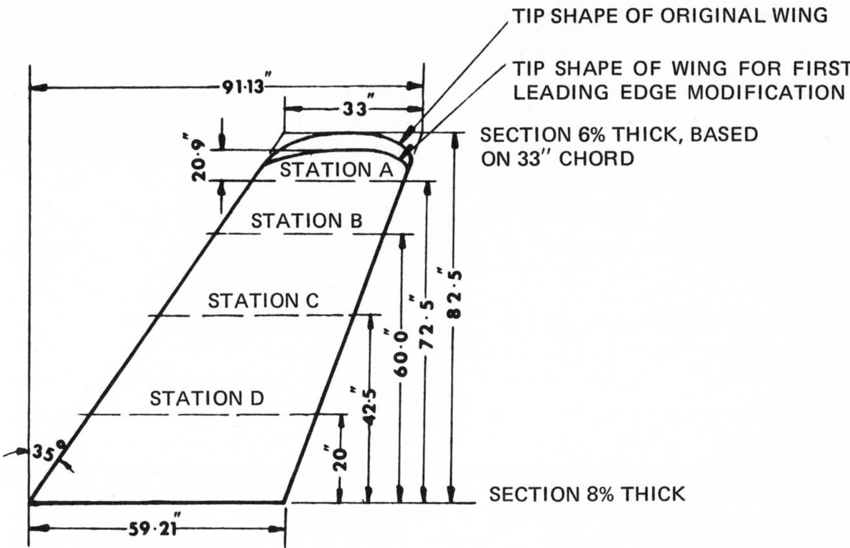

FIG. 3.1 Planform of wing without chord extension (*from Wallis, 1965*).

3. Turbulent separation from the critical region just downstream of the leading edge.
4. Turbulent separation from the nose which is associated with buffeting and a large-scale vortex flow.
5. Turbulent separation that moved forward from the trailing edge.
6. Nose separation on the lower surface at a small angle of attack.

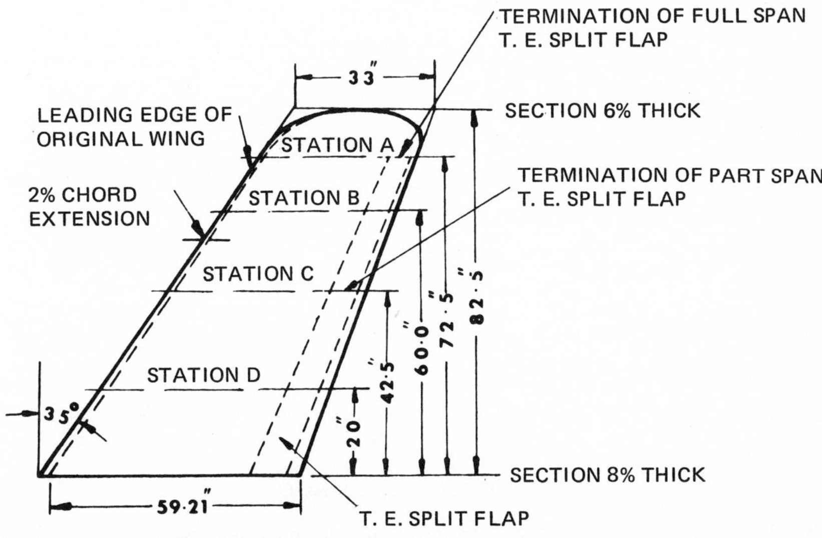

FIG. 3.2 Planform of wing with chord extension for second droop nose (*from Wallis, 1965*).

The Wallis investigation included the possibility of leading edge separation control by proper design of the leading edge droop nose shape and the control of separation in general by trailing flaps, air jets, and vortex generators. The droop nose may be considered as a possible alternate to leading edge flaps. Two different kinds of droop noses (Figs. 3.3 and 3.4) were tested at four spanwise stations.

The first droop nose was based on the NACA 220 mean camber line which is characterized by a camber restricted to the first 15 percent of the chord

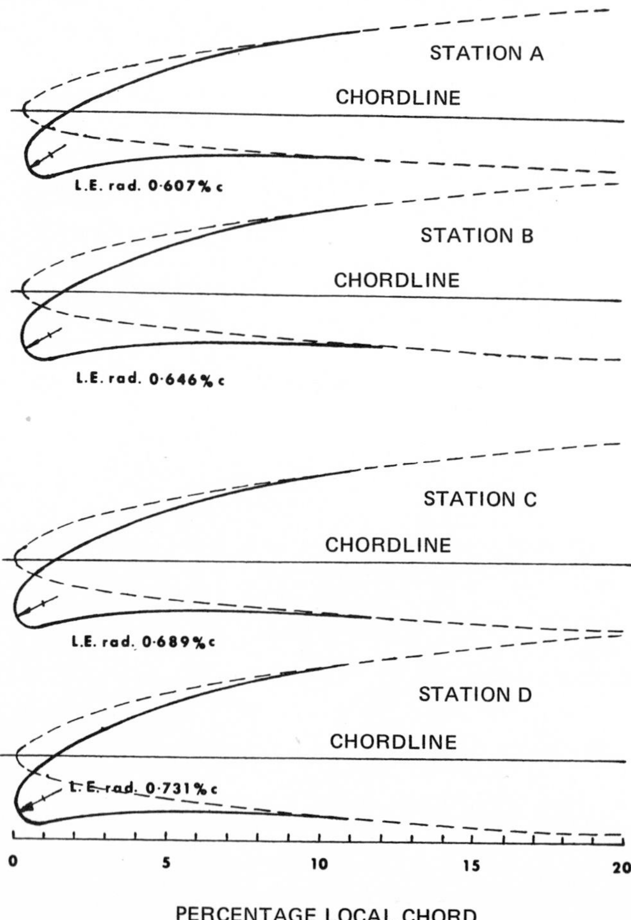

FIG. 3.3 Comparison of first modification with original nose (*from Wallis, 1965*).

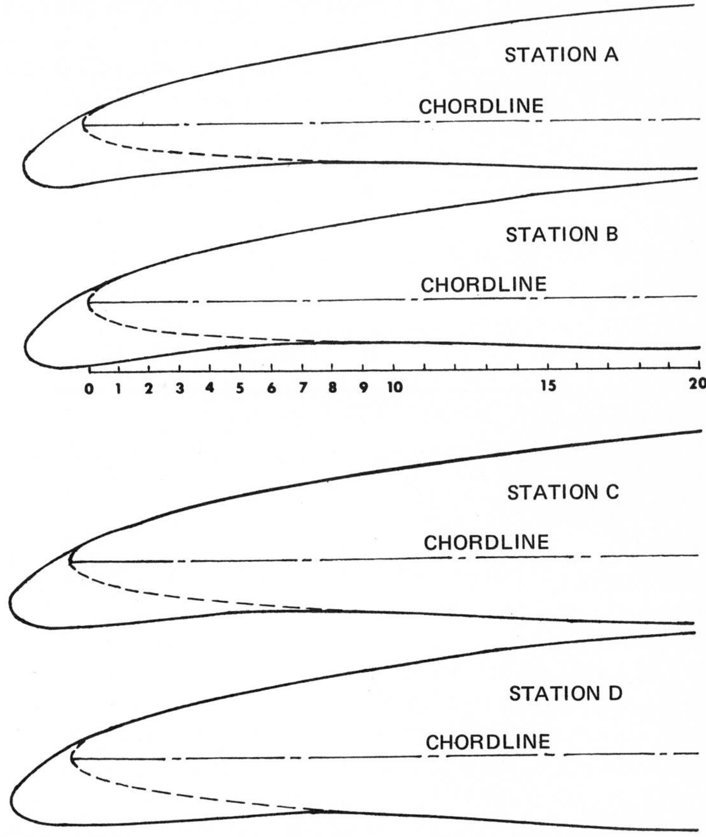

FIG. 3.4 Comparison of second modification with original nose (*from Wallis, 1965*).

and a curvature that increases as the leading edge is approached. At the leading edge, the tangent to the camber line makes an angle of over 20 deg with the basic chord line. The second droop consisted of a 2-percent chord extension forward of the leading edge and was designed so that the new nose shape faired into the existing contour but was completely external to it.

The Victor bomber has a Fowler-type trailing edge flap, but it was impractical to reproduce this type of flap on the model. Hence, the effect of flaps was simulated by a 20-percent chord split flaps, deflected at 45 deg relative to the chord line. In addition, an air jet was provided for separation control. Two jets were used, lower surface jets for the suppression of laminar separation and upper surface jets for the control of turbulent separation over the forward portion of the wing. Vane-type generators were mounted on both the lower and the upper surfaces.

Before presentation of the control techniques applied, let us study the flow phenomena that occurred on the wing. Because of the adverse pressure gradient near the leading edge, a laminar separation bubble formed near the leading edge. With an increase in angle of attack and an appropriate Reynolds number, the separation line reattached to the surface along a line that was less than 1 percent of the chord downstream from the separation line. As the angle of attack increased further, flow around the airfoil suddenly broke down, causing nose stalling (this is dependent on Reynolds number). At low Reynolds number, the onset of separation with increasing angle of attack appeared to be related to the breakdown of the momentum transfer mechanism at the downstream edge of the bubble. At moderate to high Reynolds numbers, turbulent separation just downstream of the closed bubble initiated collapse of the nose flow. For intermediate Reynolds numbers, both phenomena played an interrelated role. Since the separation problem is closely related to the phenomenon of transition (which is dependent on various factors such as free-stream turbulence, surface roughness pressure gradient, compressibility, etc.), large variations in the critical Reynolds number are inevitable.

The control technique differed for these different flows since the changing nature of the flow separation phenomena at the nose could strongly influence flow instability for Reynolds numbers within the critical range. This instability may be reduced by fitting a droop nose, but the resulting hysteresis is a penalty. As used here, hysteresis means that two alternate types of flow can be presented at a given angle of attack. The hysteresis can be fully understood when a detailed study is made of the interaction of separated and attached flows. The initial onset of flow separation at the nose can be induced to occur at the appropriate inboard position by providing spanwise variations of the nose droop and/or boundary layer control. After stall initiates, adequate control over the outer wing flow should be arranged to ensure satisfactory longitudinal and lateral stability characteristics. However, as experimental data indicate, the application of trailing edge flaps cannot be expected to alter the nature of stall development. More details of flow control by means of the various devices are described in the following pages.

3.3.1.1 Leading-edge droop

Leading-edge droop offers several advantages as a method for flow control:

1. It greatly reduces the suction peak for a given C_L or angle of attack α (see Fig. 3.5).
2. It reduces the total pressure recovery in the nose region at a given $C_{p\min}$ (see Fig. 3.6).
3. Because of the effect indicated in Item 2, the turbulence that initiates flow separation at the nose is delayed to a greater $C_{p\min}$.

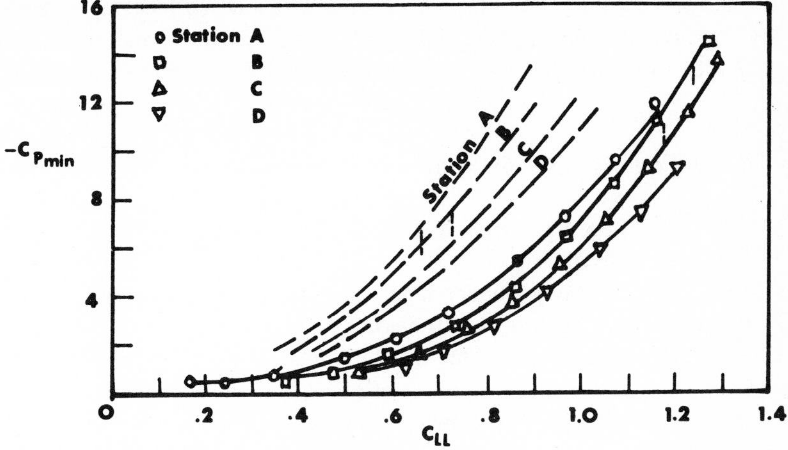

FIG. 3.5 Minimum pressure coefficient versus local lift coefficient (*from Wallis, 1965*).

4. The higher local velocity (i.e., the higher Reynolds number) associated with this increased suction peak tends to restrict the size of the laminar separation bubble and hence minimizes boundary layer thickening.

When stall occurs, however, the flow pattern changes considerably with leading-edge droop. There will be some buffeting and loss of lift near the tip.

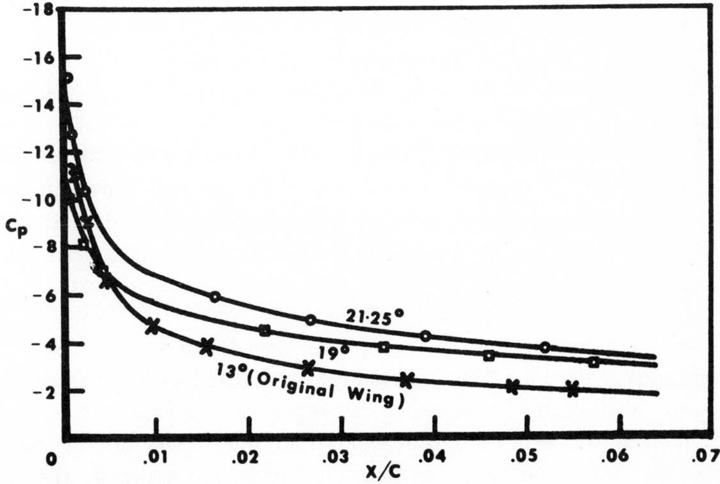

FIG. 3.6 Upper surface pressure distribution near nose at station *c* for first droop nose; parameter is incidence angle (*from Wallis, 1965*).

The inboard extremity of the vortex flow will now be located in the region downstream of about the 15-percent chord position and not in close proximity to the leading edge.

3.3.1.2 Air jets

Air jets are useful for the control of flow separation in the following ways:

1. They delay flow separation at the nose.
2. They reduce hysteresis.
3. They stabilize the boundary layer flow on the nose in a certain range of angles of attack (say, up to 20 deg).
4. They delay separation moving from the rear.
5. They regulate the spanwise onset of separation and the ease with which the tip flow is held over a worthwhile range of angle of attack.

The control of flow by an air jet is attractive because of the simplicity of its installation and operation and the effectiveness of the method at both low and high speeds. The effect of the lower surface air jet device is similar to high Reynolds numbers at the laminar separation bubble. It provides a marked gain in the angle of attack at which the general flow separation at the nose first appears on the wing. However, Wallis found that a large amount of boundary layer control was needed to increase the separation lift coefficient for the case of flow separation at the nose by, say, 0.1 or 0.2. This is due to the steepness of $C_{p_{\min}}$ on the curves plotted versus local lift coefficient C_{LL}. Hence, although air jets can greatly improve the handling properties of an aircraft, they are not an alternative to leading edge controls for substantially increasing lift.

3.3.1.3 Trailing-edge flap

The flow around the flapped part of the wing modifies that around the outboard portion of the wing in such a manner that the effective incidence and thus the local lift are increased.

The trailing-edge flap is not a useful device either for control of flow separation at the nose or for reduction of hysteresis. Flow separation occurs at the nose and develops relatively independently of the flap arrangement, and the degree of hysteresis is not altered by the addition of a part-span flap.

3.3.1.4 Vortex generator

Wallis mounted a vane-type vortex generator on the lower surface to safeguard against the marked hysteresis associated with the second droop nose during flight tests. In addition, an upper surface vane-type vortex generator suppressed the shock-induced separation, but it was not expected to exercise any major influence on the separation properties at the nose of the wing.

3.3.1.5 Wing fence

Wallis also conducted a series of experiments wherein a leading-edge boundary-layer fence was used on the second droop nose model (Fig. 3.4) to cause the following flow phenomena:

1. The early initiation of separation at the leading edge on the inboard side of the fence.
2. An increment of boundary-layer stability on the nose just outboard of the fence.
3. A delay in the process by which the inboard flow separation contaminates the outer wing flow.

3.3.1.6 Wing design technique

Wallis (1965) has indicated that when designing a wing, special consideration should be given to shaping the leading edge to enable appropriate control of stall.

With turbulent separation just downstream of the leading edge that initiates the stall (provided the flight Reynolds number is high enough), the magnitude and gradient of the pressure recovery on the airfoil nose are the most important variables to be considered for design purposes. $C_{p\,\mathrm{min}}$ is a measure of the former, and the latter is fixed by contouring the nose. Since the magnitude of $C_{p\,\mathrm{min}}$ measures the pressure recovery required, this quantity is one of the most important variables that govern flow separation at the nose. By controlling the spanwise variation of this parameter for a given wing lift coefficient, it is possible to initiate leading-edge separation in the desired location. Note that the relationship between $C_{p\,\mathrm{min}}$ and local lift coefficient C_{LL} for a given airfoil section is—to a first order—not influenced by any three-dimensional effect. At a given wing incidence, the section with the highest $C_{p\,\mathrm{min}}$ is usually the most susceptible to leading-edge separation. Separation is imminent when $C_{p\,\mathrm{min}}$ reaches a value that is essentially constant for a given type of section.

On the critical leading-edge section, the design technique is to provide sufficient reserve of $C_{p\,\mathrm{min}}$ for the outboard wing section at the nose stall angle of attack to counter any tendency for separation to engulf the outer regions.

Since the main outcome of using the trailing-edge flaps is the introduction of bodily displacement of $C_{p\,\mathrm{min}}$ versus C_{LL} along the lift axis, the magnitude of this displacement for the representative flap is a factor to be considered in design.

When leading-edge flow conditions are ineffectual in providing a measure of flow control over the outward spread of separation, then a boundary layer fence may be extremely useful.

It is suggested that more attention should be paid to the forward turbulent separation and hysteresis problem because these may provide the key to a satisfactory wing stall.

3.3.2 Separation Control by Provoking Separation

As mentioned previously, a spike is sometimes mounted in front of a blunt body exposed to a supersonic stream in order to provoke laminar as well as turbulent flow separation along the spike length. Such a spike not only reduces drag (although not so effectively as at zero angle of attack) but also increases lift and, correspondingly, changes the moment at angles of attack. Hence, a spike is an effective control device. It is also practical in that it is handy and simple compared to variable geometry skirts, for example. However, its applicability is somewhat limited unless the unfavorable effect of flow oscillation on the probe can be eliminated. Such oscillation on the probe may cause aerodynamic disturbances.

If the separated flow reattaches downstream on the body surface, then a higher pressure and a larger rate of heat transfer are to be expected locally at the reattachment region.

A spike is usually made of a solid material. (An air jet from the forward stagnation of a blunt body produces an aerodynamic effect similar to that obtained from a spike.) Flow over the spiked body depends on the length, diameter, and shape of the spike, the shape of the blunt body, and the variation of Reynolds number, angle of attack, etc.

The effect of spike length at larger angles of attack is sketched in Fig. 3.7 for four cases: very short, very long, and just shorter and just longer than the critical length for minimum drag. This sketch represents interpretations of schlieren photographs with special emphasis on the formation of vortices. As indicated, the flow phenomena are rather complex, and accordingly, the interpretations are tentative in nature.

Note that for the conditions illustrated in Fig. 3.7, nonsymmetric flow causes an additive complication in the form of longitudinal vortices associated with the lift on the leeside of the spike. The critical length for minimum drag depends on Reynolds number, Mach number, angle of attack, and ratio of spike diameter to nose diameter, etc. When a spike exceeds this critical length, the separation moves suddenly downstream along the spike and becomes unstable, oscillating rapidly back and forth along the spike. Such oscillation may persist as long as the boundary layer on the spike remains laminar. With the very short spike, the conical flow started at the tip of the spike but when the angle of attack was increased, the flow over the nose became asymmetric without changing the general form of the flow pattern. For the spike that was just longer than critical length for minimum drag, lift vortex sheets formed on the forepart of the spike. However, these sheets were modified by a conical separation on the lee side of the spike, according to their strength and position relative to the spike.

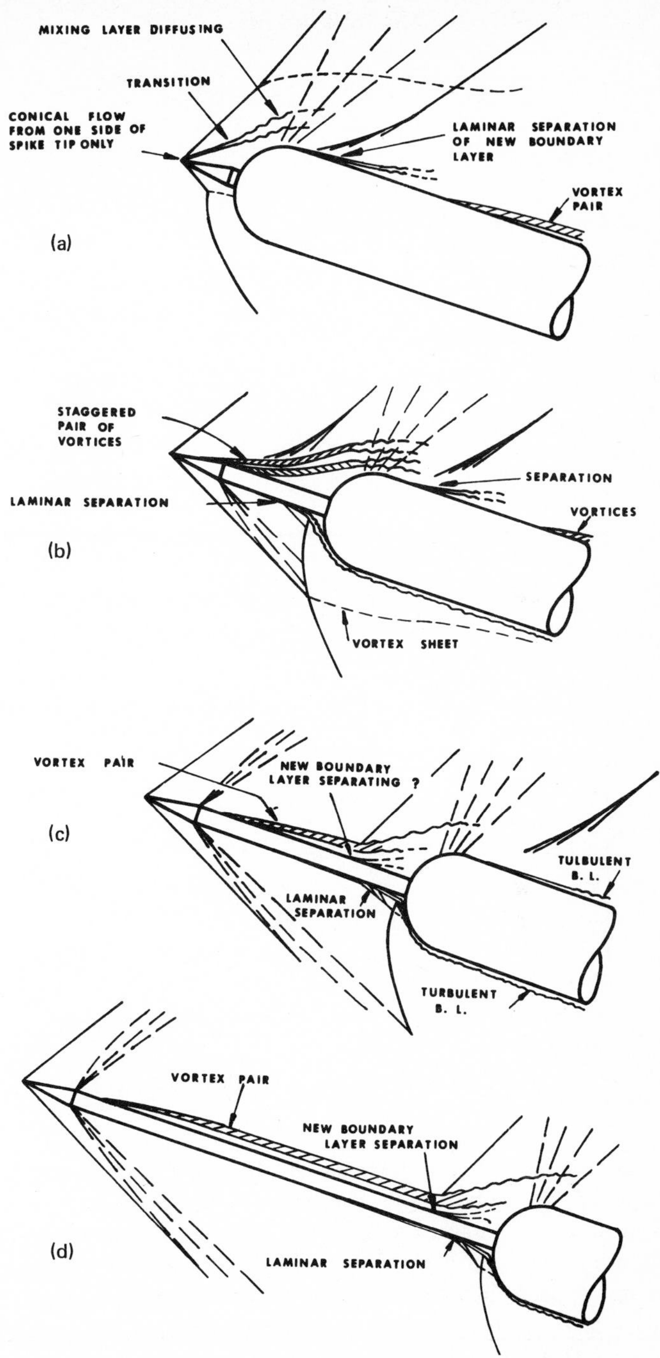

FIG. 3.7 Flow at higher angles of incidence: (*a*) very short spike (*b*) spike just shorter than the critical length (*c*) spike just longer than critical length (*d*) very long spike (*from Hunt, 1958*).

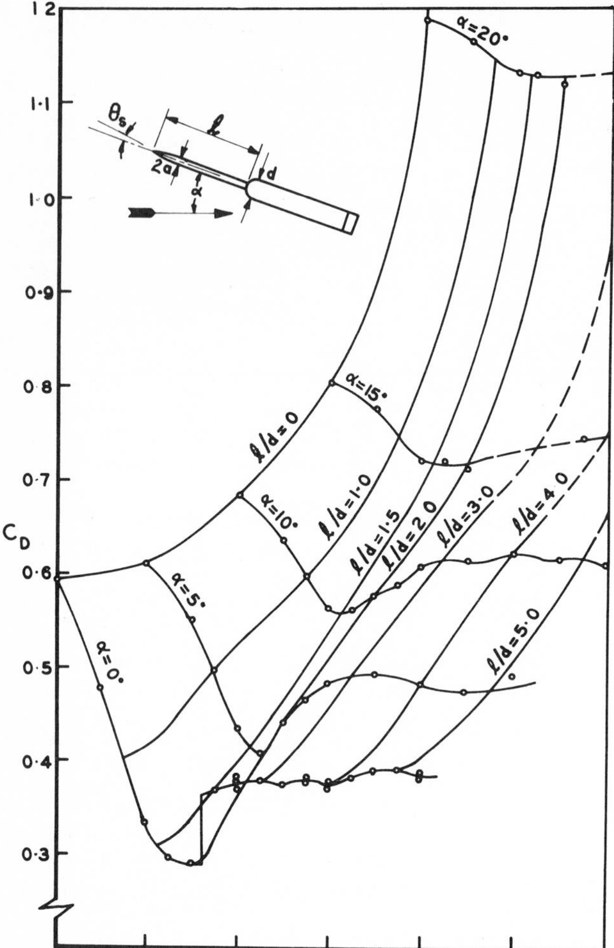

FIG. 3.8 Effect of spike length and angle of attack on drag ($M_\infty = 1.81$, $2a/d = 0.2$, $\theta_s = 10°$, $Re_d = 310{,}000$; *from Hunt, 1958*).

Figures 3.8 and 3.9 show the effect of spike length and angle of attack on drag and lift, respectively. The value of C_D is corrected from the measured value to reduce drag to what would exist had free-stream static pressure acted uniformly over the base of the model. The value of C_L is corrected from the measured value by accounting for the lift increment on the model mounting elbow and the component of the base pressure.

It is clear from Fig. 3.8 that a large reduction in drag at zero angle of attack can be achieved by attaching a spike. The reduction of drag is less pronounced with an increase in angle of attack. However, a longer spike is more effective in reducing drag at the same angle of attack. The lift is nearly independent of spike length, but it increases with an increase in angle of attack.

Figure 3.10 shows the normal force and center of pressure location with flat-nosed spikes of various lengths and angles of attack. (The location was measured by a three-component, strain-gage balance.) The symbol A_f refers to the area of the flat-faced portion of the spike nose with respect to body-nose area in percent.

It is seen from Fig. 3.10 that with an increase in angle of attack, the normal force increases rapidly but the center of pressure does not vary appreciably. Although the normal force is larger with than without a spike,

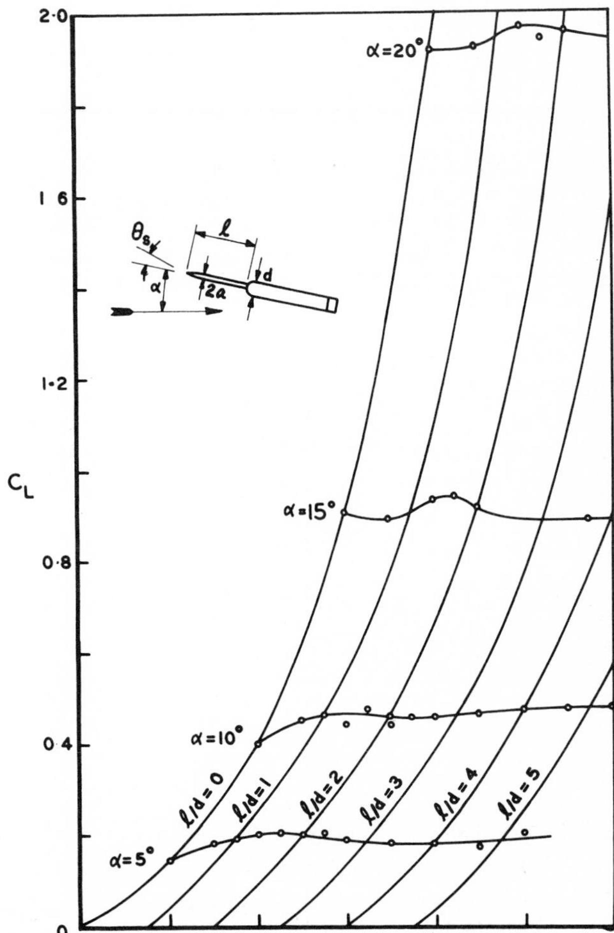

FIG. 3.9 Effect of spike length and angle of attack on lift ($M_\infty = 1.81$, $2a/d = 0.2$, semiangle of tip of spike is 10 deg, $Re_d = 310{,}000$; *from Hunt, 1958*).

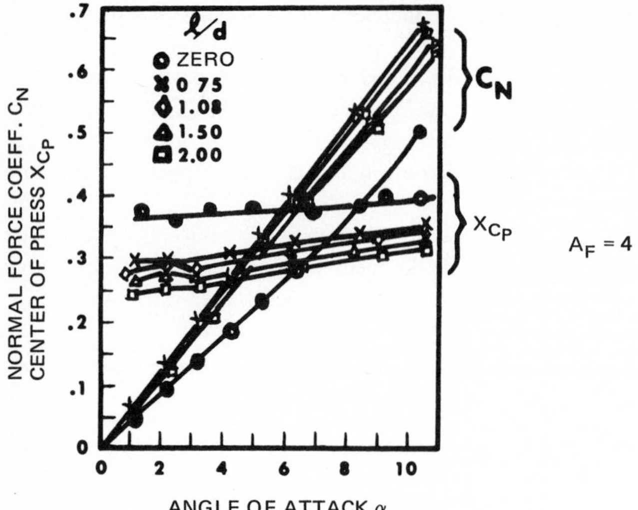

FIG. 3.10 Normal forces and center of pressure location on a body with flat-tipped spikes of various lengths (solid symbols indicate center of pressure, $M_\infty = 2.5$, $Re_d = (0.57 \pm 0.02) \times 10^6$; *from Album, 1961*).

the effect of varying the spike length is small; the normal force is almost equal for spikes of different lengths.

In Fig. 3.11 the pitching moment measured by the strain-gage balance is plotted versus angle of attack. The symbol X refers to distance from model nose to a point along the body longitudinal axis in fraction of body length. It is evident that the spikes increase the pitching moment and that longer spikes increase the moment. If the flow changes from laminar to turbulent, then drag suddenly increases as shown in Fig. 3.12.

The transition occurred at approximately $l/d \approx 1.7$. The effects of Mach number on C_D is also presented in Fig. 3.12. Here C_D is defined by

$$C_D = \frac{D}{qA} + \frac{p_B - p_a}{q}$$

where q is the free-stream dynamic pressure,
D is the total drag measured by the balance,
A is the cross-sectional area of the body,
p_B is the base pressure, and
p_a is the static pressure of the ambient free stream.

The shape of the spike tip affects the drag. For example, an arrow-shaped windshield tip reduces the drag more than does a sharp ogive tip.

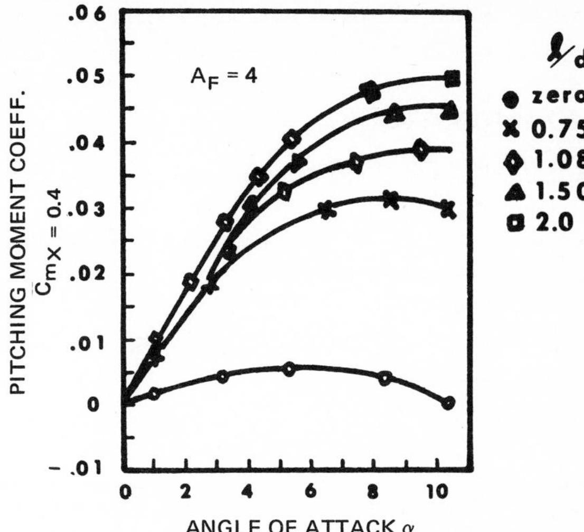

FIG. 3.11 Pitching moment characteristics (about $X = 0.4$) of bodies with flat-tipped spikes at various lengths ($M_\infty = 2.5$, $Re_d = (0.57 \pm 0.02) \times 10^6$; *from Album, 1961*).

The position of separation can be controlled by changing the spike length because there is a critical length for the jump of separation point which depends on Mach number, diameter of spike, etc. With an increase in angle of attack, the separation point at the tip of the spike jumps to a retarded position and back to the tip again with decreasing angle of attack.

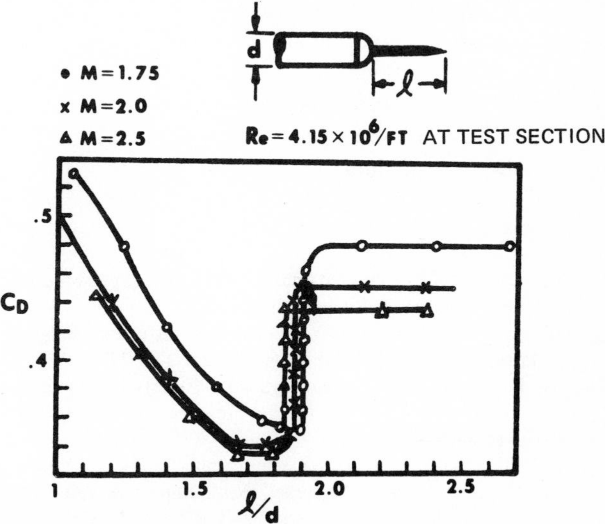

FIG. 3.12 Effect of Mach number on a body with a nose spike (*from Daniels and Yoshihara, 1954*).

The unfavorable oscillation on the spike may be eliminated by changing the separated laminar flow to turbulent or by providing a hemisphere nose shape.

NOMENCLATURE

A	Cross-sectional area of body
a	Radius of spike
A_f	Area of the flat-faced portion of spike nose with respect to body nose area in percent
C_D	Drag coefficient
C_{DP}	Equivalent pump drag coefficient per unit span
C_L	Lift coefficient
C_{LL}	Local lift coefficient
C_m	Pitching moment coefficient
C_N	Normal force coefficient
$C_{P\min}$	Minimum pressure coefficient
D	Total drag measured by balance
d	Diameter of after body
l	Spike length
P	Power required by suction compressor
p_a	Static pressure of ambient free stream
p_B	Base pressure
q	Stream dynamic pressure
u	Streamwise velocity component
u_∞	Flight velocity
X	Distance from model nose to a point along the body longitudinal axis in fraction of body length
y	Distance perpendicular to wall
α	Angle of attack
η_a	Efficiency of propulsive system
θ_s	Spike tip half angle
ρ	Density of fluid

REFERENCES

Album, H. H. (1961). "Spiked blunt bodies in a supersonic flow," U.S. Air Force Office of Scientific Research Report AFOSR 307.

Brown, A. C. et al. (1968). "Subsonic diffusers designed integrally with vortex generators," *J. Aircraft*, Vol. 5, No. 3, pp. 221–229.

Daniels, L. E. and H. Yoshihara (1954). "Effects of the upstream influence of a shock wave at supersonic speeds in the presence of a separated boundary layer," WADC Tech. Report 54-31.

Hunt, G. K. (1958). "Supersonic wind-tunnel study of reducing the drag of a bluff body at incidence by means of a spike," RAE Report AERO 2606.

Wallis, R. A. (1965). "Wind tunnel studies of leading edge separation phenomena on a quarter scale model of the outer panel of the Handley Page Victor wing with and without nose droop," Ministry of Aviation, R&M 3455 (AD 813 973).

SUPPLEMENTARY REFERENCES

Chen, R. T. N. et. al. (1973). "Development and evaluation of an automatic departure prevention system and stall inhibitor for fighter aircraft, Tech. Rept. AFFDL-TR-73-29, Air Force Flight Dynamics Lab., U.S. Air Force Systems Command, Wright-Patterson AFB, Ohio.

Hoak, D. E. (1964). "Pressure measurements for Mach five flows over winged re-entry configurations with aerodynamic controls," RTD-TDR-63-4179, Pts. I & II, Air Force Flight Dynamics Lab., U.S. Air Force Systems Command, Wright-Patterson AFB, Ohio.

Hoak, D. E. and W. J. Klotzback (1964). "Pressure and heat transfer measurements for Mach 8 flows over a blunt pyramidal configuration with aerodynamic control," FDL-TDR-64-2, Pts. I, II, III, & IV, Air Force Flight Dynamics Lab., U.S. Air Force Systems Command, Wright-Patterson AFB, Ohio.

Hoak, D. E. and W. J. Klotzback (1964). "Pressure and heat transfer measurements for March 21 flows over a blunt pyramidal configuration with aerodynamic control," FDL-TDR-64-120, Air Force Flight Dynamics Lab., U.S. Air Force Systems Command, Wright-Patterson AFB, Ohio.

Lachmann, G. V. (1962). "Boundary layer control for low drag," *Aircraft Eng.*, Vol. 34, No. 397, pp. 68–69.

Marchman, J. F. and J. N. Uzel (1972). "Effect of several wing tip modifications on a trailing vortex," *J. Aircraft*, Vol. 9, p. 684.

Taylor, C. R. (1969). "Flight test results of a trailing edge flap designed for direct-lift control," NASA-CR-1426, National Aeronautics and Space Administration, Washington, D.C.

chapter 4

PREVENTION AND DELAY OF SEPARATION WITHOUT THE USE OF AUXILIARY POWER

INTRODUCTORY REMARKS

This chapter describes techniques that control separation by geometrical design of the main body configuration, by additive geometrical configurations which increase the level of turbulence, and by screens which suppress oscillation in separated regions.

The techniques that use no auxiliary power are most effective when applied to turbulent flows. They exploit the natural characteristics of turbulence so that natural mixing may cause sufficient pressure recovery to keep the boundary layer flow from stagnating under the action of rising pressure. The distribution of pressure or velocity over the wetted body surface may be obtained by potential flow analysis. The refined solution can be obtained by modifying the configuration of the body surface, i.e., adding the displacement thickness to the solid surface.

The treatment of geometrical design of the main body includes the study of an optimum technique to produce a zero skin friction surface, the investigation of flow separation control techniques on airfoils and a discussion of airfoil design techniques. A very low drag is potentially achievable by taking advantage of the shortness of pressure rise obtainable under the zero friction condition on an airfoil; it can be utilized to increase the proportion of the chord over which the pressure gradient is favorable and thus delay transition. The reason is shortened chord downstream of transition where the zero friction control could be utilized immediately after the transition.

4.1 CONTROL BY ZERO SKIN FRICTION

Based on the criterion of separation, a geometry for the main body may be determined over which the skin friction remains continuously zero in the direction of the flow path. Under that condition, the flow is almost at the point of separation but still remains attached on the body surface. This gives not only the lowest drag but also a very low heat transfer rate because convective heat transfer is closely related to skin friction. At the condition of separation, the local pressure gradient is the largest. Therefore, for a given initial boundary layer, the flow with zero friction appears to achieve the specified possible pressure rise in the shortest possible distance and probably with the least dissipation of energy.

4.1.1 Prediction of Zero Skin Friction

In formulating a method for the prediction of zero skin friction, Stratford (1959a and 1959b) set up the procedure for numerical computation and designed a wall surface on which the skin friction is zero. His method is based on the equations of motion and the analysis of turbulence, utilizing either dimensional analysis or mixing length theory. For convenience, he divides the boundary layer in the region of rising pressure into two distinct layers, outer and inner. The pressure rise causes a lowering of the dynamic head in the outer layer and the losses due to shear stress are considered almost the same as for flow on a flat plate. In the inner layer, on the other hand, the inertia forces are small so that the velocity profile is distorted by the pressure gradient until the latter is largely balanced by the transverse gradient of the shear stress. For the prediction of zero friction, the outer layer solutions are matched with those of the inner layers. In his formulation, Stratford (1959a) considered the simplified pressure gradient as shown in Fig. 4.1, the

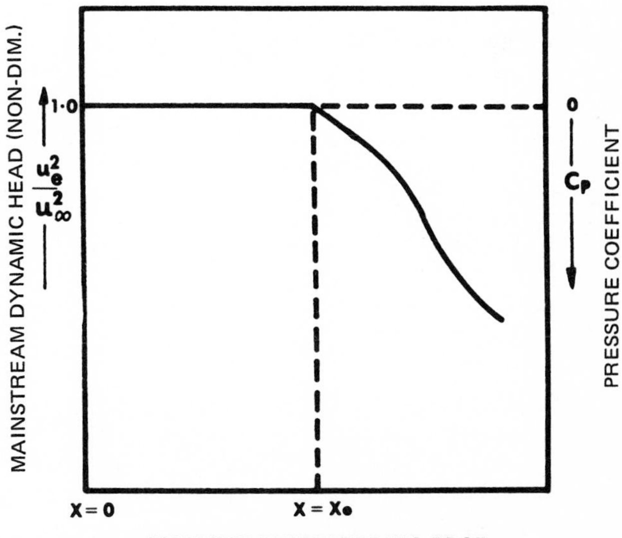

FIG. 4.1 Pressure distribution considered in analyzing zero skin friction (*from Stratford, 1959a*).

consequent development of the boundary layer in a sudden pressure gradient (Fig. 4.2) and the behavior of the boundary layer downstream of separation (Fig. 4.3).

The constant pressure region (Fig. 4.1) may represent the favorable pressure gradient region of an airfoil because that favorable region can be replaced by an equivalent length of a corresponding flat plate. Since a favorable pressure gradient produces a thinner boundary layer than does a zero pressure gradient, an imaginary length–the so-called equivalent length of a flat plate–is somewhat shorter than the actual distance from the forward stagnation point to the position of the maximum velocity of potential flow. The equivalent length for turbulent flow x_{eq} measured from the leading edge can be computed from the following equation extending the integral to the position

$$\frac{x_{eq}}{L} = \int_0^{x/L} \left(\frac{u_e}{u_{e\max}}\right)^{8.17} d\left(\frac{x}{L}\right)$$

where $u_{e\max}$ is the maximum velocity of potential flow and L is the reference length. Some authors have used a value smaller than the 8.17 given above, e.g., 5.

4.1.1.1 The outer layer

If the pressure rise is rapid, the shear forces in the outer portion of the boundary layer are small compared to either the inertia forces or the pressure gradient. The pressure rise causes a lowering of the dynamic head and the losses due to shear stress may be considered almost the same as for flow on a flat plate. Hence, Stratford (1959a) considers a flow condition of outer layer as follows: up to $x = x_0$, the velocity profile is unchanged; at $x > x_0$, a

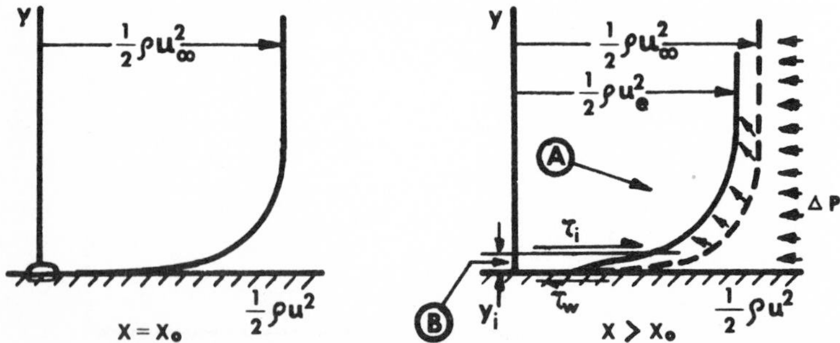

FIG. 4.2 Development of a boundary layer in a sudden pressure gradient (*from Stratford, 1959a*). At $x = x_0$ the profile is unchanged except at $y = 0$. Just downstream of x_0 there is a general lowering of velocity in the outer layer (A) and a change of shape in the inner layer (B).

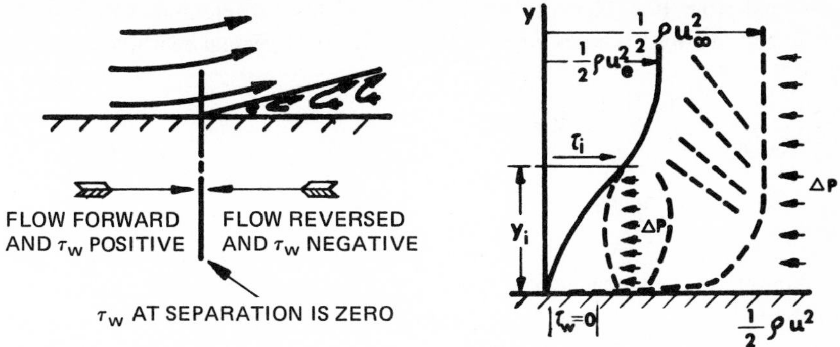

FIG. 4.3 Boundary layer behavior at the separation position (*from Stratford, 1959a*). The flow reaches the separation condition of zero skin friction when the backward force $y_i\,\Delta p$ can only be adequately balanced by the shear stress difference $(\tau_i - \tau_W)$ if τ_W is zero.

constant pressure prevails, i.e., $p = p'$. The prime denotes the condition downstream of $x = x_0$. The total pressure may be assumed approximately constant throughout, thus

$$\frac{1}{2}\,\rho u^2(x, \Psi) = \frac{1}{2}\,\rho u'^2(x, \Psi) - (p - p_0) \qquad (\Psi \geqslant \Psi_i) \tag{4.1}$$

where $\Psi = \int_0^y u\,dy$, $(\Psi \geqslant \Psi_i)$ and the condition $\Psi \geqslant \Psi_i$ refers to an application limited to the outer region of the velocity profile (Ψ_i is the value of Ψ at the edge of the inner layer). The reduction in the value of the local dynamic head $\frac{1}{2}\,\rho u^2$ from its initial value $\frac{1}{2}\,\rho u_\infty^2$ is due to three effects: the rise in static pressure, the viscosity between $x = 0$ and x_0, and the viscosity downstream of x_0. The first two of these three effects are included in Eq. (4.1). The third effect, which is likely to be relatively small for the outer part of the boundary layer at separation, is allowed for approximation. The velocity in the boundary layer without pressure gradient u' is evaluated as follows:

$$\frac{u'}{u_\infty} = \left(\frac{y'}{\delta'}\right)^{1/n} \tag{4.2a}$$

where

$$\delta' = \left[(n+1)\,\frac{n+2}{n}\right]\theta' \tag{4.2b}$$

$$\theta' = 0.036 \times R_e^{-1/5} \tag{4.2c}$$

The value of n varies slightly with Reynolds number but is usually close to 7. The general solution for the outer region of the boundary layer may be obtained from Eqs. (4.1) and (4.2). If Eq. (4.1) is differentiated with respect to Ψ and $u(\partial/\partial\Psi)$ is replaced by $\partial/\partial y$, the form of general solution is

$$\left|\frac{\partial u}{\partial y}\right|_{(x,\,\Psi)} = \left|\frac{\partial u'}{\partial y'}\right|_{(x,\,\Psi)} \qquad (\Psi \geqslant \Psi_i) \tag{4.3}$$

4.1.1.2 The inner layer

In the outer layer, the action of the pressure rise is interpreted as a direct reduction in the dynamic pressure head along each streamline and the back pressure force is balanced by the inertia forces. In the inner layer, however, the effect of fluid inertia is too small for this mechanism to be possible; the inertia forces at the wall are zero and thus the pressure forces are balanced by the shear force. In the inner layer there is a transition between fluid at the wall, for which the pressure force is entirely balanced by the shear force and fluid in the outer layer, where the pressure force simply causes a direct reduction of dynamic head. Stratford (1959a) obtained the inner layer solution in the form of Eq. (4.4) by considering dimensional arguments which yielded the same result as that of mixing length theory:

$$\frac{1}{2}\,\rho u^2 = \frac{2}{(0.41\beta)^2}\,\frac{\partial p}{\partial x}\,y(\tau_w = 0 \qquad y < y_i) \tag{4.4}$$

The value of 0.41 is the von Karman constant for a flat plate and β is an empirical factor that depends on the value of the second derivative:

$$\begin{aligned} \beta &= 0.66\left(\frac{d^2p}{dx^2} < 0\right) \\ \beta &= 0.73\left(\frac{d^2p}{dx^2} \geqslant 0\right) \end{aligned} \tag{4.5}$$

The relevant value of d^2p/dx^2 is to be evaluated immediately prior to separation.

4.1.1.3 The joint solution

At the joint between the inner and the outer layers, continuity is required in Ψ, u, and $\partial u/\partial y$. At the joint, the values of Ψ and $\partial u/\partial y$ are equal for the inner and outer layers. From Eq. (4.3), Ψ and $\partial u/\partial y$ for the inner layer at the joint are equal to Ψ (or Ψ') and $\partial u'/\partial y'$ for the corresponding point on the comparison profile. Equating $\Psi(\partial u/\partial y)^3$, calculated respectively from Eqs. (4.4) and (4.2*a*), the joint solution at the separation point is

$$\left(\frac{y'}{\delta'}\right)^{(2n-4)/n} = \frac{3(0.41\beta)^4}{(n+1)(n\delta'\,dC_p/dx)^2} \qquad (\Psi = \Psi_i) \tag{4.6}$$

A further property of the joint

$$\frac{u^2}{u'^2} = \frac{3}{n+1} \qquad (\Psi = \Psi_i) \tag{4.7}$$

is obtained by comparing $u^2/(\Psi \partial u/\partial y)$–calculated from Eq. (4.4)–with the corresponding quantity calculated from Eq. (4.2*a*). The separation condition may be derived from Eqs. (4.1) and (4.2*a*) as

$$C_p = \left(\frac{y'}{\delta'}\right)^{2/n} \frac{1-u^2}{u'^2} \qquad \Psi \geqslant \Psi_i \qquad C_p \leqslant \frac{1-u^2}{u'^2} \tag{4.8}$$

where $C_p = (p - p_0)/(\frac{1}{2}\rho u^2)$.

Using the values at the joint as given by Eqs. (4.6) and (4.7) and substituting Eqs. (4.2*b*) and (4.2*c*) for δ' leads to

$$(2C_p)^{1(n-2)/4}\left(x\frac{dC_p}{dx}\right)^{1/2} = 1.06\beta(10^{-6}\,\text{Re})^{1/10}\left(C_p \leqslant \frac{n-2}{n+1}\right) \tag{4.9}$$

where Re is Reynolds number based on the local value of x and peak velocity u_{e_0}. The value of n is evaluated by the expression $n = \log_{10} \text{Re}_s$. Here $\text{Re}_s = x_s u_\infty/\nu$ (x_s is the distance to the position of separation).

4.1.2 Design of a Surface for Zero Skin Friction

Stratford (1959b) had little difficulty in successfully designing a 3-ft-long wind-tunnel wall along which the turbulent incompressible skin friction was effectively zero. The wind tunnel had an 8-in^2 working section, a 9:1 contraction ratio, and a nominally 2:1 two-dimensional diffuser that led to a fan via a transition length. The air speed in the working section was 55 ft/sec. Figure 4.4 shows the top surface of the test section which produced zero skin friction.

Stratford (1959b) also gives the design procedure for such a continuous zero skin friction wind tunnel. He first estimated the pressure distribution for the design from available evidence on boundary layer separation. Then using this pressure distribution, he obtained the design geometry by applying corrections based on physical arguments as to the shape that would be predicted from one-dimensional flow theory. Such a design is only approximate because the pressure distribution is affected by the boundary layer growth. Thus, he included provision for varying the divergence and the shape of the opposite wall, in conjunction–if necessary–with building up the surface of the test wall.

The Stratford (1959b) experiments were made on a very low aspect ratio wall and precautions were required to keep the flow two-dimensional. Thus

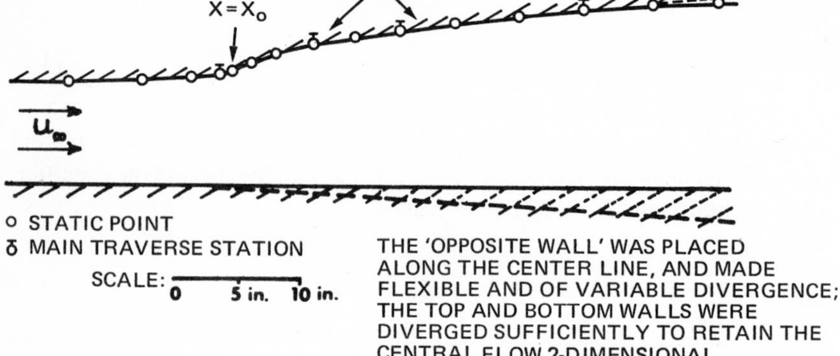

FIG. 4.4 Design of the test section for zero skin friction (*from Stratford, 1959b*).

care must be taken in applying his results to designs for higher aspect ratios. Correction of pressure distribution near the discontinuity at $x = x_0$ was necessary because the boundary layer suddenly thickened at this point of discontinuity due to the steep pressure gradient.

Figure 4.5 shows the spanwise variation of the total head. Along a considerable part of the span, the total head was equal to the total head at the inlet; this is very similar to that used in his theoretical analysis (see Fig. 4.1).

Stratford (1959b) attempted to design a wall such that the given pressure rise could be obtained in the shortest possible distance. The wall was to have zero skin friction continuously, but it appears that he did not obtain a completely zero skin friction on the wall. An attempt to vary the pressure rise to provide continuous zero skin friction for the flow was not successful.

The shortest possible pressure rise for a given initial boundary layer is obtained when the hump in the dynamic head close to the wall is reduced to zero and the dynamic head becomes linear at all positions along the wall. Figures 4.6 and 4.7 indicate the idealized theoretical dynamic head profiles and the momentum thickness at separation for $\text{Re}_s = 10^6$, both as a function of C_{ps}. Figure 4.8 gives the pressure distribution for flow with zero skin friction along flow path at $\text{Re}_0 = 10^6$, and Fig. 4.9 shows the momentum thickness of the boundary layer at separation as a function of C_{ps}. The comparisons of Figs. 4.7 to 4.9 confirm the analytical predictions.

It was stated in Chap. 1 that a criterion for separation such as that of von Doenhoff and Tetervin (see Sec. 1.2.2.1, *The von Doenhoff and Tetervin procedure*) was the condition of zero skin friction. But the strict definition of zero skin friction is not applicable for the continuous zero skin friction surface since separation may not occur. However, a quasi-zero skin friction condition, which is very close to the strictly zero skin friction case, can be used for the design of a continuous nearly zero skin friction surface. The

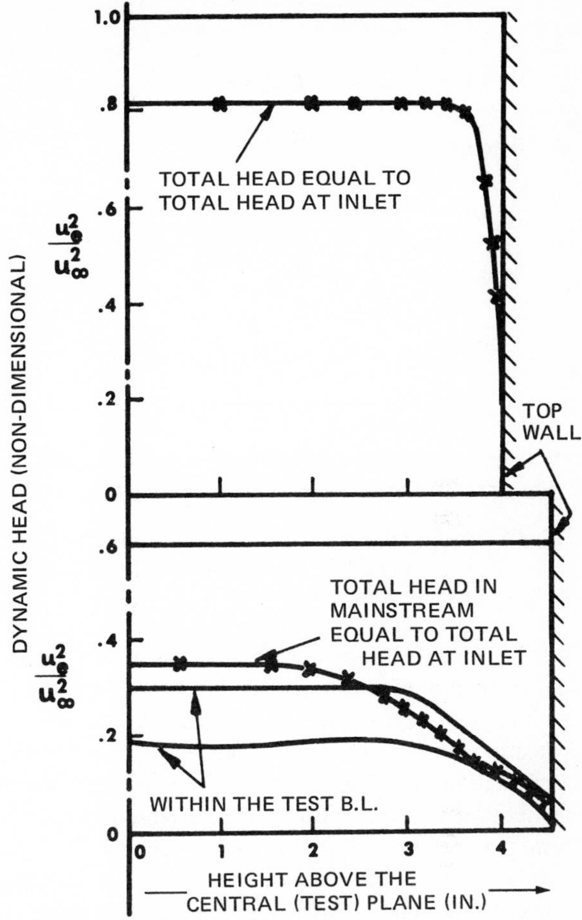

FIG. 4.5 Spanwise variation of total head (*from Stratford, 1959b*).

pressure distribution as predicted by the condition of skin friction that is very close to zero is compared in Fig. 4.10 to that given by Stratford and to experimental data reported by Clauser (1954).

A fixed value of $H = 2.3$ was used for the von Doenhoff-Tetervin equation (1943). Stratford (1959b) found that although the von Doenhoff-Tetervin equation is reasonably accurate for predicting separation when the profile shape changes rapidly, there appears to be no evidence in support of their empirical auxiliary relations if a boundary layer near separation has a steady profile shape. Therefore, the derivation of the pressure distribution by the von Doenhoff-Tetervin equation is not appropriate for the continuous zero friction case as evidenced in Fig. 4.10. The two experimental pressure

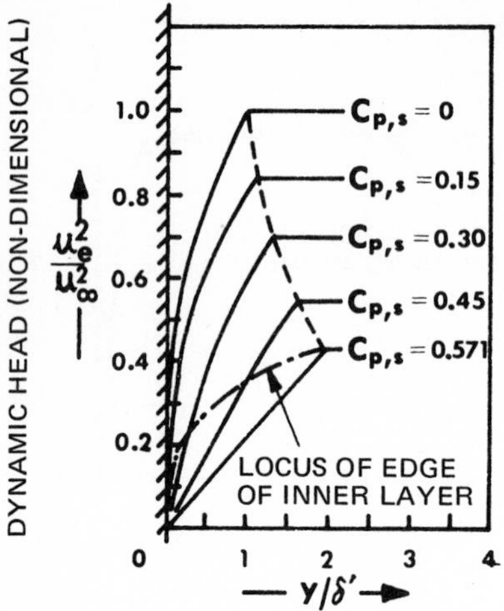

FIG. 4.6 Idealized theoretical dynamic head profiles (at separation $Re_0 = x_0 u_\infty / \nu = 10^6$, δ' = thickness of the flat plate comparison profile; *from Stratford, 1959a*).

distributions of Clauser (1954) included in that figure gave a constant shape in the outer part of the boundary layer downstream of an initial settling length. At the first experimental condition, the boundary layer was far from separation, but at the second one, the skin friction closely approximated that at separation, although it was larger than zero. Hence, the Stratford pressure distribution may be considered as a reasonable extrapolation to zero friction from two pressure distributions of Clauser. The difference in the Reynolds number between these data is not significant enough to cause any appreciable effect. It was not possible to compare the velocity profiles $(u - u_e)/u^*$ used

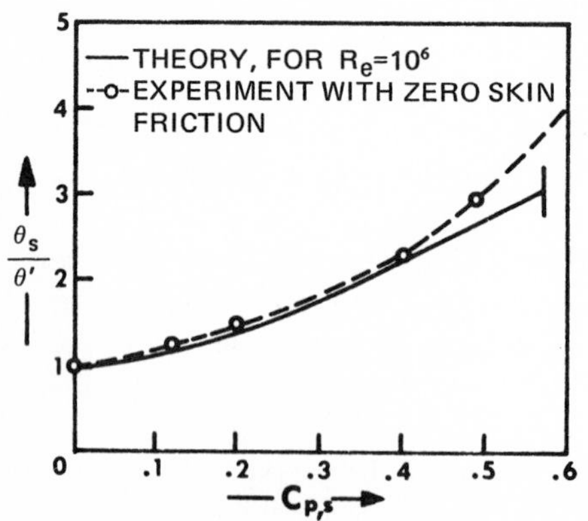

FIG. 4.7 Momentum thickness of the boundary layer at separation (*from Stratford, 1959a*). θ' = momentum thickness of the comparison profile).

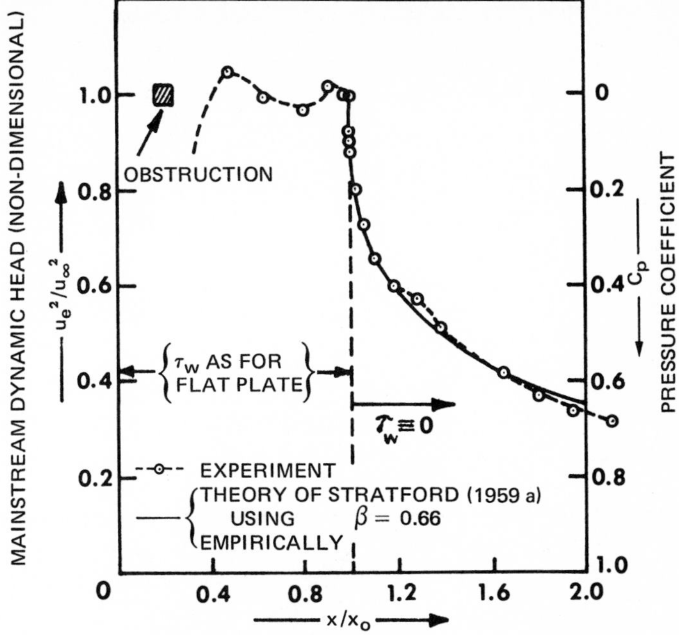

FIG. 4.8 Pressure distribution for flow with zero skin friction ($Re_0 = 10^6$). The initial region (a length of about $\frac{1}{3}x_0$) was replaced by an equivalent obstruction, and the equivalence, including indirectly the normality of the turbulence level, was checked by boundary layer traverse (*from Stratford, 1959b*).

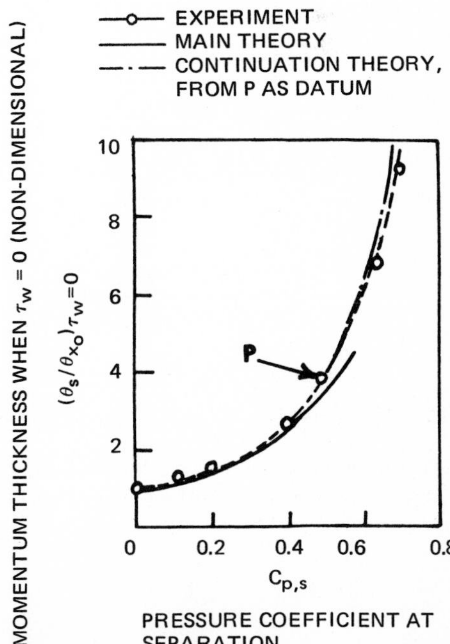

FIG. 4.9 Momentum thickness of the boundary layer (*from Stratford, 1959b*).

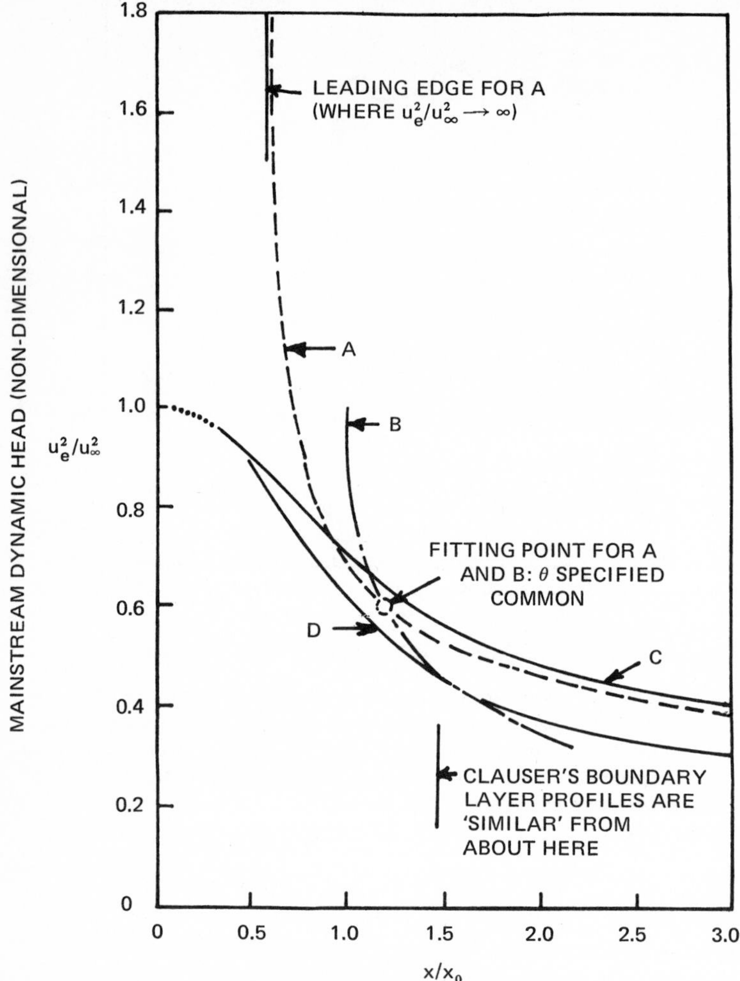

x = distance from the equivalent leading edge
(except for curve A)

A ----Theoretical, Von Doenhoff & Tetervin (1943), using H = const.
H = const. = 2.3
B —·— Stratford (1959b), Fig. 4
C —— Clauser's 1st experiment (1954)
D —— Clauser's 2nd experiment (1954)
(C and D have been appropriately scaled in x)

FIG. 4.10 Comparison of Stratford and other pressure distributions (*from Stratford, 1959b*). Clauser first and second experiments (C and D) were appropriately scaled in x; x is the distance from the equivalent leading edge except for Curve A. The theoretical data of von Doenhoff and Tetervin (A) used H = constant = 2.3.

by Clauser because as skin friction approaches zero, $1/u^* = 1/\sqrt{\tau_w/\rho}$ becomes infinite. Wortmann (1955, 1957) used the Truckenbrodt method (1952) to obtain pressure distribution for a constant factor H. Transition precluded an exact comparison, but the steepest pressure rise used by Wortmann appeared to be intermediate between the experiments of Clauser and Stratford.

The Speidel (1955) test of an airfoil designed by Wortmann also indicates a more gradual pressure rise than that of the Stratford experiment. Some of the difference in the rates for pressure recovery probably occurred because of difference in the Reynolds numbers employed.

4.2 PREVENTION OF INCOMPRESSIBLE FLOW SEPARATION ON AIRFOILS AND WINGS

Flow separation may lead to stall, a phenomenon that accounts for 25 to 50 percent of the fatal accidents on civil aircraft. In his summary of flow separation on airfoils, Carrow (1954) categorized various types of separation, with emphasis on transition. Here the separation of incompressible flow on airfoils and wings is considered within the framework of general separation phenomena, and particular attention is given later to the initial separation. Reference is made to the Chappell (1968) investigations which lead to the estimation of initial separation on airfoils and wings exposed to subsonic flow stream. The prevention of separation is then possible if the flight conditions are adjusted so that they do not reach the estimated values of initial separation.

4.2.1 Flow Separation Phenomena on Airfoils and Wings

The phenomenon of flow separation depends on the shape of airfoils and wings of constant symmetrical, cambered, twisted, and nonconstant section. The available investigations on nonsymmetrical sections are rather limited, but there is no fundamental reason to doubt that the more thorough investigations of constant symmetrical sections cannot be extended to other shapes. Accordingly, the logical step is to present the initial separation over a constant symmetrical section as the basic problem. However, information on common separation phenomena occurring for all shapes of airfoils and wings may be useful in order to estimate the initial separation.

For simplicity, consider only aerodynamically smooth airfoils or wings with no arbitrary fixing of transition at the leading edge. The effect of fixed transition will be described additionally. As shown in Fig. 4.11*a* and 4.11*b*, two types of boundary layer separation are generally possible, namely, laminar separation at or near the leading edge and turbulent separation from the trailing edge, depending on camber, ratio of thickness to chord, leading edge radius, and Reynolds number. As indicated in Fig. 4.11*c*, separation from the leading edge of swept wings is usually manifested in the form of a vortex sheet with its axis inclined at an angle to the free-stream direction.

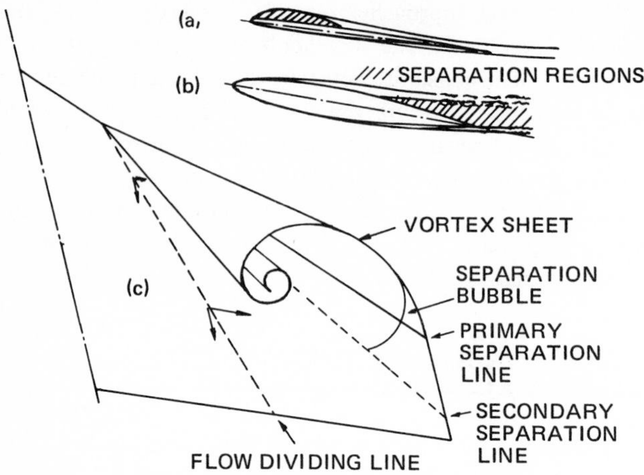

FIG. 4.11 Types of separation (*from Chappell, 1968*).

Laminar separation is usually accompanied by a subsequent reattachment following transition in the free shear layer to the turbulent state, forming a region of local separation or bubble. The laminar separation bubble is either "short" or "long" depending on the magnitude of Reynolds number (based on velocity and momentum thickness of boundary layer at separation) and pressure gradient. The type of bubble is predictable; among the available analyses, that of Gaster (1966) predicts the most precise result.

If the bubble is short and the angle of attack is then increased, the separation point moves forward to a region of increasing surface curvature. Eventually turbulent reattachment of the full shear fails to take place, and there is a consequent sudden loss of lift (Fig. 4.12*a*) and increased drag.

This bursting of the laminar separation bubble is called leading-edge stall, but in order to avoid ambiguity with the three-dimensional case, Chappell (1968) classified this phenomenon as short-bubble stall. Since separation involving formation of a short bubble can occur on most conventional airfoil sections of moderate thickness to chord ratio in the range of $0.09 < t/c < 0.15$, specific attention is necessary to prevent the disaster following breakaway of a short bubble.

If the bubble is long, then an increase in angle of attack produces a progressive rearward movement of the reattachment, thus increasing the length of the bubble until it coincides with the trailing edge. The separation is reached at about this angle of attack; any further increase results in a gradual reduction of lift, as shown in Fig. 4.12*b*. This separation process can also occur on most conventional thin airfoils in a range of t/c up to about 0.09 and is usually called thin-airfoil stall (Chappell, 1968, refers to it as the *long-bubble stall*). Aerodynamically, whereas the formation and development of a long bubble has a considerable adverse effect on drag via the pressure distribution, the existence of a short bubble has a negligible effect right up to the harmful stall condition.

The turbulent separation from the trailing-edge as illustrated in Fig. 4.12*c* is characteristic of most conventional thick airfoil sections in a range of t/c greater than 0.12. An increase in angle of attack produces a gradual forward movement of the point of separation and a steady and gradual decrease in lift.

It is possible for both short-bubble and trailing-edge separation to exist on the same families of airfoils at the same time over a certain range of Reynolds numbers. The former generally starts to develop at a lower angle of attack than does the latter; see Fig. 4.12*d*. This combination of separation consequently displays characteristics of both the short-bubble and the trailing-edge separation, with the possibility of either a semirounded lift-curve peak followed by an abrupt decrease in lift or a relatively sharp lift-curve peak followed by a relatively rapid decrease in lift. This combined type of separation is effectively an intermediate type which may or may not occur

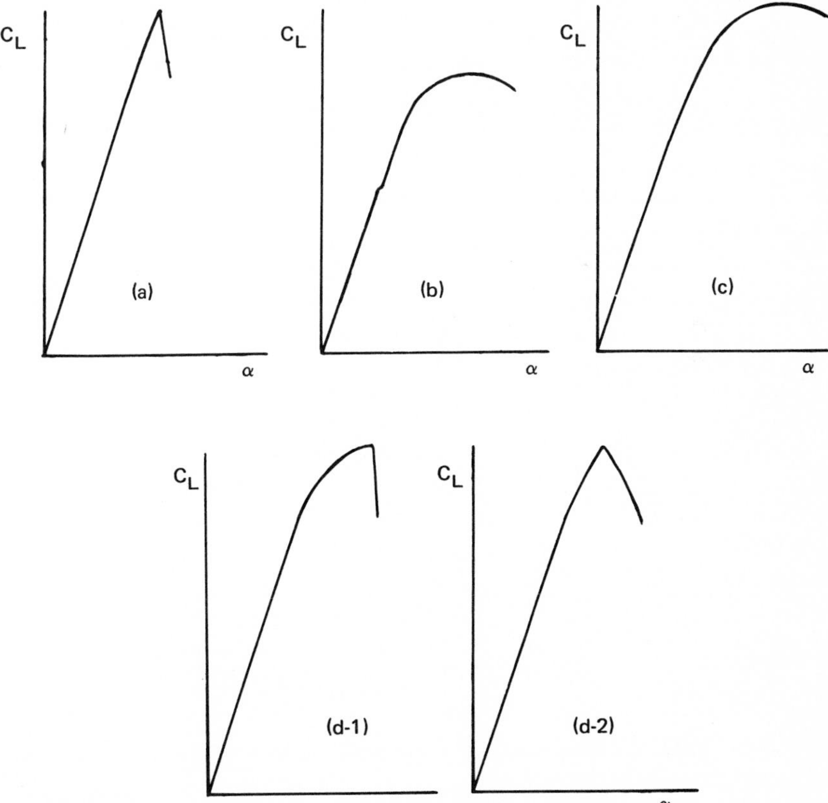

FIG. 4.12 Types of airfoil stall: (*a*) short-bubble stall; (*b*) long-bubble stall; (*c*) trailing-edge stall; (*d*) combined stall (*from Chappell, 1968*).

when the thickness/chord ratio of a given family of airfoils is decreased at constant Reynolds number and the type of separation subsequently changes from the trailing-edge type to the short-bubble type.

The boundaries for the various types mentioned can be distinguished by relating the Re_c (the Reynolds number based on the streamwise values of chord and velocity) to the quantity $z_{1.25}/c$, where $z_{1.25}$ is the upper surface ordinate of the airfoil at 1.25 percent chord and c is the chord length. These boundaries are shown in Fig. 4.13.

This correlation was shown by Gault (1957) after careful study of the aerodynamic characteristics of over 150 different airfoils, both symmetrical and cambered, over a range of Reynolds numbers from 0.7×10^6 to 25×10^6. Figure 4.13 is a good guide for classifying initial separation on

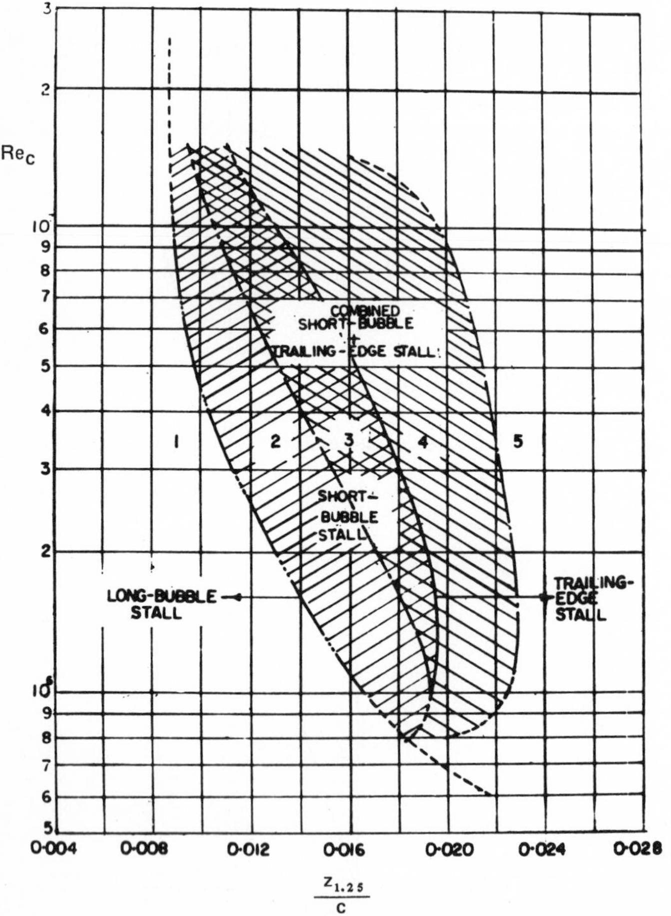

FIG. 4.13 Boundaries for various types of airfoil stall (*from Chappell, 1968*).

airfoils, not only for the pure types (Fig. 4.12*a*–4.12*c*) but also for the combined type (Fig. 4.12*d*) because separation is initiated either from the leading edge or on the trailing edge.

Now consider the stall of wings. The boundaries of various types of separation can be made for the initial stall by plotting $(\mathrm{Re}_{ct})_n$ versus $(z_{1.25}/c)_n$ as shown in Fig. 4.14 for wings with constant symmetrical sections. The symbols Re_{ct} refer to Reynolds number based on the velocity and chord at the wing tip, and the subscript n indicates the normal direction to the leading edge. Note that Re_c and $z_{1.25}/c$ were used for airfoils (Fig. 4.13) whereas $(\mathrm{Re}_{ct})_n$ and $(z_{1.25}/c)_n$ are used for plan wings (Fig. 4.14). Chappell (1968) formulated $(\mathrm{Re}_{ct})_n$ as

$$(\mathrm{Re}_{ct})_n = \mathrm{Re}_{c_s} \cos \Lambda_0 \frac{(c_t)_n}{c_s} \tag{4.10}$$

where

$$\frac{(c_t)_n}{c_s} = \frac{3\lambda}{2} \frac{1+\lambda}{1+\lambda+\lambda^2} [\cos \Lambda_0 + \sin \Lambda_0 \tan (\Lambda_0 - \Lambda_{tr})] \tag{4.11}$$

Here c_t is the wing tip chord, c_s is the aerodynamic mean chord in the streamwise direction, Λ is the wing sweep back angle, and λ is the wing taper ratio. The subscripts 0 and *tr* respectively refer to wing leading and trailing edges. Furthermore,

$$\left(\frac{z_{1.25}}{c}\right)_n \doteqdot \left(\frac{z}{c}\right)_s [\cos \Lambda_0 + \sin \Lambda_0 \tan (\Lambda_0 - \Lambda_{tr})]^{-1} \tag{4.12}$$

where the subscript s refers to the streamwise direction. The quantity $(z/c)_s$ corresponds to $(x/c)_s$ given by

$$\left(\frac{x}{c}\right)_s \doteqdot 0.0125 [1 + \tan \Lambda_0 \cdot \tan (\Lambda_0 - \Lambda_{tr})] \tag{4.13}$$

and x is the general coordinate station.

Wings that exhibited leading-edge initial separation fell in regions 1 and 2, while those that exhibited initial trailing-edge separation fell in regions 2, 3, and 4.

The effects of boundary-layer outflow on highly swept wings might be expected to result in the shift of the boundaries indicated in Fig. 4.14. Lack of sufficient data on three-dimensional flow makes it impossible to prove this shift of or to substantiate the correlation precisely for initial separation type. There is a particular need for much more experimental data on wings that exhibit trailing edge or combined separation.

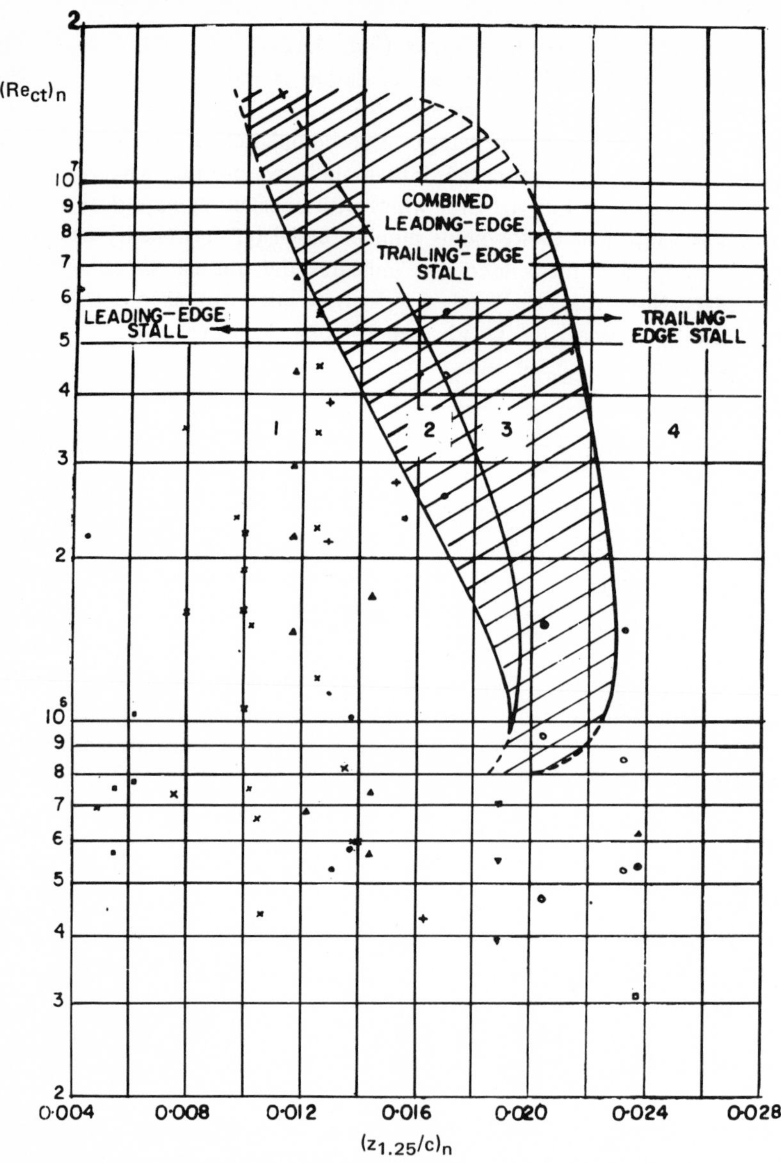

FIG. 4.14 Initial stall characteristics of plane wings with constant symmetrical sections (*from Chappell, 1968*).

4.2.2 Estimation of Initial Separation on Airfoils and Plane Wings

The initial leading- and trailing-edge separations may be estimated from lift coefficient C_L and angle of attack α. The Chappell (1968) correlation of experimental data for plane wings with constant symmetrical sections is utilized for this purpose.

4.2.2.1 Leading-edge initial separation

The correlation parameters include Reynolds number, leading-edge radius, section thickness, and all effects of planform geometry. Chappell (1968) proposes that a Reynolds number based on velocity and leading-edge radius is a logical parameter which combines these effects; both are measured normal to the wing leading edge at the tip:

$$(\mathrm{Re}_{\rho t})_n = \mathrm{Re}_{c_s} \cos \Lambda_0 \left(\frac{\rho}{c}\right)_n \frac{(c_t)_n}{c_s} \tag{4.14}$$

where ρ is the wing section leading-edge radius and

$$\frac{(\rho/c)_n}{(\rho/c)_s} \doteqdot \{\cos \Lambda_0 \left[\cos \Lambda_0 + \sin \Lambda_0 \tan (\Lambda_0 - \Lambda_{tr})\right]\}^{-1} \tag{4.15}$$

This parameter $(\mathrm{Re}_{\rho t})_n$ should be correlated with α_k, C_{Lk}, or some function that provides the best fit to the experimental data. The subscript k refers to the end of the linear variation of force and moment characteristics. The value of α_k is affected by leading-edge sweep. Since the simple sweep theory may not be sufficient for the complex problems associated with separation in three dimensions, pure empirical correlations are needed to find the relation between α_k and the leading-edge sweep.

For plane wings and wing-body combinations of constant symmetrical sections, the values of $\alpha_k \cos \Lambda_0$ can be correlated to $(\mathrm{Re}_{\rho t})_n$ with a strong dependence of $(t/c)_c$; see the available experimental data shown in Fig. 4.15. These data were for wings of up to 10-percent streamwise thickness to chord ratio, and the transition position was unknown except for a few cases in which transition was fixed by a boundary layer trip near the leading edge.

The preferred methods for estimating values of α_k are (1) surface flow visualization (for vortex separation) and (2) onset of nonlinearities in the $C_D \sim C_L{}^2$ plot. They are also obtainable by means of a pressure plot over the tip chord (collapse of leading-edge suction peak for leading-edge separation and loss of trailing-edge pressure for trailing-edge separation) and of the onset of nonlinearities in lift and pitching moment characteristics ($C_L \sim \alpha$ and $C_{m\frac{1}{4}} \sim \alpha$).

Note particularly that bubble separation at the leading edge of two-dimensional wings can be included in the same correlation as vortex separation at the leading edge of three-dimensional wings (Fig. 4.15). Chappell (1968) suggests that the data of Fig. 4.15 may be used to predict the spanwise position of the part-span vortex on a tapered swept wing at any angle of attack, at least for spanwise stations outside the influence of the wing vortex or the fuselage. Furthermore, such prediction of spanwise position of the

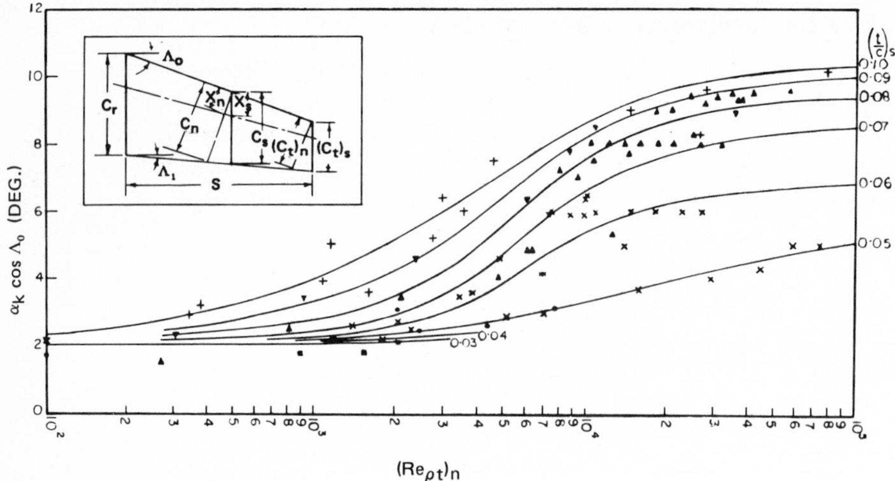

FIG. 4.15 Boundaries of leading-edge separation for plane wings with constant symmetrical section (*from Chappell, 1968*).

part-span vortex should be restricted to wings with straight leading edges since curvature promotes a more rapid movement of the vortex.

Table 4.1 indicates the ranges of parameters from available experimental data used to establish the correlation of Figs. 4.15, 4.16, and 4.17 for predicting leading-edge initial separation. Caution must be exercised in using parameters well outside these ranges or in using unconventional sections. This is also true for Table 4.2 which is given later.

Figures 4.16 and 4.17 show that 90 percent of the data for α_k and C_{Lk} are predicted within an accuracy of ±20 percent at low C_{Lk} values (less than 0.6). Above values of about 0.6, the data are mostly two-dimensional, and the percentage of scatter is reduced, thus enabling the C_{Lk} to be predicted within an accuracy of ±0.1 percent. There are many possible explanations for the overall scatter, not the least of which is aeroelastic distortion on thin wings at high sweep angles and aspect ratios; under high dynamic pressures, these can easily introduce errors of 20 percent in angle of attack (at the wing tip). Another source of error lies in the value of the leading-edge radius assumed for the wing section. The theoretical radius has been used appropriately, but the actual radii for the model were not known. Also, in a few cases which exhibited small divergence, there was some difficulty in deciding the exact point of divergence of the $C_D \sim C_L{}^2$ plot from the linear.

4.2.2.2 Trailing-edge initial separation

Chappell (1968) proposed to correlate $\alpha_k \cos \Lambda_0$ with $(Re_{ct})_n$, given by Eqs. (4.10) and (4.11). However, a better correlation was found by plotting

TABLE 4.1 Ranges of parameters of available experimental data used to establish the correlations for predicting leading-edge initial separation and stall (incompressible flow)

Symbol	$(t/c)_s$	AR	λ	Λ_0 deg	Λ_{tr} deg	M	$Re_c = \times 10^{-6}$
□	0.03	3.0	0 to 0.4	53.1	0	0.6	3.0 to 4.8
○	0.04 to 0.05	3.0 to 3.9	0 and 0.2	38.6 to 60.0	−5.1 to 60.0	⩽0.2 and 0.8	2.0 to 7.0
X	0.051 to 0.06	1.97 to 4.0	0.01 to 1.0	32.6 to 63.0	0 to 57.1	0.1 to 0.8	1.5 to 15.0
X	0.06		Two-dimensional section			0.1 to 0.2	1.9 to 6.8
Δ	0.07		Two-dimensional section			0.1 to 0.2	1.5 to 6.0
Δ	0.075		Two-dimensional section			0.1 to 0.2	1.5 to 6.0
Δ	0.075 to 0.078	2.03 to 5.5	0.32 to 0.53	46.6 to 60.0	34.5 to 39.8	0.1 to 0.5	1.0 to 18.0
Δ	0.08		Two-dimensional section			0.1 to 0.2	1.5 to 6.0
∇	0.085 and 0.089	2.33 to 5.16	0.35 and 1.0	45.0 and 55.0	32.0 and 45.0	0.1 and 0.2	0.6 to 1.4
∇	0.09		Two-dimensional section			0.2	5.8
+	0.10	2.23 to 4.3	0.14 to 0.39	44.0 to 50.2	0 and 26.6	0.1	0.6 to 11.3
+	0.10		Two-dimensional section			0.1	0.6

Source: Chappell (1968).

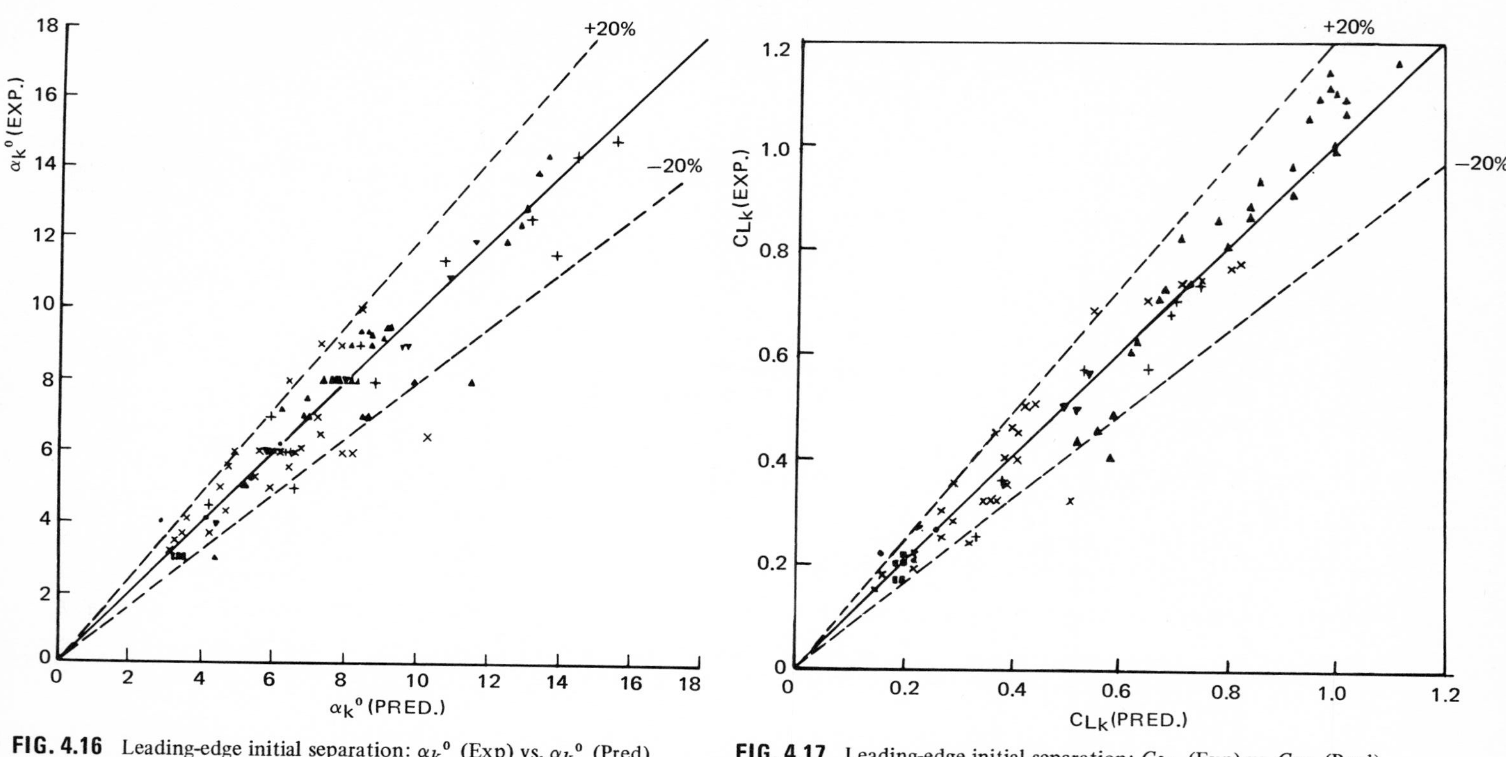

FIG. 4.16 Leading-edge initial separation: $\alpha_k{}^0$ (Exp) vs. $\alpha_k{}^0$ (Pred) (*from Chappell, 1968*).

FIG. 4.17 Leading-edge initial separation: C_{Lk} (Exp) vs. C_{Lk} (Pred) (*from Chappell, 1968*).

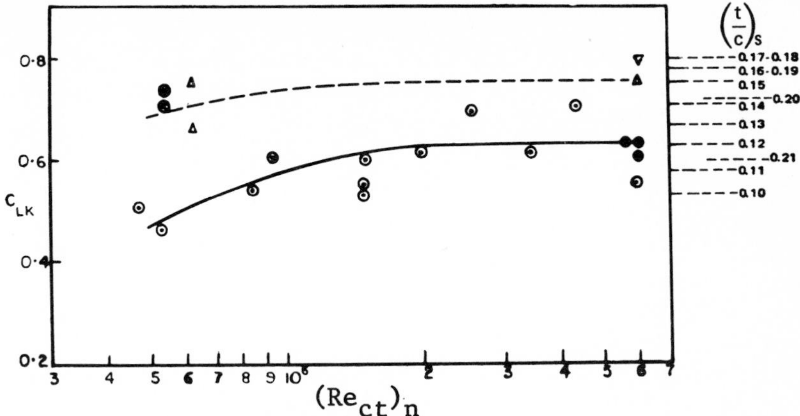

FIG. 4.18 Boundaries of trailing-edge initial separation for plane wings with constant symmetrical section (*from Chappell, 1968*).

C_{Lk} directly against $(\text{Re}_{ct})_n$ (as shown in Fig. 4.18) with the parameter $(t/c)_s$. This is used in predicting values of C_{Lk} and α_k for trailing-edge initial separation.

Figures 4.19 and 4.20 compare the values for C_{Lk} and α_k predicted by using Fig. 4.18 to the available experimental data listed in Table 4.2.

Figures 4.19 and 4.20 were drawn for $(t/c)_s = 0.12$ because most of the available test data were that value. However, there were quite broad ranges for other parameters within 15 plotted points for $(t/c)_s = 0.12$, e.g., aspect ratios from 3 to 10, sweepback angles from 35 to 45 deg, and some two-dimensional data. The fixing of transition point by means of a boundary layer trip (e.g., a band of uniform roughness) could have more effect on turbulent than on laminar separation. In this regard, it should be noted that some of the tests with turbulent separation utilized transition fixed by roughness whereas others were free transitions. This effect of transition position could be a dominating factor where scatter is concerned (±15 percent for C_{Lk} and α_k).

4.2.3 Estimation of Separation on Three-dimensional Wings

All of the types of boundary-layer separation discussed in Sec. 4.2.2 are possible for three-dimensional subsonic flow over wings. However, because of the complex nature of flow separation on wings, it is not feasible to classify the characteristics of either the overall separation or the section separation in the same simple relation of C_L versus α as shown for the airfoils in Fig. 4.12. It *is* possible nevertheless to divide wing separation into leading-edge, trailing-edge, or a combination of both types.

On the basis of physical arguments, it is reasonable to expect that initial separation types are closely allied in both two- and three-dimensional flows

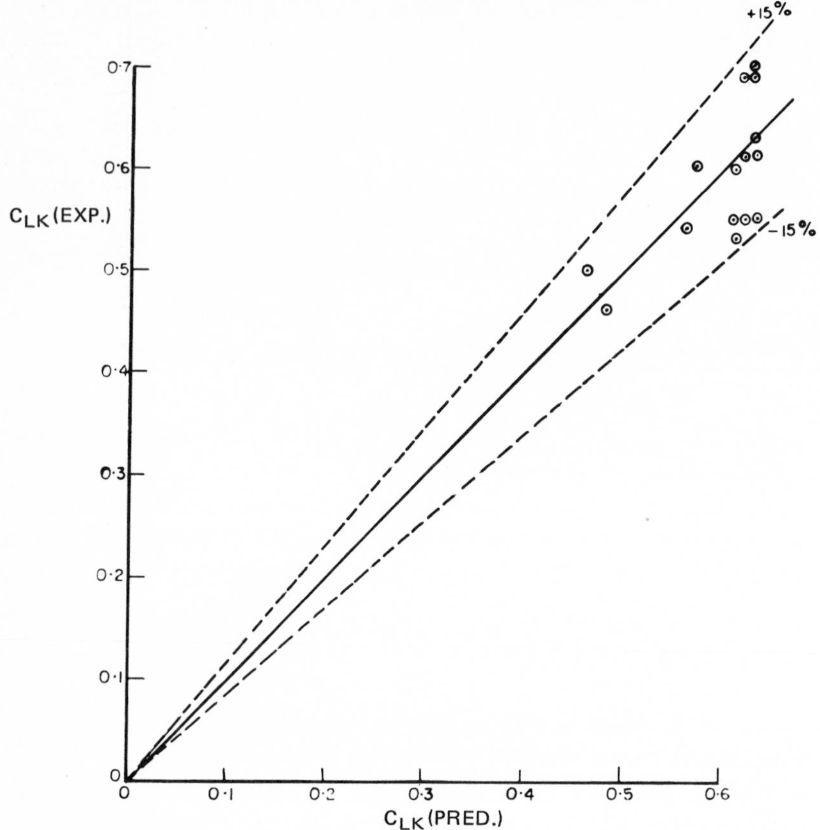

FIG. 4.19 Trailing-edge initial separation: C_{Lk} (Exp) vs C_{Lk} (Pred); $(t/c)_s = 0.12$ only (*from Chappell, 1968*).

and that not only a streamwise but also a spanwise spread of separation should be considered for the three-dimensional case. For wings that exhibit combined separation, initial separation will generally occur from one edge only; the second type will occur at a later stage in the flow development as angle of attack is increased.

In practice, the separation phenomenon for the body shape used for flow conditions in engineering is three-dimensional. Unfortunately both analytical and experimental investigations are scarce for three-dimensional separation on wings and even more so for propellers. Hence the present development of a proper prevention technique is limited in scope.

To shed more light on the complex three-dimensional boundary layer problem, separation phenomena are now considered for three commonly used wing shapes: unswept, swept, and delta types.

4.2.3.1 Particular features of flow and its separation on three-dimensional wings

In practice, the three-dimensional boundary layer involving secondary flow or crossflow may become a knotty problem. Crossflow may occur even on rectangular wing if it is placed at an angle of yaw or composed of airfoil with susceptibility for separation; see Jones (1947).

If the yawed or swept wing is of very high aspect ratio, then certain boundary layer and separation phenomena are determined independently by the crosswise component of velocity. The effect of sweepback is to increase the area of stable laminar flow over the wing and to decrease the lift coefficient at which flow separation occurs. Although at a particular angle of attack, flow may not separate on an unswept wing, it may do so on a sweptback wing; see Fig. 4.21.

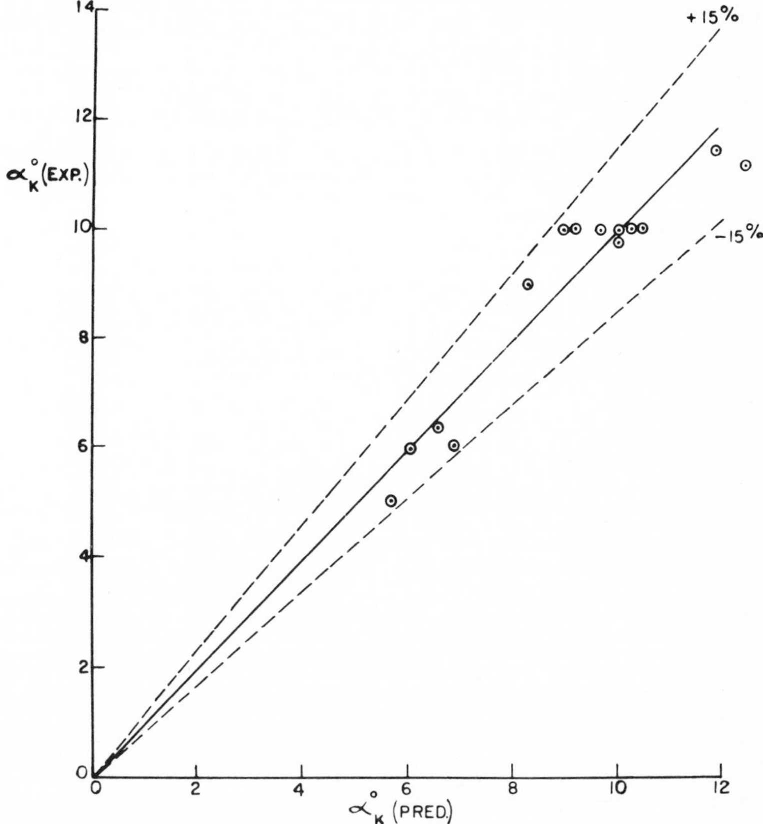

FIG. 4.20 Trailing-edge initial separation: $\alpha_k{}^\circ$ (Exp) vs. $\alpha_k{}^\circ$ (Pred); $(t/c)_s = 0.12$ only (*from Chappell, 1968*).

TABLE 4.2 Ranges of parameters of available experimental data used to establish the correlations for predicting trailing-edge initial separation and stall (incompressible flow)

Symbol	$(t/c)_s$	AR	λ	Λ_0 deg	Λ_{tr} deg	M	$Re_c = \times 10^{-6}$
□	0.075	1.5	1.0	60.0	60.0	⩽0.2	1.24
○	0.12	3.0 to 10.1	0.45 to 1.0	36.5 to 46.3	31.2 to 45.0	0.1 to 0.3	1.1 to 10.0
○	0.12		Two-dimensional sections			⩽0.2	1.5 to 6.0
⊗	0.13	4.5	1.0	30.0	30.0	⩽0.2	0.72
△	0.15	3.0 and 6.0	1.0	0	0	⩽0.2	0.6
△	0.15		Two-dimensional sections			⩽0.2	6.0
▽	0.18		Two-dimensional sections			⩽0.2	6.0
⊕	0.21		Two-dimensional sections			⩽0.2	6.0

Source: Chappell (1968).

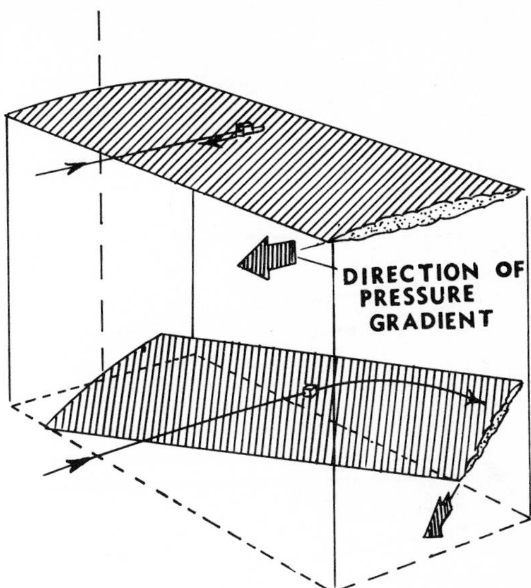

FIG. 4.21 Effect of yaw on the paths of particles in a separated boundary layer (*from Jones, 1947*).

In the case of an oblique wing, the resultant pressure gradient is at right angles to the long axis of the wing. Both this pressure and components of the viscous stress distribution lying in this direction are determined by the crosswise component of velocity. Hence, the circumstances that lead to flow separation over the straight wing are reproduced if the crosswise component of motion of the sweepback wing is the same as that of the unswept wing. This phenomenon is helpful in explaining the adverse effect of sweepback on the lift and the drag of a wing. According to the two-dimensional theory, if a wing with reverse boundary layer flow and maximum lift at $C_L = 1.4$ is yawed 45 deg, there would be separation accompanied by a fully developed lateral motion of the boundary layer at $C_L = 0.7$. At 60 deg yaw, the predicted maximum lift coefficient would be only 0.35. Wind-tunnel observation of the boundary-layer flow over sweptback wings agree qualitatively with these predictions in regard to flow separation, but the expected loss in maximum lift does not occur.

The two-dimensional theory is applicable up to separation and apparently the effect of separated flow downstream of that point is important to the lift.

Figure 4.22 clearly demonstrates the occurrence of crossflow over the wing of a Messerschmidt Me 109. Airflow breakaway over the wing caused this plane to bank abruptly. At first the disturbance was restricted to the wing root, but it spread very fast–and on the side only–to the wing tip. Suddenly and without any warning, the aircraft was thrown over on its side, leaving the

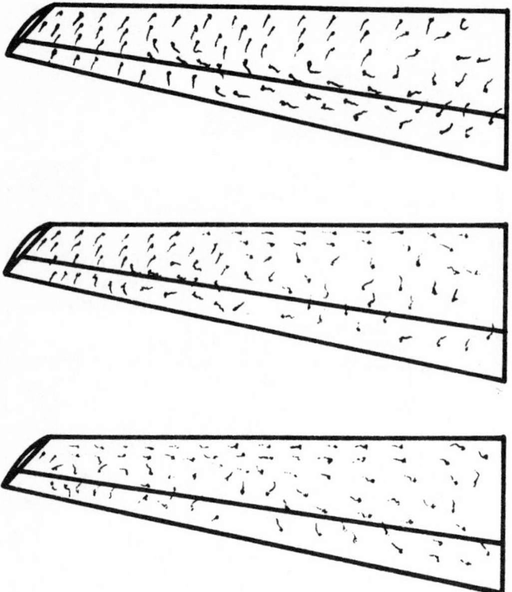

FIG. 4.22 Reconstruction of three tuft diagrams (at intervals of 0.5 seconds) showing the striking flow processes during stalling of a Messerschmidt Me 109 with an experimental twisted wing but without slots. The position of the silk thread shows the rapid sideways spread of the initial disturbance, accompanied by strong cross flows in the boundary layer (*from Liebe, 1952*).

pilot helpless to prevent the banking. The sideway flow of the boundary layer caused a rapid loss in lift. The formation of crossflow at the leading edge is demonstrated in Fig. 4.22 by tufts running at right angles to the flight direction. The undesirable premature breakaway of the crossflow in the outer region (arising from the outward boundary layer flow) is a pronounced feature of flow over the sweptback wing. For a swept wing, the pressure on the suction surface in the neighborhood of the nose is greatly reduced compared to that downstream because of the displacement of the wing section; see Fig. 4.23. Thus, the fluid particles in the boundary layer move along toward the lower pressure region and form a strong crossflow toward the tip.

In order to control stall, Dommasch et al. (1961) proposed the following techniques for all wing configurations. A favorable stalling characteristic may be provided by using washout (aerodynamic twist of the wing whereby the effective angle of attack of the tip sections is made less than that of the root sections) or by using spoilers which produce an early flow collapse of the root. The stalling characteristics of very smooth and laminar flow wings can be improved by roughening the leading edge of the tip section to give the effect of a higher Reynolds number.

4.2.3.2 Unswept wing

The flow over an unswept wing may be considered as roughly two-dimensional insofar as types of initial separation are concerned. Rectangular wings with constant spanwise sections tend to separate at all sections simultaneously. Possibly there is a slight bias towards root separation, but this is difficult to assess from the available flow visualization studies conducted by Goett and Bullivant (1938). In the case of rectangular wings, trailing-edge separation appears to start from the inboard sections as reported by Goett and Bullivant (1938). This effect is simply explained by the fact that a spanwise loading calculation indicates the occurrence of maximum load at the center section; if the section maximum coefficient is assumed constant, initial separation forms at this station. It is more difficult to delay stall near the tips for a tapered wing than for a rectangular wing because the lower Reynolds number at the tip favors early tip stall (Dommasch et al., 1961). Because of the large downwash at the very tips, the tips themselves do not normally stall, but the sections just inboard do.

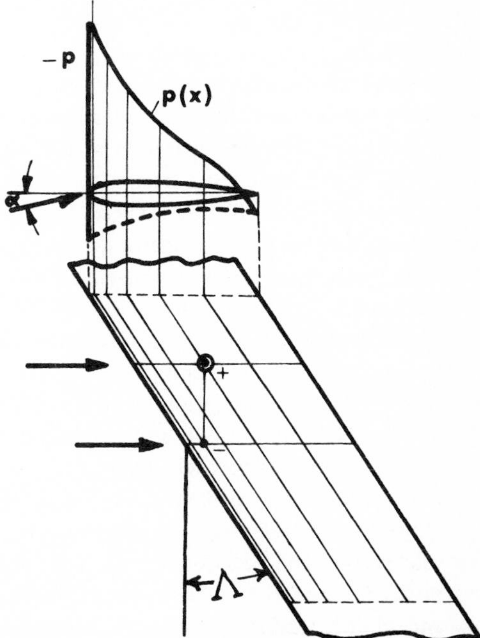

FIG. 4.23 Cross flow in the boundary layer on a swept wing; curves of constant pressure (isobar) are shown on the suction side of wing (*from Schlichting & Truckenbrodt, 1967*).

4.2.3.3 Swept wings

Swept wings are provided for modern aircraft that fly at high speeds. One attributable effect is a change in the spanwise distribution of downwash; this causes the load on a wing of given aspect ratio and taper ratio to be concentrated further outboard. Therefore, flow separation occurs for the outboard sections before the inboard sections. In addition, boundary layer outflow from the root toward the tip produces thicker boundary layers at the tip sections than at the root. Thus trailing-edge separation at the tips is favored for both tapered and untapered swept wings (Graham, 1951). Rogers et al. (1960) found that leading-edge separation for swept, untapered wings slightly favored the tip section initially. Hall and Rogers (1960) showed that a small leading-edge radius at the tip of a tapered swept wing is also conducive to initial separation in this region.

For wing-body combinations, the body might induce initial separation at the wing root for unswept tapered wings, especially if fillets are poorly designed. But for wings with appreciable sweep, the effects of boundary layer outflow and load distribution almost certainly preclude such a possibility.

Note that at a certain angle of attack, swept wings exhibit the leading-edge vortex type of separation often accompanied by a separation bubble (Fig. 4.11*c*). The vortex sheet forms initially close to the wing tips and spreads inboard as angle of attack increases. The formation of leading-edge vortices for plane wings of constant symmetrical section is affected by Reynolds number, section thickness to chord ratio, leading-edge radius, and leading-edge sweep, When there is full-chord separation at the wing tip, the lift curve slope reaches a maximum and thereafter begins to decrease towards stall as the angle of attack increases still further. These flow changes are also reflected in the development of pitching moment for wings with trailing-edge sweeps of about the same magnitude as the leading-edge sweep.

The leading-edge vortex formation may be delayed by increasing Reynolds number and leading-edge radius (Garner & Cox, 1961). It may be conjectured that the increase of leading-edge sweep could be detrimental, causing leading-edge vortex separation at lower angles of attack. However, Rogers et al. (1960) found that this is not necessarily so because vortex separation can be delayed on the sweptback wing until the equivalent two-dimensional angle of attack is about 50 percent greater than that for the airfoil. It is postulated that the strong lateral flow in the boundary layer of the swept wing causes this alleviation. The location of transition may affect the leading-edge vortex separation because if the transition occurs early enough (say, forward of the 5 percent chord station), then the possibility of laminar separation at the leading edge would be eliminated.

4.2.3.4 Delta wings

Figure 4.24 shows the basic pattern of flow over a slender wing at an angle of attack. A vortex is formed by rolling up of the shear layer which separates from

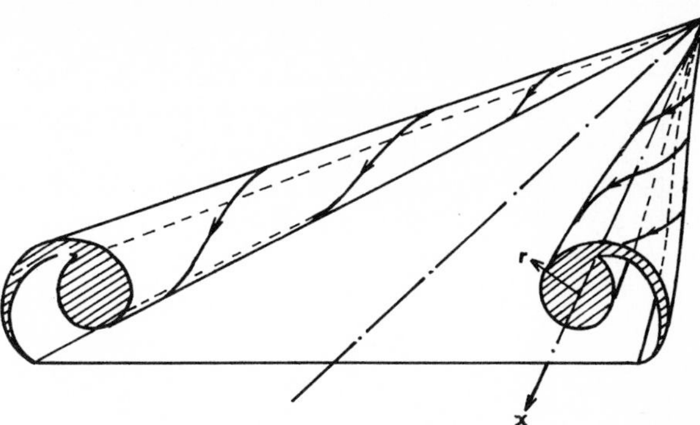

FIG. 4.24 Basic pattern of flow over a slender wing at incidence (*from Hall, 1959*).

the leading edge of a slender wing at angles of attack. Hall (1959) used a simplified model of the vortex (chosen with the aid of experimental results) to analyze this vortex flow. He assumed that if the rotational character of the flow is accounted for, no vortex sheet needs to be considered and viscous diffusion may be neglected. He further assumed that the vortex flow is axially symmetric and incompressible, and that the velocity field is conical.

Hall determined the vortex structure by solving the Euler equations of motion. Since the rotational nonviscous flow is described by the Euler equations of motion and may be called an Euler vortex, its structure is continuous. It differs from the potential line vortex in that it can be generated and its strength can be varied along its length by the natural spiralling in of vorticity.

Figure 4.25 is a reproduction of part of a total head survey by Harvey (1958) over a 20 deg delta wing at both incidence and yaw. The leeward vortex shown there is a good example of a flow pattern without complication from secondary separation. The closed contour, indicated by a broken line, roughly indicates the region of spiralling flow; this is the area considered to be covered by the vortex and it is considerable. Note that before the vortex is formed, the shear layer becomes fairly diffuse and cannot be distinguished after less than one convolution of the spiral. The gradients of total head in the vortex are small compared with those near the leading edge or in the boundary layer over the wing. This figure shows a progressive departure from axial symmetry as the distance from the axis is increased.

The high axial velocities and low pressures within the vortex shown by the Hall analysis are in general agreement with experiment. In the vortex, the pressure p invariably decreases with decreasing $\xi = r/x$, where x and r are cylindrical coordinates (x is taken along the axis of symmetry with origin at the wing apex). This occurs because the gradient $\partial p/\partial \xi$ is proportional to v^2; the symbol v refers to circumferential velocity component. Since for conical

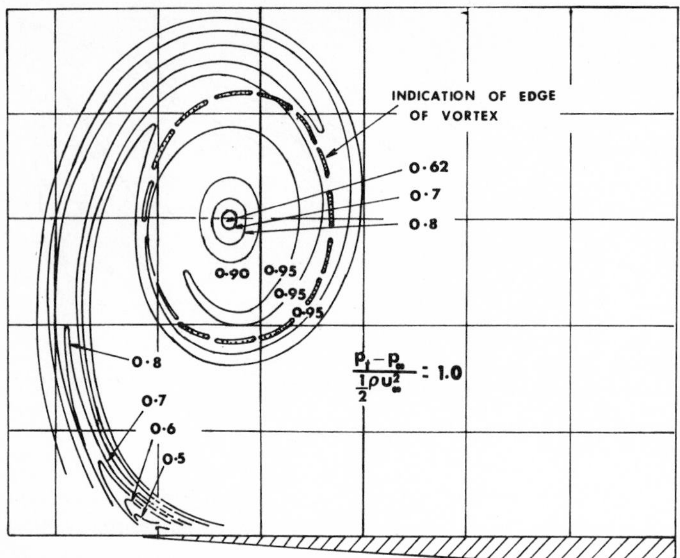

FIG. 4.25 Total head survey over a 20-degree delta wing at 15-degree incidence and 10-degree yaw (*from Harvey, 1958*; also *Hall, 1959*).

pressure distribution $\partial p/\partial x = P = 0$ and $w/u < \xi$ (u and w are the axial and radial component of velocity, respectively), the pressure must decrease along the spiralling path of a fluid element in the vortex. Thus the fluid element is continuously accelerated. The greater the magnitude of $\partial p/\partial \xi$ or of v^2 and the greater the difference $(\xi - w/u)$, the larger the acceleration.

A positive pressure gradient, for which P is positive, reduces u and w and increases v and the magnitude of the pressure difference $(p - p_0)$, which has a negative value. A negative pressure gradient has the reverse effect where p_0 is the static pressure at the outside edge of the vortex.

The structure of flow within the vortex is markedly sensitive to changes at the edge in the geometry of the spiralling streamlines and in the pressure distribution.

4.2.4 Effects of Separation on Drag and Moment

The effects of separation on drag and moment are now described in order to supplement the relations presented in Secs. 4.2.1–4.2.3 for maximum lift and angle of attack at the separation of flow on airfoils and wings.

The drag of conventional airfoils depends mainly on friction; this is determined by the transition position, i.e., the length of laminar and turbulent flow. With an increase in angle of attack, a slight increase of drag follows. However, drag rises rapidly in the region of $C_{L\,max}$ due to the stall. Since both pressure drag and friction drag decrease with increasing Reynolds number, the airfoil drag is affected by Reynolds number. As shown in Fig. 4.26, the characteristics of a laminar airfoil are different from those of conventional airfoils.

In the limited region of small lift coefficients, the drag of a laminar airfoil is constant and independent of angle of attack provided the Reynolds number is sufficiently large to prevent laminar separation. Of course, the drag of a laminar airfoil is much less than that of an ordinary airfoil in this region. The moment coefficient C_m is essentially constant and independent of angle of attack in a larger region of up to 6 deg for this laminar airfoil.

Although no separation occurs at subsonic and supersonic speeds with this same airfoil, a dangerous separation may occur at a transonic region where the flow is mixed, i.e., flow of subsonic and supersonic speeds; see Fig. 4.27. This figure indicates incompressible flow (Fig. 4.27*a*), compressible flow which does not reach the sonic limit (Fig. 4.27*b*), and the formation of compression shock when the flow exceeds the sonic limit (Fig. 4.27*c* and 4.27*d*), and the typical pressure distribution and flow field at supersonic speed (Fig. 4.27*f* and 4.27*g*). The behavior of flow in the immediate neighborhood of the velocity of sound is not yet fully understood. It is known from observations that the process from subsonic to supersonic flow is continuous but the reverse process causes a compression shock, as may be observed on the Laval nozzle. Since

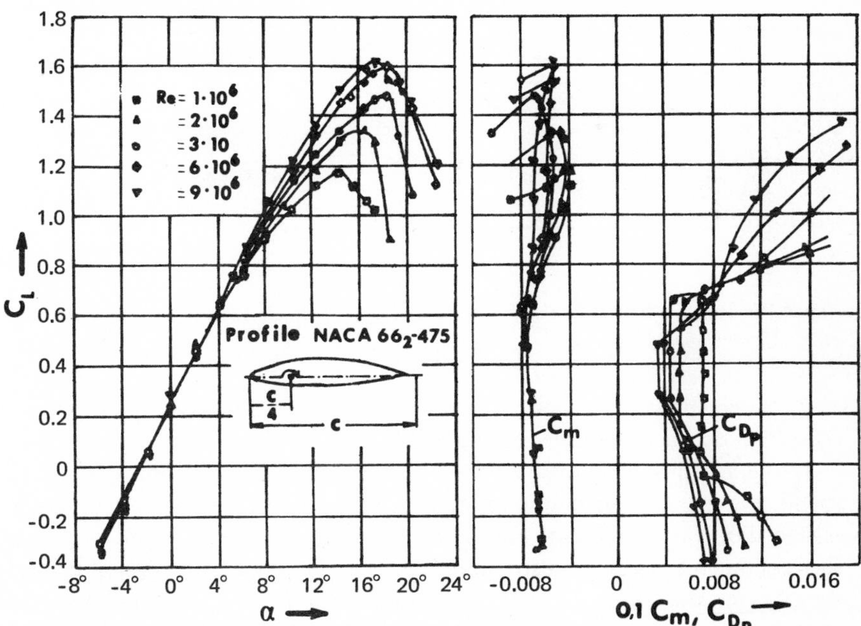

FIG. 4.26 Three component measurements for NACA laminar airfoil at various Reynolds numbers (*from Loftin & Smith, 1949*).

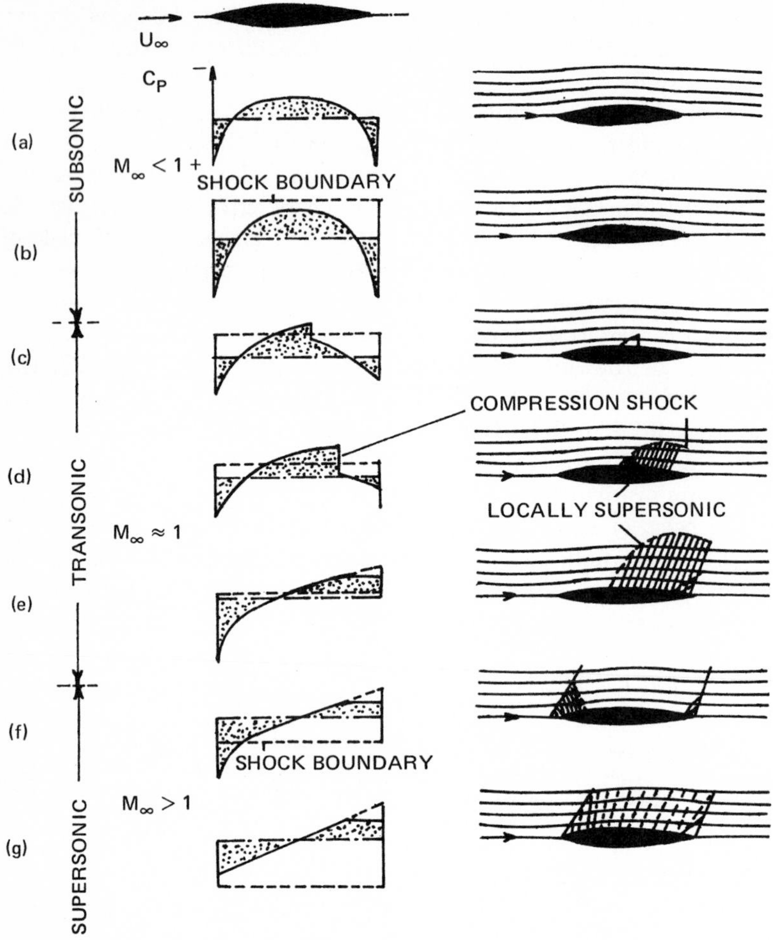

FIG. 4.27 Pressure distribution and flow field of a biconvex airfoil in the transonic region (*from Quick, 1951*).

such a compression shock causes a sudden pressure rise downstream, separation occurs as indicated in Fig. 4.28.

If the maximum velocity on the entire airfoil is less than the velocity of sound, the pressure distribution is continuous at the free-stream Mach number regime. For the test case of Fig. 4.28*a*, until M_∞ reaches approximately 0.6, the continuous pressure distribution prevails. But, at $M_\infty > 0.7$, the local velocity on the airfoil becomes larger than the velocity of sound and the pressure downstream of the pressure minimum rises abruptly. Such pressure jumps (compression shocks) become greater with increasing Mach number. Such a compression shock is dangerous because of the subsequent flow separation. The relation of C_L versus M_∞ is shown in Fig. 4.29 and the

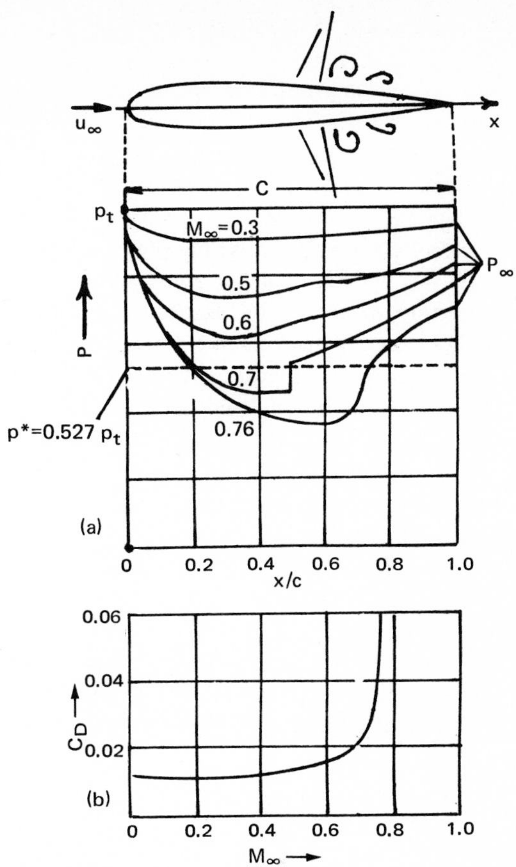

FIG. 4.28 Measurements on a NACA 4412 airfoil at subsonic speeds and $\alpha = 0$ (*from Stack et al., 1938*).

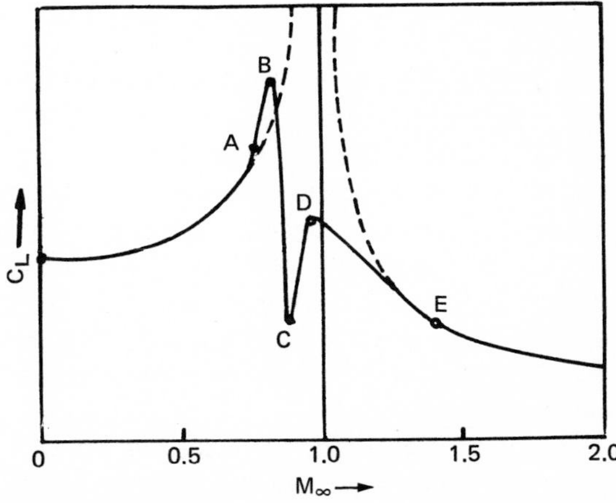

FIG. 4.29 Lift coefficient of a wing as a function of Mach number (*from Holder, 1964*).

position of compression shock as well as velocity distribution is shown in Fig. 4.30. In these figures, symbols A-E refer to the following:

A–At $M_\infty = 0.75$: No compression shock occurs because the velocity is less than the sonic velocity on most of both surfaces of the airfoil; where the velocity is greatest, it is only slightly above sonic.

B–At $M_\infty = 0.81$: On the upper surface, the local velocity on the upstream portion of the airfoil exceeds the sonic velocity and causes compression shock at 70 percent of the airfoil chord. On the lower surface, the subsonic speed prevails everywhere. The lift increases with Mach number until this condition (point B) is reached.

C–At $M_\infty = 0.89$: On the lower surface, the local velocity exceeds the velocity of sound over a large extent of the airfoil, thus causing a compression shock at the downstream position on the lower surface. The velocity distribution varies strongly and lift is greatly reduced.

D–At $M_\infty = 0.98$: The compression shock on both the upper and lower surfaces is weaker than at $M_\infty = 0.89$ and this shock is formed near the trailing edge of the airfoil. The lift is larger than for condition C.

E–At $M_\infty = 1.4$: Pure supersonic flow is reached with the typical velocity distribution. The magnitude of lift may be predicted by the linearized theory of Ackeret (1925).

4.2.5 Factors that Affect Control

Many factors can affect techniques for control of separation. On aircraft, for example, these include various operational problems such as positions and configurations of flaps and tailplanes, propeller slipstream, protuberance, and interaction phenomena resulting from different parts of the aircraft. Although it is beyond the scope of this section to investigate all these complex problems, a few pertinent factors are presented here, e.g., Reynolds number and nose shape.

Reynolds number has already been mentioned (Secs. 4.2.1-4.2.3) as one of those factors that affect separation or stall. The phenomenon of stall on an airfoil is closely related to the separation bubble that forms between the laminar separation point and the turbulent reattachment point. With an increase in Reynolds number, the reattachment point moves upstream until it reaches the laminar separation point, i.e., the length of the separation bubble becomes zero. The maximum lift rises rapidly with increasing Reynolds number and is affected by two factors that are superimposed. An increase in Reynolds number affects lift favorably at a fixed angle of attack because the separated flow region is reduced. The angle of attack of the wing can be increased with increasing Reynolds number until separation finally occurs. At a very large Reynolds number, transition takes place before laminar separation, and the transition point moves upstream with increasing Reynolds number, thus increasing the length and thickness of the turbulent boundary

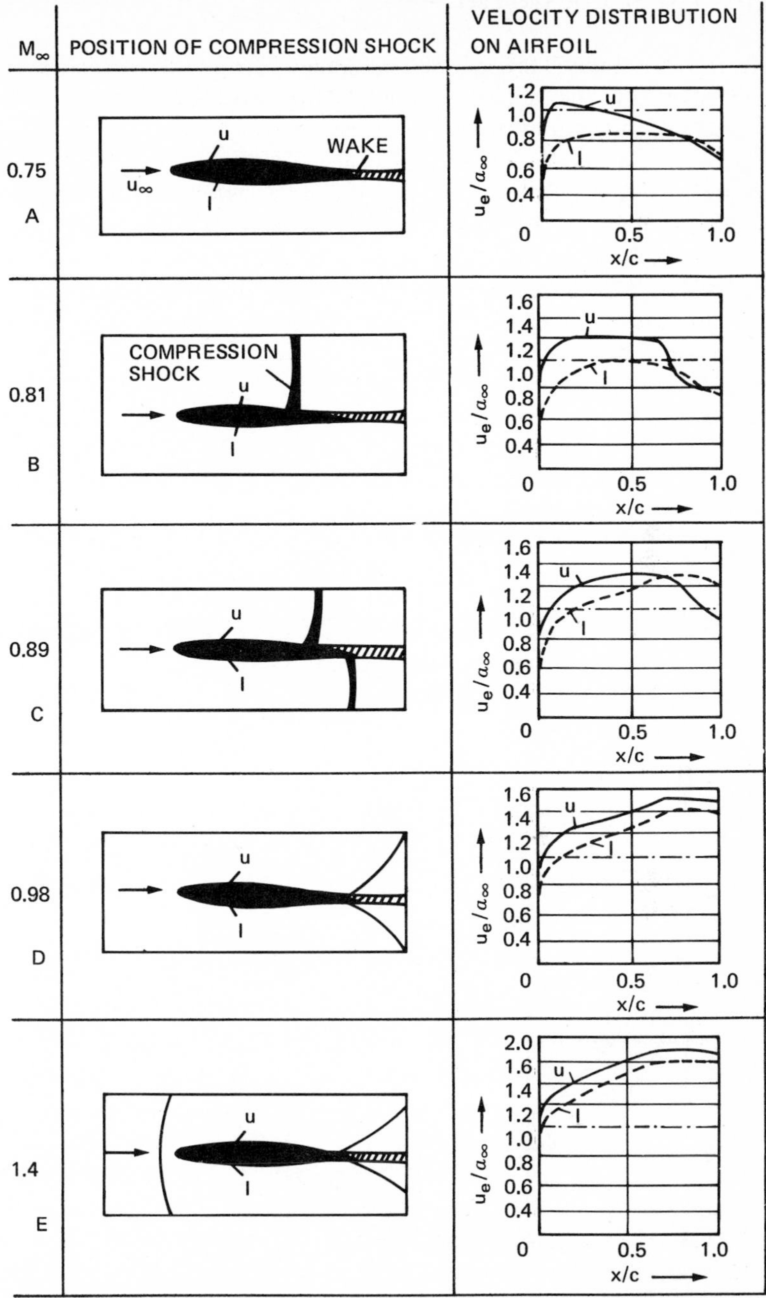

FIG. 4.30 Transonic flow over an airfoil at various Mach numbers for a 2-degree angle of attack (points A–E correspond to the lift coefficient of Fig. 4.29, a refers to velocity of sound; *from Holder, 1964*).

layer. As a result, the circulation around the profile decreases, i.e., the maximum lift can again decrease at very large Reynolds numbers. The example of variation of pressure distribution with Reynolds numbers shown in Fig. 4.31 demonstrates that when the Reynolds number varies, various separation characteristics occur for the same airfoil at the same angle of attack. At $Re = 1 \times 10^5$ and 4.5×10^5, for a thick airfoil (NACA 4412) and a thin airfoil (NACA 64A006) the pressure distribution is similar as shown in Fig. 4.32. This indicates that a low angles of attack, a similar type of separation occurs on a thin airfoil as on a thick airfoil, although the separation for the thicker airfoil occurs at a larger (although still low) angle of attack.

If the Reynolds number is increased to 1.8×10^6, then a short laminar separation bubble of length $0.005c$ is formed on the nose of the NACA 4412 airfoil, and turbulent flow separation occurs at $x/c = 0.40$. At a still larger

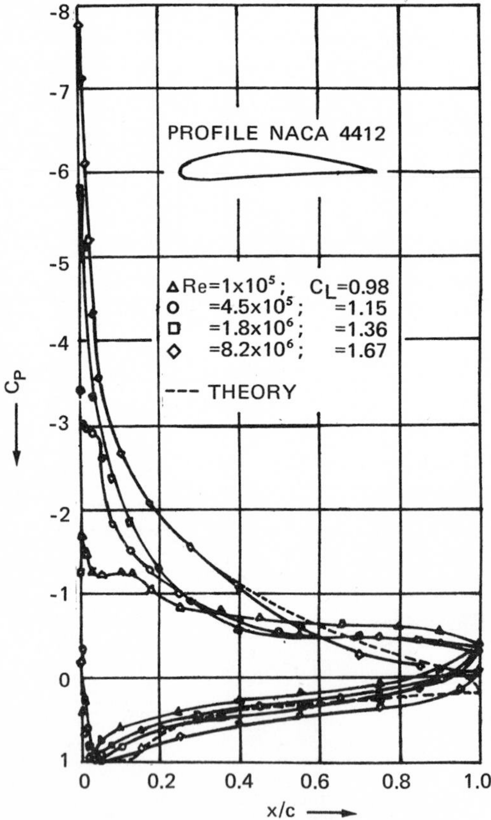

FIG. 4.31 Effect of Reynolds number on an airfoil pressure distribution at a large angle of attack (NACA 4412 airfoil, $\alpha = 16$ deg.; *from Pinkerton, 1936*).

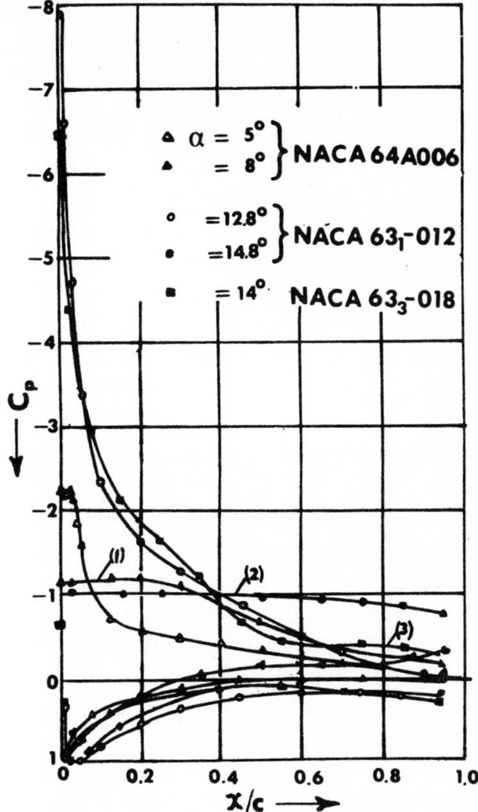

FIG. 4.32 Measured pressure distribution at Reynolds number 5.8 × 10⁶ for NACA-6-series airfoils with various separation characteristics in the region of maximum lift (*from McCullough & Gault, 1951*): (1) separation on a thin airfoil, (2) laminar separation at nose of airfoil, (3) turbulent separation.

Reynolds number, $Re = 8.2 \times 10^6$, the flow over the whole airfoil is attached. A further increase in Reynolds number has no practical effect on the pressure distribution as may be predicted by the modified potential theory of Pinkerton (1936); see Fig. 4.31.

Figure 4.32 shows the measured pressure distribution of three representative airfoils of NACA-6 series with various separation characteristics in the region of maximum lift at a Reynolds number of 5.8×10^6. The separated region is characterized by a region x/c within which $C_p = 0$. It is clearly seen that with thicker airfoils such as NACA 63_1-012 and 63_3-018, C_p drops more sharply in the nose region than with the thinner NACA 64A006 airfoil. A drop in C_p is noticeable for NACA 64A006 in the nose region at an angle of attack of $\alpha = 5$ deg, but separation occurs at the leading edge at $\alpha = 8$ deg and then spreads downstream. For the thicker airfoil, NACA 63_1-012, a very steep drop in C_p occurs in the leading edge region at $\alpha = 12.8$ deg; at $\alpha = 14.8$ deg, laminar separation occurs at the nose and suddenly breaks down, spreading the separated region over the whole suction side. The rapid reduction of lift follows at an angle of attack larger than that corresponding to $C_{L\,max}$. For the thickest airfoil, NACA 63_3-018, turbulent separation occurs as the trailing edge is approached. There is a sharp peak in C_p on the suction side for an angle of attack larger than that corresponding to $C_{L\,max}$. However, as angle of attack is further increased, the separation region spreads upstream toward the nose, reducing the lift continuously.

Figure 4.33 illustrates the effect of the boundary layer on the pressure distribution as the angle of attack is increased. The full curve indicates pressure distribution predicted by potential theory. With increased angle of attack, the separation occurs on the suction side near the trailing edge (Fig. 4.33*b*) and if the angle of attack is increased further, the separation point moves upstream. A closed vortex is formed in the separated region. At an angle of attack larger than that corresponding to $C_{L\,max}$, the separation point moves to the nose and forms a long separation bubble; the flow reattaches downstream as indicated in Fig. 4.33*c*.

The most important geometrical parameter for separation at large angles of attack (and relating to the maximum lift) is the shape of the nose because its shape determines the pressure distribution in the neighborhood of the nose. Figure 4.34 gives $C_{L\,max}$ (at a definite Reynolds number 6×10^6) as a

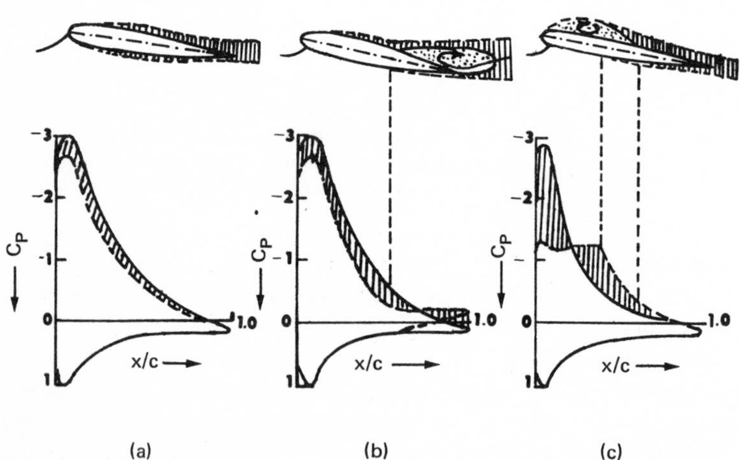

FIG. 4.33 Variation of pressure distribution on airfoil with increase in angle of attack: (*a*) unseparated flow; (*b*) rear separation; (*c*) leading-edge separation and long bubble (qualitative distributions: first inviscid approximation and experiment; *from Thwaites, 1960*).

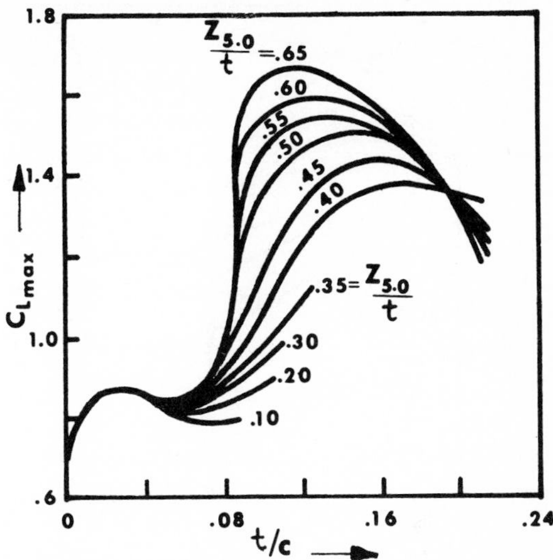

FIG. 4.34 Maximum lift at Re = 6 × 10^6 as a function of thickness ratio t/c and nose radius $z_{5.0}/t$ ($z_{5.0}$ is z evaluated at $x/c = 0.05$; *from Nonweiler, 1955, 1956*).

function of thickness ratio of t/c where t is the thickness of the airfoil and c is its chord length. Already as mentioned in Sec. 4.2.1, Chappell (1968) uses the ratio of the profile coordinate at $x = 0.0125c$ and the chord ($z_{1.25}/c$) to characterize the nose shape whereas Thwaites (1960) uses the ratio of the profile coordinate at $x = 0.05c$ and the thickness ($z_{5.0}/t$).

It is seen from Fig. 4.34 that $C_{L\,max}$ is not affected by the nose radius of a very thin airfoil but that for airfoils of moderate thickness, it increases with an increase in $z_{5.0}/t$. Gault (1957) was the first to use the parameter of profile coordinate ($z_{1.25}$ of the suction side of the profile at $x/c = 0.0125$) in an attempt to eliminate the dependence of Reynolds number in a universal diagram region for various separation phenomena. Figure 4.35 includes measurements on about 150 smooth airfoils in a low turbulence level wind tunnel.

A sharp leading edge or a very small nose radius airfoil (for $z_{1.25}/c < 0.009$ at all Reynolds numbers) exhibit the special characteristic of thin-airfoil stall. At small angles of attack, separation occurs at the nose followed by reattachment. The boundary layer at reattachment is neither typically laminar nor turbulent, but the turbulent characteristics prevail as the trailing edge is approached. With an increase in angle of attack, the reattachment point moves downstream so that the separated region becomes larger and the lift decreases correspondingly. When the separated flow region is extended over the whole suction surface, then the value of C_L decreases with an increase in angle of attack.

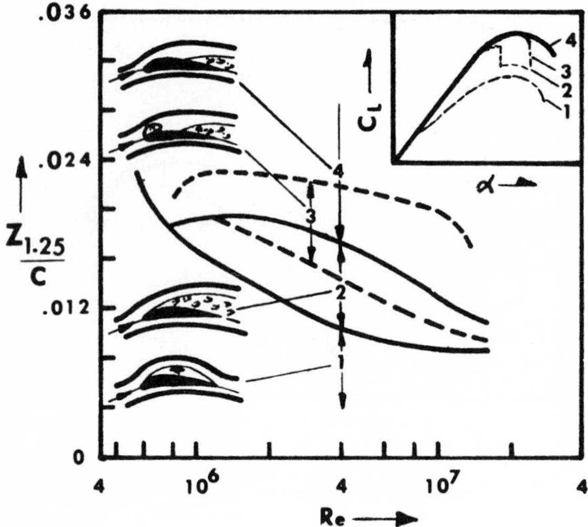

FIG. 4.35 Flow separation on a profile as a function of Reynolds number and nose radius (separation on a thin airfoil (1), laminar separation on airfoil nose (2), combination of laminar and turbulent separation (3), and turbulent separation (4); *from Gault, 1957*).

A fundamentally different separation characteristic is found for a wing of moderate thickness, average nose radius, and relatively strong leading edge curvature. For this shape, the rapid pressure rise downstream of the profile nose causes laminar separation at a large angle of attack. Transition takes place in the separated flow and is followed by turbulent reattachment downstream. The size of the resultant laminar separation bubble decreases with increasing angle of attack because the transition as well as the turbulent reattachment point approach the separation point, and both move toward the leading edge. Finally, the laminar boundary layer in the immediate neighborhood of the nose separates where the curvature of the surface is so large that transition cannot affect reattachment. As previously mentioned, this phenomenon is known as laminar leading-edge stall. It is characterized by a sudden, rapid reduction of lift. For most airfoils with thick profiles ($t/c > 0.15$) and large nose radii, the flow downstream of laminar separation reattaches at large angles of attack. In this case, the maximum lift is affected by two mutually influencing phenomena, the so-called combined leading-edge and trailing-edge stall. These are the spread of the laminar separation bubble at the nose and the occurrence of turbulent separation at the trailing edge which moves upstream with increasing angle of attack. The maximum lift depends on which of these two phenomena is the dominant one. The separation bubble may disappear if the profile is thick and strongly cambered and if the Reynolds number is very large (natural transition takes place at large Reynolds numbers). The turbulent boundary layer separates at the trailing edge, causing the so-called trailing-edge stall. For this case, the separation point moves upstream continuously with increasing angle of attack and the lift drops gradually rather than suddenly. The design of the nose shape is discussed in Sec. 4.3.

4.3 DESIGN CONSIDERATION IN PREVENTING FLOW SEPARATION ON AIRFOILS AND WINGS

Some pertinent and practical aspects will now be presented for the design of airfoils, wings, and airfoil flap combinations which may consist of a part of the wing. A large portion of this section refers to design consideration of two-dimensional airfoils exposed to incompressible flow.

4.3.1 Airfoils for Incompressible Flow

The pressure distribution on an airfoil and the behavior of the boundary layer at various angles of attack are the governing factors to be considered in preventing stall without the use of auxiliary power. The particular features of the droop nose (discussed previously) are studied here.

Wortmann (1961) utilized three NACA airfoils (0012, 64-012, and 66-012) to investigate the method of design for low drag airfoils. (His study was not primarily concerned with separation.) All had the same ratio of maximum thickness to chord (12 percent). These NACA airfoils had been designed to give a velocity distribution which was selected with primary emphasis on desired boundary layer characteristics. The boundary layer could be kept laminar within a region of slight favorable pressure gradient; thus the location of maximum velocity is an important item to be considered. If the peak suction occurs far aft, then a decrease in frictional drag may be expected.

Airfoil 0012 is an older type of wing section whereas the two others are NACA 6 series which have their peak suction further aft. The extensive favorable pressure gradient of the low-drag airfoils stabilizes the laminar boundary layer and delays transition until the region of highest velocity is reached. However, a design which locates the peak suction further aft is no panacea because the adverse pressure gradient becomes steeper beyond the peak and the load on the boundary layer (which is now turbulent within this region) increases correspondingly. Finally as the peak suction is moved aft, the pressure drag increases at a higher rate than the frictional drag decreases. Therefore, for a given thickness and a given Reynolds number, there is an optimum aft limit of the chordwise position of the peak suction with respect to drag.

The dotted line in Fig. 4.36 indicates the velocity distribution of the three airfoils at a positive angle of attack of 1.3 deg. Note that peak suction occurred at the leading edge and that the subsequent fall in velocity

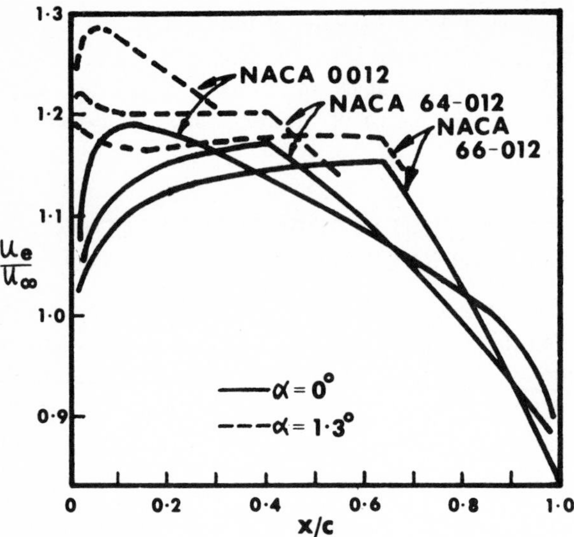

FIG. 4.36 Velocity distributions for three 12-percent-thick NACA airfoils at two angles of attack (*from Wortmann, 1961*).

destabilized the laminar boundary layer. Thus the transition point jumped far forward and caused a sudden increase in drag. This occurrence of this process at negative angles of incidence causes the characteristic indentation in the curve of profile drag of the section called the "low-drag range" or the "low-drag bucket." Older airfoils (e.g., the 0012) generally cause no irregularities in such a process because the transition position changes gradually. Therefore, it may be stated that the width of the low-drag bucket is closely related to the minimum drag. Figure 4.36 indicates that the 66-012 airfoil had both the smallest drag (because peak suction occurs further aft) and the narrowest drag bucket.

If the peak suction is located far aft, then the favorable pressure gradient becomes flatter at increased angle of attack, and rapidly becomes unfavorable compared to the initially steeper pressure gradient. On the other hand, if a certain aft position of the peak suction is maintained, then the low-drag bucket widens with increasing airfoil thickness, but the minimum drag also increases.

Almost irrespective of airfoil type, a given width of the low-drag bucket for these NACA low-drag airfoils was always associated with a given drag within the range. No advantage may be claimed until this interdependence is altered, i.e., until a smaller drag is obtained with the same range or a wide low-drag range is achieved with the same drag.

With an increase in Reynolds number, the laminar boundary layer becomes less stable; thus the width of the low-drag bucket is reduced. The low-drag bucket disappears completely at Reynolds number (based on chord length and free-stream velocity) larger than 10^7.

The minimum drag generally decreases with increasing Reynolds number as a result of two opposing effects. An increase in Reynolds number reduces the friction drag, and because the boundary layer thins, the already small pressure drag becomes still smaller. On the other hand, the transition point moves upstream due to the decreasing stability of the laminar boundary layer and thus increases minimum drag. Both effects are approximately equal for Re of the order 20 to 30 million.

The suitability of an airfoil for high subsonic speeds is determined by its critical Mach number. Of the three airfoils with equal thickness ratio shown in Fig. 4.36, the 66-012 will have the largest critical Mach number, not only at a single angle of attack but also within a whole range of angles of attack. The velocity distribution suitable for achieving a substantial region with laminar flow is also helpful in achieving a high critical Mach number. When the angle of attack is increased to the point when the velocity peak is at the leading edge, then the critical Mach number is rapidly reduced. If the critical Mach number of a low-drag airfoil is plotted against lift, then a characteristic indentation similar to the low-drag bucket is obtained.

Walz (1944) studied the effects of contour changes for concave and convex tails of symmetrical airfoils sections. He found that at $Re = 3 \times 10^7$, the turbulent boundary layer was thinner on an airfoil with a concave tail and cusp than on an airfoil with a convex tail end. Neither shape parameter indicates that was favorable for preventing separation; in fact there was the danger of separation at the trailing edge of the airfoil with the convex tail end.

Wortmann (1955) had also studied the behavior of the turbulent boundary layer in an adverse pressure gradient region. He paid particular attention to the velocity distribution for which the relevant turbulent boundary layer showed the least tendency to separate, even at small Reynolds numbers. This is equivalent to stipulating that the shape parameter must remain constant throughout the flow. Hence, contrary to conventional usage, his starting point for airfoil design was not velocity distribution but rather specifically stipulated characteristics for the boundary layer. The corresponding velocity distribution over the airfoil were obtained through trial and error by using the approximate method of Truckenbrodt (1952). These velocity distributions not only prevented separation, but the boundary layer for which they were developed was more favorable than that for the velocity distributions of the NACA sections.

Two results of these calculations are shown in Fig. 4.37. With velocity distribution 1 (which corresponds to a NACA airfoil), the shape factor H and the momentum thickness θ rapidly increased near the trailing edge. On the other hand, with velocity distribution 2, the magnitude of H and θ remained practically constant and the boundary layer thickness at the trailing edge was smaller than with the NACA section.

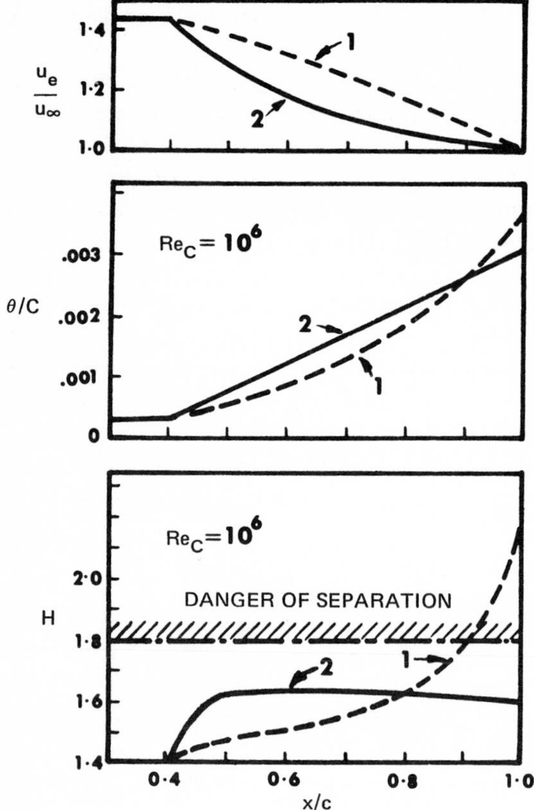

FIG. 4.37 Development of the turbulent boundary layer for two different velocity profiles ($Re_c = 10^6$, θ = momentum thickness, H = shape parameter; *from Wortmann, 1961*).

Wortmann (1955) also considered the design procedure for an airfoil by inverting the calculation of the boundary layer. He started from a given shape parameter and calculated the shape of concave velocity distribution as a function of Reynolds number and chordwise position of transition. In determining an optimum velocity distribution, it should be possible to calculate the development of $\theta(x)$ for different shape parameters and Reynolds numbers. However, Wortmann obtained analytically inconclusive results because the approximate methods he used for computing the turbulent boundary layer were too inaccurate in the special case of the concave velocity distribution. Therefore, he later suggested the use of experimental results before attempting to design by analytical calculation.

The positive velocity gradient on the NACA low-drag sections changed abruptly into a more or less steep negative velocity gradient and at this point, the formation of separation bubbles was observed, even at high Reynolds numbers on the order of 10^7. Laminar separation occurred with subsequent reattachment on the upper surface of the airfoil after the boundary layer became turbulent. The formation of a separation bubble thickens the initial boundary layer. Wortmann (1957) observed that when this separation bubble was removed by forced transition (coinciding with the point of separation) an increase in drag occurred on the NACA 64-418. With a concave velocity distribution which initially had an extremely steep gradient, he obtained a distinct reduction in drag by moving the point of transition upstream and preventing the occurrence of separation bubbles.

The best solution would be the occurrence of transition prior to separation and an unchanged transition point. The steeper the decrease of velocity in the transition region, the more urgent is the need for such a control of transition. Such a steep decrease in velocity is found on the upper surface at large angle of attack, i.e., in the region of maximum lift.

Figure 4.38 illustrates the unpleasant effects of a separation bubble that extends into a region of a steep negative velocity gradient. This figure shows the lift-drag curve of a 20-percent thick airfoil for which the positive velocity gradient changed directly into the steep negative gradient associated with concave velocity distributions. At the test conditions of $Re = 1.5 \times 10^6$, a separation bubble formed when the trip wires were smaller than the critical roughness height. Although the size of this bubble was small, it caused a very

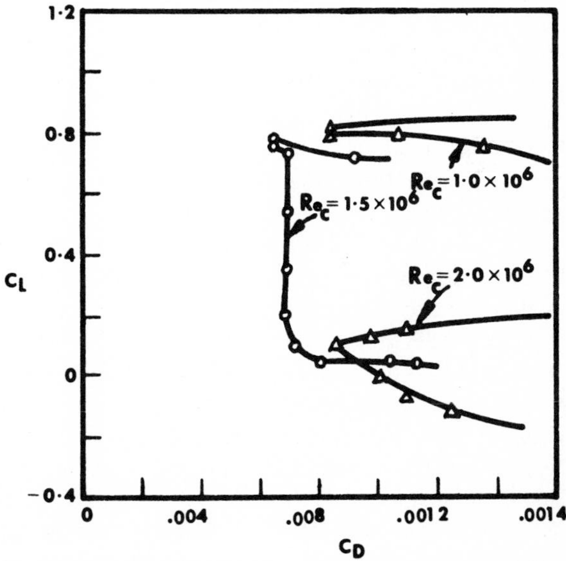

FIG. 4.38 Lift-drag curve of the 20-percent-thick aerofoil FX3 for two Reynolds numbers (at $Re_c = 1.5 \times 10^6$, trip wires prevent the formation of a separation bubble; at $Re_c = 1.0 \times 10^6$, the same wires are no longer effective and a separation bubble about $0.03c$–$0.05c$ wide develops at $x = 0.7c$; *from Wortmann, 1961*).

large increase in drag. The turbulent boundary layer aft of the bubble remained completely attached to the trailing edge.

Wortmann (1957) found that it is possible to control transition by inserting a transition element (a so-called "instability range") between the positive and the negative velocity gradients rather than connecting them directly. Laminar separation may be prevented by a flat negative velocity gradient in this instability range. However, a high degree of laminar boundary layer instability may result. The velocity distribution is given by $u_e = Cx^m$. This is known for laminar flow over a wedge near the forward stagnation and results in similar solutions of the boundary layer equations. The symbols C and m are constants. No laminar boundary layer separation occurs for $m > -0.091$.

Recall that if the angle of attack is sufficiently large, the laminar boundary layer will separate from the upper surface of the leading edge of the airfoil in the form of a local separation bubble. But a turbulent boundary layer will separate near the trailing edge and form a wake downstream. With a further increase in angle of attack, the separation point of the turbulent boundary layer will move toward the leading edge, limiting the maximum lift. But after a critical angle has been exceeded, the laminar separation bubble will suddenly transform into a long separated region that extends over the whole airfoil chord, causing a sudden loss of lift. The laminar separation bubble remains short and innocuous if $\mathrm{Re}_\delta^* > 400$–$500$. Since each bubble that occurs within a steep negative velocity gradient exerts a very unfavorable influence on the development of the turbulent boundary layer, it is necessary to design the airfoil so that the laminar separation bubble does not occur at the nose. In other words, care should be taken with the velocity distribution over the upper surface so that an instability range will form near the leading edge at high angles of attack. The points of instability and separation of the laminar boundary layer should be sufficiently apart so that transition occurs near the separation point.

The separation points are illustrated in Figs. 4.39 and 4.40. The solid lines in these figures refer to the NACA 63-015 section and the dotted lines to a modification of this airfoil within $0.002 \leqslant x/c \leqslant 0.060$.

Figure 4.39 shows the velocity distribution on the upper surface for 0- and 20-deg angles of attack. The dotted velocity distribution also shown for $\alpha = 20$ deg is determined by $u_e = Cx^m$ for a constant shape parameter in the vicinity of the separation boundary. Figure 4.40 gives results for the boundary layer calculation. Note that for the NACA 63-015 airfoil, separation of the laminar boundary layer had already occurred at $x/c = 0.016$. For this case, $\mathrm{Re}_\delta^*/\sqrt{\mathrm{Re}_c} = 0.8$ with $(\mathrm{Re}_\delta^*)_{\mathrm{crit}} = 400$, the ensuing separation bubble would be innocuous for Reynolds numbers $\mathrm{Re}_c = u_\infty c/\nu \geqslant 0.25 \times 10^6$ and would be followed by an attached turbulent boundary layer. With the modified velocity distribution, separation moved downstream of $x = 0.06c$. Then $\mathrm{Re}_\delta^* = u_e \delta^*/\nu$ was nearly doubled and the critical Reynolds number was lowered to $\mathrm{Re}_c \approx 0.06 \times 10^6$.

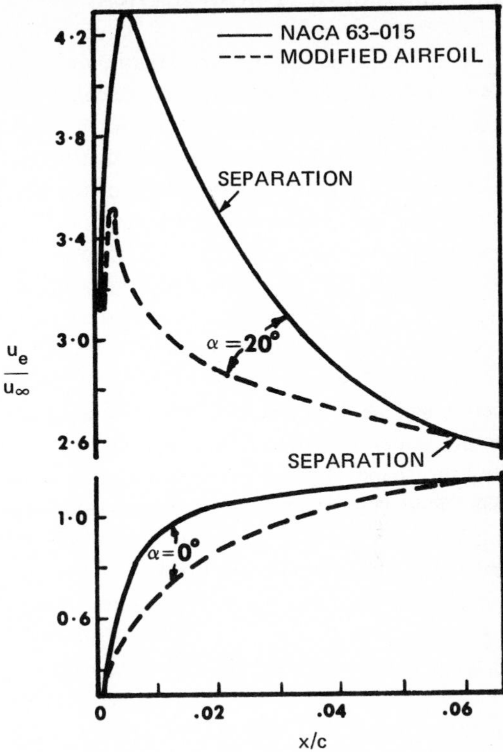

FIG. 4.39 Velocity distribution near the leading edge of an NACA 63-015 section and for a modified section (the lower curves are for $\alpha = 0$ deg; *from Wortmann, 1961*).

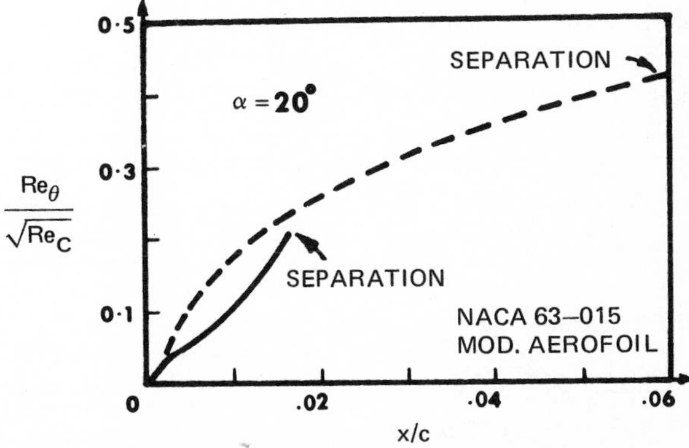

FIG. 4.40 Development of the boundary layer thickness with laminar flow near the leading edge of an NACA 63-015 section and for a modified version (angle of incidence $\alpha = 20$ deg; in each case, calculation is taken as far as the separation point; *from Wortmann, 1961*).

Since the velocities increase rapidly with an increase in angle of attack, it is essential that the nose shape be calculated accurately. The flow control is based on the velocity distribution, for example, that given in Fig. 4.39. It is immaterial whether such a velocity distribution is obtained by choice of the thickness distribution, by the shape of camber line, or a combination of both. Thus even with thinner airfoils, it is possible to obtain velocity distributions similar to that shown in Fig. 4.39. However, with decreasing thickness, it will be more difficult to prevent the unfavorable velocity distribution at small angles of attack on the lower side of the airfoil.

If the flow separates upstream of the trailing edge, then it will be of interest to investigate how the velocity distribution obtained by potential flow theory will be affected. Separation at the trailing edge will cause all velocities on the upper surface of the wing nose to be lowered at about the same rate; thus the shape of the instability range remains practically unchanged because that range is relatively insensitive to small changes of flow conditions. One might therefore conclude that the insertion of an instability range is a practical method for flow control.

As mentioned previously, the nose shape is a very influential factor in leading-edge separation. The example of so-called "droop nose" described in Chap. 3 is one solution to prevent separation at subsonic flow speed. Another example is sketched in Fig. 4.41.

The criterion for laminar leading-edge separation on the droop nose is a velocity ratio $u_{es}/u_{e\max}$ of 0.94 to 0.95; here u_{es} is the potential velocity at the separation point and $u_{e\max}$ is maximum potential velocity. This criterion was given by Owen and Klanfer (1953) following extensive investigations of leading-edge separation on NACA 63-009, 63-012, 64A 006, 65-3-018, 66-2-516, 66 3-018 airfoils and wedges (wedge angles from 5 to 12 deg) for a range of Re_δ^* from 400 to 4883 and angles of attack from 0 to 10 deg. Figure 4.42 indicates the relationship between velocity at separation and maximum velocity.

A measurement by Gyorgyfalvy (1959) indicates that the droop nose sketched in Fig. 4.41 yields $u_{es}/u_{e\max} = 0.95$ at an air speed of 36–54 mph

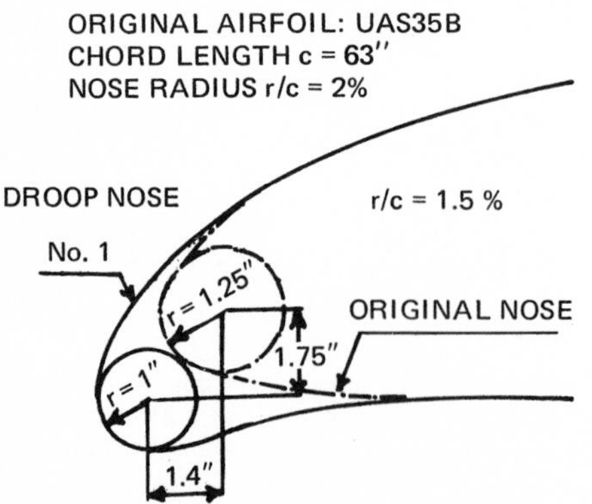

FIG. 4.41 Model of droop nose (*from Gyorgyfalvy, 1959*).

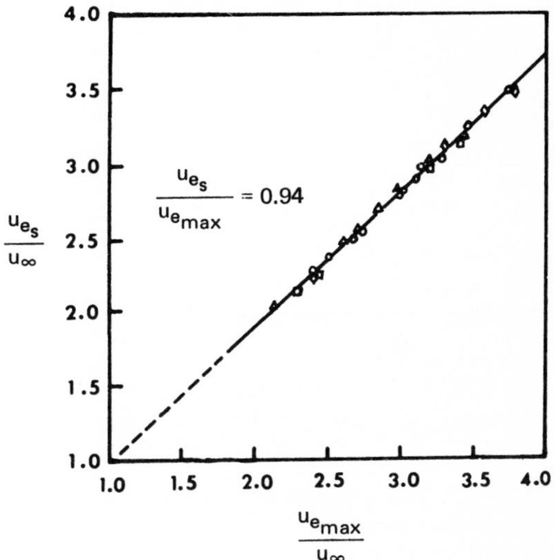

FIG. 4.42 Relation between velocity at separation and maximum velocity (*from Owen & Klanfer, 1953*).

and without flap deflection. The critical ratio of $u_{e_s}/u_{e_{max}}$ is independent not only of flap deflection but also of angle of attack and airfoil shape. However, this simple criterion of velocity ratio may not be valid if the centrifugal effect becomes important on the curved surface that provides high lift. Gault (1955) utilized a critical value of 0.94 for this ratio (instead of 0.95) for two-dimensional airfoils. He indicated that this ratio can also be used qualitatively for thin sweptback wings. Gyorgyfalvy (1959) confirmed this 0.94 separation criterion by measurements on the original and on a modified form (the "droop nose") airfoil.

Figure 4.43 shows the somewhat different configurations of leading-edge shape designed by Cornish (1965). His design objective was to obtain the maximum lift coefficient by a combination of airfoil shape and blowing. The first attempts to avoid laminar leading-edge separation on the Piper L 21 resulted in greatly increased camber at the leading edge as indicated in Fig. 4.43. This modification was only moderately successful in that it achieved only a slight increase in maximum lift coefficient. A more effective technique is shown in the lower portion of Fig. 4.43. It proved adequate for lift coefficients above 5.0 and applied to the Cessna L-19. In this case, the radius of the leading-edge circle was increased and the center of the circle was

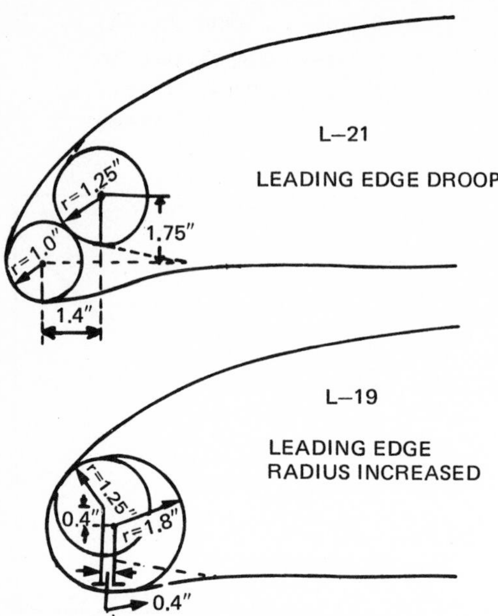

FIG. 4.43 Leading-edge modifications for a wing used with blowing (*from Cornish, 1965*).

dropped until the circle was again tangent to the upper surface. It is interesting to note that at a lift coefficient of 5.0, the stagnation point on Cessna L-19 was at about 20–25 percent aft of the leading edge on the under surface of the wing. This leading-edge modification has been also used for Piper L-21, Auster MA-4, and Fokker S-12. However, their minimum speeds (dictated by maximum lift coefficient) were dictated by reason other than laminar separation at the leading edge.

Flow separation on wings or airfoils can be controlled by flaps, by a leading edge device, or by their combined application. So far as the author knows, investigations of flap which may be related to separation are mostly for $C_{L\,max}$. The use of flaps enables the $C_{L\,max}$ to be increased, and this also increases the range of angles of attack at which flow separation can be avoided.

However, a large portion of the increment in $C_{L\,max}$ attainable with Fowler flaps comes from the increase in chord due to flap extension beyond the airfoil trailing edges rather than from an improvement in stalling performance (McCormick, 1967).

Cornish (1958) used tufts for flow visualization when he measured $C_{L\,max}$ and studied the unorthodox flow separation phenomenon on a flap. The flaps had been carefully faired to a modified Cessna L-19 plane; tufts on the their upper surfaces indicated whether flow was attached on the flaps. However, the measured value of $C_{L\,max}$ was lower than for attached flow. Subsequent

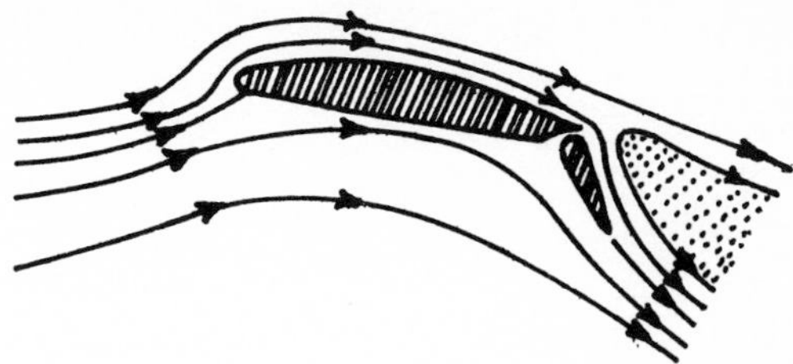

FIG. 4.44 Flow separation away from a flap (*from Cornish, 1958*; also *McCormick, 1967*).

observations showed that although the flow was attached at the surface of the flaps, it was separated a short distance away from the flaps and was not turned by them; see Fig. 4.44. Since the control of separation by flaps has not been investigated directly to any extent, this technique will not be discussed.

4.3.2 Flaps on Delta Wings at Supersonic Speeds

Experiments by Maki and James (1957) and by James and Maki (1957) have shown that the control of the leading-edge separation with slaps enables flap effectiveness and the maintenance of aircraft longitudinal stability to high angles of attack. It should be noted that the critical angle of attack to cause flow separation can be increased by deflecting the nose flap on a swept wing. The general nature of the stall is not substantially affected as indicated by the tuft visualization study of Eyre and Butler (1967).

Keyes and Ashby (1967) presented an important aspect of flow separation related to the hinge moment characteristics of a trailing-edge flap on a swept delta wing at supersonic flight. Their test was conducted on a 75 deg swept delta wing at $M_\infty = 6$, $Re = 4.03 \times 10^6$ (based on the root chord of the delta wing) for 30 and 90-deg angles of attack and flap deflections (δ) of 0 to 30 deg. Two types of leading edges (blunt and sharp) and two (0.2 and 0.3) flap aspect ratios A_{fl}/A_w were tested; A_{fl} is the flap planform area and A_w is the delta wing planform area. Figure 4.45 gives the configurations of the delta wing and the flap.

In the area of the wing-flap junction, the boundary layer was transitional up to 60-deg angle of attack and laminar at larger angles from 60 to 90 deg. Boundary-layer separation occurred on the wing and flap for a flap deflection angle of 30 deg up to an angle of attack of approximately 50 deg. A typical pressure distribution and schlieren photograph is shown in Fig. 4.46. Separation occurred at some deflection angle below 30 deg for the delta-wing flap combination at $\alpha = 30$ deg and $\delta = 30$ deg, but not at 20 deg. A slight separated region existed at $\alpha = 40$ deg and $\delta = 20$ deg.

In order to investigate the angle of attack related to flow separation, let us digress here and discuss some of the flight characteristics of a delta wing as it

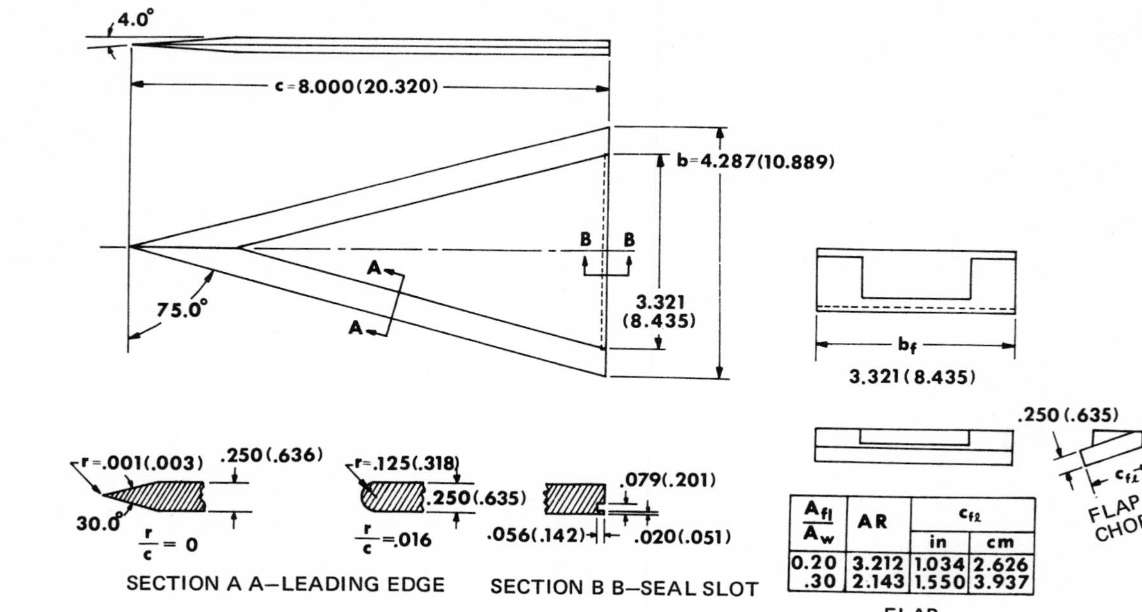

$\frac{A_{fl}}{A_w}$	AR	$c_{f\ell}$ in	$c_{f\ell}$ cm
0.20	3.212	1.034	2.626
.30	2.143	1.550	3.937

FIG. 4.45 Model configuration of delta wing and flap used in tests (all dimensions are in inches; equivalent values in centimeters are shown in parentheses; *from Keys & Ashby, 1967*).

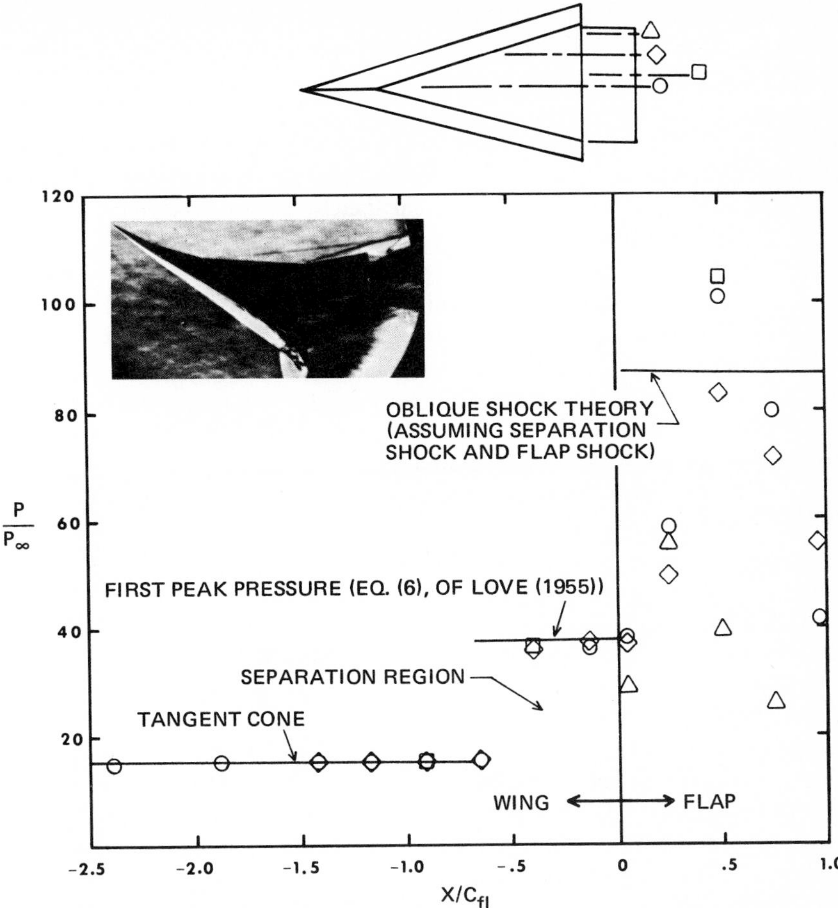

FIG. 4.46 Pressure distribution and region of separation on delta wing flap ($\alpha = 30$ deg, $\delta = 30$ deg, $A = 2.143$, $r/c \approx 0$, and $Re_\infty = 3.36 \times 10^6$; *from Keyes & Ashby, 1967*).

would be used for the application proposed by Keyes and Ashby (1967). A delta wing that has some advantage as a reentry vehicle over the complete reentry trajectory may reenter the atmosphere at a high angle of attack (80 to 90 deg) ballistic trajectory. Because the maximum lift coefficient for a delta wing occurs at an angle of attack near 50 deg, most of the trajectory could be flown in the moderate angle of attack range of 30 to 60 deg.

The hinge-moment coefficient of a trailing-edge flap on a delta wing is generally determined by the local flow condition. It is affected by the angle of attack, flap deflection angle, bow-shock/flap-shock interaction, and type of boundary-layer separation (laminar, transitional, or turbulent). Hence, the flow separation effect on the hinge moment of a flap must be studied along with the other factors just described.

At low angles of attack and the test conditions of Keyes and Ashby (1967) the bow shock/flap-shock intersection occurred downstream of the trailing edge of the flap. As the angle of attack is increased, this intersection moved closer to the flap and eventually moved in front of the flap. As shown in Fig. 4.47, an expansion or a compression wave was reflected from this intersection.

When this intersection occurs as shown, the reflected wave can influence the pressure distribution on the flap and thereby the hinge moment. As the angle of attack is increased, the reflected compression wave probably increases the pressure level at the trailing edge of the flap, shifting the center of pressure of the flap rearward. This shift in center of pressure would cause an increase in the hinge moment. As α is increased further, the influence of the reflected wave would probably move forward on the flap, but the details of the flow are difficult to determine because of the very complex flow field. The abrupt change in the hinge moment caused by the reflected wave at the bow-shock/flap-shock interaction amounted to approximately 40 percent for $\delta = 20$ deg and could cause dangerous problems for operational control.

The separation, along with the influence of the bow-shock/flap-shock interaction, tends to keep the hinge moment coefficient nearly constant at the lower range of angles of attack ($\alpha = 30$–37.5 deg for $\delta = 30$ deg). Separation also tends to delay the onset of subsonic flow on the flap since the angle of the separation "ramp" is less than that of the flap-deflection angle.

Figure 4.46 also shows the measured and calculated values of pressure on the wing and on the flap in the separated region. Calculated and measured maximum pressures on the flap were only in fair agreement. Tangent-cone theory was used to calculate the pressure level on the wing. The first peak pressure in the separated region was computed by using Eq. (6) of NACA TN 3601 (Love, 1955) for turbulent flow. The pressure level on the flap was evaluated by assuming that the flow passes through both the separation shock and the flap shock. The good agreement of the measured and computed first peak pressure in the separated region indicates that the boundary layer at the wing-flap junction was close to being fully turbulent.

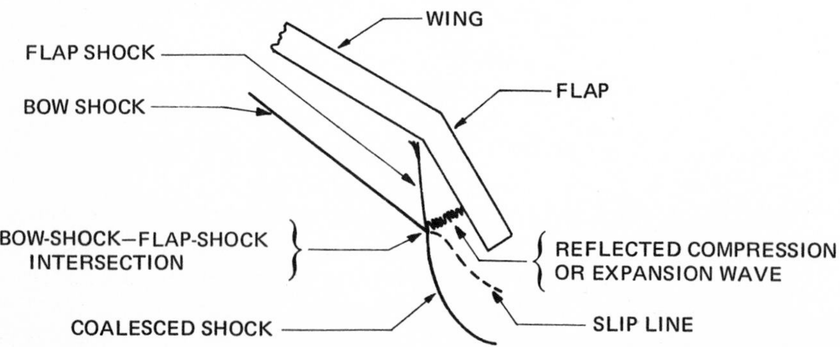

FIG. 4.47 Intersection of the bow-shock and flap-shock (*from Keyes & Ashby, 1967*).

The boundary-layer separation over the flap was not extensive at or below a deflection angle of 20 deg for any angle of attack. The boundary-layer separation at a deflection angle of 30 deg, in conjunction with the shock intersection effects, tended to reduce the hinge moment.

The experimental data of Keyes and Ashby (1967) were obtained at only one Reynolds number. However, known differences in the extent of separation and in the rise of peak pressure for laminar, transitional, and turbulent boundary layers, enable the conclusion that the type of boundary layer can strongly influence the hinge-moment characteristics at deflection angles large enough to cause boundary layer separation.

4.3.3 Boundary Layer Fences

Measurements on swept wings by Jones (1947) and Jacobs (1952) indicate that boundary layer thickness increases as the tip is approached. This can cause premature separation and subsequently the very abrupt roll that is much feared for on sweptback wing aircraft. Techniques for controlling the onset of separation at the outer wing and the undesirable roll include the use of a boundary layer fence. To be fully successful for unswept wings, a boundary layer fence must be able to isolate the incipient breakaway process and restrict it to a narrow space. Examples of shapes of boundary layer fences are shown in Fig. 4.48.

The dimensions of the boundary layer fence should be such that it is mounted upstream of the low pressure area and projects above the boundary layer. Thus its height should be higher than the probable thickness of the boundary layer; a medium height that is one-half the thickness of the airfoil has been found most effective.

Queijo et al. (1954) have shown that the static longitudinal stability can be improved by using a chordwise wing fence on a 35-deg sweptback wing, by lowering the horizontal tail, or by a combination of both. With the slats retarded and at moderate angles of attack, the static longitudinal stability can be improved by placing chordwise wing fences at a spanwise station of about 73 percent of the wing semispan from the plane of symmetry. The nose of the fence must extend slightly beyond or around the wing leading edge. With slats extended, the static longitudinal stability characteristics can be appreciably improved by placing a chordwise fence at a spanwise position of

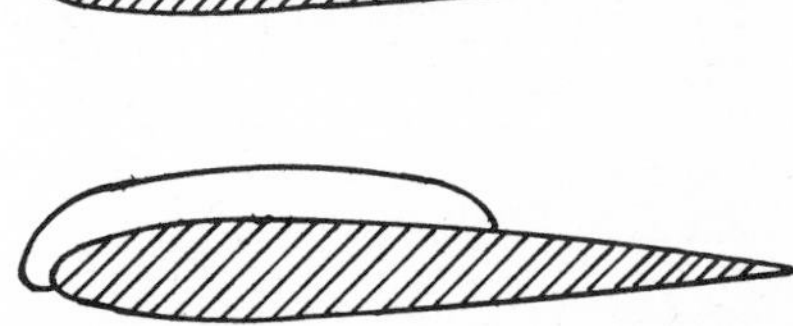

FIG. 4.48 Shapes of boundary layer fence used in the first flight of the Messerschmidt Me 109 without slots (*from Liebe, 1952*).

approximately 36 percent of the wing semispan from the plane of symmetry.

4.3.4 Propellers

Himmelskamp (1945) investigated flow phenomena on a rotating propeller blade by evaluating the local lift coefficient of the propeller blade from the measured pressure distribution, as shown in Fig. 4.49. Wind-tunnel measurements of a fixed wing are included in the figure for purposes of comparison. It is seen that maximum C_L values strongly increased in the neighborhood of the hub amounting to 3.2 compared to 1.4 for a fixed wing. The shift of the separation toward the larger angles of attack is due to the action of Coriolis forces in the boundary layer; this causes additional streamwise acceleration similar to a pressure drop. The centrifugal force also acts favorably on the boundary layer by retarding separation. Experiments on separation from rotors (Velkoff et al., 1969) were discussed in Chap. 2, Sec. 2.1.10.

4.3.5 Wing and Propeller Combinations

The flow downstream of a propeller affects separation on the wing. This effect is shown in Fig. 4.50 and compared with the separation behavior of a wing without a propeller. The cross-hatched areas indicate the separated regions. The left-hand column shows the separation behavior on the wing with no propeller. In this case, separation occurs along the trailing edge starting at wing center. With increasing angle of attack, the separated region spreads out over the entire wing span. The right-hand side shows the separation behavior

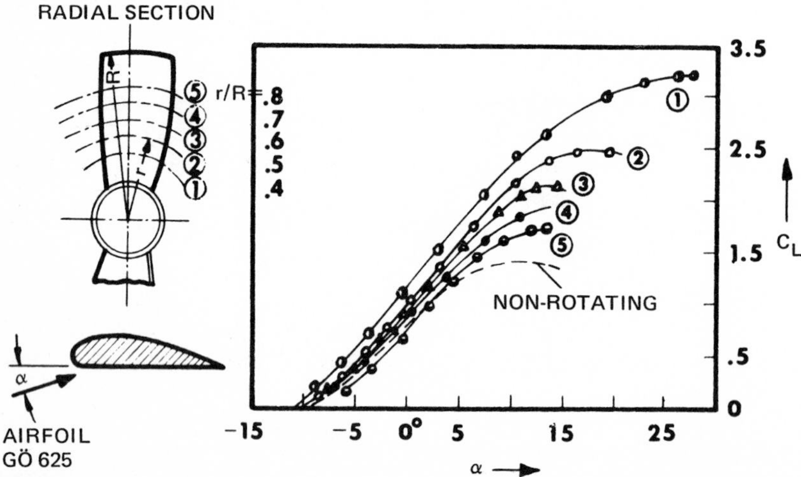

FIG. 4.49 Local lift coefficient at various sections of a rotating propeller blade (*from Himmelskamp, 1945*).

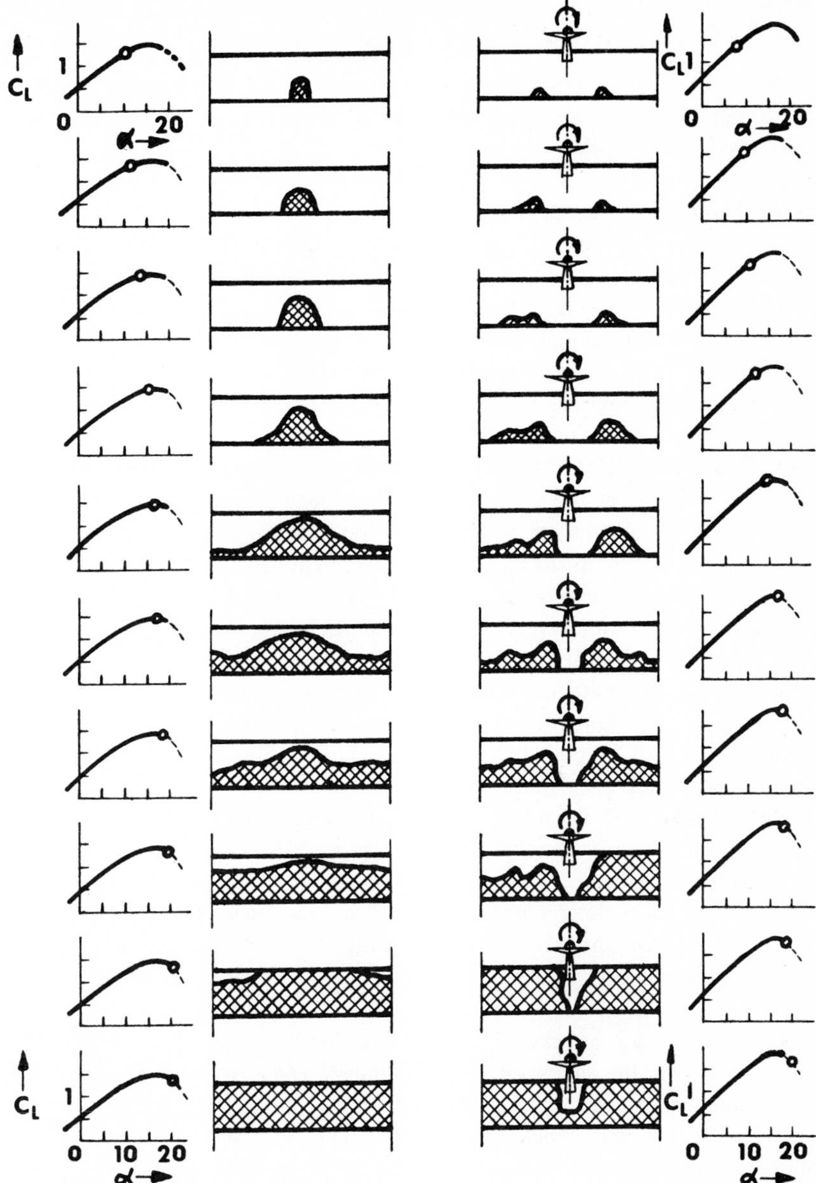

FIG. 4.50 Influence of oblique blowing on behavior of separation (*from Weinig, 1940*).

as affected by propeller downstream flow. In this case, the separation starts on the trailing edge but at the rim of propeller tip flow. It then spreads out toward the wing tip and toward the leading edge with increasing angle of attack, but the flow remains attached in the region immediately downstream of the propeller.

4.4 SHOCK-INDUCED FLOW SEPARATION

In practice, separation induced by a shock is undesirable; if possible, therefore, it should be eliminated at the design stage. However, it is necessary to consider the following facts (Pearcey, 1961). Although the separation introduces direct losses into the boundary layer flow, at the same time it reduces the wave drag by softening the foot of the shock itself and retarding its rearward movement. Therefore, the reduction in boundary layer drag resulting from prevention of the separation might easily be offset by an increase in wave drag. Hence, prevention of separation could actually accelerate the drag rise at a fixed angle of attack; it may increase the lift and (although unlikely) may reduce the lift/drag ratio.

Consider the physical phenomenon of shock-induced separation (see Fig. 1.7, Chap. 1). It is caused by the inability of the retarded boundary layer air to negotiate the adverse pressure gradients imposed by the shock waves. The pressure gradients in the external flow may become very large even for weak shocks, but the shocks are softened at a body surface by interaction with the boundary layer and so do not become severe enough to cause separation until the magnitude of the pressure rise increases substantially. The pressure rise imposed on the boundary layer by the shock wave thickens the stream tube of low velocity in the inner (subsonic) part of the layer. This thickening deflects the outer part of the layer and the external flow outward, away from the wall, generating a band of compression waves in the supersonic flow. These waves start in the supersonic part of the boundary layer and propagate out into the external flow.

With the increase in overall pressure resulting from the stronger shock, the gradients in this more gradual pressure rise and the degree of deceleration in the inner strata increase progressively until the stage is reached at which forward flow is no longer possible at the surface itself. The inner strata then leave the surface and continue their progressive increase in the degree of outward deflection in the outer parts of boundary layer and in the external flow.

Separation may be caused at the foot of a shock or upstream of the impingement of the shock onto the boundary layer. When it does occur, the interaction can spread appreciably both upstream and downstream from the point at which the shock itself exists just outside the boundary layer. Shock-induced separation may be divided into three regions: (*a*) at the upstream end, the region of relatively sharp pressure rise near the separation point itself; (*b*) at the downstream end, the region of relatively sharp pressure rise associated with reattachment; and (*c*) at the dead air under the shear layer (which may have a substantial depth), the region between (*a*) and (*b*) where the pressure rise is relatively slow.

The two-dimensional models used by Pearcey (1961) are sketched in Fig. 4.51. He paid particular attention to the case of a plane oblique shock

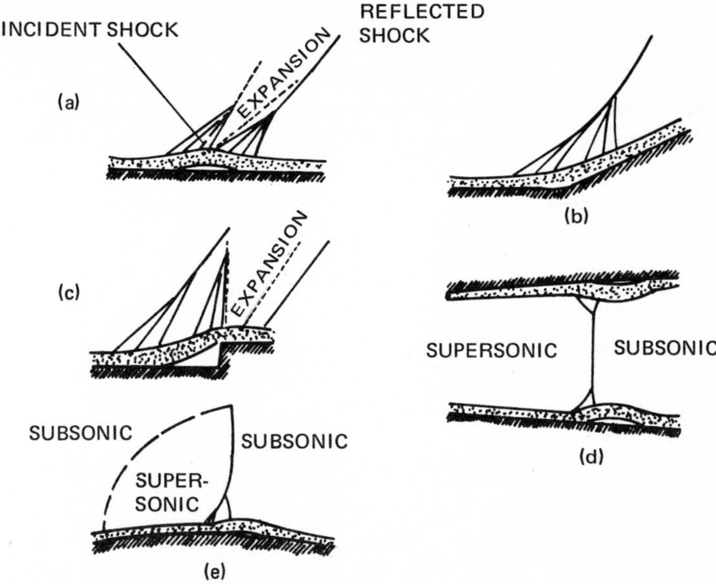

FIG. 4.51 Types of interactions between boundary layer and shock wave: (*a*) incident shock; (*b*) compression corner; (*c*) step; (*d*) normal shock (duct); (*e*) normal shock (airfoil) (*from Pearcey, 1961*).

generated in the stream and reflected from a flat wall (incident shock, Fig. 4.51*a*) and the case of a plane oblique shock generated at the corner of a flat wall with an abrupt change of shape (compression corner, Fig. 4.51*b*). The first case, or model, simulates flow in a supersonic diffuser while the other case simulates flow at the concave hinge of a control flap or flow at the trailing edge of a straight wing at supersonic speeds. Furthermore, these models provide a useful basis for discussing the effects of certain fundamental parameters, such as the state of boundary layer, Reynolds number, Mach number, heat transfer, surface curvature, etc., and their influence on the design requirements for preventing separation.

We may then proceed to the case of a shock generated by a step in the wall, simulating, for example, a spoiler (Fig. 4.51*c*), and that of a normal shock in a duct (Fig. 4.51*d*) which is important for supersonic intakes and diffusers. The more complex case of the shock wave terminating the local region of nonuniform supersonic flow imbedded in subsonic flow past a wing (Fig. 4.51*e*) may be developed from these relatively simple examples by considering similarities and differences.

The pertinent formulas and data which may be used for calculations are discussed and their applications indicated for the body shape of practical engineering concern. As mentioned in Chap. 1, the prediction of shock-induced separation may be simplified by introducing the concept of free interaction. Free interaction is defined as the fully developed separated flow whose characteristics are not affected by downstream characteristics.

As mentioned in Chap. 1, for laminar flow in free interaction over a flat plate, the pressure rise to separation with no heat transfer is given by Gadd (1957) as

$$C_{p_s} = \frac{1.13}{\sqrt{\beta_0 \mathrm{Re}_x{}^{1/2}}} \tag{4.16}$$

where $\beta_0 = \sqrt{M_0{}^2 - 1}$ and subscript 0 refers to a region upstream of the interaction.

Chapman et al. (1958) empirically determined the pressure rise to separation for turbulent flow in free interaction over a flat plate with no heat transfer. Pearcey (1961) plotted its mean value as a function of M_0, as shown in Fig. 4.52; there $\tilde{c}_{f_0}$ is the ratio of c_{f_0} at a given Re_{x_0} to the corresponding value at $\mathrm{Re}_{x_0} = 10^6$ and the subscript 0 refers to the region upstream of the interaction in uniform flow.

The Hakkinen (1959) criterion for an incident shock which causes laminar separation has been presented in Chap. 1. For a wholly laminar interaction, the overall pressure rise for which separation first occurs (that is, for which separation is incipient) was formulated experimentally by Hakkinen et al. (1959) as exactly double the pressure rise up to the separation point in a free interaction, i.e.,

$$C_{p_{\text{incipient}}} = 2\dot{C}_{p_s} \tag{4.17}$$

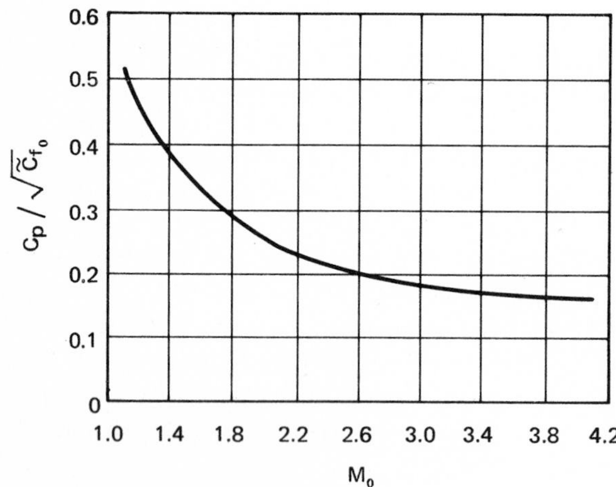

FIG. 4.52 Pressure rise to the separation point in free interactions (turbulent boundary layers on flat plates with zero heat transfer—from an empirical mean curve; *from Chapman, et al., 1958*).

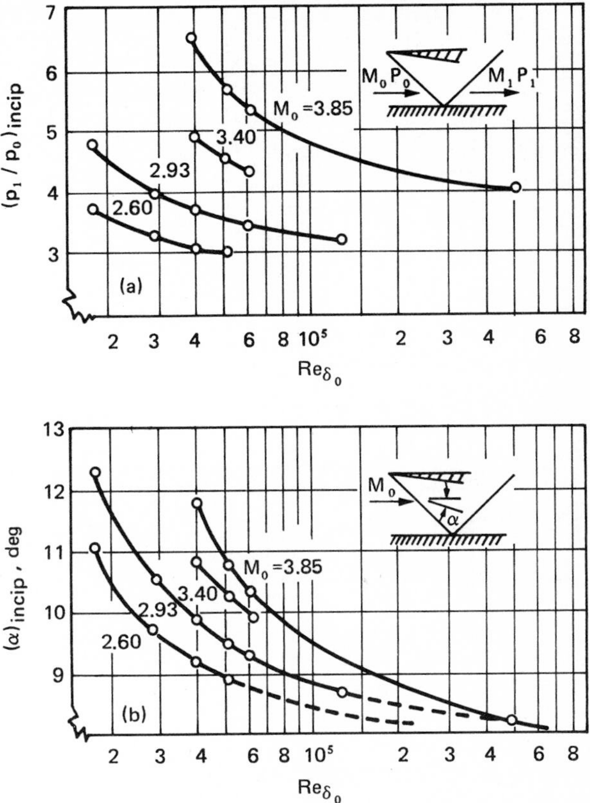

FIG. 4.53 Conditions for the occurrence of separation for incident shocks on turbulent boundary layers (*from Kuehn, 1959*).

Kuehn (1959) experimentally determined the corresponding shock strength, flow deflection, and compressive turning for the occurrence of turbulent shock-induced separation; see Figs. 4.53 and 4.54. As indicated there, the incident shock and the compression corner give almost identical results until M_0 becomes large. Furthermore, the pressure ratio to provoke separation increases from a value near $M_0 = 2$ (that is, about 30 percent greater than the pressure ratio for separation in a free interaction) to a value at $M_0 = 4$ that is more than twice the pressure ratio to separation.

The six body shapes shown in Fig. 4.55 were selected by Pearcey (1961) to demonstrate the prevention of shock-induced separation by practical design features and without the use of auxiliary power. For the diffuser with one plane wall exposed to supersonic flow, Fig. 4.55*a*, point *A* is a compression corner that causes an incident shock on the opposite wall at *B*. The design problem is that the separation must be prevented at point *B* because there the

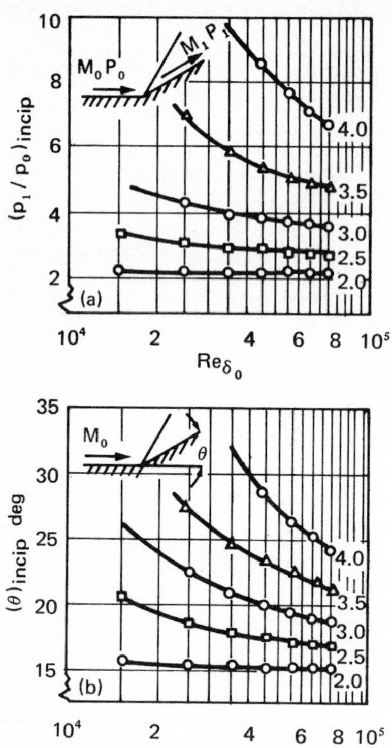

FIG. 4.54 Conditions for the occurrence of separation at compression corners with turbulent boundary layers (*from Kuehn, 1959*).

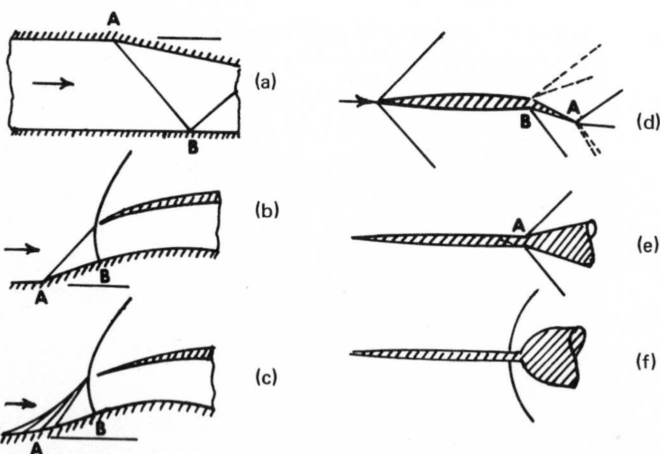

FIG. 4.55 Designs to avoid shock-induced separation or its effects: (*a*) diffuser with one plane wall; (*b*) side intake with compression corner; (*c*) side intake with curved compression; (*d*) aerofoil with deflected flap; (*e*) probe with conical afterbody; (*f*) probe with bluff afterbody (*from Pearcey, 1961*).

imposed pressure rise amounts to double that at A. For the case of a supersonic side intake with compression corner, Fig. 4.55b, design conditions at compression corner A and at the normal shock B must be carefully considered. A curved compression, Fig. 4.55c, may ultimately be preferred. For an airfoil with flap, Fig. 4.55d, the compression occurs at the corners of A and B. This case can be studied with the previous three because the flow with an oblique shock at the trailing edge of an airfoil (deflecting the flow from the surface on to the dividing streamline) is equivalent to the flow at a solid compression corner; see Henshall and Cash (1955).

Figure 4.55e shows a static probe or spike mounted to a conical center body. The compression corner that involves oblique shock is located at A. For this and many other similar cases, the flow deflection and the pressure rise should be kept below the values as given by either Eq. (4.17) or by Figs. 4.53 and 4.54, depending on whether the boundary layer is likely to be laminar or turbulent upstream of the interaction. For normal shock waves impinging on turbulent boundary layers, the design should keep the upstream Mach number below about 1.3 (the corresponding value for laminar layers has never been properly established but it is necessarily very low). For cases where the conditions for incipient separation are not clearly known, a safe but probably conservative design is one in which the imposed pressure rise is less than the pressure rise for separation in a free interaction; see Eq. (4.16) and Fig. 4.52. For a blunt body with a spike, Fig. 4.55f, or a step, there must always be some separation, and the most relevant design consideration would probably be to keep the datum holes or other point of interest clear of the separated flow. The Bogdonoff (1955) experimental results for steps can be referred to for guidance.

If prevention of separation is a main concern, as, for example, for supersonic side intakes, then beyond certain stages, it becomes more important than the maintenance of laminar flow. Separation may be prevented by fixing transition (in particular, immediately upstream of the shock), but an increase in skin friction is to be expected and artificial transition is not always easy to achieve at any prescribed point in supersonic flow at low Reynolds number. (A similar conflict but in the opposite sense may also arise when early separation is provoked in order to reduce skin friction in aerodynamic heating. This conflict is likely to be strongest where some form of distributed suction or cooling is applied in order to delay transition to turbulent flow.)

Sweepback and small aspect ratios are very helpful in preventing shock-induced separation. The latter approach relies on the general spanwise relief in the stream tube constriction at the local sonic speed to slow up the development of local velocities on the surface. The combination of low aspect ratio and sweepback yields further delay in the onset of separation and relief from the adverse effects of the shock waves by virtue of the reduced Mach number component normal to the isobars on the wing. The reader is referred to Pearcey (1961) for an extensive and detailed presentation.

4.5 AUGMENTATION OF THE TURBULENCE LEVEL

A technique is often used whereby the flow energy is increased by guide vanes and the turbulence level is augmented by fixed vortex generators and screens. A guide vane may energize the retarded flow by acceleration of fluid particles. A decrease in the magnitude of the adverse pressure gradient can be realized by placing a fixed vortex generator in the region of adverse pressure gradient and by forcing mixing in the boundary layer. Since displacement thickness and momentum thickness change when a vortex generator is introduced, its effect can be formulated by the boundary layer behavior. An almost endless variety of vane types and fixed devices can be conceived to provide for forced mixing and a number of different shapes of such devices are used in practice. The investigations of Kamiyama (1967) and those of Schubauer and Spangenberg (1960) are good sources of information.

4.5.1 Guide Vanes

The term *vanes* can be interpreted in terms of the principal function they achieve. Slats and vanes are often relative concepts that depend on the dimensions of the device in relation to those of the entire body. If they are comparable to the size of the body, then the system may be referred to as either slats or vanes. For illustration the following case is considered: When the dimensions of a number of staggered and closely arranged slats are comparable to those of the airfoil for which they are provided, then the arrangement may result in slotted vanes as shown in Fig. 4.56. Note that great variations of slotted vane designs have little effect on the curves shown in that figure. They intersect the solid vane curve at A; the magnitude of C_D for B, C_1, and C_2, indicates that the increase of C_L beyond A is achieved at less viscous drag for slotted vanes. The tip vortices (induced drag) must be considered for finite aspect ratios. It is also readily seen that the slotted vane type of airfoil is effective in substantially increasing the lift provided by an unslotted airfoil at the same drag.

Kamiyama (1967) used conformal mapping techniques to analyze the effect of plate vanes inserted in subsonic flow within a two-dimensional channel with uniform flow u_{e_1} entering the bend. He evaluated the effect of turning vanes on the wall pressure of a band by using the Green theorem to find the wall pressure distribution of a prescribed bend. Figure 4.57 shows the configuration of the bend and the inserted turning vane.

The pressure coefficient C_p is defined by

$$C_p = \frac{p - p_1}{\frac{1}{2}\rho u_{e_1}^2} = 1 - \left(\frac{u_e}{u_{e_1}}\right)^2$$

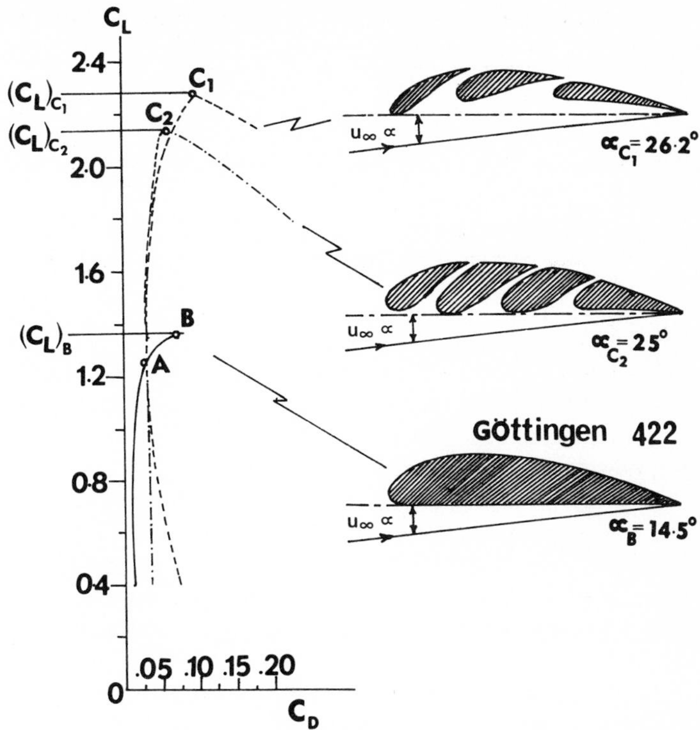

FIG. 4.56 Effect of slots on force coefficients (angles shown for aspect ratios of 6, polar curves shown for infinite aspect ratios, slotted vane designs by G. Lachmann; *from Frey & Vasuki, 1966*).

where the subscript 1 refers to the condition at the entrance to the channel and u_e is flow velocity.

Further evidence of the validity of prediction for bends with and without vanes is given in Table 4.3. Note that the data given there also show the effectiveness of vanes.

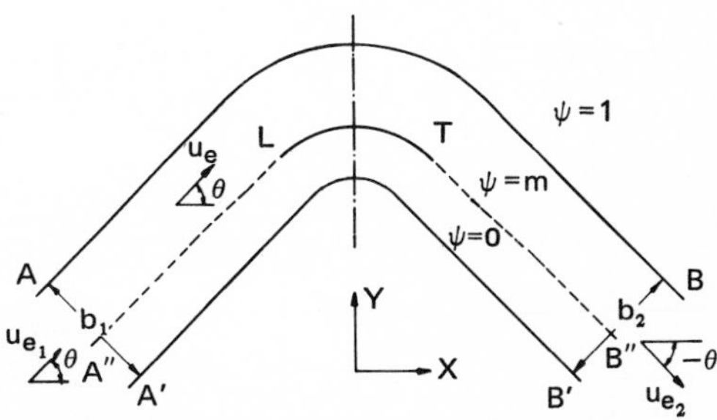

FIG. 4.57 Physical plane for a bend and a turning vane (*from Kamiyama, 1967*).

TABLE 4.3 Comparison of predicted and experimental values for surface pressure distribution

	One plate at $r_p = \sqrt{r_i \cdot r_o}$		Without plate	
Guide plate	Theory	Experiment (u_{e_1} = 8 to 12 m/s)	Theory	Experiment (u_{e_1} = 11.34 m/s)
$C_{p\min}$	−0.756	−0.780	−1.040	−1.056
$C_{p\max}$	−0.278	−0.280	0.465	0.456

Source: Kamiyama (1967).

The turning vane was located at a radius of $r_p = \sqrt{r_i \cdot r_o}$; here r_i is the inner radius, r_o is the outer radius, and p is the channel width. As seen in Fig. 4.58, the difference between predicted and measured values of C_p for a 90 deg conventional bend was only about 3 percent. The difference becomes larger downstream of separation.

The effect of a number of vanes on the surface pressure is given in Fig. 4.59. Two single turning vanes were placed at a radius of $r_{p_1} = \sqrt[3]{r_i^2 \cdot r_o}$ and $r_{p_2} = \sqrt[3]{r_i r_o^2}$.

It is clear from Figs. 4.58 and 4.59 that the presence of a turning vane reduces the absolute value of $C_{p\min}$ and that an increase in the number of vanes gives a further reduction.

Figure 4.60 indicates the extent to which the inner radius affects both flow velocity and pressure. If the radius becomes smaller, the effect of turning vanes on the pressure or velocity may be greater.

It is seen from this investigation that vanes are effective in energizing the flow by accelerating it and that higher flow velocity may be achieved by using vanes with a sharp curvature. Furthermore, vanes are an effective means of guiding the flow in a desired direction.

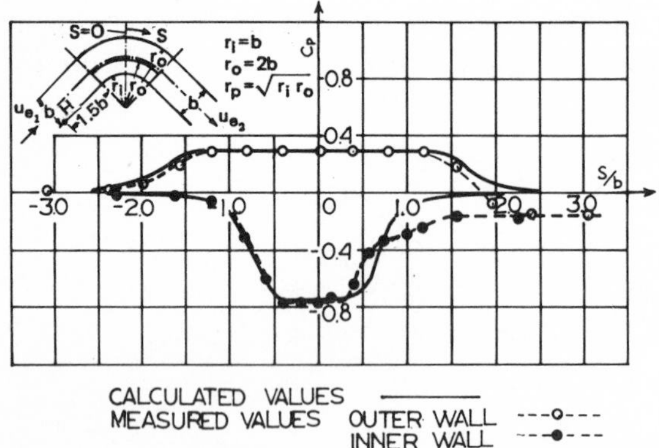

FIG. 4.58 Comparison of calculated and measured values of surface pressure distribution (*from Kamiyama, 1967*).

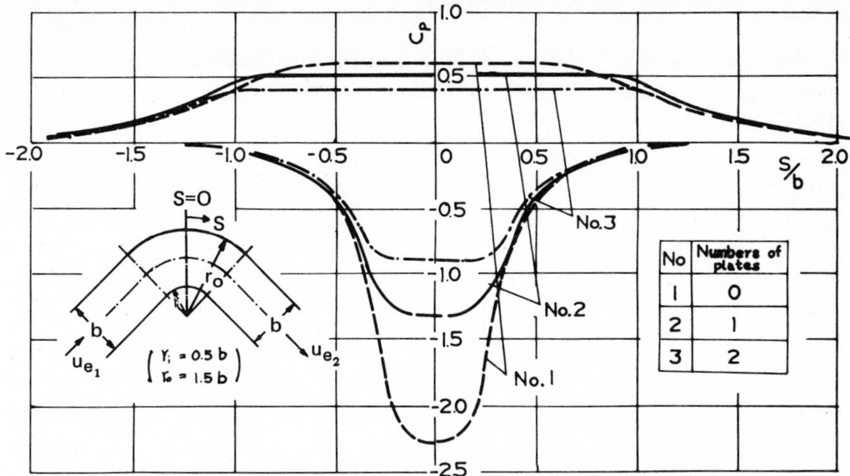

FIG. 4.59 Effect of number of plates on the surface pressure distributions of a bend (*from Kamiyama, 1967*).

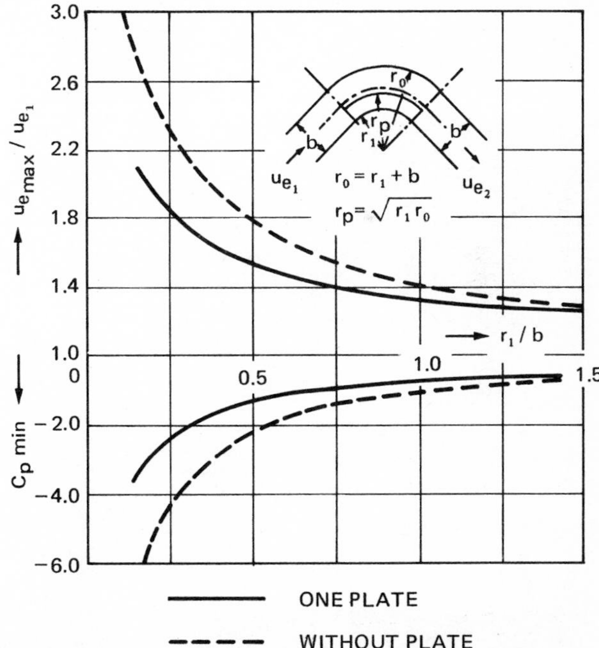

FIG. 4.60 Relations between inner radius (r_i/b) and maximum velocity ($u_{e_{max}}$) or minimum pressure coefficient ($C_{p_{min}}$) at a bend (*from Kamiyama, 1967*).

Frey and Vasuki (1966) used flow visualization techniques to study the effects of thin vanes on flow separation in a rectangular channel; see Fig. 4.61, 4.62, and 4.63.

The flow pattern shown in Fig. 4.61*a* indicates that, to a great extent, the redistribution of flow is basically independent of the shape of the elbow diffuser or of the expansion ratio. The main flow detaches from both sides and streams toward the outer boundary. The width of the flow is 0.7*b* (*b* refers to inlet width) at 1.25*b* from the corner. The center of a large, almost circular vortex (0.6*b* radius) is present at this distance. The vortex provides flow for replacement of fluid entrained in the fluid boundary eddies; the particular shape of this vortex at this expansion ratio contributes to a more rapid redistribution of flow. After separation, the width of the main flow (52) is approximately equal to or smaller than that of the inlet. The standing enforced vortex system (89) becomes more elongated in the flow direction if the expansion ratio increases above 2 or 3.

Figure 4.62*b* indicates that the installation of vanes enables a quick redistribution of flow. Auxiliary vanes, consisting of elementary vanes, achieve a quick flow redistribution throughout. The flow is first strongly diverted and then expanded. The outer detached zone augments the expansion of the flow by means of the vortex (for replacement fluid).

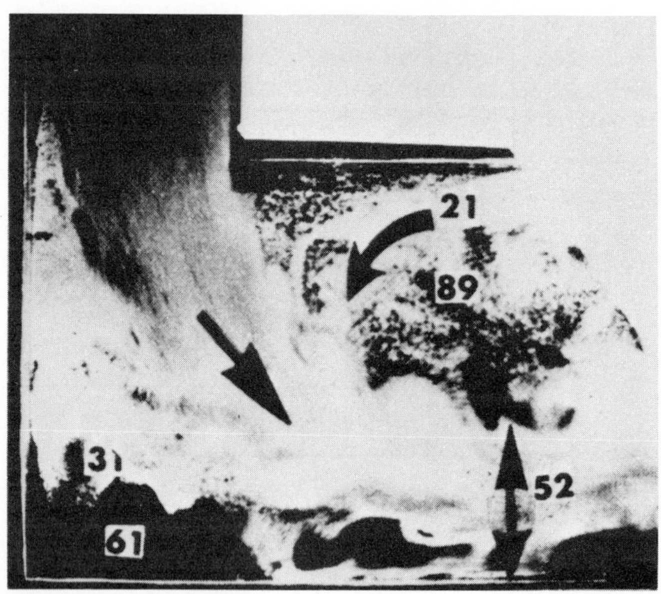

(a)

FIG. 4.61 Flow control in an elbow diffuser: (*a*) main stream detaches from both sides and streams toward the outer boundary; (*from Frey & Vasuki, 1966*).

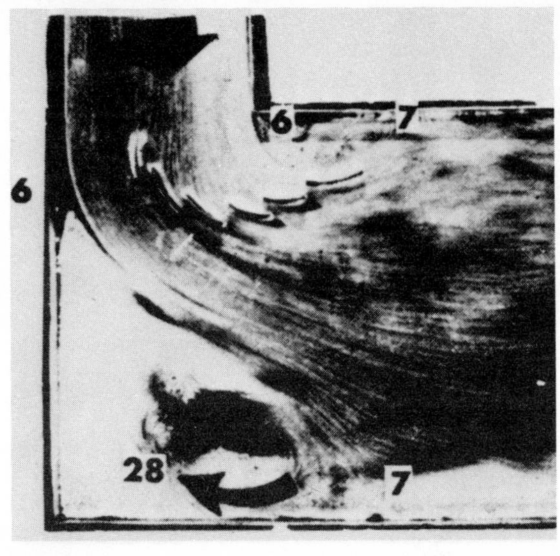

(b)

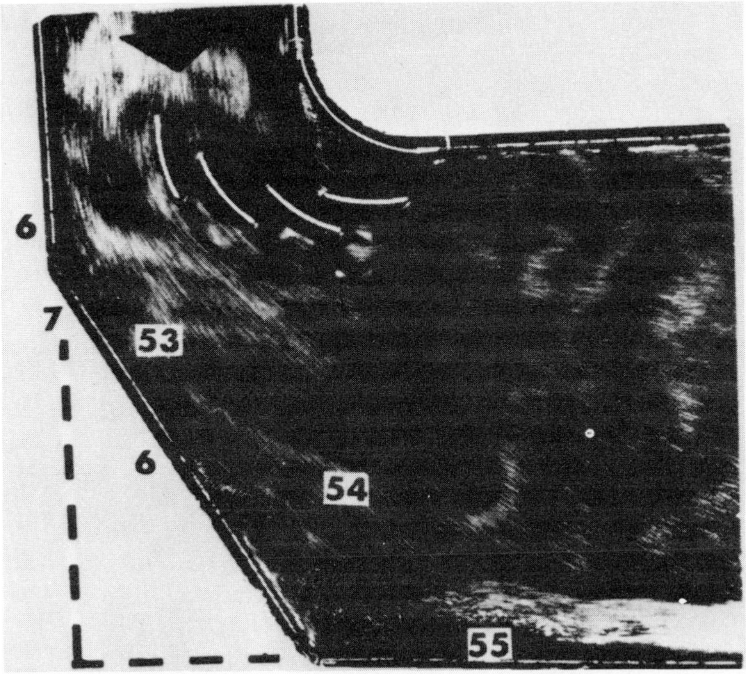

(c)

FIG. 4.61 Flow control in an elbow diffuser: (*b*) auxiliary vanes (seven elementary vanes) achieve quick flow redistribution throughout; (*c*) auxiliary vanes (four elementary vanes) supplemented by the boundaries of the wall achieve quick flow redistribution (*from Frey & Vasuki, 1966*).

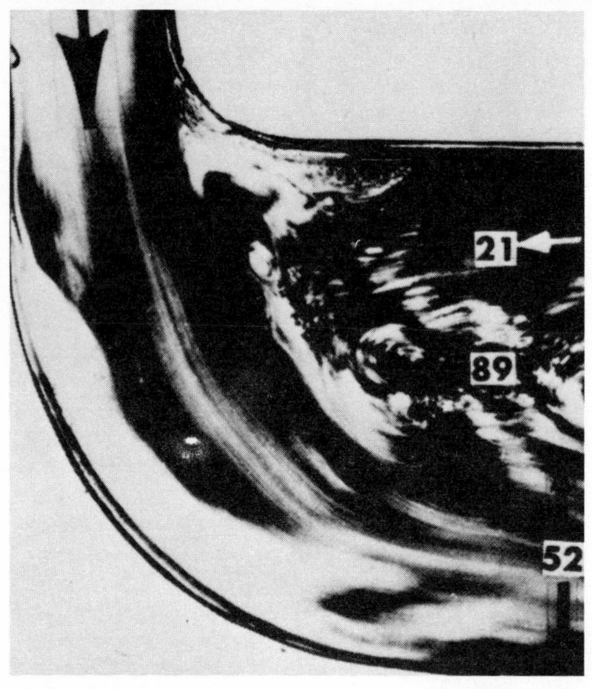

(a)

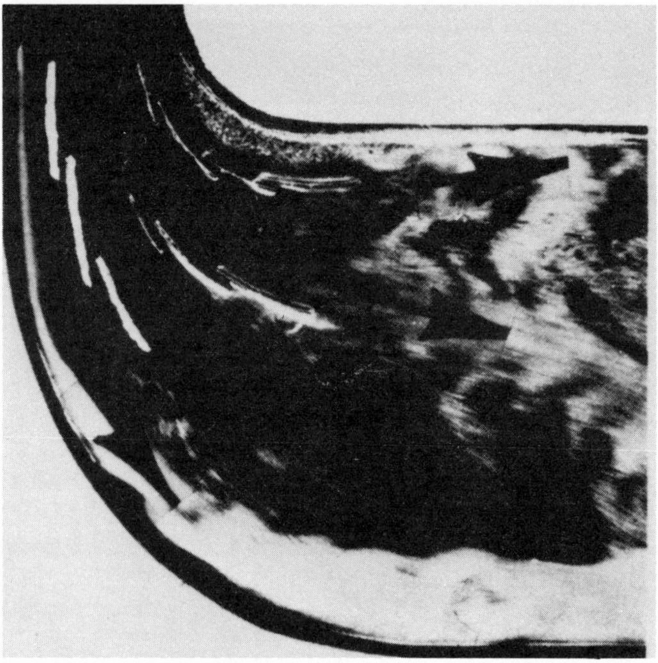

(b)

FIG. 4.62 Flow separation in a curved elbow diffuser: (*a*) redistribution of flow by a large oblong vortex; (*b*) redistribution of flow by three rows of vanes (*from Frey & Vasuki, 1966*).

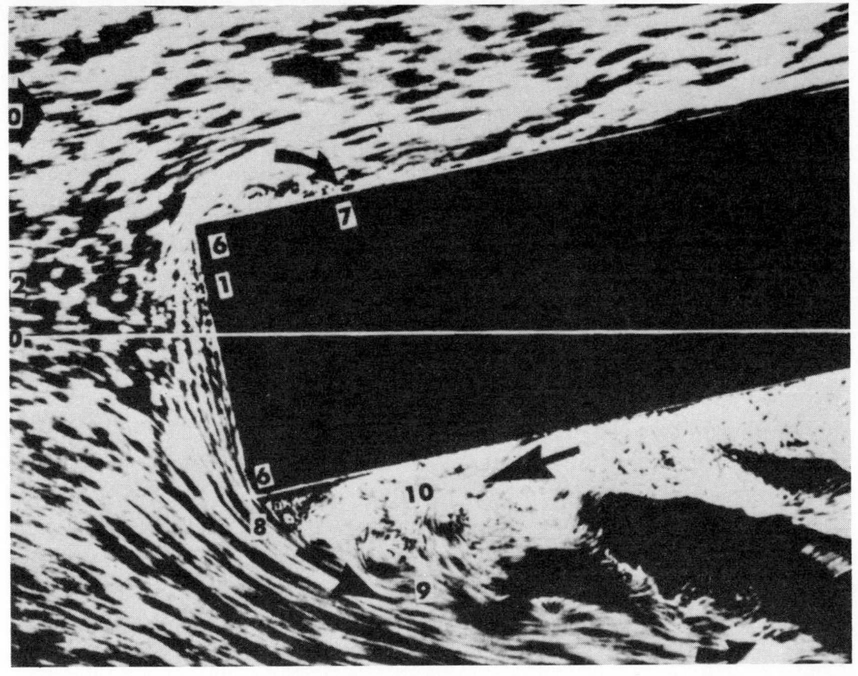

(a)

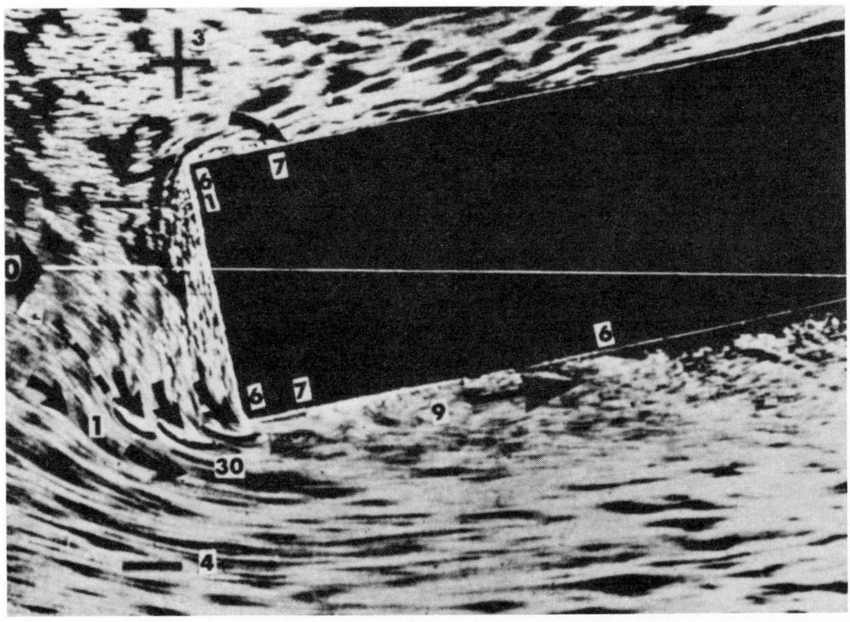

(b)

FIG. 4.63 Flow separation around a bridge pier: (*a*) unstable vortex zones at sharp corners (large detached zone at lower corner, small zone of separation at upper corner); (*b*) three vanes divert the flow around the sharp corner (bridge pier model oblique to flow; sharp-edged flat leading edge; *from Frey & Vasuki, 1966*).

Figure 4.61*c* demonstrates that the standing vortices can be reduced if the auxiliary vanes and the boundaries of the wall supplement each other. The auxiliary vanes, consisting of four elementary vanes, are sufficient for quick flow redistribution by using $r_i/b_i = 0.6$ (r_i refers to inner radius) and deflecting the corner boundary at 30 deg as shown. The detached vanes and boundaries should be developed together and not independently. The design shown in Fig. 4.61*c* is superior to those of Figs. 4.61*a* and 4.61*b* for such applications as the heat transfer on the sedimentation in apparatus for drying powders, etc. The boundary of the wall (53) can be improved (54 and 55) by reducing the overloading at the lowest part of a precipitation chamber beginning at 55; see the dashed lines on Fig. 4.61*c*.

The design data for the curved elbow diffuser shown in Fig. 4.62 are $r_i/b_i = 0.75$ and $r_0/b_i = 3$ (r_0 refers to outer radius). The effective flow width is approximately constant through the elbow diffuser. In Fig. 4.62*a*, reverse flow is provided by a large oblong vortex at the inner side. This type of vortex requires a substantially larger redistribution zone (compare with Fig. 4.61*a*). For the same design data, Fig. 4.62*b* demonstrates the improved control provided by three rows of vanes. There flow redistribution is similar to that in a rectangular channel; compare with Figs. 4.61*a*–4.61*c*.

Guide vanes are effective not only for flow control within a channel or diffuser surrounded by walls but also for external control around a body submerged in a stream. To illustrate, consider the two-dimensional visualization of flow around the model of a bridge pier model as shown in Fig. 4.63. Unstable vortex zones of different length exist at each corner. The flow around the sharp corners is effectively diverted by three vanes (see point 30 in Fig. 4.63*b*) although their effects are weak at the upper corner. Auxiliary vanes create flow around the corner and attached flow along the suction side. The zone of detached flow at the upper corner is reduced and the stagnation point is moved closer to the upper corner. To some extent, the flow around the vanes and through the slots of the auxiliary vane is a function of the local Reynolds number.

The vanes may be placed at smaller or larger angles of attack and at large or small distances from each other in parallel, in series, or at positive or negative stagger angles. The positive stagger angle creates strong flow diversion, and the negative stagger angle creates strong turbulence. Physically, this different behavior is merely a matter of the direction of the slots. According to the pressure difference, they guide the flow tangentially or opposite to the direction of flow around the vane set. Geometrically, this different behavior means that the same device can be made to produce either type of flow through an appropriate choice of the angle of attack. The set of vanes greatly reduces the oncoming turbulence through the action of the slot and the rapid changes of velocities throughout the velocity field. Furthermore, the vane set substantially damps out oncoming oscillations.

4.5.2 Vortex Generators

Vortex generators in a variety of shapes have found wide application for eliminating separation on wings and on diffusers. The vortex generated is used to forcefully enhance the natural mixing of turbulent flow. To provide insight into the effects of vortex generators on flow, the physics of forced mixing in the boundary layer will be reviewed.

4.5.2.1 Forced mixing in the boundary layer

The boundary layer formed on a flat plate was investigated by Schubauer and Spangenberg (1960) in a special wind tunnel wherein a variety of adverse pressure gradient could be obtained and the effects of forced mixing by various types of vortex generators could be studied in a two-dimensional turbulent boundary layer.

Section 4.1 discussed control techniques for exploiting the natural turbulence and keeping the fluid from stagnating under the action of the opposing pressure, i.e., by proper design of the body shape to enable natural turbulent mixing of slowed fluid near the wall with faster fluid in the outer region of boundary layer. This effect of natural mixing is often expressed as a turbulent shear stress and the agency responsible is likened to an "eddy viscosity." This eddy viscosity is on the order of 100 times greater than ordinary viscosity in a turbulent boundary layer and on the order of 1000 times the effect of mixing on the mean flow field in jets and wakes. It may be studied by observing the behavior of displacement and momentum thickness, δ^* and θ, and the shape factor, $H = \delta^*/\theta$, without becoming involved with the detailed flow process. Although stationary three-dimensional flow patterns will be encountered when forced mixing is applied, consider for simplicity the incompressible two-dimensional von Karman momentum equation:

$$\frac{d\theta}{dx} = \frac{\theta}{q}\frac{dp}{dx}\left(\frac{1}{2}\frac{\delta^*}{\theta} + 1\right) + \frac{\tau_w}{2q} \tag{4.18}$$

Representative average values of δ^* and θ are obtained from the two-dimensional momentum equation (4.18) by taking an average over a sufficient spanwise-distributed pattern. This equation is an expression for the growth of θ with respect to x in terms of the forces exerted on the flow due to the pressure gradient and wall friction. Furthermore, the retardation of the flow by the forces causes δ^* to increase also. However, the manner in which the value of δ^* depends on the amount of fluid through which the retardation is distributed can be illustrated by the following examples. For the case of zero pressure gradient, the value of $H = \delta^*/\theta$ is 2.6 for laminar flow, but only 1.3 in the transition to turbulent flow. Inasmuch as θ remains virtually unchanged by transition, this means that δ^* is decreased to one-half its original value by the mixing action of turbulence. Therefore, it may be said that mixing affects δ^* directly. It is known that H is reduced when the adverse pressure gradient is reduced; hence the reduced adverse pressure gradient and the increased rate of mixing are expected to have similar effects. If H is held at some fixed value, then the increased rate of mixing yields a larger pressure gradient. On the other hand, if H is increasing in the flow direction and the rate of mixing is increased, then either there is a decrease in dH/dx or else it is held the same for a larger value of the pressure gradient. Since the criterion of incipient flow separation is given by a fixed value of H, say, larger than 2, an increase in the rate of mixing can be expected to make the pressure recovery higher before separation occurs, particularly where the pressure gradient is larger. Only qualitative estimates of this kind are possible because any reduction of H below its separation value normally brings skin friction into the picture. For the turbulent mixing by eddy motion comprising bulk movements of fluid, the mixing weakens as the wall is approached and finally disappears, leaving only molecular viscosity in a laminar sublayer at the wall.

Thus, the increase in skin friction of turbulent flow is attributed to a velocity increase near the wall rather than to a uniform, across-the-layer increase in eddy viscosity. This phenomenon carries with it a compensating feature, namely, a shaping of the velocity profile toward lower values of H, even less than the value for laminar flow.

4.5.2.2 Shapes for vortex generators

The natural turbulent mixing is an ideal one; the induced mixing brought about by the velocity differences that already exist within the boundary layer is applied by one portion of the fluid on another, and no momentum loss is charged against the natural mixing process. To match this ideal case, then, mixing should not be achieved by using up the momentum of the mean flow. Rather the velocity field should be rearranged to produce steeper gradients within which turbulent mixing motions are generated and across which they transfer momentum. A "pure" rearrangement implies guiding the fluid to new positions, with no residual motions to require an induced drag from the mixing devices. However, such an ideal condition cannot be realized because no fluid-handling device can escape the drag that arises from frictional effects; hence, if residual motions contribute enough to the mixing, their suppression would not be warranted. The vortex generator is an example of the case where the derived mixing depends on induced streamwise vortices. Now the role of vortex generator can be easily understood because such vortices may be regarded as a rearranging mechanism that is installed in the flow and whose influence persists for a considerable distance downstream.

Although the concept of promoting self-mixing by simply rearranging flow was adhered to in the experimental investigations of Schubauer and Spangenberg (1960), the forms designed to minimize induced drag (by cancelling residual vortex motion) generally exhibited as much or more drag than those that produce strong vortex trails. Furthermore, it was found that mixing on a coarse scale by a relatively large, widely spaced device was far more effective than fine-scale mixing. Under those conditions, multiple rows were less effective than a single row of devices properly spaced and properly stationed.

Figure 4.64 illustrates the test results for a fixed single-row arrangement. The spacing and position for maximum delay of separation and maximum pressure recovery were found by trial and error. The height of the devices was on the order of the boundary layer thickness, but this was not varied during the attempt to find the optimum position. The drag of the devices was measured in their single-row or with an identical device on each side. Rotating devices or moving agitators were not tested. The reasoning behind the devices and their intended actions were described by Schubauer and Spangenberg (1960) as follows:

Device (A), called the simple plow, is intended to part the boundary layer and guide the outer flow toward the wall into the furrow so produced. The motion thus starts and continues downstream in the form of a pair of trailing vortices. The shielded plow (B) is device (A-2) modified by the addition of enough side shielding to effectively eliminate the vortices. However, these elaborate additions merely increased the drag of the devices. Accordingly, the shielded plow (B) was abandoned in favor of a scoop (C) which incorporated some shielding and was easier to construct. The twist interchanger (D) and tapered fin (G) represent other ideas for overturning the flow without intentionally involving residual vortex motions. In practice, it was not possible to reduce drag by eliminating trailing vortices, and since this amounted to stopping the action started by the device, its effectiveness was reduced without compensative effects.

The triangular plow (E) is a less refined version of (A) and has much the same action, and ramp (F) is simple (E) used in reverse. By virtue of being in the nonuniform velocity field of the boundary layer, the dome (H) is designed to generate vortices without adding much wetted area. Device (J) represents a conventional vortex generator of the flat-plate type with a trapezoidal shape and is arranged in pairs to produce counterrotating vortices. The shielded sink (K) falls in a separate category because suction is employed. If the air is withdrawn through a hole in the wall, the air above that entering the hole would be deflected toward the wall. The effect produced is not unlike that of a plow, except that now the boundary-layer flow passes into the hole instead of being pushed aside. The primary aim was to obtain mixing with a decrement in momentum thickness rather than the increment which characterized external devices.

Effects of vortex generators on the pressure distribution on a wind-tunnel wall In order to establish a basis for investigating the effect of forced mixing, Schubauer and Spangenberg (1960) used various adverse pressure distributions (Fig. 4.65) to study the boundary layer. Each distribution was sufficient to

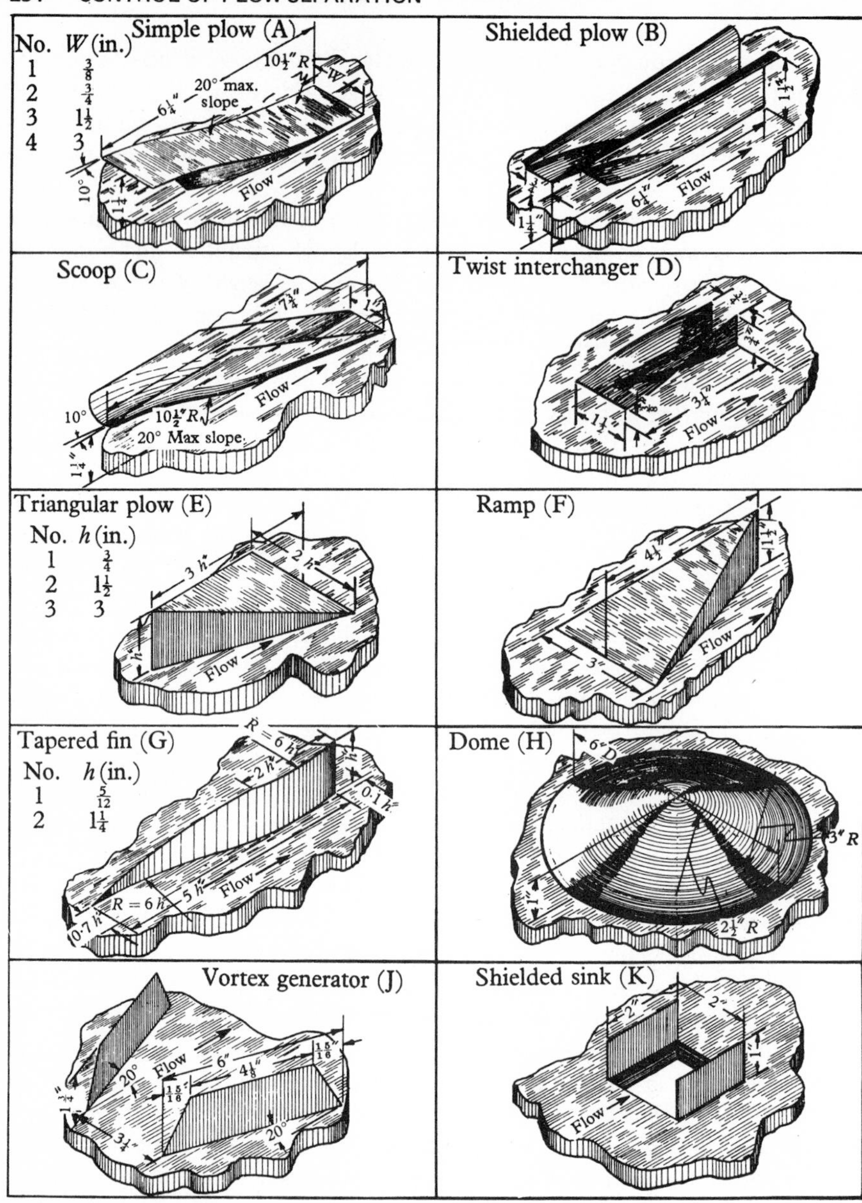

FIG. 4.64 Mixing devices (an angle of attack of about 10 deg was necessary to avoid laminar flow separation on the upper surface of devices (A), (B), and (C); scale effects were not investigated; these might impair the effectiveness of any of the devices; *from Schubauer & Spangenberg, 1960*).

cause eventual separation of the boundary layer on the wind-tunnel wall. Considerable effort was required to attain a two-dimensional flow.

In all cases the pressure was held constant up to the position, namely, 4 ft from the leading edge where the original of x was taken. The pressure was then made to rise as indicated in Fig. 4.65, where notations A, B, C, D, E,

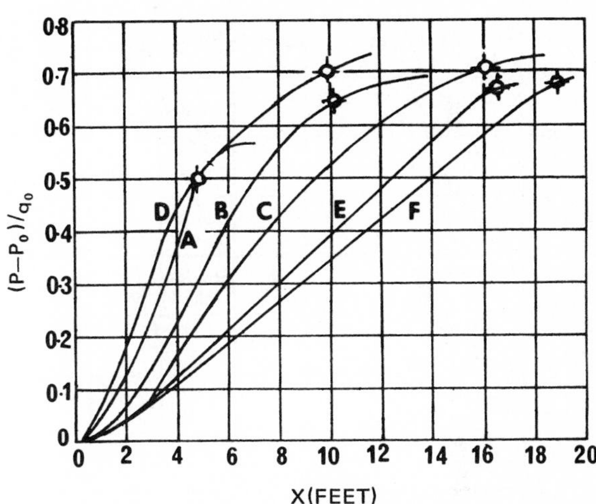

FIG. 4.65 Pressure distributions resulting from mixing devices (the origin of x is taken as 4 ft. from the leading edge; the pressure is uniform for $-4 \leqslant x \leqslant 0$; *from Schubauer & Spangenberg, 1960*).

and F are not referred to in the generator types in Fig. 4.64. Symbols p and q refer to the static pressure at the wall and the dynamic pressure, respectively. Subscript 0 indicates free-stream conditions at $x = 0$.

Distribution A was set up first in order to provide a condition of early separation and low-pressure recovery to which forced mixing could be applied with some expectation of improvement; see Fig. 4.66. Distribution B and C were next established in order to match as closely as possible the range of pressure recovery obtained with mixing devices applied to condition A. Since skin friction may become important, the investigation was extended to include distributions D, E, and F. Forced mixing was applied only in the case of distribution A.

Relative performance of various devices The summary of mixing characteristics given in Table 4.4 is useful in judging the relative performance of the various devices.

Starting from a separation point at 4.83 ft (pressure distribution A of Fig. 4.65 without a mixing device) and a pressure-recovery coefficient $(p - p_0)/q_0$ at separation of 0.5, the effect of the several devices may be compared by running down the columns in Table 4.4 headed x_s, $(p_s - p_0)/q_0$, and $\Delta p_s/q_0$. From the drag of the individual members of a set of devices, the effective increment in momentum thickness at the trailing edge of a row of devices was calculated according to the formula

$$\Delta\theta_{tr} = \frac{D}{2Lq_{tr}}$$

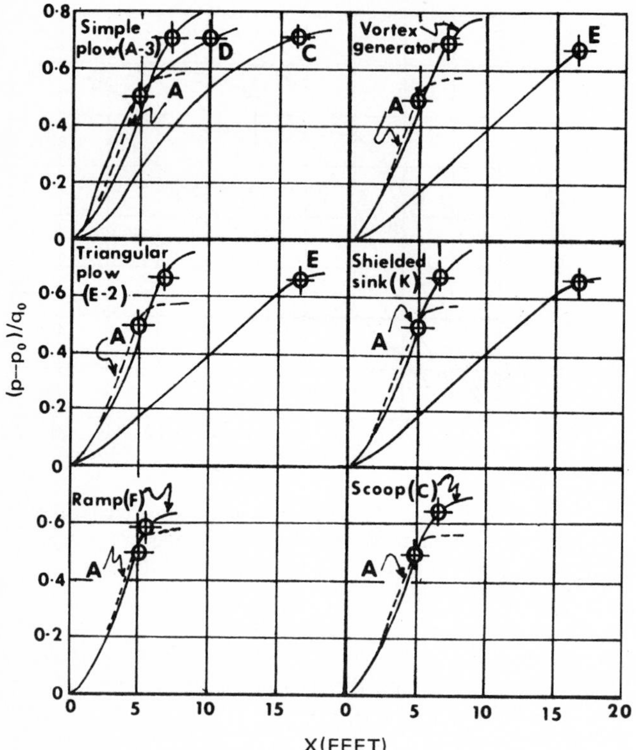

FIG. 4.66 Typical effect of mixing devices on pressure distributions (Distributions A, C, D, E, are the same as those given in Fig. 4.65; *from Schubauer & Spangenberg, 1960*).

where D is the drag of one member, L is the cross-stream spacing between devices, and q_{tr} is the free-stream dynamic pressure at the trailing edge of the devices. The fractional increase is expressed by $\Delta\theta_{tr}/\theta_{tr}$ where θ_{tr} is the momentum thickness at the same position in the absence of the mixing device. The quantity $\Delta p_s/q_0$ may be regarded as a measure of accomplishment and $\Delta\theta_{tr}/\theta_{tr}$ as the price paid for it. The suction quantity was fixed at 5.6 percent of the boundary-layer flow for the shielded sink (K); this produced the decrease $-\Delta\theta_{tr}$ as shown in Table 4.4. Note that its effect on the separation point and pressure recovery places makes the shielded sink one of the more effective devices.

Effect of mixing on boundary layer development In order to learn what was happening to the flow downstream of A-3, C, E-2, F, J, and K devices (Fig. 4.64), Schubauer and Spangenberg (1960) carried out a detailed velocity survey of the boundary layer and pressure distribution. They were particularly

TABLE 4.4 Summary of mixing characteristics

Device	L	x_{tr}	x_s	$(p_s - p_0)/q_0$	$\Delta p_s/q_0$	D/q_{tr}	$\Delta\theta_{tr}$	$\Delta\theta_{tr}/\theta_{tr}$
None	—	—	4.83	0.50	0.0	0.0	0.0	0.0
Simple plow (A-1)	6	0.52	6.33	0.61	0.11	0.0010	0.0010	0.112
Simple plow (A-2)	4.5	2.60	7.00	0.71	0.21	0.0015	0.0020	0.126
Simple plow (A-3)	8	2.60	7.25	0.71	0.21	0.0045	0.0034	0.214
Simple plow (A-4)	8	3.85	6.83	0.71	0.21	0.0100	0.0075	0.274
Shielded plow (B)	—	0.52	—	—	—	0.0025	—	—
Scoop (C)	6	1.48	6.67	0.65	0.15	0.0023	0.0023	0.200
Twist interchanger (D)	3	1.94	5.75	0.58	0.08	0.0016	0.0032	0.242
Triangular plow (E-1)	6	1.02	6.42	0.66	0.16	0.0015	0.0015	0.147
Triangular plow (E-2)	6	2.87	6.75	0.67	0.17	0.0050	0.0050	0.289
Triangular plow (E-3)	6	4.08	7.92	0.75	0.25	0.0240	0.0240	0.780
Ramp (F)	6*	2.87	5.75	0.59	0.09	0.0037	0.0037	0.214
Tapered fin (G-1)	2	0.17	6.42	0.52	0.02	—	—	—
Tapered fin (G-2)	4	0.52	6.00	0.62	0.12	0.0033	0.0050	0.562
Dome (H)	6	0.50	5.92	0.60	0.10	0.0032	0.0032	0.364
Vortex generator (J)	6.75	3.80	7.17	0.69	0.19	0.0097	0.0086	0.323
Shielded sink (K)	6†	2.67	6.75	0.67	0.17	—	−0.0030	−0.176

Source: From Schubauer & Spangenberg, 1960.

Note: L and x_{tr} respectively indicate optimum spacing (in inches) and position (in feet); x_s is the mean position (in feet) of separation over the central 45 in. of span. $\Delta p_s/q_o$ is the increase in pressure coefficient caused by a particular device. It is equal to $[(p_s - p_o)/q_o] - 0.50$. D/q_{tr} is given in square feet and $\Delta\theta_{tr}$ in feet.

*The optimum value of L was 3 in., giving $x_s = 6.08$ ft. A value of 6 in. was used because the investigators were interested in comparing devices (F) and (E-2).

†Optimum L and x_t not determined.

interested in determining the average state of the flow in terms of an average δ^* and an average θ (the average was taken across the span of the devices). They found that the mechanics of mixing were about the same for all the devices tried, namely, an induction into the boundary layer of currents of higher velocity usually accompanied by streamwise vortices. Furthermore, an approximately constant percentage increase or decrease in θ was maintained throughout the subsequent course of the boundary layer. However, this rule does not apply unless the more gradual pressure rise has essentially the same form as that of the steep rise to which forced mixing is applied.

Figure 4.67 gives an example of a set of velocity profiles at various cross-stream positions (z/L) and various distances x along the wall behind the simple plow (device A-3). Here δ is the boundary-layer thickness for the same station in the absence of mixing devices. The two sets of dashed curves

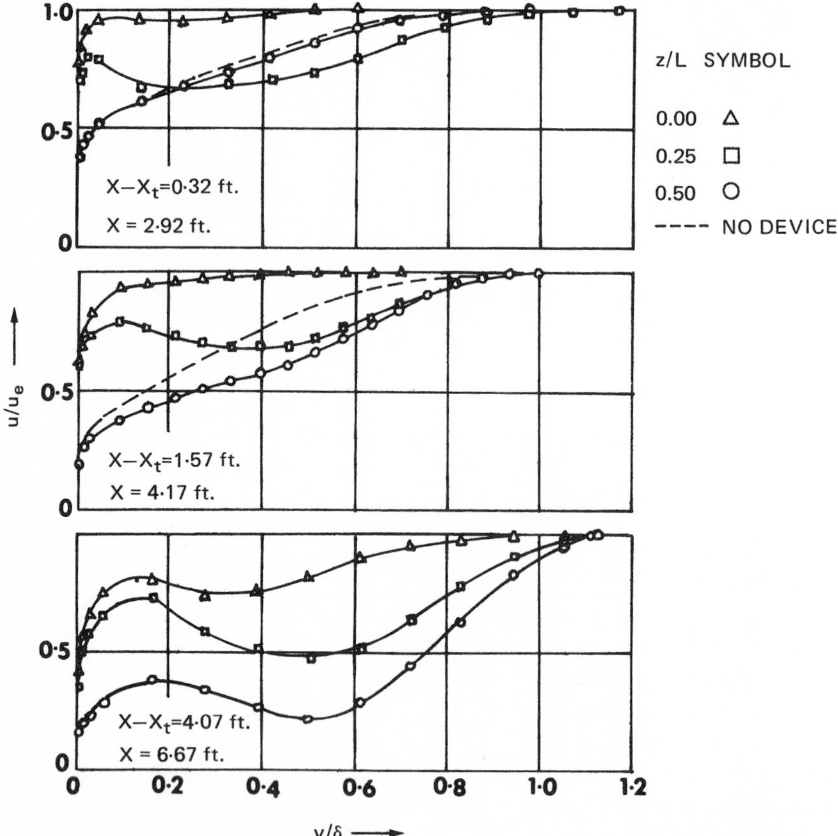

FIG. 4.67 Mean-velocity profiles aft of simple plow (A-3) for conditions given in Table 4.4. Here δ is the boundary layer thickness at the same station (x) with mixing devices absent (*from Schubauer & Spangenberg, 1960*).

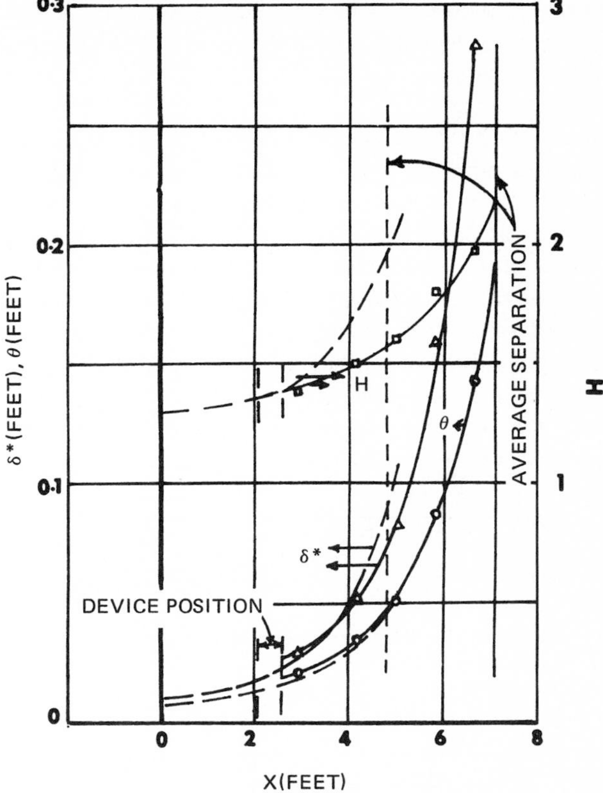

FIG. 4.68 Effect of forced mixing on separation and average thickness and shape parameters (here δ^* and θ are mean values derived by averaging local values over a sufficient span to be equivalent to a two-dimensional case; *from Schubauer & Spangenberg, 1960*).

indicate the velocity profiles for the condition without the mixing device. With variations in degree, the example in Fig. 4.67 is typical of the velocity patterns of the other devices. One exception was the ramp (F); it showed a similar pattern, but the higher flow rate was displaced to the region between devices. From Fig. 4.67 and similar velocity profiles, local values of δ^* and θ were calculated, and the average of these was taken to obtain parameters for an equivalent two-dimensional boundary layer. The experimental results for the six devices enable certain common behavior patterns to be observed under the operating conditions given in Table 4.4 and generalities derived. Figure 4.68 indicates the typical change in boundary layer development brought about by installing vortex generators, in this case the simple plow (A-3).

The performance of the devices was generally similar although not all were as effective as (A-3) in improving the pressure distribution over that attainable without devices. The drag of device A-3 resulted in an increase in $\Delta\delta^*$ and $\Delta\theta_t$ placing its curves above those without devices. The symbols $\Delta\delta^*$ and $\Delta\theta_t$ refer to the increment of the displacement thickness and momentum thickness at the trailing edge of the device. However, subsequent increases in δ^* and θ were rather slower, and the solid curves approached or were below the dashed curve continuing with the characteristic steepening rise to the delayed separation point. For the shielded sink (K), $\Delta\theta_t$ was negative, and step in δ^* and θ was down, and the curves were entirely below the dashed curves. The values of H were reduced but at separation, they reached about the same value as found for the natural layer. This value ranged from 1.75 to 2.3 for an average of 2. These values of H are somewhat lower than those quoted in the literature. This is attributed to the fact that they apply to the upstream extreme of a fluctuating separation point.

Figure 4.69 compares boundary-layer development when forced mixing by various devices was used to attain a given pressure recovery and, alternately,

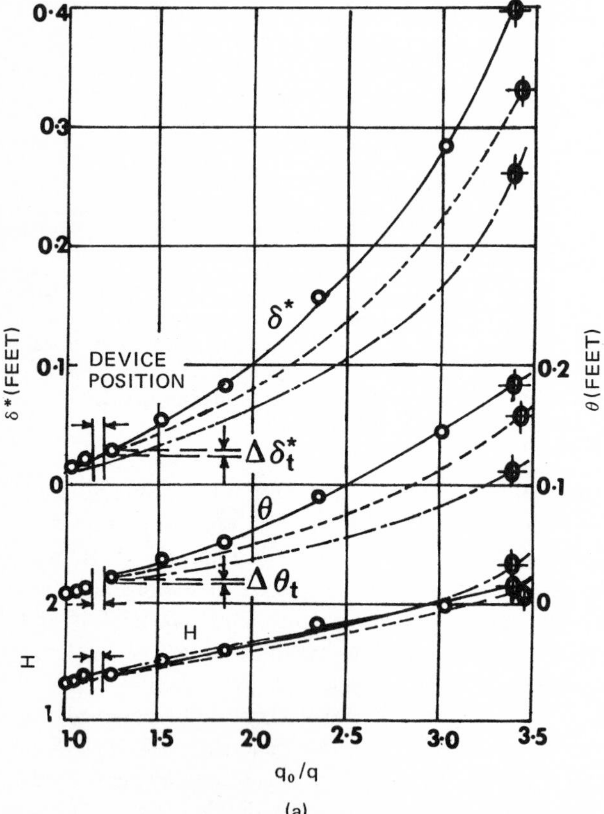

FIG. 4.69 Development of boundary layer by forced mixing and by adjusting pressure distribution: (*a*) performance of the simple plow (A-3) (*from Schubauer & Spangenberg, 1960*).

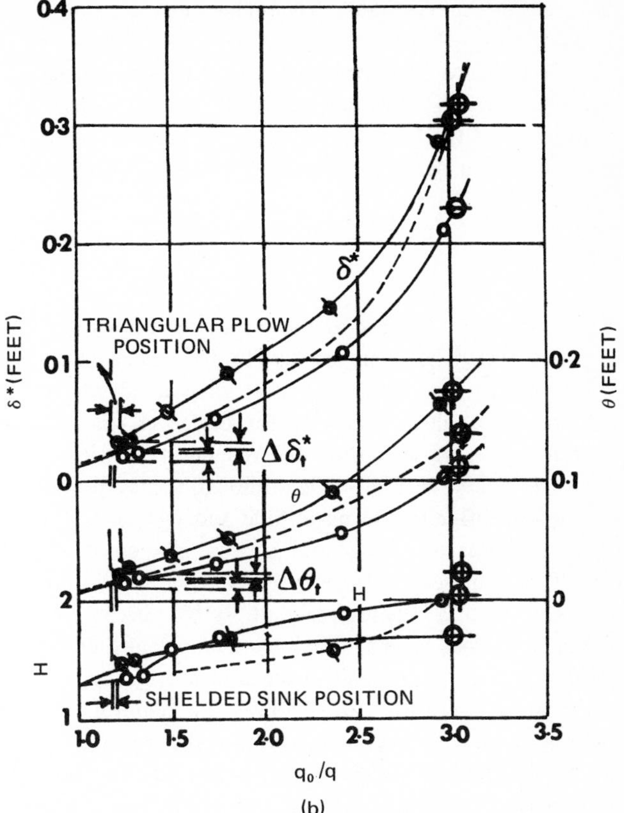

FIG. 4.69 Development of boundary layer by forced mixing and by adjusting pressure distribution: (*b*) performance of the triangular plow (E-3) and the shielded sink (K) (*from Schubauer & Spangenberg, 1960*).

when the recovery was attained by adjusting the pressure distribution alone.

The performance of the simple plow (A-3) is illustrated in Fig. 4.69*a* by comparing the values of δ^*, θ, and H for forced mixing with those of pressure distribution C and D. Note that the drag of the device caused a step-up in $\Delta\delta_t^*$ and $\Delta\theta_t$. This lifted the curves for δ^* and θ aft of the "device position" above the corresponding curves for C and D. The principal feature to be pointed out here is that the θ curve for forced mixing lies above that for pressure distribution C by an almost constant percentage that is approximately equal to that given by $\Delta\theta_t/\theta_t$ in Table 4.4. Although the data for pressure distribution B and C were available only for this comparison, this feature was present to some extent in all cases. Note further that the H curves generally lay close together. These tendencies indicate that skin friction played a negligible role and that $\Delta\theta_t$ at the device affected the subsequent development by causing a constant percentage increase in θ as mentioned previously.

The pressure recovery obtained with the triangular plow (E-2) and the shielded sink (K) are compared with pressure distribution E in Fig. 4.69*b*. Inspection of the θ curves reveals that the value of $\Delta\theta_t$ at the devices affected the subsequent development of the boundary layer according to

$$\frac{\theta}{\theta_i} = \left(\frac{q_i}{q}\right)^{(H/2+1)}$$

where subscript i refers to the initial condition, the special case $\tau_w = 0$, and $H =$ constant. The value of $\Delta\theta_t/\theta_t$ for device (E-2) is given in Table 4.4 as 28.9 percent; at separation the increase above the dashed curve is 27.2 percent. The value of $\Delta\theta_t/\theta_t$ for device (K) is given in Table 4.4 as -17.6 percent, while at separation the reduction is -19.4 percent. A similar relation for δ^* is lacking in Fig. 4.69*a*, although it may prevail for other cases.

Devices differ in the magnitude and sign of $\Delta\delta_t^*$ and $\Delta\theta_t$ and, accordingly, their responses to a decreased pressure gradient differ. When $\Delta\theta_t$ is positive, a penalty is imposed on forced mixing. This is not so much because of the force on the devices themselves but rather because the resulting percentage momentum loss is magnified by being deposited in a developing boundary layer. On the other hand, when $\Delta\theta_t$ is negative, the same source provides a dividend. A negative $\Delta\theta_t$ involves the removal of fluid with a momentum deficiency and involves some additional losses from fluid handling; hence not all of the dividend is clear profit. However, the obvious advantage of forced mixing is the saving in diffuser wall length that is possible for a given pressure recovery. With pressure distribution E, e.g., attainment of a pressure recovery coefficient of 0.67 requires the adverse pressure region to be 16.6 ft long. When either the triangular plow (E-2) or the shielded sink (K) is used, a length of only 6.8 ft is required for the same pressure recovery coefficient, a saving of 9.8 ft. A similar comparison between pressure distribution C and the simple plow (A-3) for a pressure-recovery coefficient is 0.7, indicates that (A-3) reduced the wall length from 16.1 to 7.3 ft, a saving of 8.8 ft.

4.5.2.3 Application of vortex generators

For convenience, the application of vortex generators is discussed with reference first to wings and then to diffusers.

Application to wings The installation of vortex generators on a wing increases maximum lift. Figure 4.70 shows the increase of lift attained when wedges were placed on a NACA 63_3-018 airfoil along a spanwise line at 0.1c from the leading edge and arranged alternately at ±22.5 deg incidence to produce a contrarotating system of vortices. The wedges extended the linear part of the lift curve and raised the stalling incidence from 14 to 20 deg, thus increasing the maximum lift from 1.33 to 1.89. With this vortex generator,

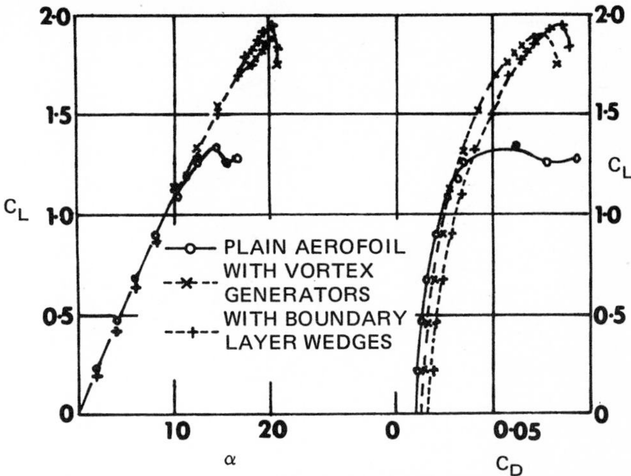

FIG. 4.70 Effect of vortex generators and boundary-layer wedges on the lift and drag of an NACA 63_3-018 airfoil (*from McCullough et al., 1951*).

the drag at $C_L > 1.1$ was less than with the plain airfoil, but at cruising incidence C_D, it was slightly greater (about 0.002). The wedge produces a slightly higher maximum lift than did a wing-type vortex generator. But as seen in Fig. 4.70, the drag of the boundary layer wedges at low incidence was twice to three times more than caused by the wing-type vortex generator (McCullough et al., 1951).

Different types of vortex generators and devices that produce vortex action can be arranged on a wing in various ways; see Figs. 4.71 and 4.72.

Application to diffusers Brown et al. (1967) successfully incorporated a wing-type vortex generator as an integral design feature of a subsonic diffuser. As mentioned previously, the principle of boundary layer control by vortex generators relies on the increased mixing between the external stream and the boundary layer that is promoted by vortices trailing longitudinally over the surface, adjacent to the edge of the boundary layer. When the arrangement is suitable, line vortices mutually interact to drive each other toward an adjacent plane wall. In this way, the fluid particles with high momentum in the stream direction are swept along helical paths toward the surface, thereby mixing with and, to some extent replacing, the retarded air at the surface. This physical process provides a continuous source of reenergization to counter the natural boundary layer retardation and growth caused by friction and adverse pressure gradient. Thus, the incorporation of a vortex generator into a diffuser designed to achieve a high pressure recovery by minimizing the energy loss and preventing separation has a high potential for success. The design technique used by Brown et al. (1967) is to pull away the wall from the vortices at such a rate that they maintain a constant distance from the wall.

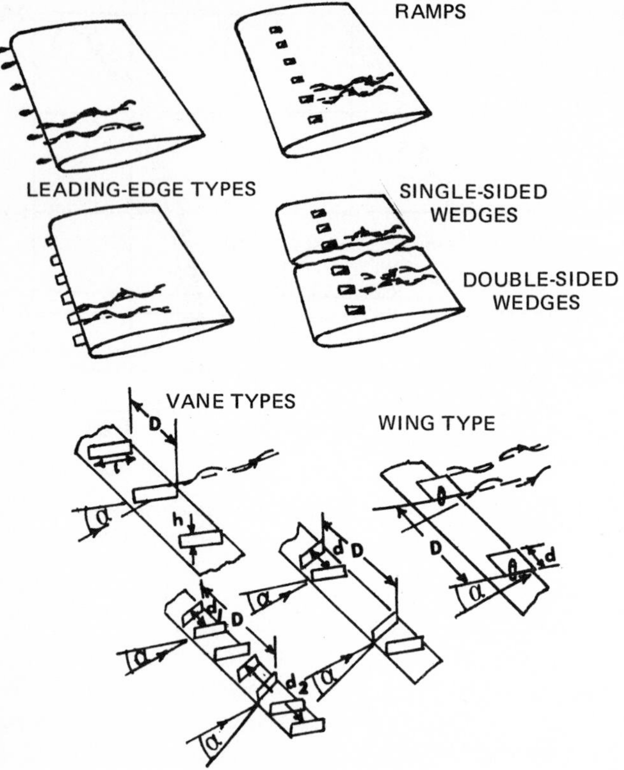

FIG. 4.71 Types of vortex generators (*from Pearcey, 1961*).

The success of this application was proved by testing two designs for simple two-dimensional diffusers. One incorporated a vortex generator and the other was a conventional high performance diffuser whose shape resembled a trumpet with constant pressure gradient. The comparison showed that both pressure recovery losses and the degree of flow uniformity may be expressed by the distortion, defined as $(p_{t_{\max}} - p_{t_{\min}}/p_{t_{\text{ave}}})$. However without a vortex

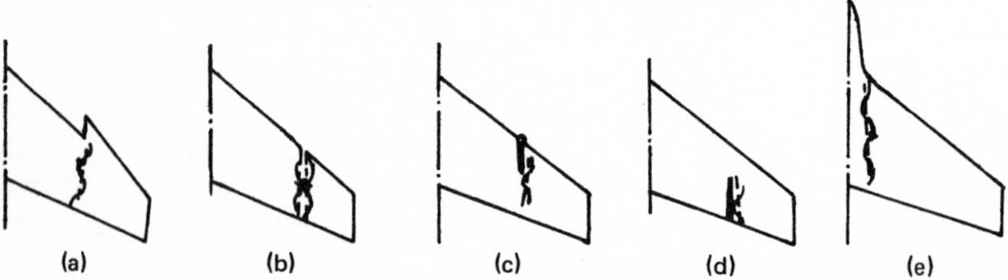

FIG. 4.72 Devices that produce vortex action: (*a*) leading-edge discontinuity; (*b*) leading-edge notch; (*c*) leading-edge fence; (*d*) rear fence; (*e*) leading-edge fairing or dorsal fin (*from Pearcey, 1961*).

generator installed, flow separated almost immediately after passing the corner in the diffuser for which the device was an integral part of the design. Separation was delayed considerably by vanes designed integrally with the diffuser, and a higher potential pressure recovery was achieved than with the conventional diffuser. It appears that the maintainance of a high vortex strength is more important than an exact match of the theoretical requirement; the analysis did not include the potential flow expansion around the corner and a real diffuser is expected to benefit both from flow expansion and from the vortex motion. Brown et al. (1967) attributed the success of the more widely spaced vanes.

Both pressure recovery losses and flow distortion were reduced by about 40 percent in the new design that incorporated a vortex generator. But when vortex generators are an integral feature of a diffuser, design mismatches carry larger penalties compared to the conventional diffuser with no vortex generator. The area of the diffuser with a vortex generator increases initially more rapidly downstream compared to the conventional diffuser.

After the successful design of this simple two-dimensional diffuser, the integrated vortex generator approach was used in a two-stage redesign of the subsonic diffuser of an inlet for a Mach 2.7 supersonic transport airplane, thus shortening the diffuser length. The desirable uniformity of the flow properties in the subsonic diffuser and the effective exchange of low momentum air near its wall with the higher momentum air outside the boundary layer was achieved by small low aspect-ratio, wing-type vortex generators. In order to ensure a substantial interchange between high and low momentum air, the vortices must run close to the edge of the boundary layer. The vortices must also be in a region where the transverse gradient of longitudinal momentum is significant. A wing-type vane whose height amounted to 1.2 boundary layer thickness, proved satisfactory as the vortex generator. However, whether vortices will remain at a required height above a wall depends on the way in which their filaments are moved by the velocity fields of their neighbors and their images. The velocity fields are determined by the initial spacing, height and sense of rotation of the vortices.

A pair of counterrotating vortices of equal strength move in a straight line in a direction normal to the plane which contains them. If a plane surface is placed parallel to the vortex plane so that the vortex planes approach the wall, then the vortices are forced away from each other. Other pairs of line vortices, parallel to the first and of similar sense, repel their neighbors, causing the vortices which approach the wall to turn and move away. This vortex pattern is sketched in Fig. 4.73.

The derivation of the paths of vortices may be given on the basis of coordinate system sketched in Fig. 4.74. Let the x-axis of a cartesian coordinate system lie in the plane of the wall and be normal to the vortex line, and let the y-axis be normal to the plane. The origin of the system is located on the wall, midway between two counterrotating vortices.

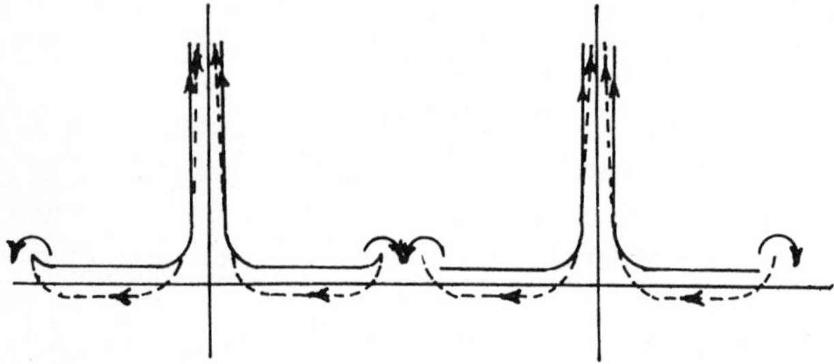

FIG. 4.73 Counterrotating vortices close to a plane (*from Brown et al., 1967*).

Analytically, the plane may be represented by reflecting the vortices in it. The vortex pair considered is one of an infinite set equally spaced a distance D apart, and all are initially at the same height above the plane. By symmetry, the motion of the complete set of vortices may be inferred from the motion of any single one. If K is denoted as the strength of a vortex, then the induced circumferential velocity at distance r from the vortex is $v = K/2\pi r$.

Assuming that all vortices have equal strength, the velocity components of the vortex in the x- and y-directions can be computed as follows:

$$\begin{aligned}\frac{dx}{dt} = \frac{K}{2\pi}\Bigg[\frac{1}{2y} &- \frac{2y}{(D-2x)^2+(2y)^2} + \frac{2y}{D^2+(2y)^2} \\ &- \frac{2y}{(2D-2x)^2+(2y)^2} + \cdots - \frac{2y}{(2x)^2+(2y)^2} + \frac{2y}{D^2+(2y)^2} \\ &- \frac{2y}{(D+2x)^2+(2y)^2} + \frac{2y}{(2D)^2+(2y)^2} - \cdots\Bigg]\end{aligned}$$

FIG. 4.74 Vortice distribution and coordinate system (*from Brown et al., 1967*).

This is reduced to

$$\frac{dx}{dt} = \frac{K}{2\pi}\left\{\frac{x^2}{2y(x^2+y^2)} + \sum_{n=1}^{\infty} \frac{16x^2y(-3n^2D^2+4(x^2+y^2)}{(n^2D^2+4y^2)[n^4D^4+8n^2D^2(y^2-x^2)+16(x^2+y^2)^2]}\right\} \tag{4.19}$$

Similarly,

$$\frac{dy}{dt} = \frac{K}{2\pi}\left\{\frac{-y^2}{2x(x^2+y^2)} + \sum_{n=1}^{\infty} \frac{16xy^2(3n^2D^2+4(x^2+y^2)}{(n^2D^2-4x^2)[n^4D^4+8n^2D^2(y^2-x^2)+16(x^2+y^2)^2]}\right\} \tag{4.20}$$

The movement of the vortices is made up of two components that result from mutual interaction between the active vortices and from the expansion of the gas above the plane as it maintains local density equilibrium. This expansion will carry the vortices with it. Equations (4.19) and (4.20) have been solved for both a fixed plane and a moving plane. The results shown in Fig. 4.73 are for the vortices located initially at $x_0/D = 0.05$ and $y_0/D = 0.08$. The results of this computation were used to design a subsonic diffuser by applying a time-distance transformation. Suppose there is a subsonic flow of velocity w in the z-direction normal to the x-y plane. If a fixed plane, parallel to the x-z plane, is imagined some distance above the initial position of the vortices, then it may be considered to flow through a two-dimensional channel. As the moving plane drops, the velocity w decreases as a function of time because of the increase in flow area. Thus, the following transformation is used:

$$z = \int_0^t w\,dt \tag{4.21}$$

Since the position of the plane is given by Eq. (4.20), which also determines the local flow area and $w(t)$, the coordinates of the wall may be determined in terms of y and z. In the design of a subsonic diffuser, the curved wall is to be terminated at or before the minimum value y is reached. Figure 4.75 shows the configuration of the subsonic diffuser that contained a vortex generator as an integral design feature (herein after referred to as "integrally designed").

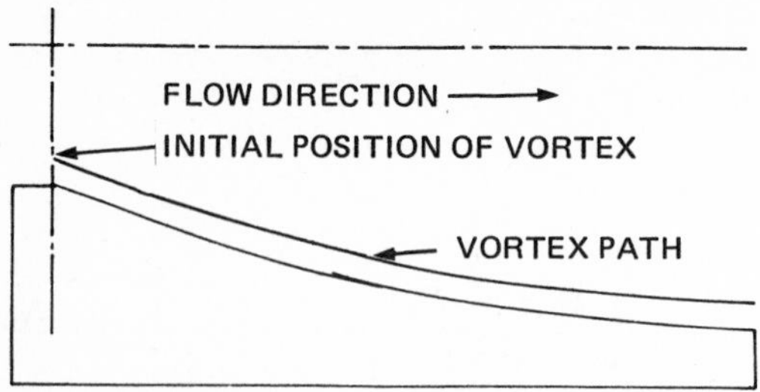

FIG. 4.75 Subsonic diffuser with vortex generator as integral design feature (*from Brown et al., 1967*).

As seen in Fig. 4.75, the integrally designed subsonic diffuser would have a very steep discontinuous initial slope, although an actual design would allow some rounding of this corner. It was assumed for this design that the natural expansion of the flow does not contribute to the effective shape of the wall. This is in direct contrast to the known design procedure for a conventional diffuser, where it is assumed that vortices are either not present or do not contribute to the diffuser expansion. The real diffuser may benefit from both effects. The difference in configuration of the two-dimensional integrally designed diffuser and the two-dimensional conventional diffuser is demonstrated in Fig. 4.76.

Now let us study in more detail the arrangement of vortex generators within the diffuser, the pressure distribution achieved, and the final subsonic diffuser design for a supersonic transport airplane. As mentioned previously, the vortex generators applied in the subsonic diffuser are rectangular wing-type vanes arranged in pairs, as shown in Fig. 4.77. The vortex generators have 6-percent thick, flat-bottomed, cambered airfoil sections designed to operate satisfactorily at high subsonic Mach numbers and a 16-deg angle of attack. The height of the vane was 20 percent larger than the

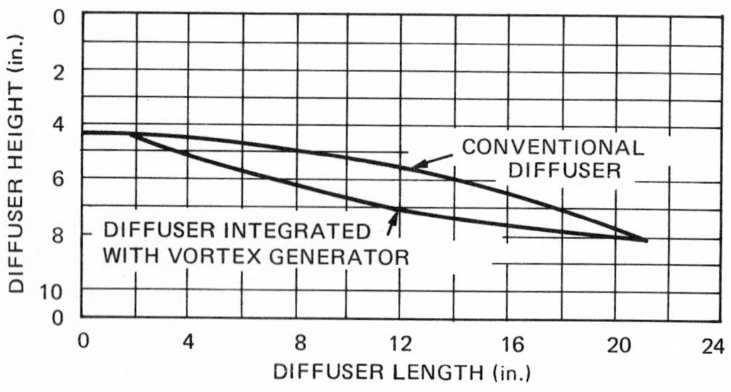

FIG. 4.76 Comparison of the two-dimensional subsonic diffusers (*from Brown et al., 1967*).

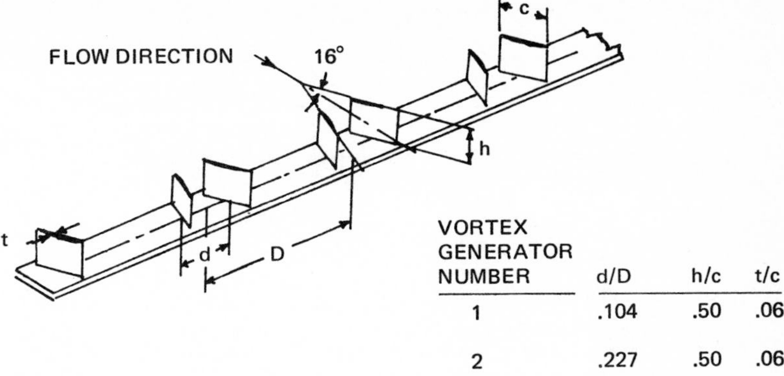

VORTEX GENERATOR NUMBER	d/D	h/c	t/c
1	.104	.50	.06
2	.227	.50	.06

FIG. 4.77 Vortex generator design configuration (*from Brown et al., 1967*).

boundary layer thickness and the chord of the vane was twice that of the height. According to slender wing theory, no increase in lift can be expected by increasing the chord length. If the pairs were arranged so that the distance between the two vanes at the midchord point amounted to one-tenth of the distance between pairs, then the airfoils would be close enough to each other to have significant lift interference and the vortex strength would be lost. Therefore, the distance between the pairs or between generators was increased so that the increased lift effectiveness counteracted the reduction in induced vortex motion toward the diffuser wall. Figures 4.78 and 4.79 show the test results for Mach 0.5 with a turbulent boundary layer of 0.27-in. thickness at the entrance.

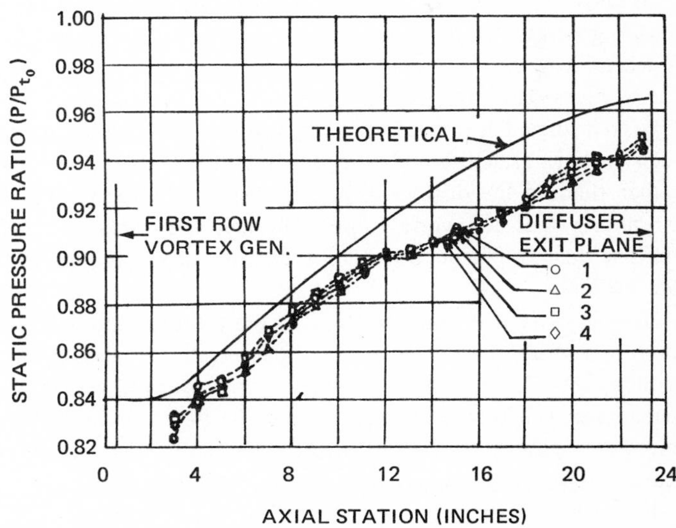

FIG. 4.78 Static pressure distribution for conventional diffuser: 1. no vortex generators; 2. $d/D = 0.104$ vortex generator; 3. $d/D = 0.224$ vortex generator; 4. wing/pylon vortex generator (*from Brown et al., 1967*).

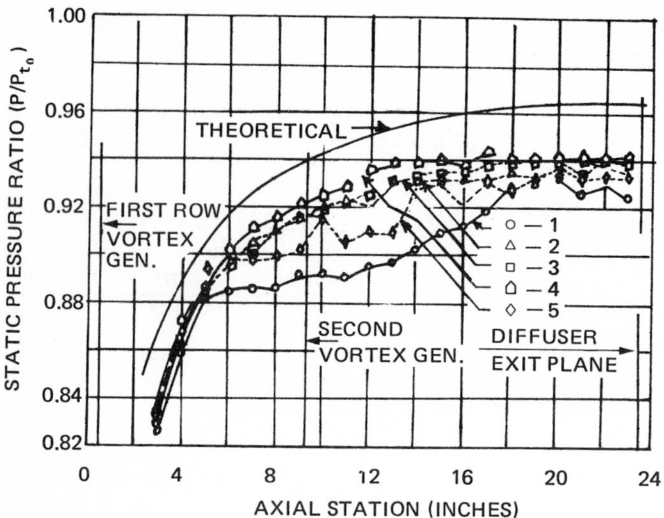

FIG. 4.79 Static pressure distribution for diffuser with integrated vortex generators: 1. no vortex generators; 2. $d/D = 0.104$ vortex generator; 3. $d/D = 0.224$ vortex generator; 4. 2nd row of vortex generator; 5. wing/pylon vortex generator (*from Brown et al., 1967*).

For the conventional diffuser involving no separation and operated efficiently at Mach 0.5, the effects of vortex generators were insignificant because their presence had almost no effect on the static pressure distribution. It is seen from Fig. 4.79 that some separation can occur even for the best vortex generator arrangement. It was therefore decided to add to the diffuser a second row about one-third of the way down to provide reinforcement to the vortices. However, it is extremely important to place the vanes so that they reinforce the vortices and do not tend to nullify them. Furthermore, their angle of attack should be optimal with respect to the local flow direction rather than to the diffuser axis. The distance between vanes was determined by noting that the vortices pass through the vanes and that at any axial station, the flow inclination to the diffuser axis is the greatest directly under the vortices.

A pair of typical contour maps of the total pressure is shown in Fig. 4.80. Both sets of measurements are for the integrally designed diffuser; the first are without a vortex generator and the second with the best single row of vortex generators. The first map indicates that the flow is essentially two-dimensional and demonstrates the characteristic effects of early separation followed by reattachment. The second map shows contours characteristically associated with counterrotating devices. The flow field is broken down into small pockets; regions with high pressure air knife down between the vanes, and low pressure air is blown out into the main stream. This type of flow distortion involves large numbers of small areas where pressure regions are lower. It is

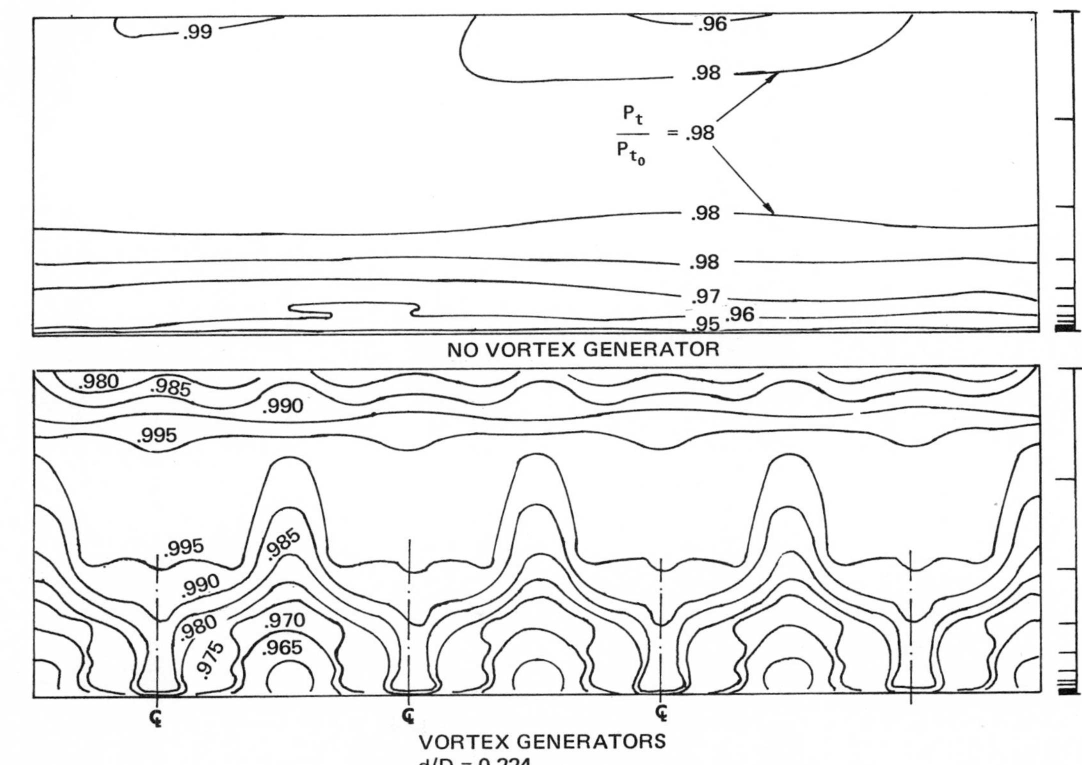

FIG. 4.80 Pitot pressure profiles at integrated diffuser exit (*from Brown et al., 1967*).

easier for the compressor to handle than the situation where there is a smaller number of larger areas with low pressure regions. Figure 4.81 illustrates the performance of the two diffusers with respect to distortion and pressure recovery.

The test results shown are consistent with the static pressure distribution. All configurations of the conventional diffuser gave about the same result, a 2-percent loss in pressure recovery and a distortion of 0.05. The distortion was calculated by using an adaptation of the General Electric method which accounts for both size and form of low pressure areas. Without vortex generator, the performance of the Brown et al. designed diffuser was little worse than the conventional one and certainly not as poor as had been expected. When the matched vortex generators were added, the performance exceeded that of the conventional diffuser; pressure recovery loss was down to 1.8 percent and distortion dropped to 0.038.

The inlet for the Lockheed supersonic transport (SST) airplane (designed by Brown et al., 1967) has a rectangular entrance and is divided into two isolated ducts by a variable geometry center body. The subsonic diffuser

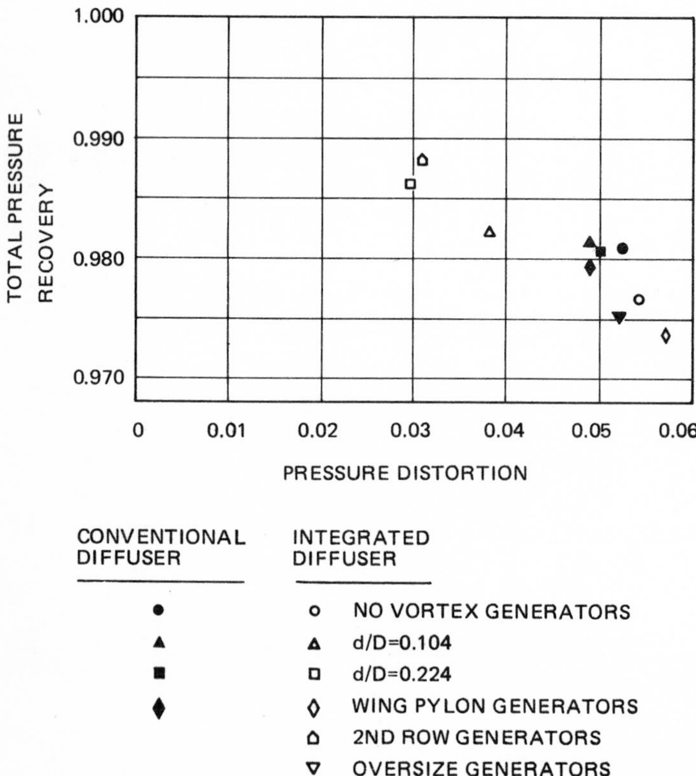

FIG. 4.81 Pressure recovery and flow distortion (*from Brown et al., 1967*).

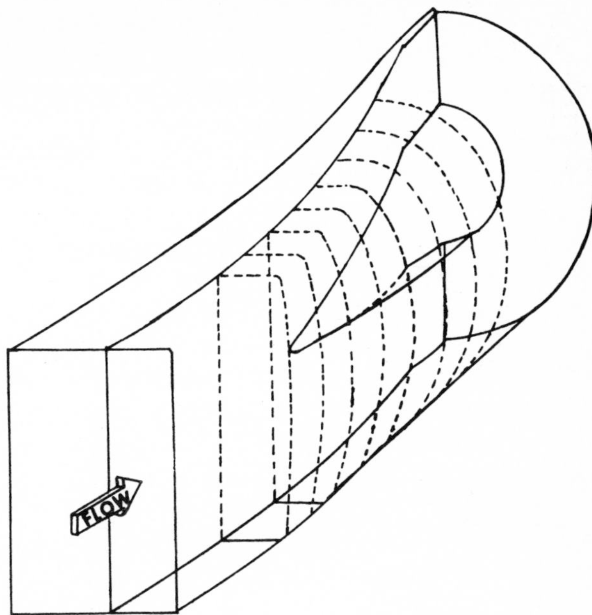

FIG. 4.82 Subsonic diffuser for Lockheed SST (*from Brown et al., 1967*).

(Fig. 4.82) enables a smooth transition from a rectangular cross section at the throat to a circular cross section at the engine surface.

The most severe adverse pressure gradient in the diffuser occurs on the center body, which has significant curvature, and at the top of the duct. The diffuser sweeps up to meet the engine which is partially buried in the wing. Consequently, when the subsonic diffuser was first shortened, there was a significant region of separated flow at the top, as shown in Fig. 4.83.

The purpose of the vortex generator design was to ensure that the line vortices remain close to the wall and do not move significantly toward the center core before they reach the engine face. This problem is more difficult for a diffuser of practical cross section than for the simpler two-dimensional one discussed previously. Hence, the additional double row of vortex generators was applied to the SST subsonic diffuser as shown in Fig. 4.84.

The positioning of the second row of vanes reflects the fact that pairs of counterrotating vortices move away from each other as they travel downstream. The separation was eliminated completely and a further improvement in performance was achieved as noted below.

Configuration	Total pressure recovery	Distortion
Clean diffuser	0.859	0.195
One row generators	0.867	0.153
Two row generators	0.877	0.122

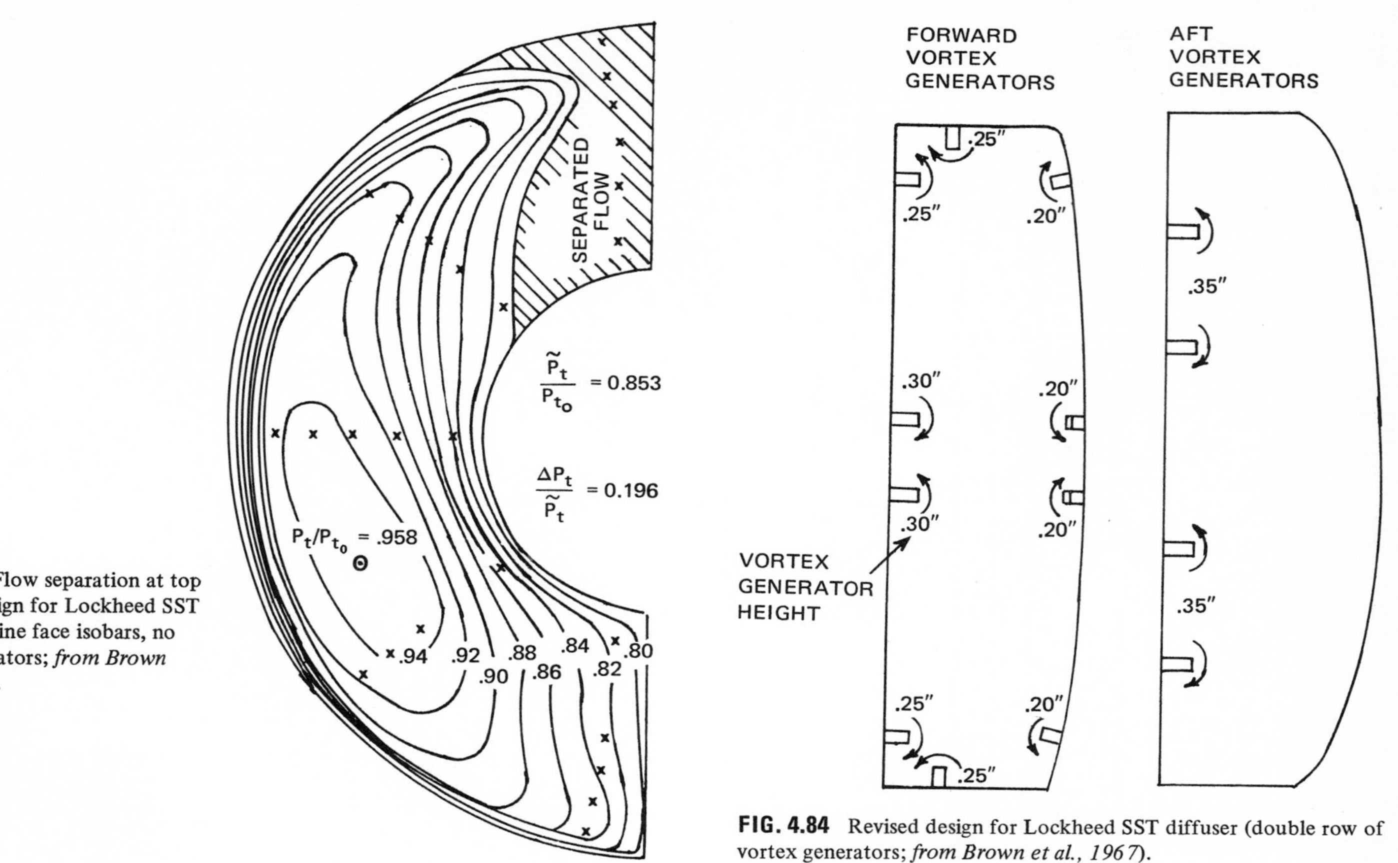

FIG. 4.83 Flow separation at top of initial design for Lockheed SST diffuser (engine face isobars, no vortex generators; *from Brown et al., 1967*).

FIG. 4.84 Revised design for Lockheed SST diffuser (double row of vortex generators; *from Brown et al., 1967*).

Pressure isobars for the diffuser with two rows of vortex generators at a pressure recovery of 0.877 are shown in Fig. 4.85.

The optimum design of the subsonic diffuser (Fig. 4.86) took advantage of the results of previous test series of the integrated design procedure. Structural and off-design performance requirements limited the amount of modification to small changes in the center body contours. Although the scope of the testing was limited, extremely gratifying results were reported (Brown et al., 1967). The operating pressure recovery was raised to 0.891 and distortion dropped to 0.095 at a free-stream Mach number of 2.65. The total pressure map showed the strong effects of the vortex generators in driving the flow into the corner (Fig. 4.87) and the generally higher levels of pressure recovery due to both more rapid subsonic diffusion and improved supersonic compression.

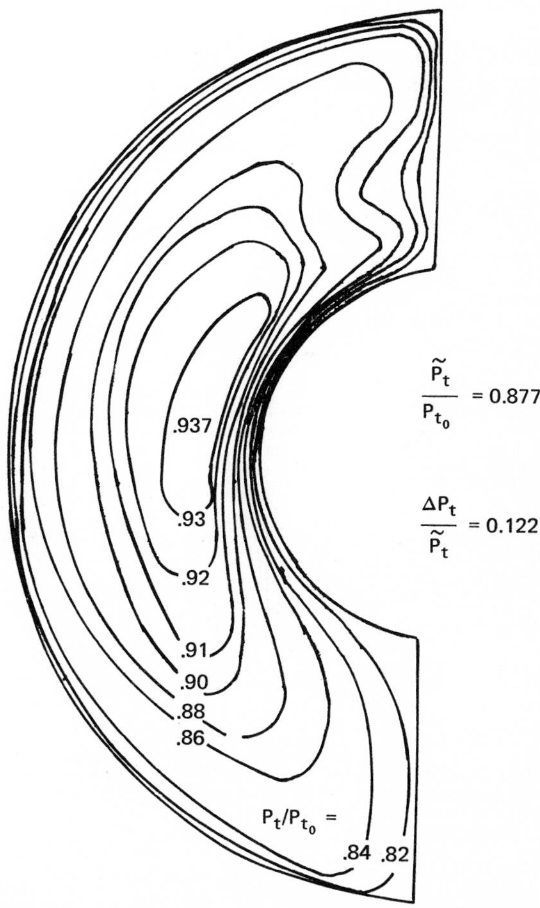

FIG. 4.85 Pressure isobars for the SST diffuser with two rows of vortex generators from engine face isobars (*from Brown et al., 1967*).

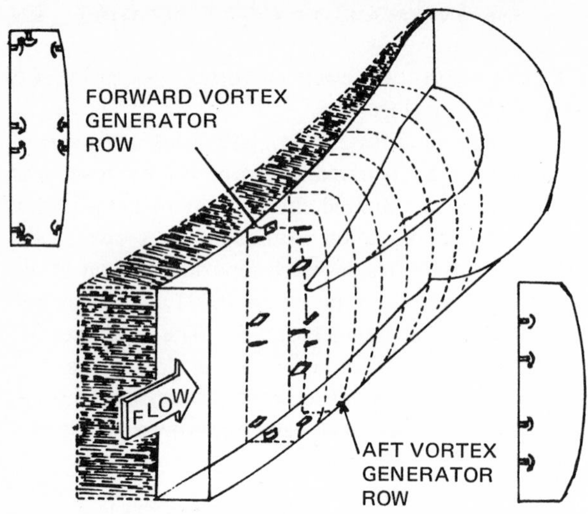

FIG. 4.86 Optimum configuration for SST subsonic diffuser with vortex generators (*from Brown et al., 1967*).

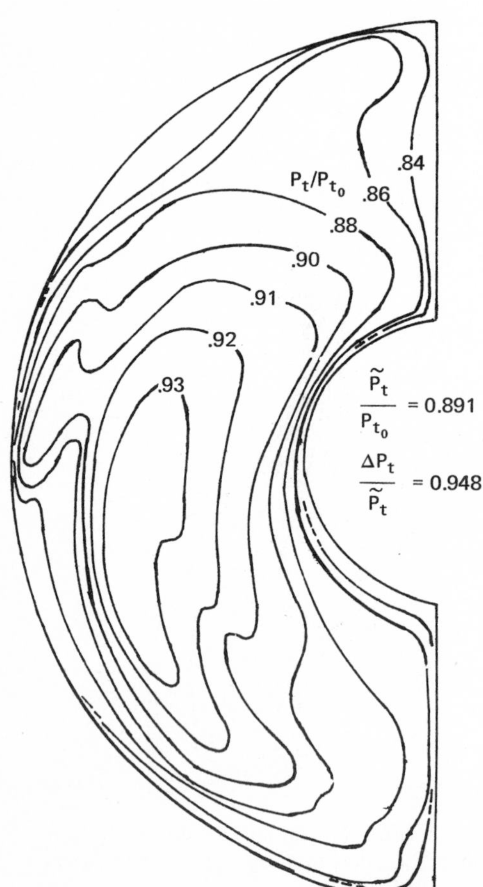

FIG. 4.87 Pressure isobars for the optimum configuration of the SST subsonic diffuser (engine face isobars; *from Brown et al., 1967*).

Figure 4.87 showed some evidence of lower pressure regions between vortex pairs, but the method of design provides some flexibility for moving the total pressure contours in a predictable manner to improve distortion acceptable by the compressor. An incorrect choice of the vortex generators in the design and arrangement could lead to significant losses in performance, but an optimum design may shorten the diffuser length about 30 percent compared to the conventional diffuser.

4.5.3 Screens

At subsonic speed, screens are useful for preventing separation, for causing reattachment of separated flows, and for increasing steadying effects. A screen is helpful when suppression of separation or a strongly steady flow are more important than efficiency. However, screens may create drag and maintenance problems.

Figure 4.88 gives an example of flow phenomena affected by a screen (Frey & Vasuki, 1966). The separated and strongly turbulent oncoming flow (57) is distributed by a screen (56). Regardless of the local flow direction immediately upstream of the screen, the flow leaves normal to the plane of the screen. Velocity differences upstream of the screen can be substantially reduced or damped out. Depending on the required uniformity of flow, more

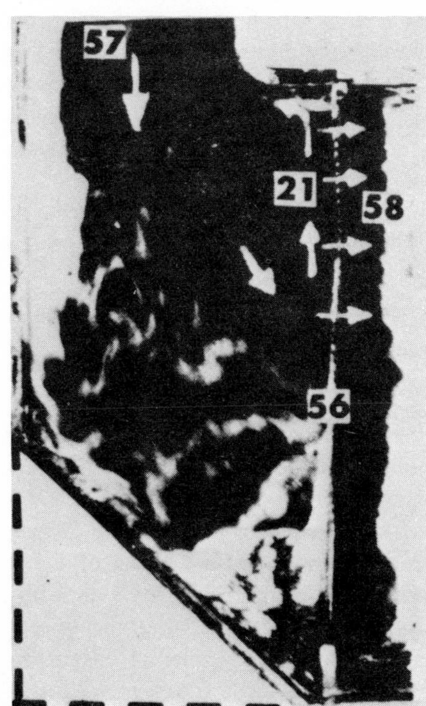

FIG. 4.88 Effect of a screen on flow phenomena (the narrow screen greatly reduces the effects of the turbulent influence and the sharply deflected inner boundary on the flow redistribution; the black-white line downstream of the screen indicates the character of the velocity distribution; *from Frey & Vasuki, 1966*).

than one screen may be used; preferably they should be of different sizes. The white patch (58) may be regarded as the velocity profile. It is obtained by instantaneously sifting a heavy band of powder on the water surface immediately downstream of the screen throughout the width (Frey & Vasuki, 1966). The use of the screen does not allow high efficiencies. A screen is frequently used as a diffuser. The efficiency of the diffuser may be expressed by:

$$E_{1.2} = \frac{P_2 - P_1}{K_1 - K_2}$$

where $E_{1.2}$ is the efficiency of the diffuser between Secs. 1 and 2, $P_2 = \int_0^{A_2} p_2 u_2 \, dA$ is the total flow of potential (pressure) energy per second across Sec. 2, $P_1 = \int_0^{A_1} p_1 u_1 \, dA$ is the same factor for Sec. 1, $K_1 = \int_0^{A_1} q_1 u_1 \, dA$ is the total flow of kinetic energy per second across Sec. 1, $K_2 = \int_0^{A_2} q_2 u_2 \, dA$ is the same factor for Sec. 2, and q is dynamic pressure.

The so-called "filling effect" of a screen is sufficiently effective to prevent separation or to cause separated flow to reattach, even for the extreme divergence of diffuser flow separation. Filling means that flow passage occurs throughout all of the available volume in the diffuser, either because of diffuser design or because of the effect of a screen. In this sense, then, filling means the absence of separation. Therefore, a *filled condition* may be defined as one in which the velocity distribution at every section of the diffuser is similar to that at the entrance. Schubauer and Spangenberg (1949) have shown that at subsonic speeds, screens divert the air flow toward the walls, thus increasing the velocity gradient and the shearing stress at the wall. Accordingly, the mechanics of the flow process through a screen are closely connected with those of the turbulent boundary layer, and the screen may prevent separation by increasing the normal velocity gradient near the diffuser wall by decreasing the pressure gradient along the wall, or by a combination of these two effects.

The filling effect depends principally on screen locations. The pressure drop acceleration of the screen is defined by $k = \Delta p/q$, where Δp is the static pressure drop across the screen. When the screen is at the most forward position, separation may be removed by increasing k and filling the diffuser fairly well upstream, but separation may still remain downstream. When the screen is arranged downstream, the diffuser is filled downstream but not upstream. If a screen with a given pressure drop acceleration is placed at the extreme end of a narrow-angle diffuser, the loss in efficiency is minimal. This enables the angle just ahead of the screen to be widened and the area ratio increased. Thus for a given area ratio and given value of k, the most efficient configuration of a diffuser and screen is that of a narrow angle terminating in a wide angle. Multiple screens in a diffuser of arbitrary shape can be as effective as a single screen in a diffuser of a special shape. It is easier to design a diffuser of simple shape providing proper choice is made for number and location of screens (Schubauer & Spangenberg, 1949).

NOMENCLATURE

AR	Aspect ratio
b	Width of channel
C	Constant
C_{DP}	Profile drag
C_m	Pitching moment coefficient
c	Chord
$\bar{\bar{c}}$	Aerodynamic mean chord
$\tilde{c}_{f_0}$	Ratio of c_{f_0} at a given Re_{x_0} to the corresponding value at $\mathrm{Re}_{x_0} = 10^6$
D	Distance between vortices
E	Efficiency
K	Strength of vortex; also kinetic energy per second
k	Pressure drop acceleration of a screen
L	Reference length; also cross-stream spacing
m	Index of power for inviscid velocity distribution; also value of stream function along turning vane
P	$P = \partial p/\partial x$; also potential (pressure) energy per second
p_t	Total pressure
p^*	Critical pressure
$\tilde{p}_t$	Area weighted total pressure
Δp_t	Difference between maximum and minimum total pressure
p_{t_0}	Total pressure at diffuser entrance
r	Cylindrical coordinate; also radius
t	Airfoil thickness
u	Axial velocity component
u^*	$u^* = \sqrt{\tau_w/\rho}$, friction velocity
v	Circumferential velocity component
W	Width
w	Radial velocity component
z	Cross stream position; also transformation shown in Equation (4.21)
$z_{1.25}$	Upper surface ordinate at 1.25-percent chord
$z_{5.0}$	Upper surface ordinate at 5-percent chord
β	$\beta = \sqrt{\mathrm{M}^2 - 1}$; also empirical factor
δ	Flap deflection angle
Λ	Wing sweep back angle
λ	Wing taper ratio
ξ	$\xi = r/x$
ρ	Leading edge radius

Superscripts

′	Condition downstream of $x = x_0$
=	Mean

Subscripts

c	Chord
eq	Equivalent
fl	Flap
i	Inner; also edge of inner layer
k	Referred to end of linear variation of force or moment
n	Normal
0	Leading edge; also outside edge of vortex; also upstream of interaction; also outer; also condition at $x = 0$; also initial condition
p	Location of vane
s	Streamwise
t	Tip
tr	Trailing edge
w	Wing plan form
1	Entrance to channel; also inlet; also condition after shock
2	Outlet

REFERENCES

Ackeret, J. (1925). "Luftkräfte an Flügeln. die mit grösserer als Schallgeschwindigkeit bewegt werden," *Z Flugtechn. Motorluftsch. 16*, pp. 72–74.

Bogdonoff, S. M. (1955). "Some experimental studies of the separation of supersonic turbulent boundary layers," paper presented at Heat Transfer and Fluid Mechanics Institute, University of California at Los Angeles.

Brown, A. C. et al. (1967). "Subsonic diffusers designed integrally with vortex generators," Paper 67-464, AIAA Third Propulsion Joint Specialist Conference, 17–21 Jul, Washington, D.C.

Carrow, D. D. (1954). "A note on the boundary layer and stalling characteristics of aerofoils," ARC Tech. Note, C. P. 174 (AD 69440).

Chapman, D. R. et al. (1958). "Investigation of separated flows in supersonic and subsonic streams with emphasis on the effect of transition," NACA Report 1356.

Chappell, P. C. (1968). "Flow separation and stall characteristics of plane, constant-section wings in subcritical flow," *J. Royal Aeron. Soc.*, Vol. 22, No. 685, pp. 82–90.

Clauser, F. H. (1954). "Turbulent boundary layers in adverse pressure gradients," *J. Aeron. Sci.*, Vol. 21, No. 2, pp. 91–108.

Cornish, J. J. (1958). "Practical high-lift systems using distributed boundary-layer control," Mississippi State University Aerophysics Department, Research Report 19.

Cornish, J. J. (1965). "Some aerodynamic and operational problems of STOL aircraft with boundary-layer control," *J. Aircraft*, Vol. 2, No. 2, pp. 78–86.

Dommasch, D. O. et al. (1961). *Airplane aerodynamics*, Third Edition, Pitman Publishing Corporation, New York.

Eyre, R. C. W., and S. F. J. Butler (1967). "Low speed wind tunnel tests on an AR 8 swept wing subsonic transport research model with B.L.C. blowing over nose and rear flaps for high lift," RAE Tech. Report 67112.

Frey, K. P. H. and N. C. Vasuki (1966). "Detached flow and control," published by the authors at Box 584, Newark, Delaware, 19711.

Gadd, G. E. (1957). "A theoretical investigation of laminar separation in supersonic flow," *J. Aeron. Sci.*, Vol. 24, No. 10, pp. 759–771.

Garner, H. C. and D. K. Cox (1961). "Surface oil-flow patterns on wings of different leading-edge radius and sweep-back," ARC CP 583.

Gaster, M. (1966). "The structure and behavior of laminar separation bubbles," in *Separated Flows*, Part II, AGARD Conference Proceedings, No. 4.

Gault, D. (1955). "Boundary layer and stalling characteristics of the NACA 63-009 airfoil sections," NACA TN 3524.

Gault, D. E. (1957). "A correlation of low-speed airfoil-section stalling characteristics with Reynolds number and airfoil geometry," NACA TN 3963.

Goett, H. J. and W. K. Bullivant (1938). "Tests of NACA 0009, 0012, and 0018 airfoils in the full-scale tunnel," NACA Report 647.

Graham, R. R. (1951). "Low-speed characteristics of a 45° sweptback wing of aspect ratio 8 from pressure distribution and force tests at Reynolds numbers from 1.5×10^6 to 4.8×10^6," NACA RM L51H13.

Gyorgyfalvy, D. (1959). "Flight research investigation of laminar separation of a high lift boundary layer control plane," Mississippi State University, Aerophysics Department Research Report 25.

Hakkinen, R. J., et al. (1959). "The interaction of an oblique shock wave with a laminar boundary layer," NASA Memo 2-18-59W.

Hall, I. M. and E. W. W. Rogers (1960). "The flow pattern on a tapered sweptback wing at Mach numbers between 0.6 and 1.6," ARC R&M 3271, Part 1.

Hall, M. G. (1959). "On the vortex associated with flow separation from a leading edge of a slender wing," RAE Tech. Note Aero 2629.

Harvey, J. K. (1958). "Some measurements on a yawed delta wing with leading edge separation," ARC 20, 451.

Henshall, B. D. and R. F. Cash (1955). "The interaction between shock waves and boundary layers at the trailing edge of a double-wedge airfoil at supersonic speed," ARC R&M 3004.

Himmelskamp, H. (1945). "Profiluntersuchungen an einem umlaufenden Propeller," Dissertation, Göttingen, 1945, Mitt. Max-Planck-Institut für Strömmungsforschung, Göttingen, No. 2, 1950.

Holder, D. W. (1964). "Transsonische Strömungen an zweidimensionalen Flügeln," *Z. Flugwiss.*, Vol. 12, pp. 285–303.

Jacobs, W. (1952). "Experimentelle Untersuchungen am schiebenden Flügel," *Ingr.-Arch.*, Vol. 20, pp. 418–426.

James, H. A. and R. L. Maki (1957). "Wind-tunnel tests of the static longitudinal characteristics at low speed of a swept-wing airplane with blowing flaps and leading-edge slats," NACA RM A57 D11.

Jones, R. T. (1947). "Effects of sweep-back on boundary layer and separation," NACA Report 884.

Kamiyama, S. (1967). "Theory on the flow through bends with turning vanes," ASME Paper 67-FE-13, presented at the Fluid Engineering Conference, 8–11 May, Chicago, Illinois.

Keyes, J. W. and G. C. Ashby (1967). "Calculated and experimental hinge moments on a trailing-edge flap of a 75% swept delta wing at Mach 6," NACA TN D-4268.

Kuehn, D. M. (1959). "Experimental investigation of the pressure rise required for the incipient separation of turbulent boundary layers in two-dimensional supersonic flow," NASA Memo 1-21-59A.

Liebe, W. (1952). "The boundary layer fence," *Interavia,* Vol. VII, No. 4, pp. 215–217.

Loftin, L. K. Jr. and H. A. Smith (1949). "Aerodynamic characteristics of 15 NACA airfoil sections at seven Reynolds numbers," NACA TN 1945.

Love, E. S. (1955). "Pressure rise associated with shock-induced boundary-layer separation," NACA TN 3601.

Maki, R. L. and H. A. James (1957). "Wind-tunnel tests of the static longitudinal characteristics at low speed of a swept-wing airplane with slotted flaps, area-suction flaps, and wing leading-edge devices," NACA RM A57A24.

McCormick, B. W., Jr. (1967). *Aerodynamics of V/STOL flight*, Academic Press, New York.

McCullough, G. B. and D. E. Gault (1951). "Examples of three representative types of airfoil-section stall at low speed," NACA TN-2502.

McCullough, G. B., et al. (1951). "Preliminary investigation of the delay of turbulent flow separation by means of wedge-shaped bodies," NACA RM A50L12.

Nonweiler, T. (1955). "Maximum lift data for symmetrical wings," *Aircraft Eng.*, Vol. 27, pp. 2–8.

Nonweiler, T. (1956). "The design of wing sections," *Aircraft Eng.*, Vol. 28, pp. 216–227.

Owen, P. R. and L. Klanfer (1953). "On the laminar boundary layer separation from the leading edge of a thin airfoil," RAE Report Aero 2508.

Pearcey, H. H. (1961). "Shock-induced separation and its prevention by design and boundary layer control," in *Boundary layer and flow control*, Vol. 2, G. V. Lachmann, ed., Pergamon Press, New York.

Pinkerton, R. M. (1936). "Calculated and measured pressure distribution over the midspan section of the NACA 4412 airfoil," NACA Report 563.

Queijo, M. J., et al. (1954). "Wind-tunnel investigation at low speed of the effects of chordwise wing fences and horizontal-tail position on the static longitudinal stability characteristics of an airplane model with 35° sweepback wing," NACA Report 1203.

Quick, A. W. (1951). "Strömmungsmechanische Probleme des Überschallfluges," *Flugwelt*, Vol. 3, pp. 68–71.

Rogers, E. W., et al. (1960). "A study of the effect of leading-edge modifications on the flow over a 50° sweptback wing at transonic speeds," ARC R&M 3270.

Schlichting, H. and E. Truckenbrodt (1967). *Aerodynamik des Fluggzeuges*, Vol. I, Springer-Verlag, Berlin.

Schubauer, G. B. and W. G. Spangenberg (1949). "Effect of screens in wide-angle diffusers," NACA Report 949.

Schubauer, G. B. and W. G. Spangenberg (1960). "Forced mixing in boundary layers," *J. Fluid Mech.*, Vol. 8, Pt. 1, pp. 10–32.

Speidel, L. (1955). "Messungenan zwei Laminarprofilen für Segelflugzeuge," *Z. Flugwiss*, Vol. 3, No. 10, pp. 353–359.

Stack, J. F., et al. (1938). "The compressibility bubble and the effect of compressibility on pressures and forces acting on an airfoil," NACA Report 646.

Stratford, B. S. (1959a). "The prediction of separation of the turbulent boundary layer," *J. Fluid Mech.*, Vol. 5, pp. 1–16.

Stratford, B. S. (1959b). "An experimental flow with zero skin friction throughout its region of pressure rise," *J. Fluid Mech.*, Vol. 5, pp. 17–35.

Thwaites, B. (1960). *Incompressible aerodynamics*, Oxford, at the Clarendon Press.

Truckenbrodt, E. (1952). "Ein Quadraturverfahren zur Berechnung der laminaren und turbulenten Reibungsschicht bei ebener und rotations-symmetrischer Strömung," *Ingr.-Arch.*, Vol. 20, No. 4. (Translated as "A method of quadrature for calculation of the laminar and turbulent boundary layer in case of plane and rotationally symmetrical flow," NACA TM 1397, 1955.)

Velkoff, H. R. (1969). "Boundary layer discontinuity on a helicopter rotor blade in hovering," paper 69-197, AIAA/AHS VTOL Research, Design and Operations Meeting, Feb., Georgia Inst. Technol., Atlanta.

von Doenhoff, A. E. and N. Tetervin (1943). "Determination of general relations for the behavior of turbulent boundary layers," NACA Report 772.

Walz, A. (1944). "Theoretische Widerstandberechnung an einem Laminarprofil mit verschiedenen Schwanzteilformen," U. M. 3131.

Weinig, F. (1940). *Aerodynamik der Luftschraube,* Verlag von Julius Springer, Berlin.

Wortmann, F. X. (1955). "Ein Beitrag zum Entwurf von Laminarprofilen für Segelflugzeuge und Hubschrauber," *Zeitsch f Wiss.*, Vol. 3, p. 333. (British MOS Translation TIL/TA 903.)

Wortmann, F. X. (1957). "Experimentelle Untersuchungen an neuen Laminarprofilen für Segelflugzeuge und Hubschranber," *Zeitsch f. Wiss.*, Vol. 5, p. 228. (British MOS Translation TIL/TA 906.)

Wortmann, F. X. (1961). "Progress in the design of low drag airfoils in *Boundary Layer and Flow Control*, Vol. 2, G. V. Lachmann, ed., Pergamon Press, New York.

SUPPLEMENTARY REFERENCES

Ashill, P. R. and D. Gilmore (1974). "Investigation of externally blown flap airfoils, with leading edge devices and slotted flaps," AGARD-CP-143, pp. (6-1)–(6-13).

Ball, K. O. W. (1971). "Flap span effects on boundary layer separation," *AIAA J.*, Vol. 9, No. 10, p. 2080.

Bley, P. and W. Ehrfeld (1971). "Determination of dissipative losses in the separation nozzle flow with pitol-tubes," Ph.D. thesis, Karlsruhe Inst. f. Kernverfahrenstechnik, Techn. Hochschule, Karlsruhe, in German.

Bradshaw, P. (1970). "Prediction of the turbulent near-wake of a symmetrical aerofoil," *AIAA J.*, Vol. 8, No. 8, p. 1507.

Brebner, G. G. (1957). "Some simple conical camber shapes to produce low lift dependent drag on a slender delta wing," ARC-CP-428.

Chang, P. K. (1960). "Laminar separation of flow around symmetrical struts at zero angle of attack," *J. Franklin Inst.*, Vol. 270, No. 5, pp. 382–396.

Chang, P. K. and W. H. Dunham (1961). "Laminar separation of flow around symmetrical struts at various angles of attack," *Schiffstechnik*, Vol. 8, No. 44, pp. 259–264.

Chang, P. K. and A. G. Strandhagen (1961). "The viscosity correction of symmetrical model lines for wave resistance analysis," *Schiffstechnik*, Vol. 8, No. 40, pp. 28–34.

Chang, P. K. et al. (1974). "Analysis for the flow field around buildings," Tech. Rept., The Catholic University of America, Washington, D.C.

Cleary, J. W. (1972). "Lee-side flow phenomena on space shuttle configurations at hypersonic speeds: Part 1, Flow separation and flow field viscous phenomena of a delta-wing shuttle orbiter configuration," NTIS HC $6.00/MF $0.95 CSCL 20M, Ames Research Center, U.S. National Aeronautics and Space Administration, Moffett Field, Calif.

de Ponte, S. and A. Baron (1974). "The effect of vortex generators on the development of a boundary layer," AGARD-CP-143, pp. (15-1)–(15-5).

Dobbinga, E. et al. (1972). "Some research on two dimensional laminar separation bubbles," N'73-14998 06-02, Technische Hogeschool, Delft (Netherlands). (Also in AGARD Fluid Dyn. of Aircraft Stalling.)

Ericsson, L. E. and J. Peter Reding (1973). "Stall-flutter analysis," *J. Aircraft*, Vol. 10, p. 5.

Eyre, R. C. W. and S. F. J. Butler (1967). "Low speed wind tunnel tests on an A.R. 8 swept wing subsonic transport research model with B.L.C. blowing over nose and flaps for high-lift," Tech. Rept. 67112, Royal Aircraft Establishment, Farnborough, England.

Gadetskiy, V. M. et al. (1972). "Investigation of the influence of vortex generators on turbulent boundary layer separation," *Uch Zap (USSR)*, Vol. 3, No. 4, pp. 22–28. (Transl. into English, NASA-TT-F-16056, 1974.)

Gartling, D. K. (1970). "Tests of vortex generators to prevent separation of supersonic flow in a compression corner," AD-734154, ARL-TR-70-44, Applied Research Lab., Texas Univ., Austin. (Avail: NTIS CSCL 20/4.)

Golovina, L. G. et al. (1972). "Flow separation in conical diffusers," NASA-TT-F-14089. (Transl. into English from *Izv. Vyssh. Ucheb. Zaved. Aviat. Tekh(Kazan)*. Vol. 14, No. 1, pp. 98–103.)

Good, M. C. and P. N. Joubert (1968). "The form drag of two-dimensional bluff-plates immersed in turbulent boundary layers," *J. Fluid Mech.*, Vol. 31, No. 3, pp. 547–582.

Hackett, J. E. and M. R. Evans (1971). "Vortex wakes behind high-lift wings," *J. Aircraft*, Vol. 8, p. 334.

Hall, M. G. et al. (1971). "Scale effects in flows over swept wings," RAE-TR-71043, Royal Aircraft Establishment, Farnborough, England.

Howell, R. H. and H. H. Korst (1971). "Separation controlled transonic drag-rise modification for V-shaped notches," *AIAA J.*, Vol. 9, No. 10, p. 2051.

Irwin, H. P. A. H. (1974). "A calculation method for the two-dimensional turbulent flow over a slotted flap," ARC-CP-1267, RAE-TR-72124, ARC-34236 (54), Aero. Res. Council, London.

Jacob, K. and D. Steinbach (1974). "A method for prediction of lift for multi-element airfoil systems with separation," AGARD-CP-143, pp. (7-1)–(7-12).

Johnson, B. and R. H. Barchi (1968). "Effect of drag-reducing additives on boundary-layer turbulence," *J. Hydronautics*, Vol. 2, No. 3, pp. 168–175.

Johnston, J. P. (1970). "Measurements in three-dimensional turbulent boundary layer induced by a swept, forward-facing step," *J. Fluid Mech.*, Vol. 42, p. 823.

Karashima, K. and K. Hasegawa (1973). "An approximate approach to base flow behind two-dimensional rearward-facing steps placed in a uniform supersonic stream," Rept. No. 501, Vol. 38, p. 2, Inst. of Space and Aeronautical Sci., Univ. of Tokyo, Tokyo.

Kaufman, L. G. et al. (1973). "Shock impingement caused by boundary-layer separation ahead of blunt fins," *AIAA J.*, Vol. 11, No. 9, p. 1363.

Kida, T. and Y. Miyai (1971). "A theory on effect of entrainment, due to jet flap which hits the ground for an aerofoil with separation," *Aero. J. Roy. Aero. Sci.*, Vol. 75, No. 723, pp. 199–201.

Levinsky, E. S. and T. Strand (1970). "A method for calculating helicopter vortex paths and wake velocities," AFFDL-TR-69-113, Air Force Flight Dynamics Lab., U.S. Air Force Systems Command, Wright-Patterson AFB, Ohio.

Lucas, R. D. (1972). "Perturbation solution for viscous incompressible flow in channels," Ph.D. thesis, Stanford University, California.

Mair, W. A. and D. J. Maull (1971). "Bluff bodies and vortex shedding–A report on euromech 17," *J. Fluid Mech.*, Vol. 45, p. 209.

Martin, J. M. et al. (1973). "A detailed experimental analysis of dynamic stall on an unsteady two-dimensional airfoil," Paper No. 702, Presented at the 29th Annual National Forum of the American Helicopter Society, May, Washington, D.C.

Maull, D. J. and R. A. Young (1973). "Vortex shedding from bluff bodies in a shear flow," *J. Fluid Mech.*, Vol. 60, p. 401.

Mayes, J. F. et al. (1970). "Transonic buffet characteristics of a 60° swept wing with design variations," *J. Aircraft*, Vol. 7, p. 524.

Merz, R. A. et al. (1973). "Effect of base mounted cylinders on a supersonic near wake," *J. Spacecraft and Rockets*, Vol. 10, No. 7, pp. 470–472.

Mirly, K. A. and B. P. Selberg (1973). "Support wire disturbances in near viscous wakes of slender supersonic bodies," *J. Spacecraft and Rockets*, Vol. 10, No. 7, pp. 474–477.

Nenni, J. P. and C. Tung (1973). "Analysis of the flow about delta wings with leading edge separation at supersonic speeds," NASA-CR-132358.

Parkinson, G. V. et al. (1974). "The aerodynamics of two-dimensional airfoils with spoilers," AGARD-CP-143, pp. (12-1)–(12-16).

Parthasarthy, K. and V. Zakkay (1969). "Turbulent slot injection studies at Mach 6," ARL-69-0066, Aerospace Research Lab., Office of Aerospace Research, Wright-Patterson AFB, Ohio.

Portnoy, H. and S. C. Russell (1971). "The effect of conical thickness distributions on the separated flow past slender delta wings," ARC-CP-1189. Avail: NTIS HMSO. London Aeron. Res. Council. (Supersedes ARC-32834.)

Pullin, D. I. (1972). "Calculations of the steady conical flow past a yawed slender delta wing with leading edge separation," IC-Aero-72-17, Imperial Coll. of Science and Technology, London.

Reding, J. P. and L. E. Ericsson (1973). "Effects of delta wing separation on shuttle dynamics," *J. Spacecraft and Rockets,* Vol. 10, No. 7, pp. 421–428.

Rom, J. and H. Portnoy (1971). "The flow near the tip and wake edge of a lifting wing with trailing edge separation," AD-734791, TAE-132, Technion-Israel Inst. of Tech. Haifa, AFOSR-72-0010-TR-SR-1.

Sato, J. (1973). "Discrete vortex method of two-dimensional jet flaps," *AIAA J.*, Vol. 11, No. 7, p. 968.

Schmitz, G. (1938). "Tragflügel mit angelenkten Klappen," *Luftfahrtforschung*, Vol. 15, No. 1/2, pp. 13–18.

Senoo, Y. and M. Nishi (1974). "Improvement of the performance of conical diffusers by vortex generators," *Trans. ASME, J. Fluids Eng.*, Ser. 1, Vol. 96, No. 1, pp. 4–10.

Sforza, P. M. and R. F. Mons (1970). "Wall-wake: Flow behind a leading edge obstacle," *AIAA J.*, Vol. 8, No. 12, p. 2162.

Smith, F. T. and K. Stewartson (1973). "Plate-injection into a separated supersonic boundary layer," *J. Fluid Mech.*, Vol. 58, p. 143.

Takahashi, H. (1971). "On the theory of free stream-lines past arbitrary obstacles," NAL-TR-247, National Aerospace Lab., Tokyo, in Japanese.

Uebelhack, H. T. (1971). "Theoretical and experimental investigation of turbulent supersonic separated flows over front steps," presented at the 4th DGLR Annual Meeting, 11–13 Oct. 1971, Baden-Baden, West Germany. (Avail: NTIS.)

Viswanath, P. R. and R. Narashima (1974). "Two-dimensional boat-tailed bases in supersonic flow," *Aeronaut. Quart.*, Vol. XXV, No. 3, pp. 210–224.

Westkaemper, J. C. and J. W. Whitten (1970). "Drag of vane-type vortex generators in compressible flow," *J. Spacecraft and Rockets*, Vol. 7, p. 1269.

chapter 5

PREVENTION AND DELAY OF SEPARATION USING AUXILIARY POWER

INTRODUCTORY REMARKS

Techniques that employ auxiliary power may be classified according to whether or not working fluid is necessary for the purpose of control. For example, it is needed in suction, blowing, or a combination of both, but not for methods that employ a moving surface. Some control techniques that utilize heat treatment require a working fluid while others do not.

For convenience, the control of flow separation by auxiliary power is divided into moving surfaces, elimination or reduction of viscosity effect, energizing the fluid motion, and a combination of the two previous items (i.e., elimination or reduction of viscosity and energizing the fluid motion).

The practical application of these control techniques requires consideration of both technical and economical feasibility. If the power required for control by a particular technique is larger than the total power saved, then it may not be practical.

5.1 PREVENTION OF SEPARATION BY MOTION OF THE SURFACE

The difference in the velocity of fluid in the free stream and at the wall causes the formation of a boundary layer. Hence, it is feasible to eliminate this difference in velocity by moving the wall in the same direction as the main flow. This energizes the low momentum boundary layer flow by the dynamic motion of the wall and thus prevents separation.

5.1.1 Flow Phenomena on a Rotating Cylinder

Prandtl (1925) showed that the separation can be prevented over a stationary circular cylinder immersed in the free stream if the cylinder is rotated with sufficient speed and in such a direction that the tangential velocity of rotation on the surface is the same as in the free stream. Then the flow field around a cylinder may be approximated by the potential flow with circulation. The effect of air circulation around a cylinder or sphere induced by rotation of the body is called a Magnus effect.

Since the lift and point of separation are affected by rotation of the body, lift enables an indirect evaluation of the effect of rotation on separation. The lift coefficient of a cylinder with circulation may be related to the separation position of a circular cylinder. This relation is affected by three different ranges of values of the circulation, as shown in Fig. 5.1 for an analysis by Thwaites (1960). With clockwise circulation $\Gamma < 2\pi u_\infty d$ (top sketch), separation occurs at point A downstream of the midpoint and $C_L < 4\pi$. But with circulation $\Gamma = 2\pi u_\infty d$ (middle sketch), no separation occurs and C_L reaches a value of 4π. With further increase of circulation (bottom sketch), an

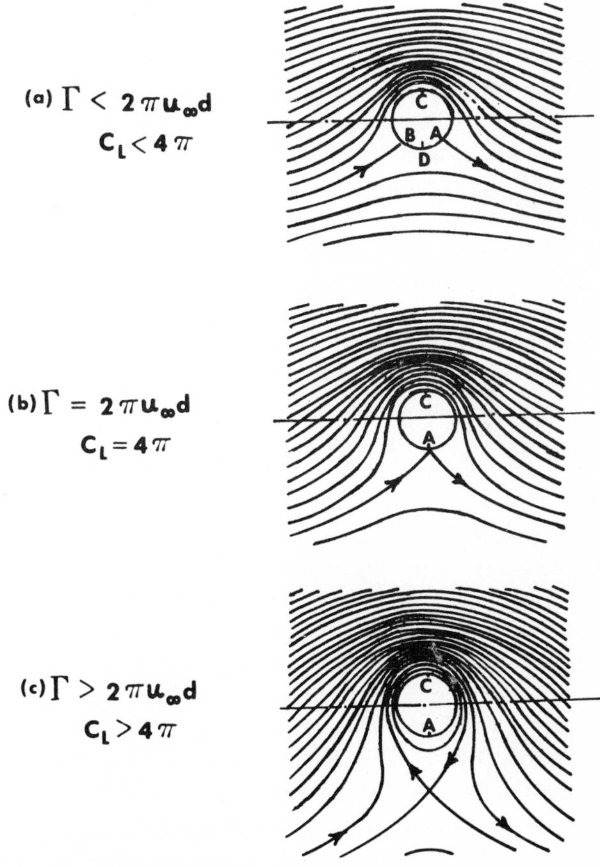

FIG. 5.1 The potential flow of a uniform stream past a circular cylinder for various values of the circulation (circulation is clockwise; *from Thwaites, 1960*).

inviscid flow pattern results in a theoretical lift coefficient greater than 4π. On the under side of the cylinder where the main flow direction and cylinder rotational direction are opposite, the effect on separation is not significant. Although Prandtl (1925) indicated that the value of C_L cannot exceed a certain value, Glauert (1959) indicates that an increase in the rotational speed of a cylinder can increase the value of the lift indefinitely.

Brady and Ludwig (1963) and Ludwig (1964) investigated in detail the occurrence of laminar flow separation on a rotating circular cylinder. The 4-in.-diameter cylinder was rotated at 2000 rpm while submerged in a free stream whose velocity was less than 30 ft/sec. The rotating cylinder was shrouded in order to impose an airfoil type of pressure gradient. One pertinent feature of the stall that occurred was the involvement of an unsteady boundary layer process because the wall and the separation point were in relative motion. There was a certain amount of unsteadiness in the boundary layer at all angular positions with negative v/u_∞ (v is the velocity component perpendicular to u_∞; u_∞ is taken as positive, and the direction toward the cylinder surface as negative). This unsteadiness gradually increased with increasing cylinder angle, but at no angular position was there a discernable jump in the magnitude of the velocity fluctuations. Such a jump could be an indication of unsteadiness of flow. Unsteady flow separation of this type differs from that of steady separation, the velocity at the separation point is not zero on the wall surface, and the formation of separated flow is not the necessary condition behind the separation. However, if the wall is in motion in either the downstream or the upstream direction, the problem involving the separation point can be reduced to a steady-flow problem. The rotating cylinder may be considered as a nearly flat wall if the separated region is expanded to a large scale.

If the wall is moving in the downstream direction, then the velocity profile in the neighborhood of separation appears like that indicated in Fig. 5.2*a*. The separation does not occur at the wall but at some height above the wall, and there would be boundary-layer flow beneath the separated region. The displacement of the separation point is downstream, varies linearly with wall velocity, and is relatively insensitive to the pressure gradient.

If the wall moves upstream, then the velocity profile is as shown in Fig. 5.2*b*. In this case it is conjectured that separation occurs at a position where there is a relatively large flattened height with a singularity at the foot, that is, one in which the velocity abruptly changes from zero near the wall to the wall velocity at the wall. The boundary layer contains a sublayer of circulating flow extending from the forward stagnation point to the separation point, and there would be no separated flow into the boundary layer.

Therefore, regardless of direction of wall movement, a criterion for laminar separation on a moving wall can be formulated by the two conditions $\partial u/\partial y = 0$ and $u = 0$ at the same point. However, experimental determination of separated flow when the wall is moving may be made more reliable by

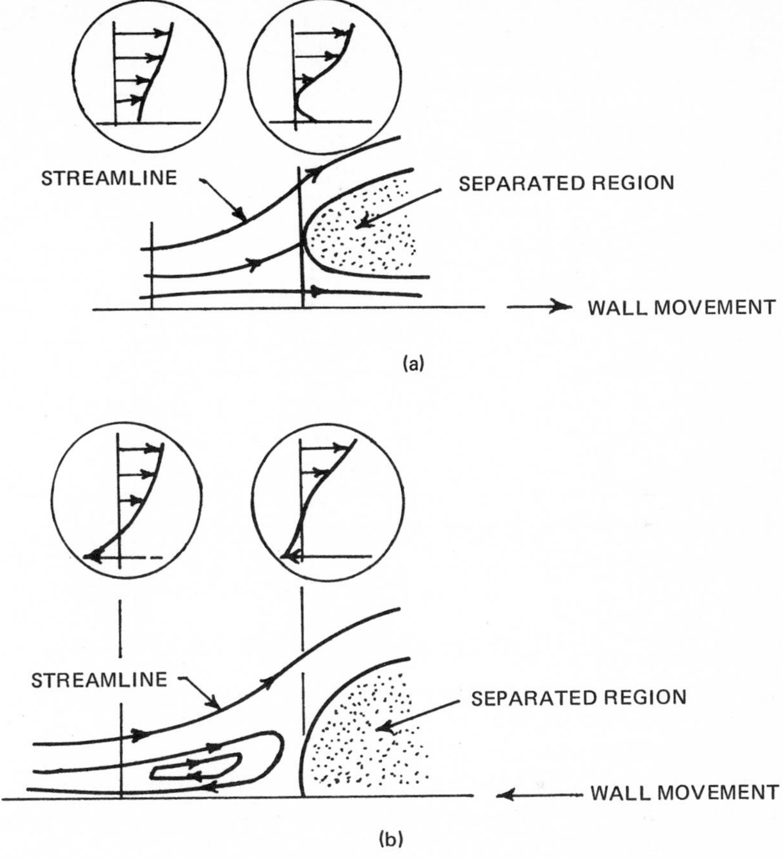

FIG. 5.2 Velocity profiles in the neighborhood of separation: (*a*) wall moving downstream (this is similar to point *A* in Fig. 5.1*a*; *from Ludwig, 1964*; (*b*) wall moving upstream (this is similar to point *B* in Fig. 5.1*a*; *from Ludwig, 1964*).

considering the behavior of the vertical velocity component in the boundary layer for the condition $u = 0$. The effect of wall velocity on the apparent position of laminar separation shown in Fig. 5.3 is based on measurement of the vertical velocity component v.

An extension of the Pohlhausen analysis, Smith (1963) gives an approximate solution that can predict the laminar separation for the case of a downstream-moving wall, but no theoretical treatment is available for predicting separation for an upstream-moving wall.

5.1.2 Practical Applications of a Rotating Cylinder

Flettner (1924) replaced the sail on a boat by a rotating cylinder and utilized the Magnus effect to move the boat forward. Since this application was found to be economically unfeasible, it will not be discussed further.

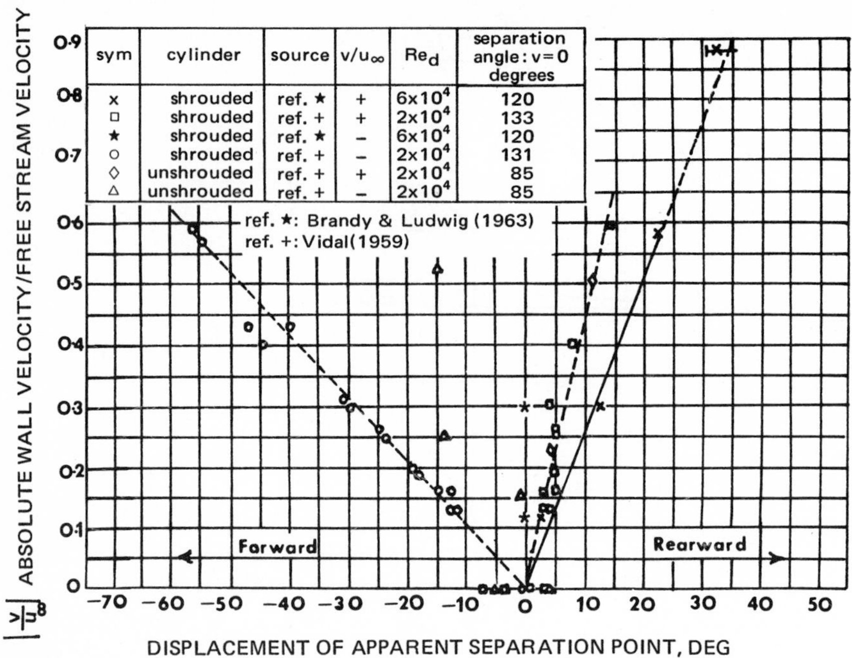

FIG. 5.3 Effect of wall velocity on apparent position of laminar separation (*from Brady & Ludwig, 1963*; v is the vertical velocity component in the boundary layer).

Alvarez-Calderon (1964) mounted a rotating cylinder that moved rearward in the direction of local flow on the upper surface of an aircraft wing. This application of a rotating cylinder to control flow separation and to improve aerodynamic performance is versatile because it can be combined with a flap, a droop nose, or leading edge slats. The method has the additional advantage of not involving energy losses or structural complications associated with the production of motion of the fluid.

Figure 5.4 shows how the boundary layer was energized by a rotating cylinder flap combination. The moving surface of the cylinder energized the low-energy boundary-layer flow impinging on the cylinder from the wing upstream of the cylinder and formed a new boundary layer downstream on the surface of the flap. Since the level of energy of this new boundary layer was high, it was possible to overcome the large adverse pressure gradient on the flap at a large angle of attack and thus prevent separation. This reenergizing process depends on the upward protrusion of the cylinder, on its peripheral speed, and on the local flap geometry.

The streamlines around the wing can be visualized by using smoke, as indicated in Fig. 5.5. Compare the extended region of separation with the cylinder stationary to the complete suppression of separation with the cylinder rotating. With the cylinder rotating, separation is controlled, and a

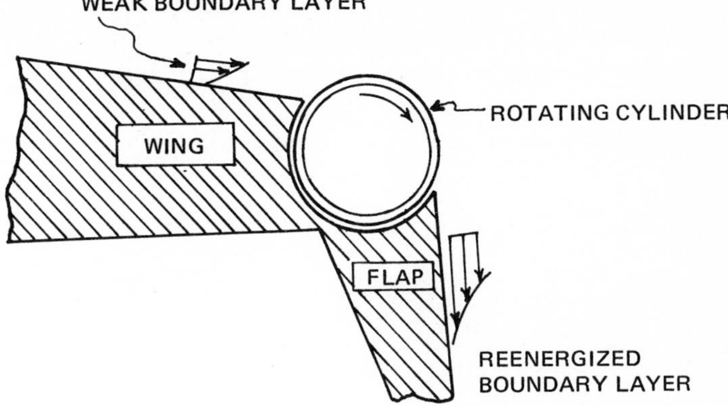

FIG. 5.4 Boundary layer control function of a rotating cylinder flap (*from Alvarez-Calderon, 1964*).

strong upwash field is induced as a low pressure region toward the leading edge of the wing. Thus the lift is greatly increased and the pitching moment reduced. Note also that the forward stagnation point is now located far back on the lower side of the airfoil.

To ensure extensive exposure of the cylinder directly to the boundary layer of the upper surface of the wing when the flap is retracted, the cylinder should be mounted on the leading edge of a plain flap with a fixed hinge as

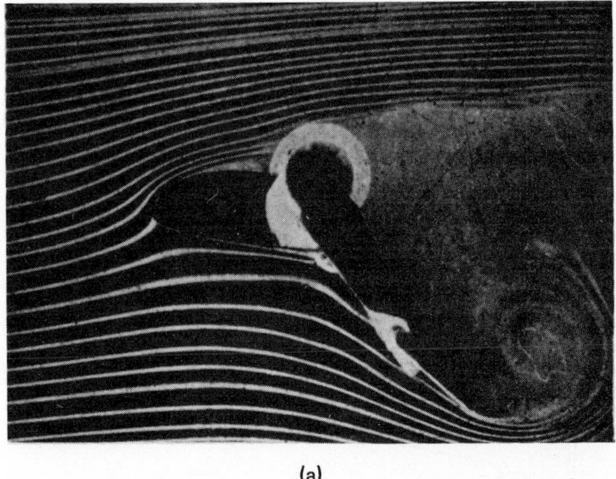

(a)

FIG. 5.5 Smoke visualization of streamlines around a rotating cylinder flap at a large flap deflection angle: (*a*) stationary cylinder (note the weak upwash on the wing and large flap turbulence typical of conventional flaps at large angles; *from Alvarez-Calderon, 1964*).

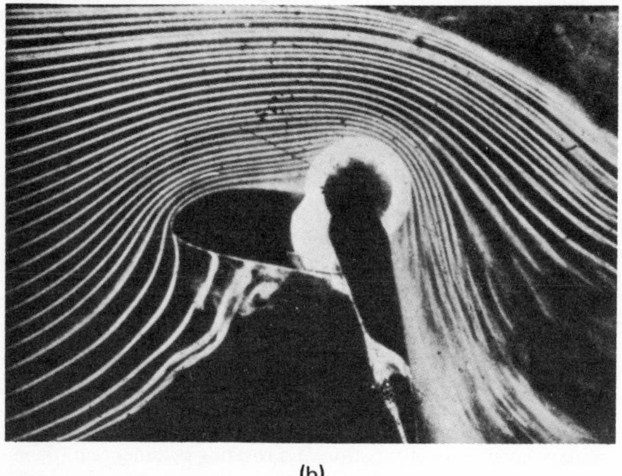

(b)

FIG. 5.5 *(continued)* Smoke visualization of streamlines around a rotating cylinder flap at a large flap deflection angle: (*b*) rotating cylinder (note the strong upwash and smooth attached flap flow; the smoke lines at the rear of the flap lose their identity due to expansion; *from Alvarez-Calderon, 1964*).

shown in Fig. 5.6. This simple installation enables the amount of cylinder protrusion into the slipstream to increase with increasing flap deflection as required. When the flap is retracted to the dashed line position, then the stationary cylinder is housed between the flap and the wing so that the undesirable protrusion of the stationary cylinder is avoided and low drag is ensured.

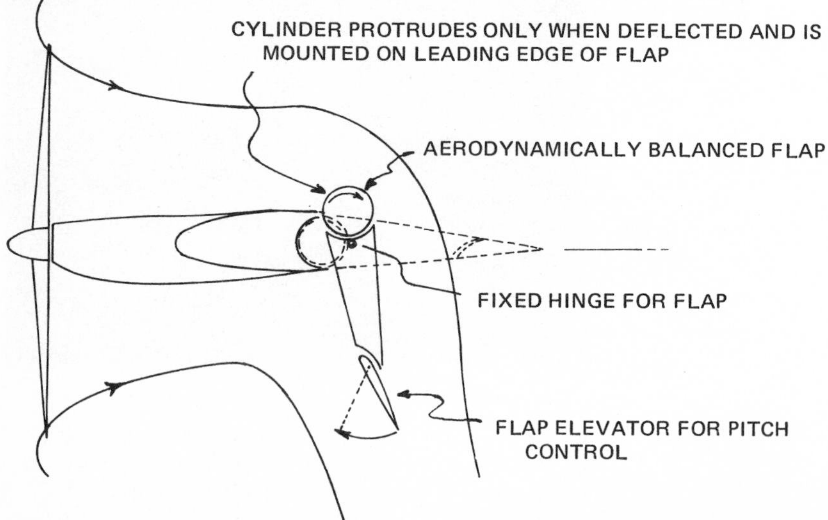

FIG. 5.6 Basic layout of a rotating cylinder flap for a deflected slipstream V/STOL application (*from Alvarez-Calderon, 1964*).

Further, the rotating cylinder flap improves the flow upstream of the cylinder and the flap area; it induces a local low pressure region and stabilizes the flow at the leading edge. This feature can be utilized to suppress leading edge separation; see Fig. 5.7.

Thus substantial increase in section lift may be obtained by mounting a rotating cylinder. The profile of NACA airfoil 23018 with a 0.40c rotating flap is shown in Fig. 5.8, and some test results for that airfoil are plotted in Fig. 5.9.

As seen in Fig. 5.9, at an unfavorably low Reynolds number of about 10^6, the highest section lift coefficient with the flap deflected and the cylinder stationary was 1.4; it was raised to 4.1 where the cylinder was rotated and the flap deflected 60 deg. Furthermore, reduced section pitching moments and thus greatly reduced flap hinge moments were obtained due to the upward protrusion of the cylinder when the flap was deflected.

A rotating cylinder may also be used in combination with propeller cross-shafting in a tilt wing type VTOL aircraft. The cross-shafting tube is installed as a high-lift, leading-edge device on the wing in order to improve flow in transition maneuvers, particularly during steep approach conditions. To avoid high drag on an exposed high-lift, leading-edge rotating cylinder system in cruise flight, it is preferable to mount the cylinder in combination

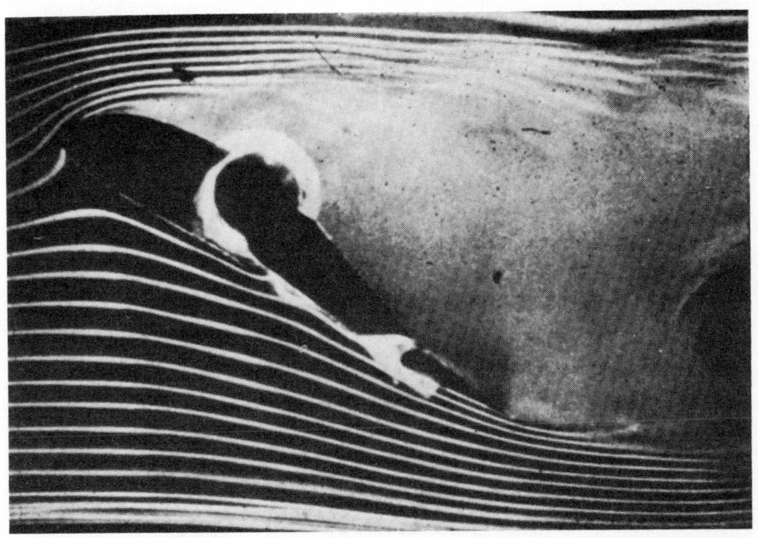

(a)

FIG. 5.7 Occurrence and suppression of leading edge separation: (*a*) stationary cylinder; (*from Alvarez-Calderon, 1964*; with a stationary cylinder, the airfoil is at a high incidence and experiences leading edge stall; with a rotating cylinder, there is attached leading edge flow and strong circulatory lift; note particularly the stagnation streamline).

(b)

FIG. 5.7 (*continued*) Occurrence and suppression of leading edge separation: (*b*) rotating cylinder (*from Alvarez-Calderon, 1964*; with a stationary cylinder, the airfoil is at a high incidence and experiences leading edge stall; with a rotating cylinder, there is attached leading edge flow and strong circulatory lift; note particularly the stagnation streamline).

with a single-pivot droop nose as shown in Fig. 5.10 or with a leading edge slat as sketched in Fig. 5.11.

Wind tunnel tests of a rotating cylinder mounted at the leading edge of a wing showed that the maximum lift coefficient is measured at about a 40-deg angle of attack, which is considerably higher than the stall angle of conventional slats. This large stall angle is a very desirable condition to increase the steep approach angles permissible for a tilt wing aircraft. Tests of a rotating cylinder system with rotational speeds in excess of 10,000 rpm

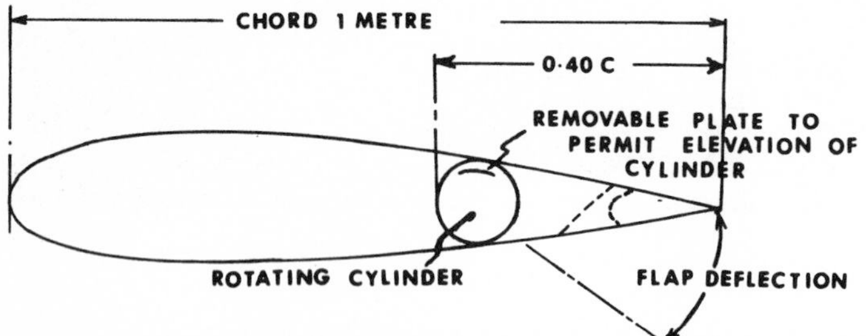

FIG. 5.8 Profile of NACA airfoil 23018 with a 0.40c rotating cylinder flap (*from Alvarez-Calderon, 1964*).

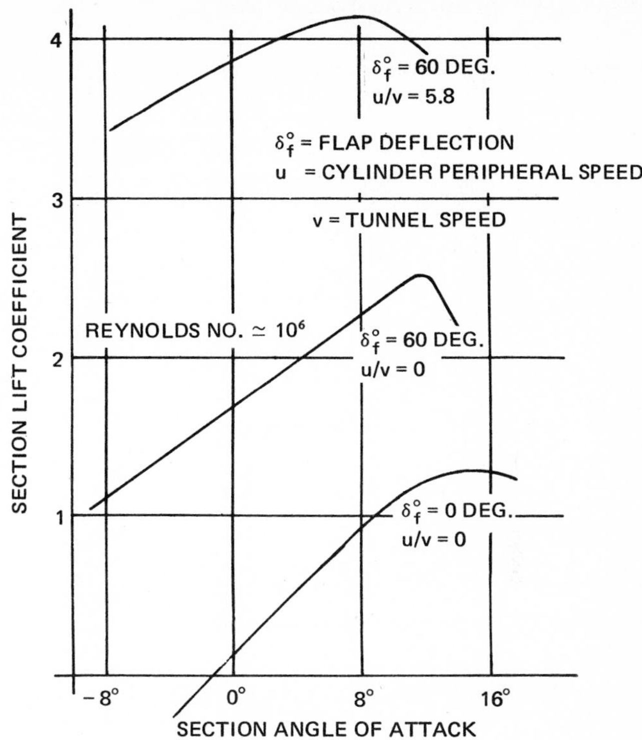

FIG. 5.9 Section lift coefficient versus angle of attack for NACA Airfoil 23018 (*from Alvarez-Calderon, 1964*; the airfoil was used with the rotating cylinder flap shown in Fig. 5.8; tests were performed at Stanford University and at the Ames Laboratory of NASA; maximum lift coefficients obtained were 1.4 with stationary cylinder and flap not deflected, 2.5 with stationary cylinder but flap deflected 60 deg, and 4.5 with cylinder rotating and flap deflected 60 deg).

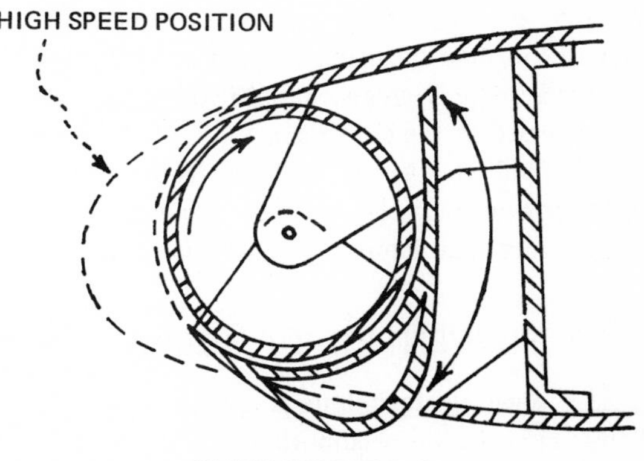

FIG. 5.10 High-lift leading-edge rotating cylinder in combination with a droop nose (*from Alvarez-Calderon, 1964*).

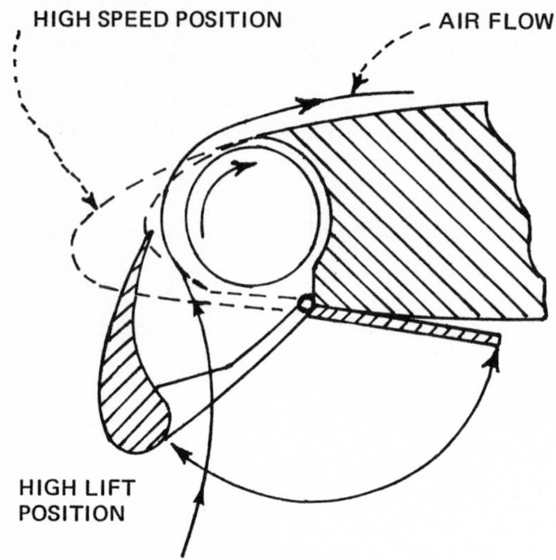

FIG. 5.11 High-lift leading-edge rotating cylinder in combination with a slat (*from Alvarez-Calderon, 1964*).

indicated that the levels of vibration and noise are low compared to those produced by the aircraft engine or propeller and lie within a range that is perfectly acceptable both structurally and operationally. They should present no discomfort to passengers.

Design studies of a rotating cylinder flap on a deflected slipstream V/STOL indicate that it offers several fundamental improvements over conventional flaps:

1. Increment of turning angle as well as turning efficiency in ground effect.
2. Pitch characteristics which are strongly and adversely dependent on flap position may be made independent of flap position. Pitch control for low speed flight is greatly simplified by using an auxiliary flap or "flapevator" immersed in the slipstream.
3. Required change of attitude from cruise to hover is greatly improved, improving tail clearance in ground effect, visibility, pilot technique in approach, and landing gear design.

5.2 PREVENTION OF SEPARATION BY SUCTION

This powerful technique is applicable to subsonic, transonic, and supersonic flows. Fluid in the neighborhood of a wall is sucked away through slots in the wall. This eliminates the viscosity effect and prevents boundary growth because a new boundary layer is formed downstream of the suction slots. Thus, by providing suction slots upstream of either the separation point or the transition region, flow separation may be prevented and laminar flow maintained to avoid turbulent flow.

This section briefly considers the combined techniques of suction and blowing. For convenience, the required quantities of suction and power are presented first inasmuch as the available pump capacity and the saved power are important factors in the practical application of the suction technique. The application of suction to prevent separation is described according to the area where suction takes place, e.g., leading edge suction, downstream suction, and trailing edge flap suction.

If the blowing of auxiliary fluid takes place in a direction perpendicular to the wall surface, then from the analytical standpoint, the analysis of suction and blowing would be similar because only the sign of the suction velocity is opposite to that of the blowing velocity. However, the analysis of suction considered here will not be repeated later in the section on blowing. The symbol v_0 refers to the velocity of suction and blowing.

The analyses of normal suction are outlined for incompressible two-dimensional and axisymmetrical flows applicable for the prevention of separation of both laminar and turbulent flows.

5.2.1 Calculation of Required Suction and Power

It is necessary to determine the minimum quantity of suction and power required so that the ideal or optimum condition of operation may be established for the economic application of suction.

5.2.1.1 Quantity of suction required

The total suction quantity Q is given by $Q = b\int_0^c (-v_0)dx = C_Q\, bcu_\infty$ where b is the wing span, c is the chord length, C_Q is the suction quantity coefficient, and v_0 is suction velocity along the outer wall directed perpendicular to wall surface. Then $C_Q = Q/bcu_\infty$. For unit span, this reduces to $C'_Q = Q/cu_\infty$. The minimum required suction for a given continuous constant suction velocity is given by

$$Q_{\min} = C_{Q\min} bcu_\infty = bs'(-v_0)_{\min} \quad \text{and} \quad C_{Q\min} = \frac{s'}{c}\left(-\frac{v_0}{u_\infty}\right)_{\min}$$

where s' is the streamwise extent of the suction zone.

Griffin and Hickey (1956) investigated the so-called critical values of the suction quantity coefficient for a flap (i.e., the minimum value of suction required to accomplish boundary layer control for the flap) for a 45-deg sweptback wing model with an aspect ratio of 6 (as shown later in Fig. 5.46).

The critical values of suction quantity are considered for area suction at a fixed angle of attack. Figure 5.12 shows a typical variation of the flap lift increment with suction quantity through area suction applied to a deflected

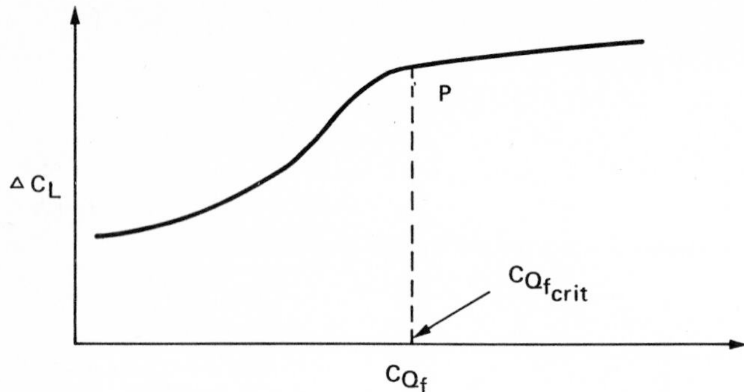

FIG. 5.12 Relationship of ΔC_L and C_{Qf} (*from Griffin & Hickey, 1956*).

flap. As indicated there, the lift increment ΔC_L due to deflection of flap increases with an increase in suction quantity coefficient of flap C_{Qf} defined by $C_{Qf} = Q_f/(u_\infty A)$ where Q_f and A are volume rate of suction flow corrected to standard atmosphere in cubic feet per second and wing area in square feet, respectively. When a point P is reached, the slope of the curve decreases and approaches a constant value. A further increase in C_{Qf} beyond point P results in relatively smaller gains in flap lift increment. Point P is also closely linked to control of flow separation. Griffin and Hickey (1956) cite tuft motion observations and static pressure distribution reported by Cook, et al. (1953) and by Kelly and Tolhurst (1954). These showed that the greater part of the flow separation on the flap has been eliminated by the time C_{Qf} reaches point P. Hence, point P may be considered as that most economical for accomplishing separation control in terms of suction quantity coefficient C_{Qf}. The value of C_{Qf} at point P is denoted as $C_{Qf_{\text{crit}}}$ and is in the 0.0006 to 0.0016 range. For practical applications, a larger value of C_{Qf} is taken (about 0.001 to 0.002) compared to the theoretical $C_{Qf_{\text{crit}}}$.

No experimental data are available on $C_{Qf_{\text{crit}}}$ for other cases of suction, for example, downstream suction, leading edge suction, etc. It may be conjectured that the same approach as used for the flap suction may be applicable for these cases.

5.2.1.2 Power required

Although the suction required for the prevention of separation may be small, a considerable loss of energy may occur when the sucked air passes through a porous material. Thus the power economy of a boundary layer with suction is an important practical item to be considered.

Pankhurst and Gregory (1952) studied the power requirement for continuous suction and its economy for the performance of an airfoil and airplane. Consider the distributed suction on the airfoil sketched in Fig. 5.13.

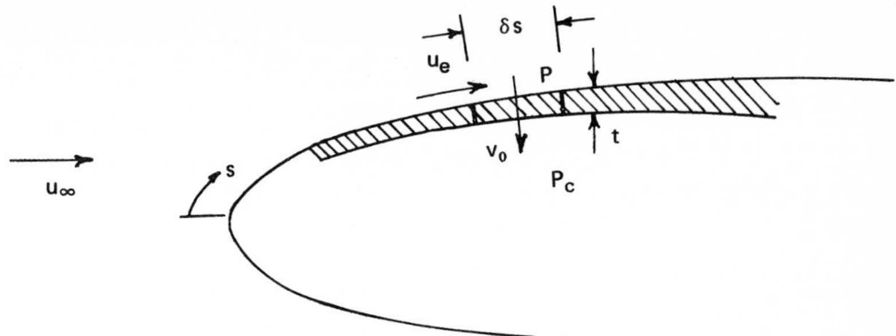

FIG. 5.13 Distributed suction on an airfoil surface (*from Pankhurst & Gregory, 1952*).

The boundary layer fluid is sucked through the airfoil surface element δs with the mean suction velocity $v_0(s)$. If κ is the resistance coefficient of the material and p_c is pressure of the inner surface, then $p - p_c = \kappa v_0 t$ where t is thickness of the wall.

It is assumed that the suction pump restores the total head of the sucked air to that of the free stream (H_0), discharging it with a velocity relative to the airplane which is equal and opposite to the forward speed of the aircraft relative to the ground u_∞, so that there is neither a sink drag nor jet thrust. If the loss in the ducting between the porous wall and the suction pump is considered (it has a small negative value), the power required for unit span of a two-dimensional airfoil is

$$P = \frac{1}{\eta_p} \int (H_0 - p_c) v_0 \, ds$$

where η_p is the efficiency of the pump. By denoting η_a as the efficiency of the propulsive system of the aircraft and c as the airfoil chord, the pump power may be formulated as an equivalent drag coefficient

$$C_{D_p} = \frac{\eta_a P}{1/2 \rho c u_\infty^2}$$

Although the value of η_a may be less than η_p, by assuming $\eta_a = \eta_p$,

$$C_{D_p} = \int \left[\frac{H_0 - p_c}{(1/2)\rho u_\infty^2}\right] \left(\frac{v_0}{u_\infty}\right) d\left(\frac{s}{c}\right) = \int \left[\frac{H_0 - p}{(1/2)\rho u_\infty^2} + \frac{p - p_c}{(1/2)\rho u_\infty^2}\right] \left(\frac{v_0}{u_\infty}\right) d\left(\frac{s}{c}\right)$$

The ideal value of C_{D_p} (denoted by $C_{D_{pi}}$) is evaluated for the ideal case of no loss across the porous material, i.e., for $p = p_c$, as

$$C_{D_{pi}} = \int \left[\frac{H_0 - p}{(1/2)\rho u_\infty^2}\right]\left(\frac{v_0}{u_\infty}\right) d\left(\frac{s}{c}\right) = \int \left(\frac{u_e}{u_\infty}\right)^2 \left(\frac{v_0}{u_\infty}\right) d\left(\frac{s}{c}\right)$$

where u_e is the velocity at the edge of boundary layer and the pressure is assumed constant through the thickness of boundary layer. Note that $C_{D_{pi}}$ varies with $\mathrm{Re}_c^{-1/2}$. When the resistance of the porpous material is considered, the "porous resistivity drag" is to be added; it is defined as

$$C_{D_{pk}} = \int \left[\frac{p - p_c}{(1/2)\rho u_\infty^2}\right] d\left(\frac{s}{c}\right)$$

Thwaites (1946) analyzed suction for rounded leading-edged and symmetrical airfoils because it was difficult to predict the effect of any type of suction on flow near a sharp edge. He determined the inviscid velocity distribution on the airfoil as a function of C_L and then applied the boundary layer momentum equation involving suction by assuming that the boundary layer is of the Blasius type. He gave the suction velocity v_0 as a parameter of $v_0/u_\infty\sqrt{\mathrm{Re}_c}$ which is dependent on C_L.

Power requirement for suction with constant inner surface pressure p_c In order to provide a constant inner surface pressure p_c, the suction chamber may be divided into several compartments. Then

$$C_{D_p} = \sum \left[\frac{H_0 - p_c}{(1/2)\rho u_\infty^2} \int' \frac{v_0}{u_0} d\left(\frac{s}{c}\right)\right] = \sum \frac{H_0 - p_c}{(1/2)\rho u_\infty^2} C_Q{'}$$

Here $\int'$ denotes the integral over each compartment, C_Q' refers to the corresponding contribution to the total suction quantity coefficient of each compartment (volume flow per unit span per unit time divided by $u_\infty c$), and the summation is over all of the compartments.

The condition for $v_0 = \text{constant}$ is $\kappa t \propto p - p_c$. For a given value of p_c in each compartment, the condition of v_0 = constant may be satisfied by varying either κ or t, or both.

Studies by Thwaites (1946) and Pankhurst et al. (1948) provide practical applications of suction to the 8 percent-thick H.S.A.V. section at a lift coefficient of 1.5. The porous surface extended from the forward stagnation point to the 0.15c position on the upper surface, corresponding to a wall distance of 0.21c. For the computations, the velocity distribution over the surface was assumed to be that of the potential flow, v_0 was assumed constant, and $(v_0/u_e)\sqrt{\mathrm{Re}_c} = 17$ at a lift coefficient of 1.5 which occurs at a 15 deg angle of attack.* The numerical results given below indicate little more

*This represents two-dimensional airfoil experimental data taken at the N.P.L. 4-ft Tunnel 2 (Pankhurst et al., 1948).

than the order of magnitude because experimental data were insufficient to enable any allowances for scale effects on the comparative figures quoted for the performance of an airfoil without suction. The suction chamber was assumed to comprise one or more compartments, with the partitions located to give minimum power requirements in each case. The pressure in each compartment was taken equal to the static pressure at the point of maximum velocity on the appropriate wall position. This is equivalent to assuming that $\kappa \cdot t = 0$ at this point and its values elsewhere may be determined for v_0 = constant. In two of the examples, corresponding compartments on either side of the middle were assumed to be interconnected to provide equal pressure since this would simplify the pumping arrangments.

Numerical examples The numerical examples for six cases calculated by Pankhurst and Gregory (1952) are as follows:

Case number and description	$C_{DP}\sqrt{\text{Re}_c}$
1. Ideal case ($p = p_c$ everywhere; zero porous resistivity)	15.3
2. One suction compartment	53.6
3. Three compartments most economically placed	25.4
4. Three compartments most economically placed (first and third interconnected)	26.4
5. Five chambers most economically placed	21.3
6. Five chambers most economically placed (first and fifth and second and fourth interconnected)	22.5

The details of numerical results for Case 5 (this yields the minimum non-ideal value of $C_{DP}\sqrt{\text{Re}_c}$) are discussed in Table 5.1 to indicate the procedure for the numerical calculations. Thus, according to the table, $C_D\sqrt{\text{Re}_c} = v_0/u_0\sqrt{\text{Re}_c} \times 1.26 = 21.3$ (C_D is drag coefficient).

TABLE 5.1 Numerical calculations of power requirements for suction

s/c	u_e/u_∞	$(u_e/u_\infty)^2$	Length of compartment	$(H_0 - p_c)/\frac{1}{2}\rho u_\infty^2 = (u_e/u_\infty)^2_{\max}$	$(u_e/u_\infty)^2_{\max} \times (s/c)$ compartment
0	0	0	0		
			0.030	1	0.030
0.030	1.0	1.0			
			0.011	8.3	0.093
0.041	2.88	8.2			
(0.053	3.87	15.0)	0.0325	15	0.488
0.01735	2.67	7.15			
			0.0365	7.15	0.261
0.110	1.96	3.85			
			0.100	3.85	0.385
0.210	1.582	2.50			
					1.26

Note: Illustrated with data of Case 5, Pankhurst and Gregory (1952).

In order to prevent the value of $k \cdot t$ from falling to zero at one point in each compartment, the thickness of the porous material could be increased everywhere by a constant amount. For example, in Case 5, if a value of $k \cdot t$ is equal to the maximum attained in the compartment with the smallest external pressure variation added—with the thinnest wall—the value of $C_D\sqrt{\mathrm{Re}_c}$ would be increased by only 16 percent, amounting to 26.1.

Now study the power requirement for an airplane weighing 10,000 lb with a wing area of 250 ft^2, a wing loading of 40 lb/ft^2, and a mean chord of 8 ft. The lift coefficient of 1.5 corresponds to a landing speed of 120 mph (150 ft/sec) at sea level, and the Reynolds number based on the mean chord is 7.66×10^6. In the calculation, three-dimensional effects are neglected and a uniform suction quantity of 48 ft^3/sec (3.7 lb/sec) is assumed. Based on these assumptions, the estimated power requirements at a lift coefficient of 1.5 and an angle of attack of 15 deg is listed below. (It is noted for reference that with zero suction, the lift coefficient amounted to 0.7 and the maximum lift was 0.87 at an angle of attack of 10.5 deg; Pankhurst & Gregory, 1952.)

Case number and description	Required horsepower
1. Ideal case	10.0
2. One section compartment	35.1
4. Three compartments; interconnected	17.3
6. Five compartments, interconnected	14.8

These data were evaluated by tests of a two-dimensional airfoil at low Reynolds numbers. Computations for power were based on the assumption that $C_Q' \propto \mathrm{Re}_c^{-1/2}$, although this assumption does need verification.

The suction head $(H_0 - p_c)$ to be produced by the pump is greatest in the compartment covering the point of the maximum velocity on the airfoil wall. At this point, the required suction head is about 40 lb/ft^2 since $k \cdot t$ has been assumed zero.

5.2.1.3 Optimum suction

The most economic suction rate may also be found from an energy point of view. Gross (1968) defined the so-called "optimum suction" as the suction rate at which there is the least loss in the boundary layer kinetic energy and in the energy necessary to pump the sucked flow to ambient static pressure and velocity. The total energy coefficient C_{Pt} is given by $C_{Pt} = C_{Ps} + C_{Pw}$. Here C_{Ps} is the suction energy coefficient or the coefficient of pumping energy, defined by $C_{Ps} = P_s/\frac{1}{2}\rho u_\infty^3 A$, where P_s is the energy required to bring suction air to ambient velocity and static pressure and A is reference area. The symbol C_{Pw} refers to the boundary layer kinetic loss coefficient defined by

$$C_{Pw} = \frac{2}{c}\left(\frac{u_e}{u_\infty}\right)^3 \theta^*$$

where

$$\theta^* = \int_0^\delta \frac{\rho u}{\rho_e u_e}\left[1 - \left(\frac{u}{u_e}\right)^2\right] dy$$

is the boundary layer kinetic energy thickness.

Since θ^* is a measure of the loss of kinetic energy in the flow due to the action of viscosity and the turbulent Reynolds stress in the boundary layer, C_{Pw} is linearly dependent on θ^*. At low suction rates, the pumping power is low but the loss in kinetic energy within the thick boundary layer is high. At high suction rates the energy loss due to the pumping is high but the loss of kinetic energy due to the thinned boundary layer is reduced. Therefore, the combined loss is minimized at some intermediate or "optimum suction" flow rate. (Examples of the "optimum suction" flow rate associated with the minimum energy loss will be given in a later section.) As mentioned previously, the characteristics of the flow affected by suction are given by δ^*, θ, and H. In practice, however, it is not always possible to measure these quantities if the flow is separated. Hence, the required or "sufficient" suction rate may be referred to the condition of fully attached flow to the surface up to the measuring rake. Gross (1968) called such a suction rate "sufficient suction." It is reasonable to assume that some basic requirement of flow is satisfied at the condition of "sufficient suction." Once a certain suction quantity is passed, the rate of change of the shape parameter $H = \delta^*/\theta$ is substantially reduced. The suction may affect the basic growth and the development of the boundary layer, causing a rapid change of boundary layer shape parameter and thickness until a certain "equilibrium" position is reached. Subsequent increase of suction serves only to thin the stable boundary layer. This suction flow rate is then the *sufficient suction.*

5.2.2 Suction in Incompressible Laminar Flow

5.2.2.1 Criteria for prevention of separation

Criteria for the prevention of separation may be conveniently grouped into two cases, continuous suction and discontinuous suction. In practice, however, an ideal realization of continuous and discontinuous suction is impossible. Because of the effects of edges, discontinuous suction (such as stepwise distribution of suction velocity) can be realized only in a very approximate manner by slots or perforation. The mathematical treatment of boundary layers with continuous suction is more advanced. The boundary layer equations for an impermeable wall also hold for a permeable wall, differing only in the condition at the wall. For the impermeable wall, the condition on the wall surface is $v = 0$, and for the permeable wall the condition on the wall surface is $v = \pm v_0$; the plus sign indicates blowing and the minus sign suction.

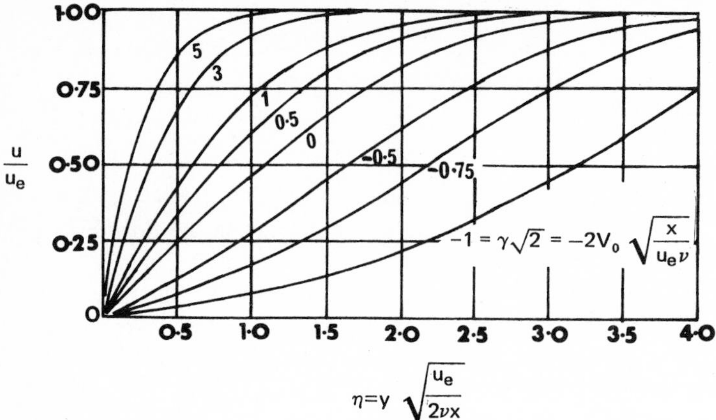

FIG. 5.14 Velocity profiles in the boundary layer of a flat plate with a suction velocity distribution $v_0 \sim 1/\sqrt{x}$ *(from Schlichting & Bussmann, 1943)*.

The various approximate methods for a permeable wall with laminar flow have been presented by Schlichting (1948), Truckenbrodt (1956), Wuest (1961a and 1961b), and Head (1961). Schlichting (1948) used the Karman momentum integral to develop an approximate method for calculating incompressible laminar boundary layers. This method has been improved by Truckenbrodt (1956). The basic idea was to approximate some auxiliary functions of the momentum equation by suitable expressions to derive the differential equation consisting of three terms. Wuest (1961b) and Head (1961) also solved the two-dimensional boundary layer equations with suction.

Consider the velocity profiles of the boundary layer as affected by suction. Some pertinent results for the boundary layer velocity profile obtained by

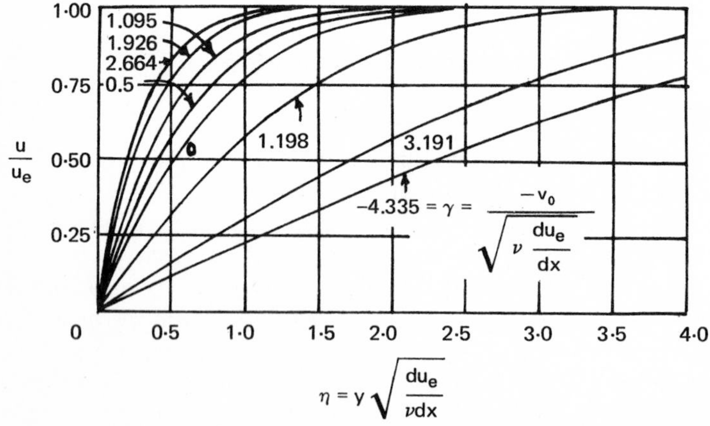

FIG. 5.15 Velocity profiles in the boundary layer at a stagnation point with constant suction *(from Schlichting & Bussmann, 1943)*.

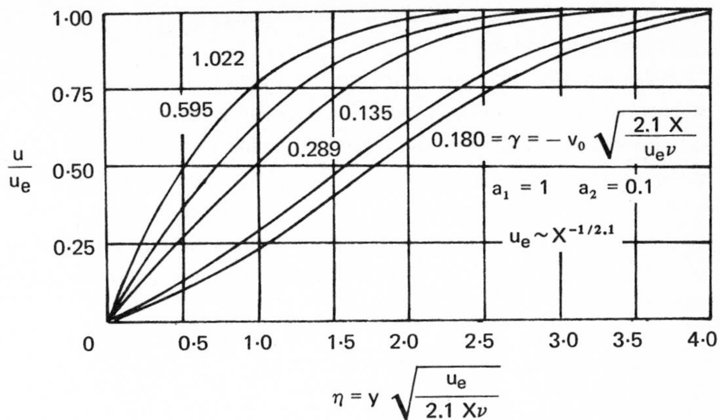

FIG. 5.16 Velocity profiles in the boundary layer of a divergent channel with suction and blowing (*from Mangler, 1944*).

similar solutions of the laminar boundary layer equation with suction are shown for a flat plate (Fig. 5.14), a stagnation point (Fig. 5.15), a divergent channel (Fig. 5.16), and a convergent channel (Fig. 5.17).

In these figures, the coefficients a_1 and a_2 are defined by

$$a_1 = \frac{1}{g}\frac{\partial}{\partial x}\left(\frac{u_e}{g}\right) \qquad a_2 = \frac{1}{g^2}\frac{\partial u_e}{\partial x}$$

and the symbol g refers to a similarity scale factor.

Although the solution for a flat plate and a convergent channel–and perhaps for a stagnation point–may not be of direct use for separation

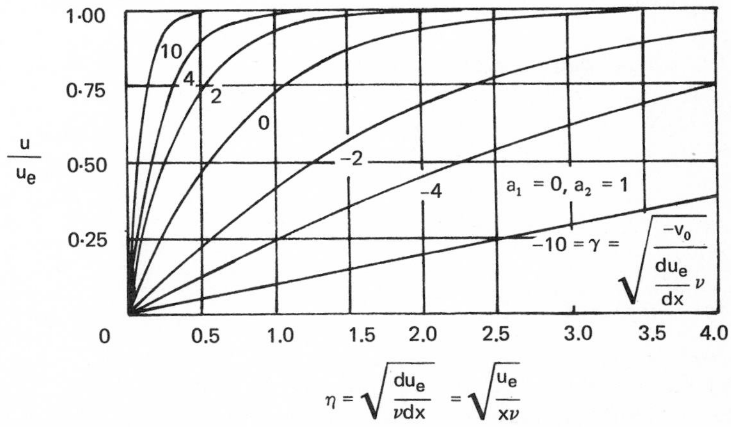

FIG. 5.17 Velocity profiles in the boundary layer of a convergent channel (sink flow) with a suction or blowing velocity $v_0 \sim 1/\sqrt{x}$ (*from Holstein, 1943*).

control, they are useful in evaluating effects of suction on the boundary layer flow.

These figures indicate that suction causes the velocity profile to become thinner whereas blowing causes it to become fuller because the momentum in the boundary layer is decreased by suction and increased by blowing. However, the velocity profile for a divergent channel (adverse pressure gradient) is not the same as for a convergent channel (favorable pressure gradient). Note that for an adverse pressure gradient in the divergent channel, the velocity profile shows an inflection point indicative of the propensity for separation.

5.2.2.2 Prediction of laminar separation with suction

Head (1961) demonstrated a method for predicting separation for a velocity distribution of the form $u_e/u_\infty = 1-(x/c)$ (where c is a chord) for three cases: (a) adverse pressure gradient with constant suction, (b) the flat plate with uniform suction following a solid entry length, and (c) on the surface of an airfoil with nonuniform suction applied in the region of adverse pressure gradient.

With constant suction The dimensionless constant suction velocity is given by

$$v_0^* = \left(\frac{v_0}{u_\infty}\right)\sqrt{\mathrm{Re}_c}$$

where $\mathrm{Re}_c = u_\infty c/\nu$ and v_0 is suction velocity. Figure 5.18 presents laminar separation with constant suction velocity as predicted by various methods. Note that all three methods use the energy equation and yield results which are almost precise, right up to separation. The compatability conditions at the wall mentioned in Fig. 5.18 are

first compatibility condition:

$$v_0\left(\frac{\partial u}{\partial y}\right)_0 = -\frac{1}{\rho}\frac{dp}{dx} + \nu\left(\frac{\partial^2 u}{\partial y^2}\right)_0$$

second compatibility condition:

$$v_0\left(\frac{\partial^2 u}{\partial y^2}\right)_0 = \nu\left(\frac{\partial^3 u}{\partial y^3}\right)_0$$

The method based on satisfying the momentum equation and the two compatibility conditions at the surface was found to be workable up to separation, but the separation it predicts is somewhat different from that given by the other methods. However, the method based on satisfying the

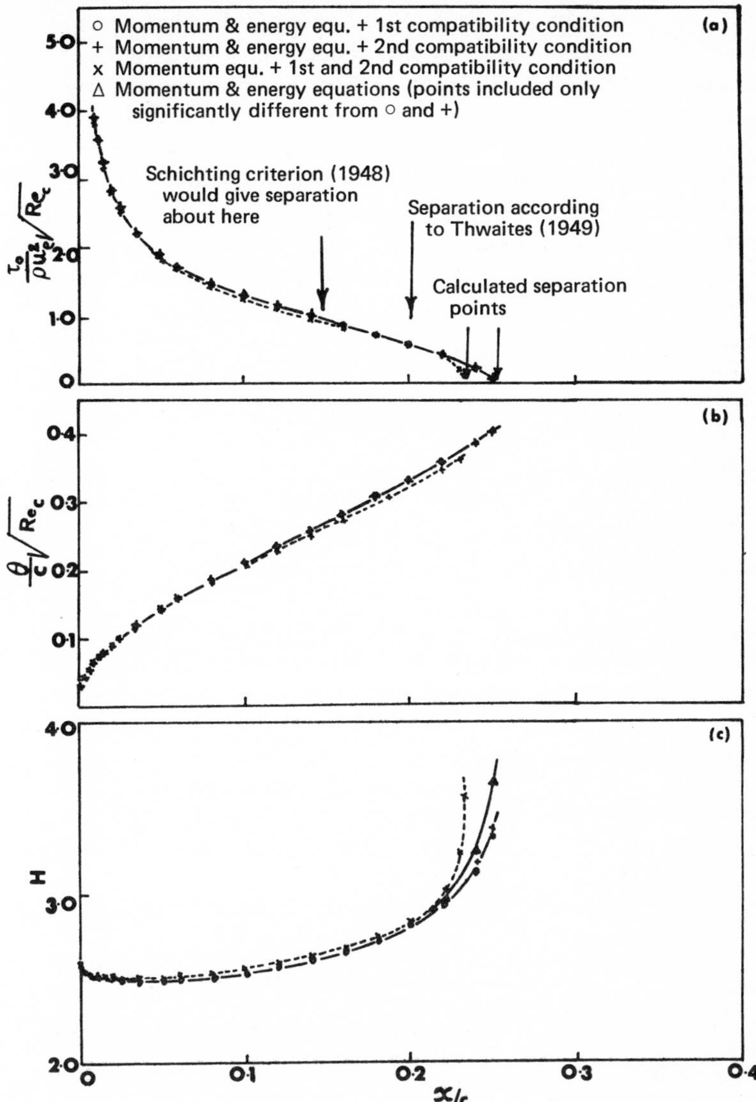

FIG. 5.18 Prediction of laminar separation with constant suction velocity (*from Head, 1961*).

momentum equation and the first compatibility condition at the wall was found to break down before separation was closely approached, and application of the Schlichting criterion (1948) unrealistically indicated separation in the region of $x/c = 0.14$. Since no exact solution for this case of suction is available for comparison, an attempt was made to show that the joint use of the momentum and energy equations may yield reasonably accurate results although no suction is involved. For such conditions (see Fig. 5.19), the computed values of skin friction for a velocity distribution of the form $u_e/u_\infty = 1-(x/c)^n$ compared favorably with those given by the exact solutions of Howarth (1938), Tani (1949), and Görtler and Witting (1957).

Head (1961) concluded from this comparison that the joint use of the momentum and the energy equations to predict separation with suctions leads to an acceptably accurate result whether or not the first or second compatibility condition at the surface is satisfied. Methods that use only the momentum equation and the first compatibility condition at the surface are unsatisfactory, but since in addition the second compatibility condition is satisfied by the use of a doubly infinite family of profiles, a workable and relatively quick method is then obtained.

With nonuniform suction Figure 5.20 shows the distribution of the potential velocity with suction applied in the region downstream of 63 percent of the chord position, where pressure rise begins (which is used in the calculations).

Figure 5.21 compares the results for this particular case as calculated by four different approximate methods and by the exact method of solution. Values of the shear stress τ_w and the boundary layer shape factor H given by all four approximate methods were very close to the pressure minimum. But downstream of this point, which is where suction is applied, the most satisfactory results were yielded by the methods that satisfy the momentum and the energy equation as well as either the first or second compatibility condition at the surface. Head (1961) has pointed out that the so-called exact

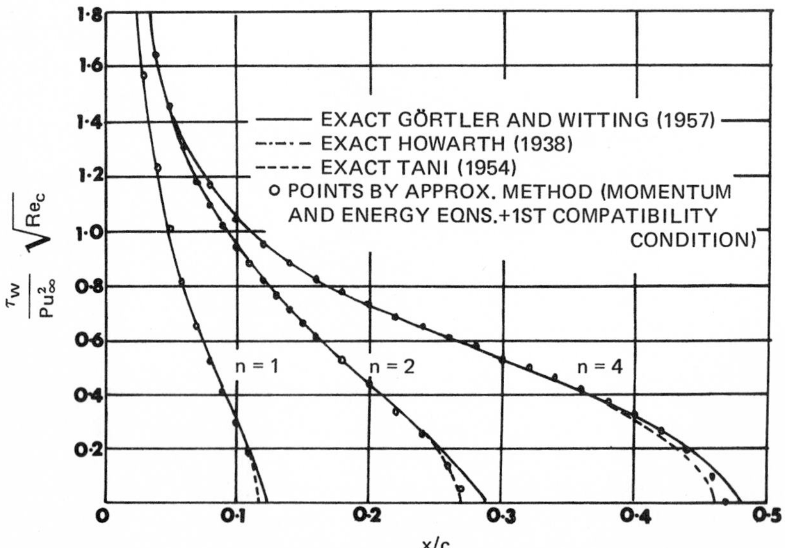

FIG. 5.19 Skin friction for $u_e/u_\infty = 1-(x/c)^n$ with no suction (*from Head, 1961*).

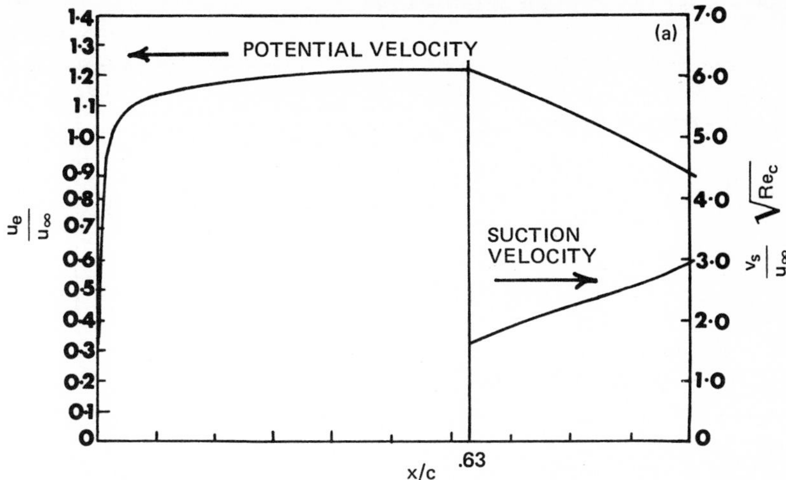

FIG. 5.20 Distribution of potential velocity and suction velocity on an airfoil surface (*from Head, 1961*).

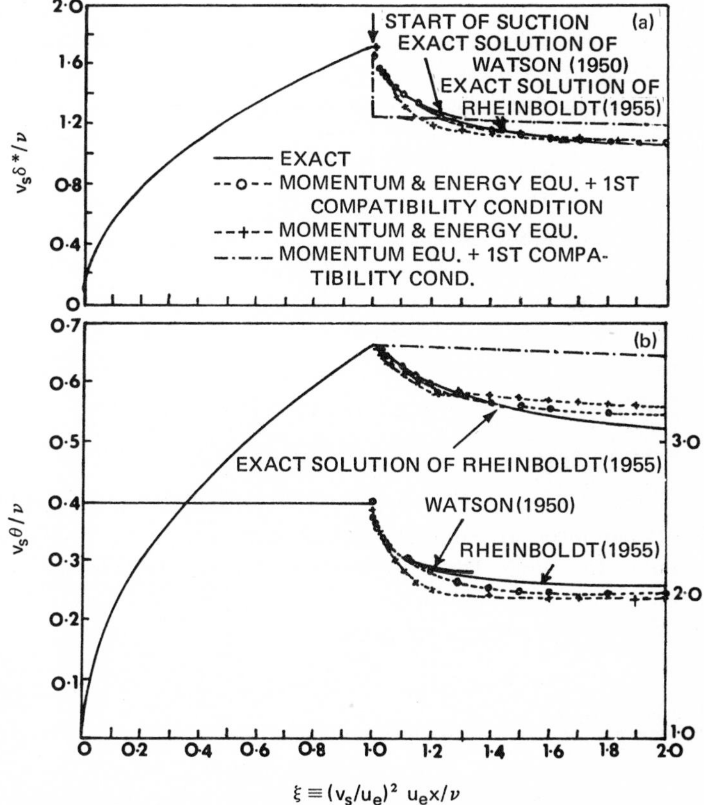

FIG. 5.21 Prediction by four approximate and one exact methods of shear stress and shape factor on an airfoil with the suction shown in Fig. 5.20 (*from Head, 1961*; he documents the Watson, 1950, solution in the form of an unpublished communication to him).

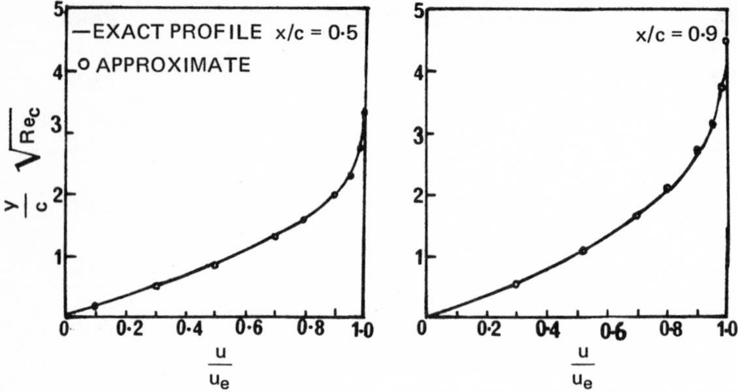

FIG. 5.22 Velocity profiles for an airfoil with the suction distribution shown in Fig. 5.20 (*from Head, 1961*).

points on Fig. 5.21 may be slightly in error because of a possible confusion of scales in the original report. The results given by the momentum equation used in conjunction with the first two compatibility conditions may also be considered to be reasonably satisfactory. However, the Truckenbrodt (1955) method yields unsatisfactory results (Head, 1961).

The momentum and energy equation with the first compatibility condition at the wall yield a close agreement with the exact solution extending to trailing edge. Thus, the velocity profiles obtained by this method for the suction distribution shown in Fig. 5.20 is in very close agreement with the exact solution at $x/c = 0.5$ and 0.9 shown in Fig. 5.22. Hence, Head (1961) recommends using the momentum and energy equation with the first compatibility condition at the wall.

5.2.3 Suction in Incompressible Turbulent Flow

5.2.3.1 Analysis of increase in lift by suction

Pechau (1958) and Schlichting (1958) studied incompressible turbulent separation involving arbitrary suction and blowing for two-dimensional and axisymmetric flows. Pertinent features of boundary layer suction to prevent separation are given by Wuest (1961a). The following forms of the momentum and the energy equations are used:

$$\frac{1}{u_e^2 R}\frac{d}{dx}(u_e^2 R\theta) + H\frac{\theta}{u_e}\frac{du_e}{dx} = \frac{\tau_0}{\rho u_e^2} - \left(\frac{v_0}{u_e}\right) \tag{5.1}$$

$$\frac{1}{u_e^3 R}\frac{d}{dx}(u_e^3 R\theta\bar{H}) = 2\int_0^\delta \frac{\tau}{\rho u_e^2}\frac{\partial}{\partial y}\left(\frac{u}{u_e}\right)dy - \left(\frac{v_0}{u_e}\right) \tag{5.2}$$

where

$$\delta^*(x) = \int_0^\delta \left(1 - \frac{u}{u_e}\right) dy = \text{displacement thickness}$$

$$\theta(x) = \int_0^\delta \frac{u}{u_e}\left(1 - \frac{u}{u_e}\right) dy = \text{momentum thickness}$$

$$H = \frac{\delta^*(x)}{\theta(x)}$$

and

$$\bar{H} = \frac{\epsilon(x)}{\theta(x)}$$

$$\epsilon(x) = \int_0^\delta \frac{u}{u_e}\left[1 - \left(\frac{u}{u_e}\right)^2\right] dy = \text{energy loss thickness}$$

For two-dimensional flow, write $R = 1$ instead of the actual radius R used for calculating axisymmetric flow. The solution requires making a proper assumption concerning the wall shear stress τ_0, given by the following formula of Ludwieg and Tillmann (1949) for the impermeable wall. For the case of suction it is necessary to adjust the proper power of $u_e\theta/\nu$:

$$\frac{\tau_0}{\rho u_e^2} = \left[\frac{0.123}{(u_e\theta/\nu)^{0.268}}\right] \times 10^{-0.678H} \qquad \text{for} \qquad 10^3 < \left(\frac{u_e\theta}{\nu}\right) < 4 \times 10^4$$

Measurements of suction and blowing by Mickley et al. (1954) and Mickley and Davis (1957) indicated that the exponent of $u_e\theta/\nu$ increases with blowing and decreases for suction approaching $-\frac{1}{6}$ although it amounts to about $-\frac{1}{4}$ for impermeable wall. Further measurements are needed for the case of suction. Pechau (1958) assumed that

$$\frac{\tau_w}{\rho u_e^2} = \frac{\alpha}{\mathrm{Re}_\theta^{1/4}} \quad \text{with } \alpha = 0.013$$

and also that $H \approx$ constant. After introducing

$$\Omega = \mathrm{Re}^{1/4} \left(\frac{u_e}{u_\infty}\right)^{5/4(H+2)} \left(\frac{\theta}{c}\right)^{5/4} \left(\frac{R}{c}\right)^{5/4} \tag{5.2a}$$

the momentum equation (5.1) becomes, in dimensionless form, with $\xi = x/c$ and $\mathrm{Re} = u_\infty c/\nu$,

$$\frac{d\Omega}{d\xi} = \frac{5}{4}\alpha\left(\frac{u_e}{u_\infty}\right)^{5/4(H+2)-1/4}\left(\frac{R}{c}\right)^{5/4} - \frac{5}{4}\mathrm{Re}^{1/5}\left(-\frac{v_0}{u_\infty}\right)\left(\frac{u_e}{u_\infty}\right)^{H+1}\left(\frac{R}{c}\right)\Omega^{1/5} \tag{5.3}$$

The integral on the right hand side of the energy equation (5.2) represents the work of the shear stress. Pechau (1958) used the following approximation:

$$\int_0^\delta \frac{\tau}{\rho u_e^2}\frac{\partial}{\partial y}\left(\frac{u}{u_e}\right)dy = \frac{\beta}{\mathrm{Re}_\theta^{1/4}}$$

where β is separation angle. Wuest (1961a) states that an exponent of $-\frac{1}{6}$ rather than $-\frac{1}{4}$ would be in better agreement with the calculations of Rotta (1952). After some calculations, we obtain

$$\bar{H}(\xi) = \frac{(u_e/u_\infty)^{H-1}}{\Omega^{4/5}}\int_0^\xi \left[\frac{2\beta}{\Omega^{1/5}}\left(\frac{u_e}{u_\infty}\right)^{(13+H)/4}\left(\frac{R}{c}\right)^{5/4} - \mathrm{R}_e^{1/5}\left(-\frac{v_0}{u_\infty}\right)\left(\frac{u_e}{u_\infty}\right)^2\frac{R}{c}\right]d\xi \tag{5.4}$$

It should be noted in Eqs. (5.3) and (5.4) that the suction terms appear only in the combination of $(-v_0/u_\infty)\cdot\mathrm{Re}^{1/5}$.

The parameter Ω vanishes at the leading edge and at a stagnation point since $\theta = 0$ and $u_e = 0$ respectively. The initial value of the shape parameter H depends on the intensity and distribution of suction.

If the suction velocity distribution is of the form $-v_0/u_\infty = A_1\xi^p$, where A_1 is a coefficient, then for bodies with stagnation point,

$$\bar{H}(\xi=0) = \left(\frac{\frac{5}{4}H + \frac{13}{4} + \frac{5}{4}}{\frac{18}{5}+1}\right)\left(\frac{2\beta}{\frac{5\alpha}{4}}\right) \quad \text{for} \quad p > \frac{3}{5} \quad A_1 \neq 0 \quad \text{or} \quad A_1 = 0$$

$$\bar{H}(\xi=0) = \left(\frac{H+2+p+1}{3+p+1}\right) \quad \text{for} \quad p < \frac{3}{5} \quad A_1 \neq 0$$

The terms in the parentheses vanish for two-dimensional flow. Further, for plates or bodies with a sharp nose,

$$\bar{H}(\xi=0) = \frac{2\beta}{\alpha} \quad \text{for} \quad p > -\frac{1}{5} \quad A_1 \neq 0 \quad \text{or} \quad A_1 = 0$$

$$\bar{H}(\xi=0) = 1 \quad \text{for} \quad p < -\frac{1}{5} \quad A_1 \neq 0$$

The relationship between $\bar{H} = \epsilon/\theta$ and $H = \delta^*/\theta$ is not greatly affected by Reynolds number; see Fig. 5.23.

Moreover, according to measurements by Wuest (1958), suction has no apparent influence on this relationship as shown in Fig. 5.24.

Therefore, Pechau (1958) assumes that turbulent separation occurs at a position where $H = 2$ or $\bar{H} = 1.58$ with suction.

5.2.3.2 Application of analysis to an airfoil

The analysis by Pechau (Schlichting & Pechau, 1959) is used in a numerical evaluation of the increase in lift by suction on the NACA 747A315 airfoil for various suction distributions (Fig. 5.25). Without suction, flow separates at the position $x/s = 0.44$ for a 12-deg angle of attack and at 0.16 for 20 deg (s is the developed length of the profile contour measured from the stagnation point). In all calculated examples, the region where suction is applied with constant inflow velocity originates at the stagnation point. The symbol a refers to the developed length of this zone.

The minimum suction velocity for attached flow has been calculated for different values of a/s ranging from 0.1 to 1.0. As shown in Fig. 5.26 the most preferable suction technique appears to be to concentrate the region of suction over the forward part of the airfoil. Maus (1968) used the Bradshaw et al. (1967) method to calculate turbulent boundary layers in decelerated flows with suction.

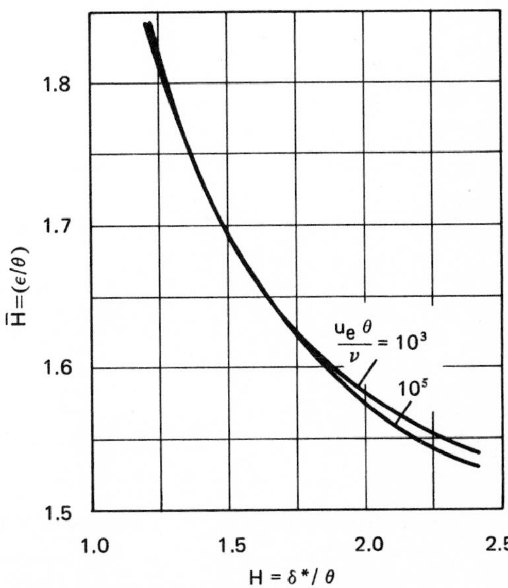

FIG. 5.23 Form parameter as a function of shape parameter in turbulent flow for two Reynolds numbers (*from Truckenbrodt, 1952*).

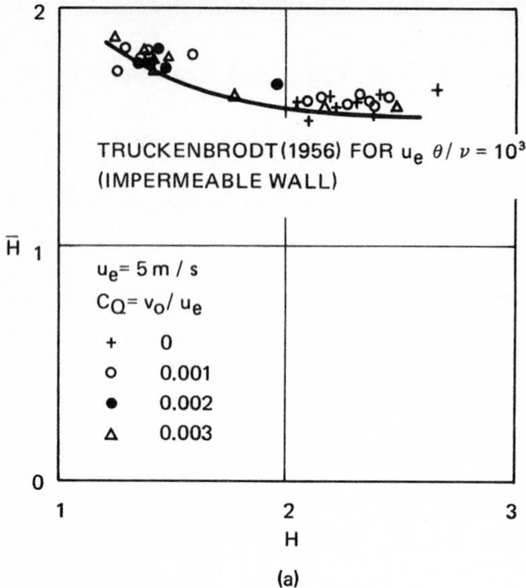

(a)

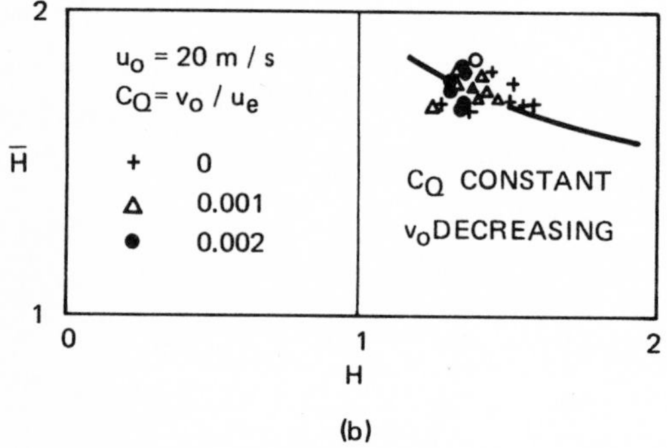

(b)

FIG. 5.24 Effect of suction on the relationship of form and shape parameters in turbulent flow on a flat plate: (*a*) suction alone; (*b*) suction with pressure rise (*from Wuest, 1958*).

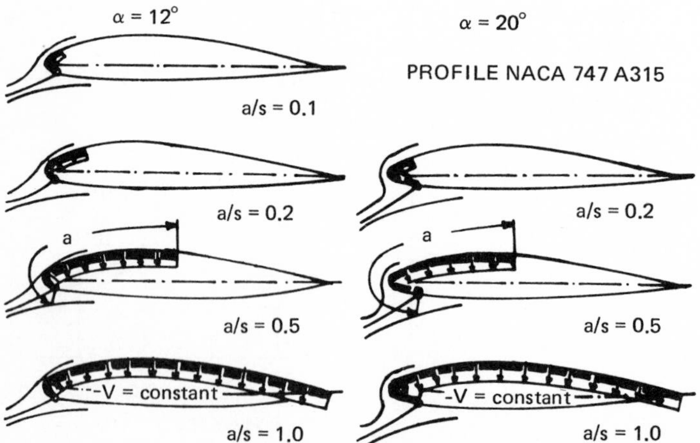

FIG. 5.25 Increase in lift of the NACA 747A315 Airfoil for various suction distributions (the different regions of continuous boundary layer suction for two angles of incidence are from Wuest, 1961a; the external streamline and the forward stagnation point were calculated by Pechau; *from Schlichting & Pechau, 1959*).

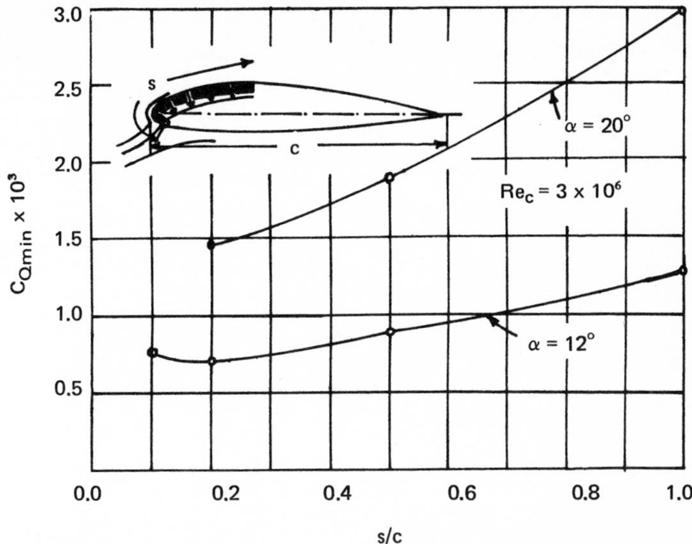

FIG. 5.26 Computed minimum suction quantities for the NACA 747A315 Airfoil as a function of the extent of the sucked region (suction with constant inflow velocity commences at the stagnation point; *from Wuest, 1961a*).

5.2.4 Application of Suction to Prevent Separation

Flow separation can be controlled by suction of the fluid flow. With an increase in flight speed, it is necessary to provide larger wing loading, large sweepback angles, and thinner airfoils. Furthermore, the largest possible lift is desired in order to satisfy the requirements for takeoff and landing. The control of separation on the upper surface of wings and flaps at large angles of attack by means of suction has proven an effective way to improve aircraft performance. It is important to select the proper technique for suction. Two will be discussed here–continuous suction and slot suction. For convenience, the application of suction to incompressible flow on an airfoil is classified into leading-edge suction, downstream suction, and trailing-edge flap suction.

Although the practical examples are described for an airfoil, the same technique is applicable to a two-dimensional subsonic diffuser because the boundary layer development and separation are governed by the external pressure distribution. Suction is effective in the region where the separation is imminent. It involves a steep pressure rise downstream of a pressure minimum. The boundary layer is generally turbulent in this region.

5.2.4.1 Leading-edge suction

Performance is strongly influenced by the region where suction is applied. In order to demonstrate this fact, Fig. 5.27 indicates the increase of lift

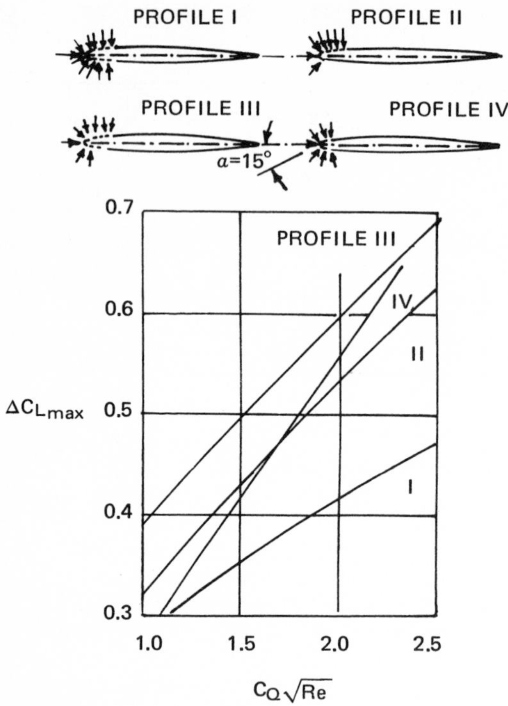

FIG. 5.27 Increase of lift achieved by four different arrangements of suction applied at the nose of a thin airfoil (continuous suction at 15-deg angle of attack; *from Pankhurst et al., 1948*, also *Schlichting & Pechau, 1959*).

achieved by four different arrangements of suction applied at the nose of a thin airfoil. Note that where equal suction areas are provided in upper and lower surfaces (Profile I), the airfoil surface yields the lowest value for $\Delta C_{L\,max}$ whereas when the suction area in the lower surface is smaller than in the upper surface (Profile III), the airfoil surface produces the highest $\Delta C_{L\,max}$.

Schlichting and Pechau (1959) give examples of leading edge suction with reference to the boundary layer characteristics (see Sec. 5.2.3.2). Figure 5.28 indicates the potential theoretical velocity distribution with no suction on an NACA 747A315 airfoil at three angles of attack (1.5, 12, and 20 deg). No turbulent separation occurred at 1.5 deg angle of attack. In the absence of suction at $\alpha = 12$ deg, separation occurred at $x/s = 0.44$ whereas at $\alpha = 20$ deg, it occurred at $x/s = 0.16$, a position farther upstream because of the pronounced suction peak in the nose regions. The symbol s refers to the wall length measured from the stagnation point to trailing edge. Since the separation occurred upstream of the forward half of the airfoil at $\alpha = 12$ and

20 deg, suction provided in the rear half of the airfoil was not effective in preventing separation. The most effective suction should be arranged in the region of the strong pressure rise downstream of the maximum velocity; this lies between x/s values of 0.1 and 0.3, as seen from the velocity distribution in Fig. 5.28.

The possible arrangement of the suction region on an airfoil at $\alpha = 12$ and 20 deg has already been presented in Fig. 5.25. It indicates that the suction zone always starts at the stagnation point. The suction velocity v_0 is over the entire suction zone, but the value of v_0 is different for each suction arrangement. Values of the boundary layer characteristics calculated by the Pechau (1958) method (see Sec. 5.2.3.1) for the suction arrangements of Fig. 5.25 are plotted in Fig. 5.29.

Since the turbulent separation point may be predicted by the value of $\bar{H} = 1.58$ (see Sec. 5.2.3), the suction velocity v_0 can be selected in such a way that the separation point lies at the trailing edge; this is indicated by the value of approximately 1.6 for $\bar{H}$ at the trailing edge, in both parts of Fig. 5.29. Furthermore, the use of suction may reduce the boundary layer momentum thickness by about 50 percent, as shown in Fig. 5.30.

The minimum suction quantity ($C_{Q\min}$) required to prevent separation as computed by the Pechau (1958) method for $\alpha = 12$ and 20 deg has already

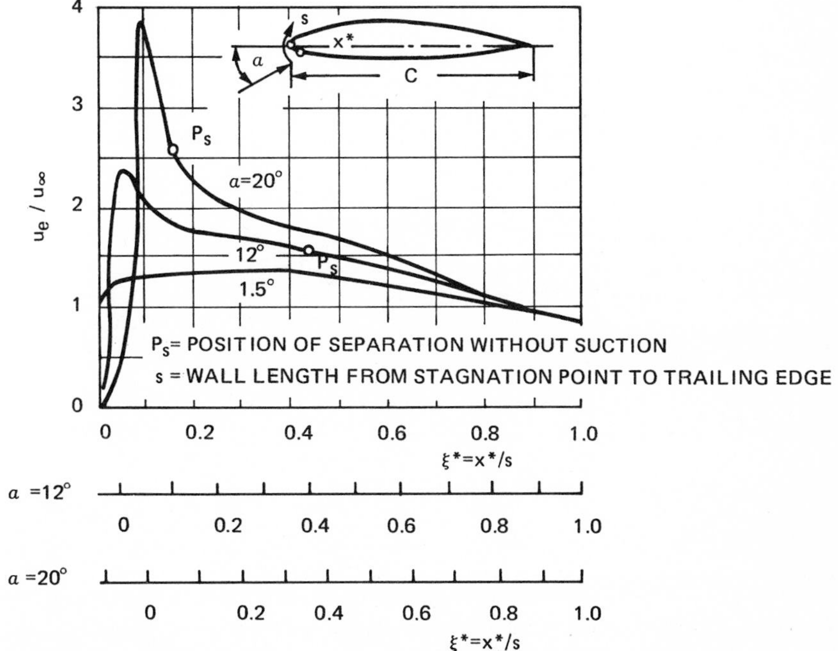

FIG. 5.28 Potential theoretical velocity distribution with no suction on the NACA 747A315 Airfoil for three angles of attack (*from Schlichting & Pechau, 1959*).

been given in Fig. 5.26 as a function of length of suction zone. Note the strong suction velocity in the nose region was more effective than suction distributed over the entire chord with small suction velocity. The most effective distribution of suction was in the forward part and ranged in length from 20 to 30 percent of the chord. Furthermore, for $\alpha = 20$ deg, more than 50 percent of the quantity may be saved if the suction is concentrated in the forward one-third of the chord length. Schlichting and Pechau (1959) also found that continuous suction through a porous wall is much more efficient that slot suction for leading-edge suction.

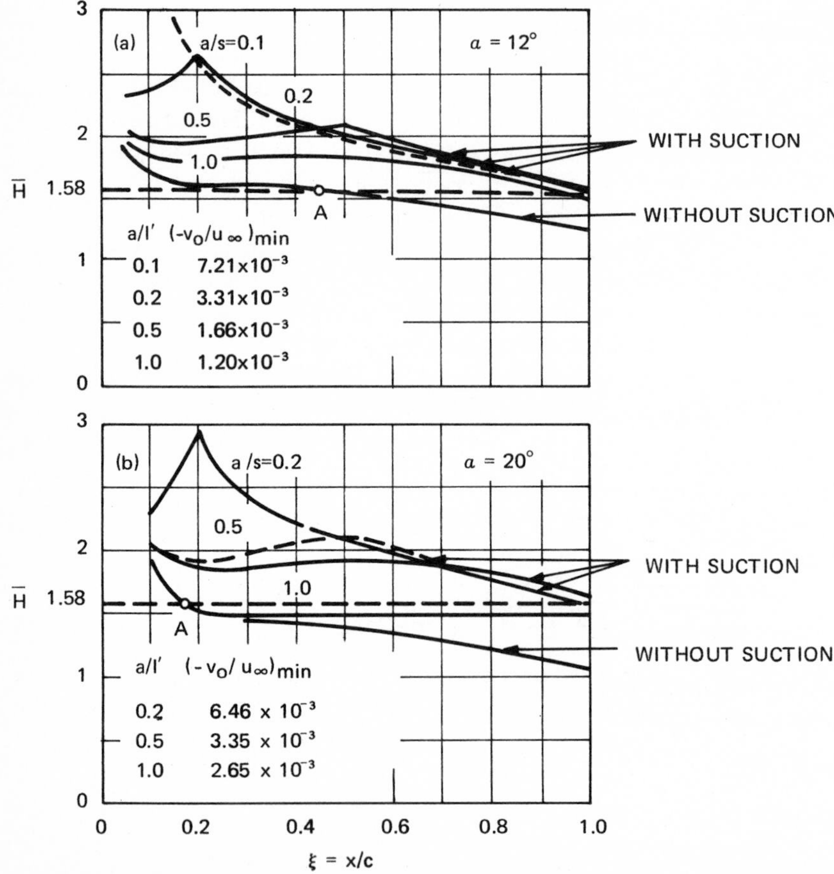

FIG. 5.29 Boundary layer characteristics of the NACA 747A315 Airfoil for different suction arrangements (the suction arrangements are those of Fig. 5.25, and the values shown were calculated by the Pechau method; Reynolds number $Re = 3 \times 10^6$; A = separation position without suction; the value of $(v_0/u_\infty)_{min}$ gives the required minimum suction velocity for prevention of separation in the corresponding suction zone; *from Schlichting and Pechau, 1959*).

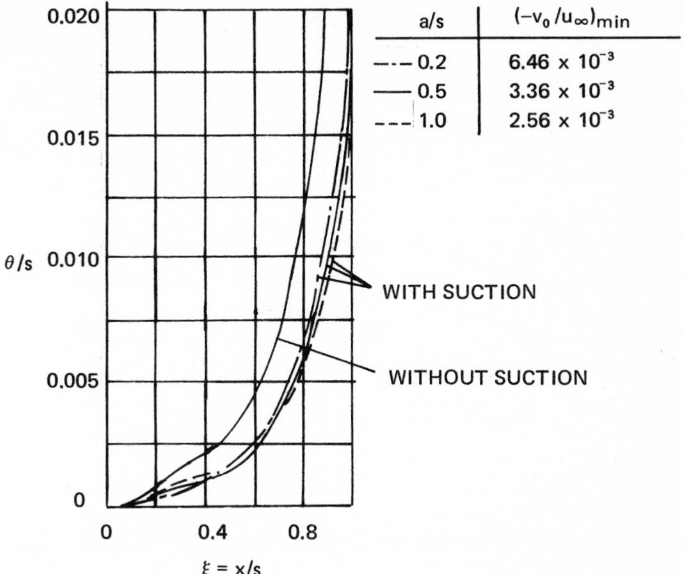

FIG. 5.30 Use of suction to reduce boundary layer momentum thickness of the NACA 747A315 Airfoil at 20-deg angle of attack (the suction regions of various lengths are for the models shown in Fig. 5.25; *from Schlichting & Pechau, 1959*).

5.2.4.2 Downstream suction

The application of suction to the surface downstream of the nose of an incompressible airfoil has been investigated experimentally by Gregory et al. (1950) and by Gross (1968). These two investigations will be discussed in turn.

Study by Gregory et al. These investigations used as their model a symmetrical airfoil (NPL 153), with a 4-ft span, 18-in. chord, and 33 percent thickness. The airfoil was fitted with a trailing edge flap and suction was provided through the porous surface from the 0.8 chord position to the trailing edge; see Fig. 5.31.

The approximate velocity distribution was a simple roof-top shape in which the velocity over the surface at zero incidence rose linearly to a maximum value at 0.8 chord and then fell linearly to the trailing edge. With no trailing edge flap, the suction provided could not prevent separation entirely. The variation of the wake drag with suction for two wind speeds is shown in Fig. 5.32 for the airfoil with and without a trailing edge flap as well as with a wire in lieu of a flap. There was no evidence of hysteresis in the movement of the separation points. The wavy form of the curves is due to slight difference in suction through the porous material; this prevented the occurrence of separation on one surface prior to the other.

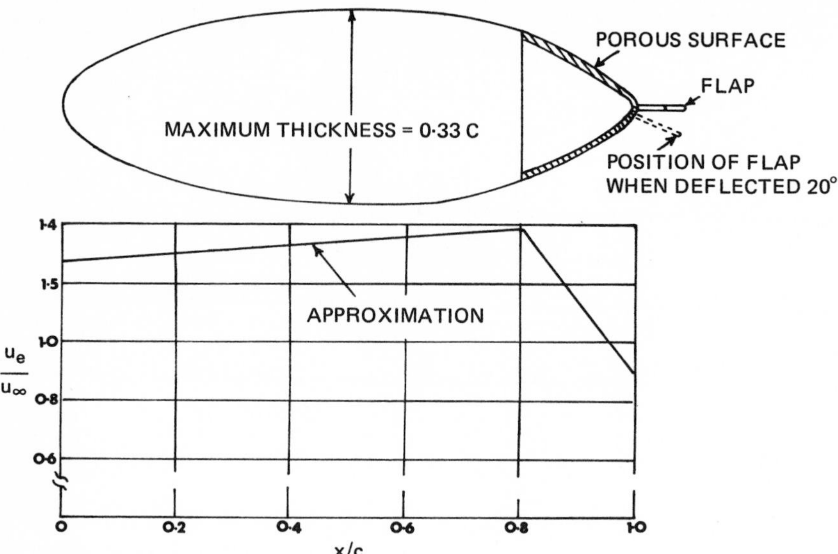

FIG. 5.31 Airfoil section (NPL 153) and velocity distribution (Approximation I) for the Gregory Model (*from Gregory et al., 1950*).

With a flap, separation was entirely prevented by the application of a sufficient amount of suction, and its boundary layer (at $\text{Re}_c = 0.58 \times 10^6$) was completely absorbed. Thus the wake drag was reduced to a value less than would have been experienced had the flap alone been exposed to the oncoming air flow. A stout 0.095-in. diameter wire, fixed along the rear of the airfoil, was effective in stabilizing the flow, and the wake drag was reduced to zero with the proper strength of suction. According to laminar boundary layer theory with distributed suction, the flow characteristics at different Reynolds number may be correlated to similar flow conditions if the parameter $C_Q'\sqrt{\text{Re}}$ is the same. Although the exact solution of $C_Q'\sqrt{\text{Re}}$ for prevention of separation is not available, the unpublished approximate solution given by Thwaites, namely $C_Q'\sqrt{\text{Re}} = 2.5$, may be used as the criterion for the prevention of separation. However, the test results at $\text{Re}_c = 0.58 \times 10^6$ showed that the value of $C_Q'\sqrt{\text{Re}}$ required to prevent separation was about 5 or double the analytical value. The largest available suction quantity, a C_Q' of 0.02, reduced the wake drag coefficient to 0.010 compared to 0.13 without suction. Streamers on a probe indicated that the flow was not two-dimensional, the boundary layer on the end fins thickened and separated due to the unnatural pressure recovery achieved by the distributed suction. There was a strong inflow from the ends of the airfoil at about the midchord position and also on the trailing edge flap.

The so-called "equivalent pump drag" was estimated for the test conditions with the small chord flaps at Reynolds numbers of 0.58 and 1.44×10^6. The

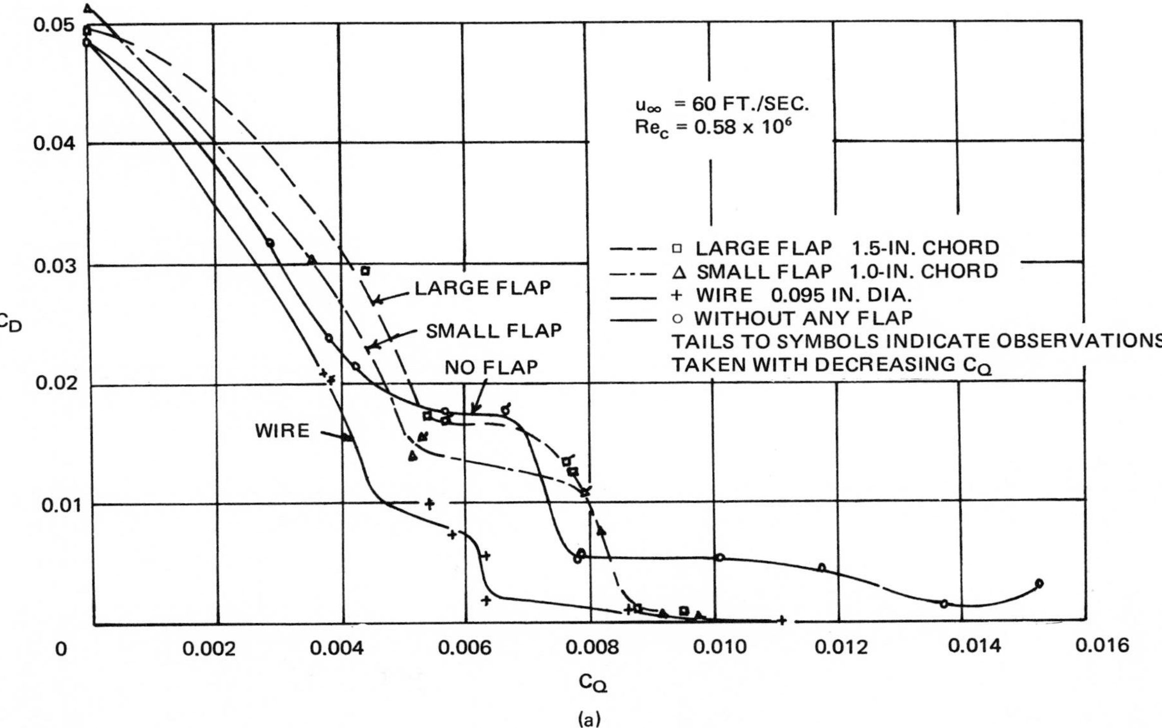

FIG. 5.32 Variation of wake drag coefficient with suction quantity coefficient for two wind speeds: *(a)* $u_\infty = 60$ ft/sec and $\mathrm{Re}_c = 0.58 \times 10^6$ *(from Gregory et al., 1950)*.

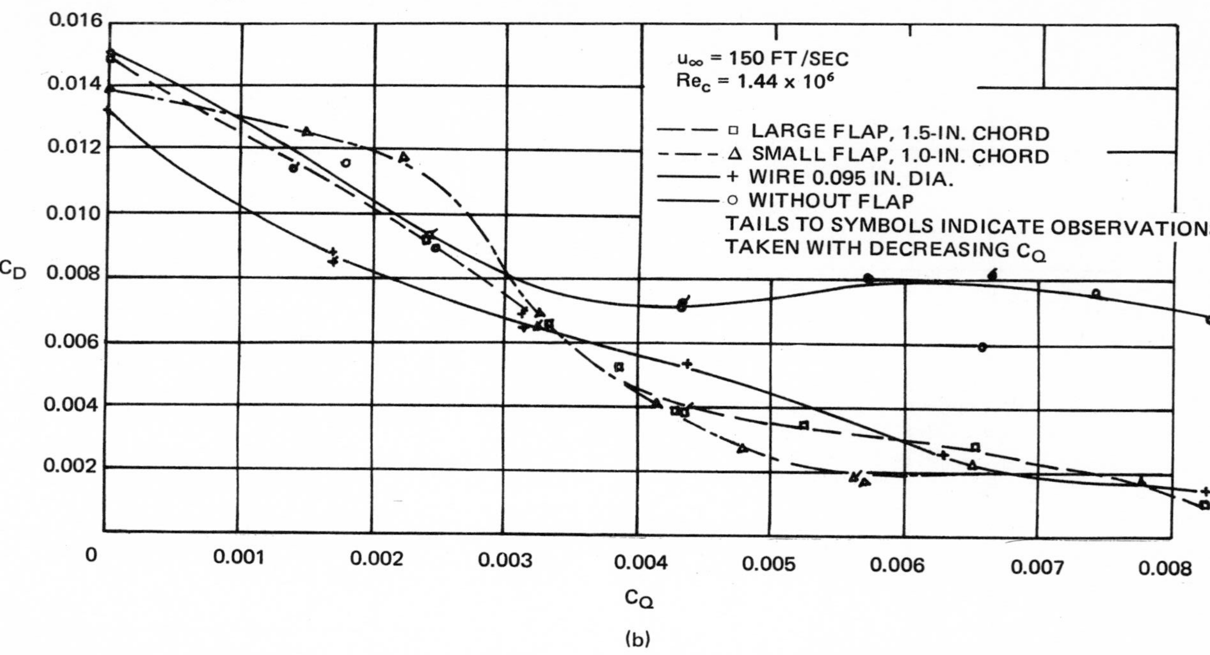

FIG. 5.32 *(continued)* Variation of wake drag coefficient with suction quantity coefficient for two wind speeds: (*b*) $u_\infty = 150$ ft/sec and $\mathrm{Re}_c = 1.44 \times 10^6$ (*from Gregory et al., 1950*).

wake drag and the effective drag coefficient for two assumed values of the pump drag are shown in Fig. 5.33. The lower curves correspond to the ideal pump drag $C_{D_{pi}}$ defined by

$$C_{D_{pi}} = \int \left(\frac{u_e}{u_\infty}\right)^2 \left(\frac{v_0}{u_\infty}\right) d\left(\frac{s}{c}\right)$$

as mentioned in Sec. 5.2.1.2.

If it is assumed that the suction chamber pressure varies to match the external pressure and the porous material resistance is neglected, then for the section tested, $C_{D_{pi}} = 1.32\, C_Q'$.

For practical purposes, the following are more realistic assumptions: a single pressure chamber with zero porous resistance at 0.8 chord and a resistance that increases toward the rear so that the normal velocity remains constant. Then a higher pump drag amounting to

$$C_{D_p} = (1.39)^2 C_Q' \approx 2C_Q'$$

may result. As indicated by the second curve, there is no need for the actual pump drag to be very much larger than this; if two section chambers are used,

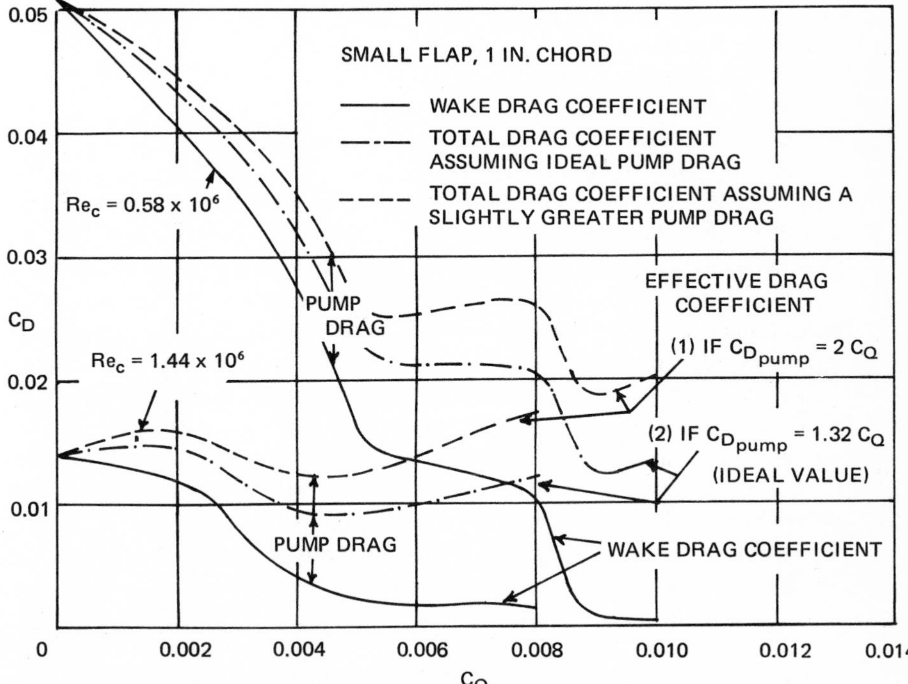

FIG. 5.33 Wake drag and effective drag coefficients for two assumed values of the pump drag (*from Gregory et al., 1950*).

the pump drag could be less. The application of suction may reduce total drag considerably at a Reynolds number of 0.58×10^6, but is less effective at the higher Reynolds number (1.44×10^6). However, the ratio of the minimum values of total drag at the two Reynolds numbers is approximately equal to the ratio of their square roots. This fact suggests that it is not unreasonable to extrapolate to higher Reynolds numbers on the assumption that $C'_Q\sqrt{\mathrm{Re}_c}$ and $C_D\sqrt{\mathrm{Re}_c}$ are constant. Such a tentative conclusion required experimental verification over a wide range of Reynolds numbers before it can be accepted as applicable.

Study by Gross The Gross (1968) investigation utilized an airfoil with a 12-percent thickness section and a 30-percent chord length flap deflected 60 deg. Turbulent flow separation was prevented by providing suction through a large number of fine slots in the pressure rise region located approximately in the forward 25 percent of chord length. It was possible to keep the turbulent boundary layer attached at ambient free-stream Mach numbers from $\mathrm{M}_\infty = 0.2$ to $\mathrm{M}_\infty = 0.315$ (corresponding to the maximum Mach numbers on the surface of the model $\mathrm{M} = 0.48$ to 0.935 and Reynolds numbers $\mathrm{Re}_c = u_\infty c/\nu = 4.1 \times 10^6$ to 6.3×10^6). The minimum observed optimum total suction flow coefficient varied from $C_{Q_t} = 1.58 \times 10^{-3}$ at $\mathrm{M}_\infty = 0.2$ to $C_{Q_t} = 1.48 \times 10^{-3}$ at $\mathrm{M}_\infty = 0.315$. The corresponding values of minimum total energy loss coefficient were $C_{p_t} = 0.013$ and $C_{p_t} = 0.022$, respectively. One-half of the pressure rise occurred within the first 15 percent of the airfoil chord and the remainder over the aft 70 percent portion.

The test results indicate that from the standpoint of minimum energy loss, it was advantageous to maintain a thin boundary layer by means of strong suction in the strongest pressure gradient of forward surface. In the downstream section, where flow deceleration was less pronounced and boundary layer became thicker, it was possible to suck with less strength while avoiding separation.

Gross (1968) considered only the forward 25 percent of the airfoil velocity distribution. It was determined by an approximate method consisting of a simple two-dimensional combination of the continuity and the Bernoulli equations modified to account for the curvature of the model and flow compressibility. The model cross-section and the theoretical velocity distribution are shown in Fig. 5.34. It also shows the theoretical velocity distribution of the representative wing of the model with and without the influence of the tunnel walls. Note that it was not possible to completely match the representative wing velocity distribution without resorting to the use of concave curvature on the test surface.

Gross used the computational method of Thompson (1965) to calculate the turbulent boundary layer development with constant suction. Figure 5.35 shows curves of H and $\mathrm{Re}_\theta/\sqrt{\mathrm{Re}_c}$ as a function of chord position x for nondimensional suction velocities of $v_0^* = -10$ and -20. Here $v_0^* = (v_0/u_\infty)\sqrt{\mathrm{Re}_c}$, v_0 is suction velocity, $\mathrm{Re}_\theta = u_e\theta/\nu$, $\mathrm{Re}_c = u_\infty c/\nu$, and c is

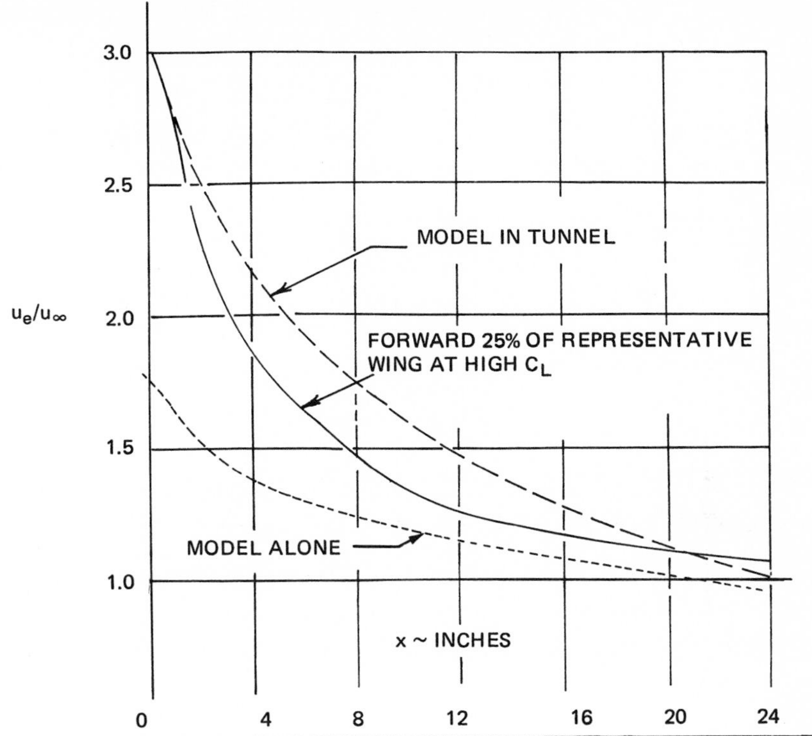

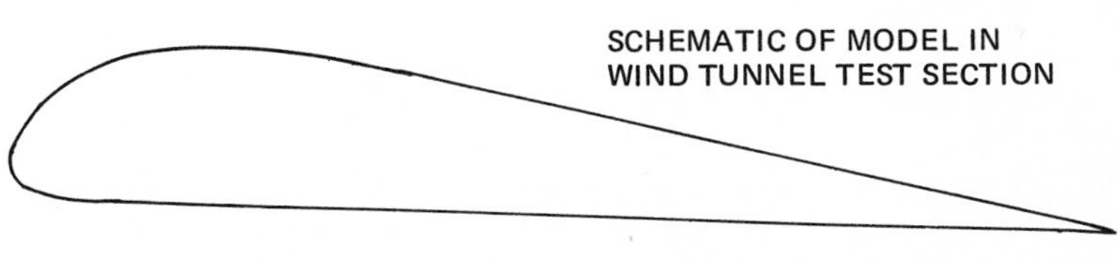

FIG. 5.34 Cross-section and theoretical velocity distribution for the Gross model (*from Gross, 1968*).

the model chord of 29.35 in. With no suction, separation was imminent at approximately 75 percent of the length of test region from the shape factor H, approaching 2.2. But the suction was sufficiently strong at a value of $v_0^* = -20$ to prevent appreciable boundary layer growth. Figure 5.36 indicates the effect of varying the chordwise suction distribution and the corresponding (theoretical) development of boundary layers. The distribution labeled in Fig. 5.36 is for a constant suction $v_0^* = -10$ and serves as the standard for

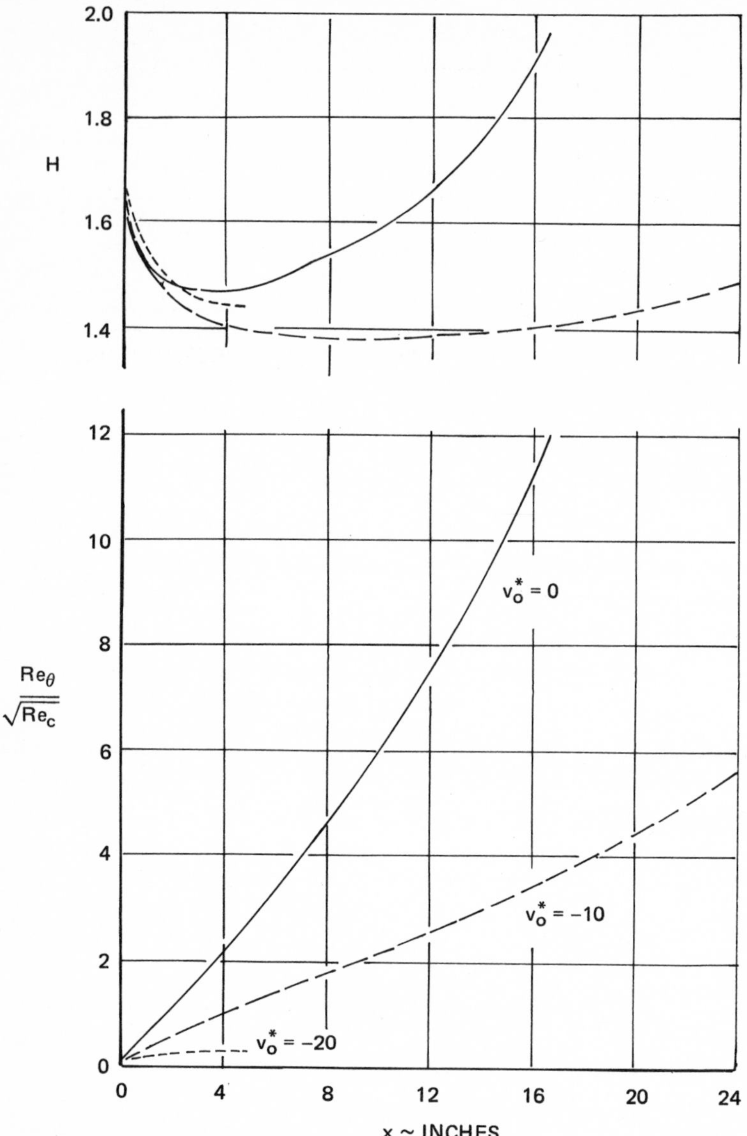

FIG. 5.35 Theoretical development of turbulent boundary layer for the Gross model with constant suction (*from Gross, 1968*).

comparison with the other types of suction. The second distribution was arranged to have the same total suction as Distribution 1 close to the trailing edge, but it had more suction near the front of the model where the boundary layer is thin. The technique whereby the boundary layer is kept thin by increased suction in the region where the flow is most strongly decelerated and allowed to thicken where the flow is less strongly decelerated is intended to reduce overall boundary layer growth for the same total suction. Distribution 3 maintained the same initial level of suction as Distribution 2, but suction was reduced at a greater rate as the flow progressed toward the trailing edge. The boundary layer was slightly thinner at the trailing edge than that with Distribution 1 for only 75 percent of the

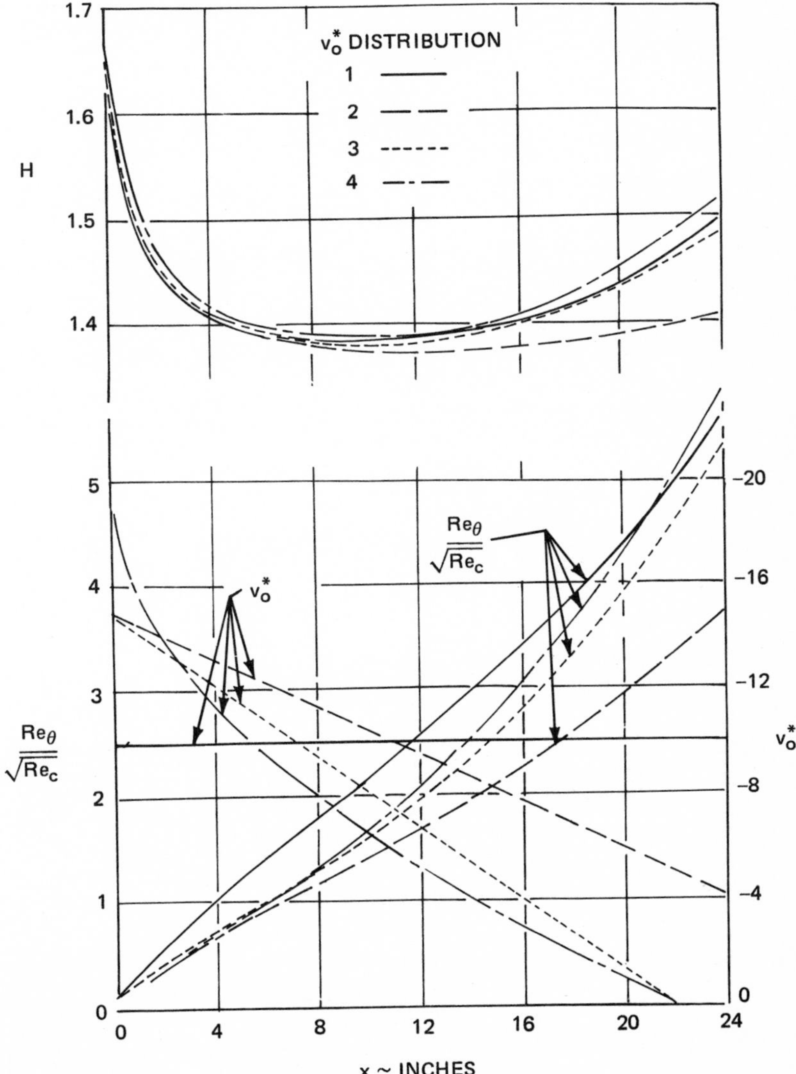

FIG. 5.36 Theoretical development of turbulent boundary layer for Gross model with different suction distributions (*from Gross, 1968*; see explanation of the various distributions given on pp. 324–325).

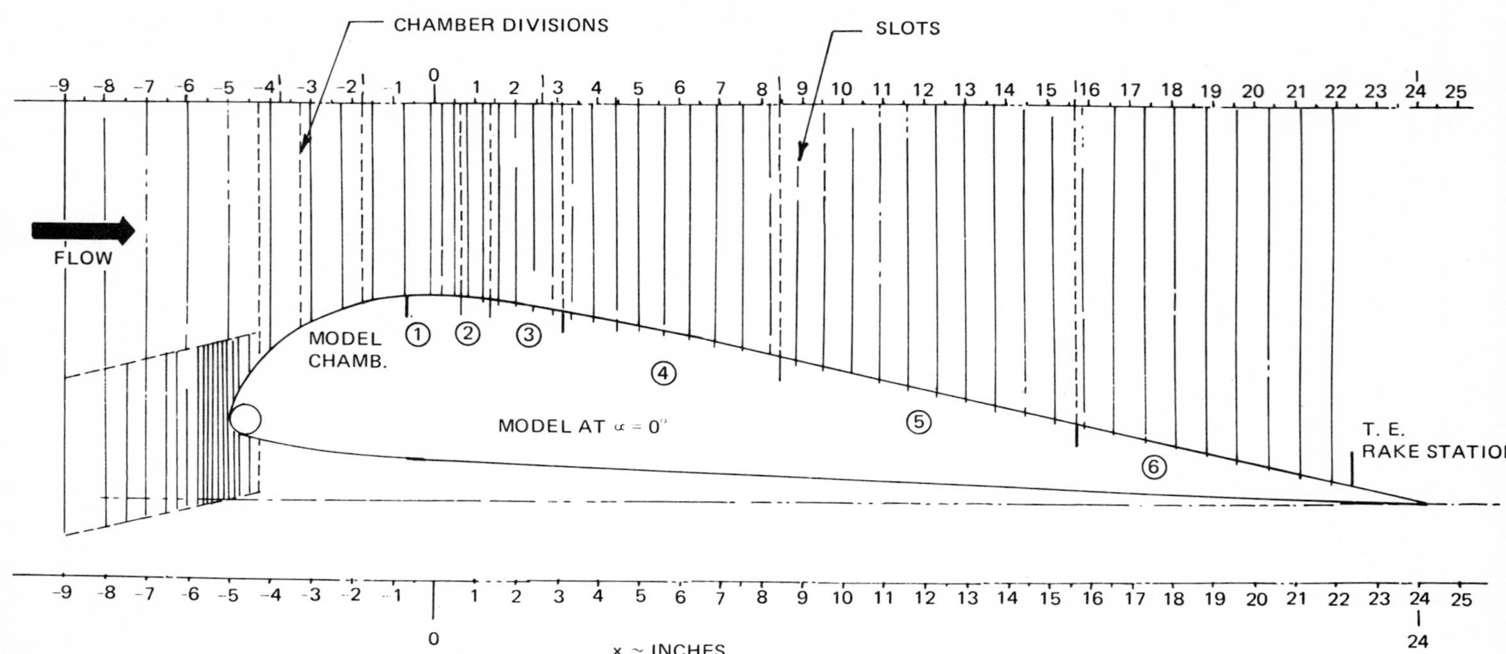

FIG. 5.37 Arrangement of the Gross model in wind tunnel test section and location of sidewall suction slots (*from Gross, 1968*).

original total suction velocity. In an attempt to carry this trend a step further, Distribution 4 was a parabolic rather than a rectangular or triangular distribution. The boundary layer thickness was approximately the same as in the case of constant suction and the total suction required was approximately 70 percent of that for Distribution 1.

Figure 5.37 is a schematic drawing of the model mounted in the wind tunnel test section. The side wall suction slot pattern as it appeared on the right side wall is also shown together with the outlines of the chambers into which the suction was drawn. Slots on the test section upper surface were continuations of the side wall slots aft of station $x = -0.1$ in. Such a model enables investigations of both boundary layer control over airfoils and separation control on a two-dimensional diffuser wall. If separation occurred when there was no suction, Gross (1968) was able to determine the minimum suction necessary to keep flow attached in the turbulent boundary layer. The change in pressure distribution due to the change of boundary layer displacement thickness with and without suction is shown in Fig. 5.38 for $M_\infty = 0.315$ and $\alpha = 0°$. Gross (1968) found that his theoretical prediction was in good agreement with experimental pressure distribution data. A similar change in pressure distribution was observed for angles of attack of 1 and 2 deg at Mach 0.2-0.325.

Figure 5.39 gives an example of the boundary layer velocity profiles affected by the total suction flow coefficient $C_{Q_t} = \Sigma\, C_{Q_s}$ where $C_{Q_s} = Q_s/\rho_\infty u_\infty A$, Q_s is the suction mass flow of an individual chamber, and A is the reference area of the model ($A = 1.9$ ft^2).

The abrupt change between the profiles shown and the profile at the lowest suction rate ($C_{Q_t} = 1.042 \times 10^{-3}$) was so pronounced that significant separation had doubtless occurred at the low suction rate.

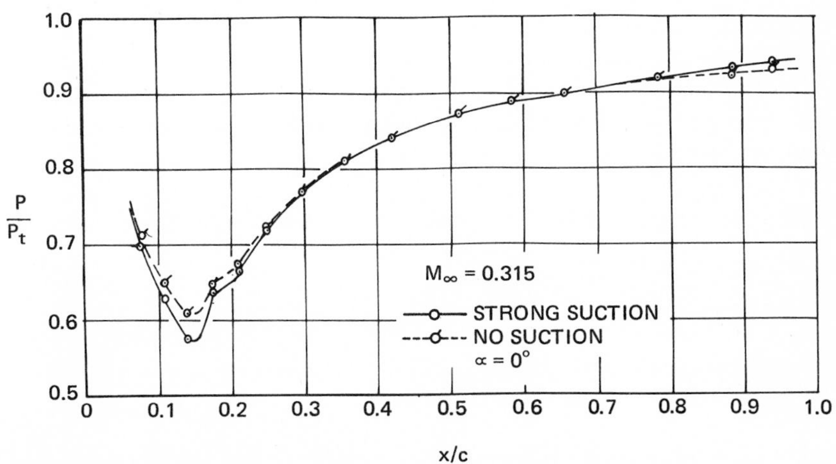

FIG. 5.38 Change of pressure distribution on the Gross model due to application of suction (*from Gross, 1968*).

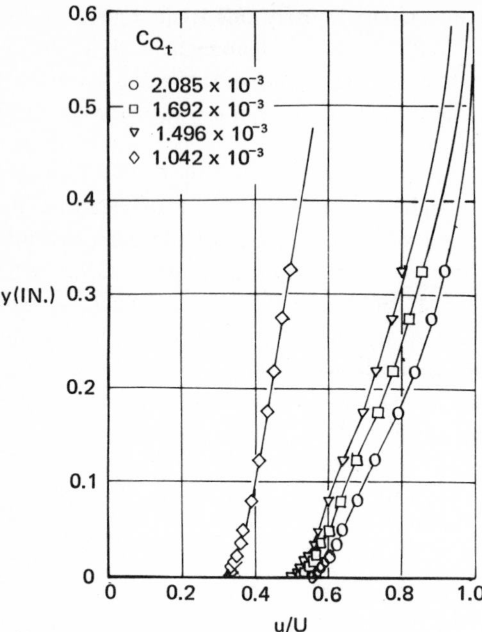

FIG. 5.39 Effect of total suction flow coefficient on boundary layer velocity profiles: representative turbulent boundary layer velocity profiles are shown for intermediate station $x/c = 0.657$, $M_\infty = 0.315$, and Distribution 3 (see Fig. 5.36); the y-scale is arbitrary and not fixed to the surface (*from Gross, 1968*).

Figure 5.40 shows the effect of suction on displacement thickness δ^*, momentum thickness θ, and shape parameter H. Note that once a certain suction quantity is passed, then the rate of change of the shape parameter H is small; thus the so-called sufficient suction point may be evaluated. Previously (see Sec. 5.2.1.3) the "sufficient" suction has been described as when the boundary layer becomes fully attached to the surface up to the measuring rake.

The following parameter is introduced for the investigation of required power:

suction power coefficient:

$$C_{ps} = \frac{P_s}{(1/2)\rho u_\infty^3 A}$$

where P_s is the energy required to bring suction air to ambient velocity and static pressure, and A is reference model area, in this case, 1.9 ft^2.

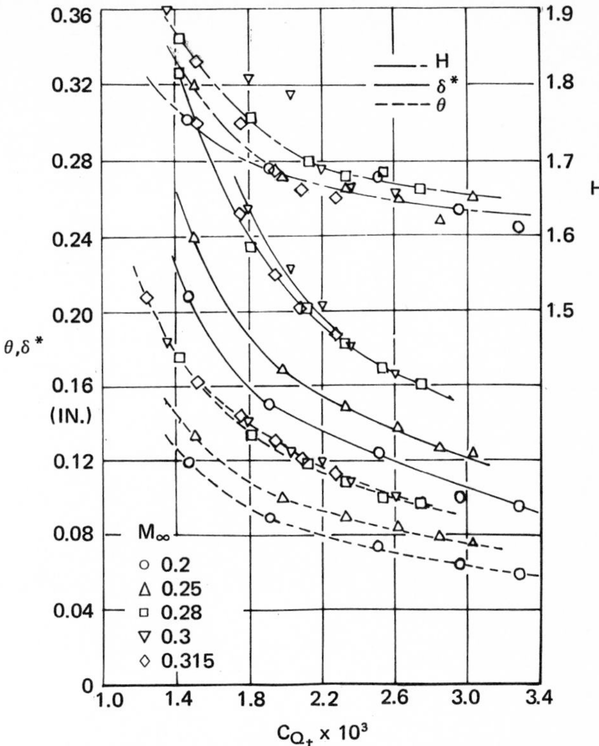

FIG. 5.40 Effect of suction flow on turbulent boundary layer momentum thickness θ, displacement thickness δ^*, and shape parameter $H = \delta^*/\theta$ (for suction distribution 3; see Fig. 5.36; *from Gross, 1968*).

Boundary layer kinetic energy loss coefficient: $C_{P_W} = (2/c)(u_e/u_\infty)^3\ \theta^*$ where c is the model chord of 27.35 in. and θ^* is the boundary layer kinetic energy thickness defined by $\theta^* = \int_0^\delta (\rho u/\rho_e u_e)\ [1-(u/u_e)^2]\ dy$. Total energy loss coefficient is $C_{Pt} = C_{Ps} + C_{PW}$.

Figure 5.41 shows the variation of suction power coefficient C_{P_S}, kinetic energy loss coefficient C_{P_W}, and total energy loss coefficient C_{P_t} as a function of total suction flow coefficient C_{Q_t} for free-stream Mach numbers $M_\infty = 0.2$–0.315. It has already been mentioned (*a*) that at low suction rates, the energy loss due to pumping power is low but the kinetic energy loss within the thick boundary layer is high and (*b*) that at high suction rates, the energy loss due to the pumping is high but the kinetic energy loss due to the thinned boundary layer is reduced. Hence, as seen in Fig. 5.41, the minimized position of combined energy loss may be located at some intermediate or "optimum suction" rate. These minimum values of $C_{P_{tm}}$ are plotted in Fig. 5.42 as a function of M_∞ and the values of C_{Qt} for minimum C_{Pt} are plotted in Fig. 5.43 as a function of M_∞.

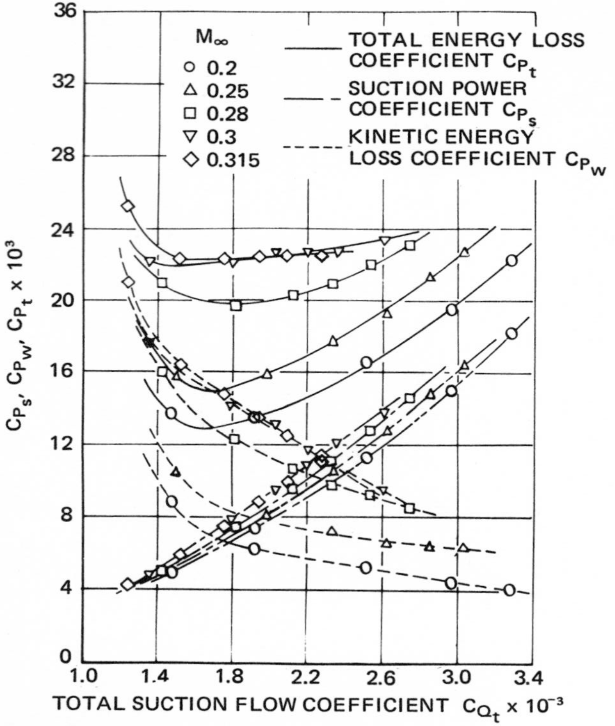

FIG. 5.41 Variation of C_{p_S}, C_{p_W}, and C_{p_t} with C_{Q_t} for several Mach numbers (for suction distribution 3; see Fig. 5.36; *from Gross, 1968*).

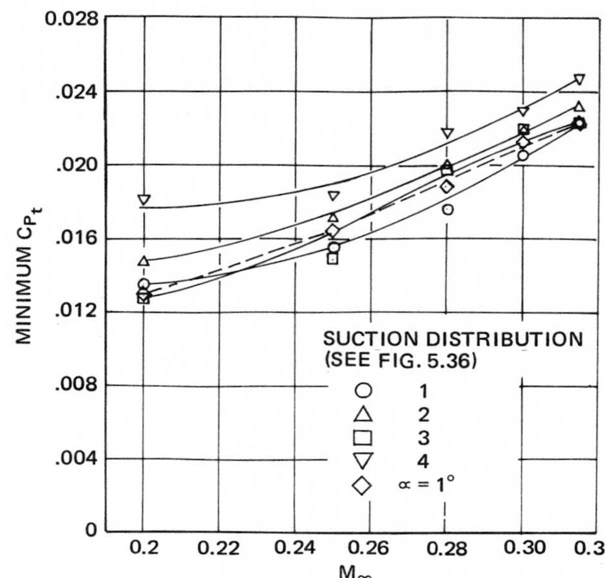

FIG. 5.42 Variation of minimum total energy loss coefficient $C_{p_{tm}}$ with Mach number (*from Gross, 1968*).

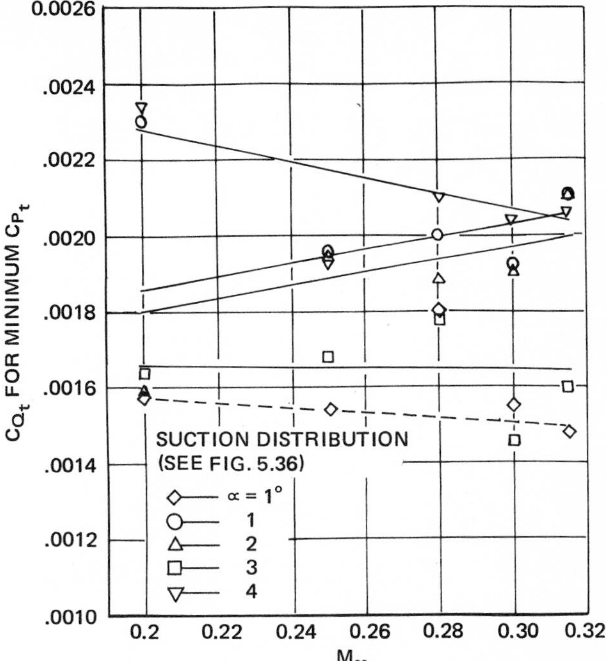

FIG. 5.43 Values of C_{Qt} for minimum C_{pt} as a function of Mach number (the lines show trends only; *from Gross, 1968*).

Figure 5.44 gives the suction distribution at the test points nearest the "optimum suction." The minimum energy loss coefficient and the corresponding optimum rate for total suction flow increased slightly with Mach number and was slightly dependent on suction distribution. The minimum observed optimum suction flow coefficient varied from $C_{Qt} = 1.58 \times 10^{-3}$ at $M_\infty = 0.2$ to $C_{Qt} = 1.48 \times 10^{-3}$ at $M_\infty = 0.315$. The corresponding values of minimum total energy loss coefficient were $C_{ptm} = 13 \times 10^{-3}$ and $C_{ptm} = 22 \times 10^{-3}$, respectively. Optimum C_{Qt} and minimum C_{pt} changed very little when the angle of attack was changed to 1 deg. Apparently the effects of the stronger pressure rise were balanced by an improvement in the chordwise suction distribution. Gross (1968) attempted to change the suction distribution to achieve the optimum condition, but he was not quite successful in doing so because the resulting suction distribution differed from the theoretical optimum. Nevertheless, the test indicated that the optimum suction distribution was such that strong suction should be provided in the upstream region to reduce the boundary layer in the region of the greatest pressure rise and allow it to grow in the region where pressure rise was less pronounced.

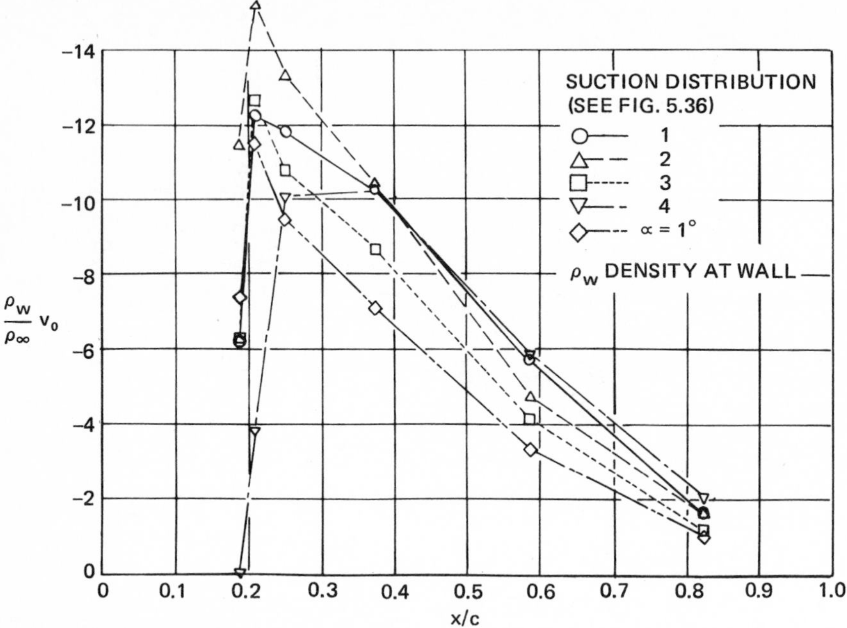

FIG. 5.44 Suction distribution at test points nearest the "optimum suction" ($M_\infty = 0.315$; *from Gross, 1968*).

5.2.4.3 Trailing-edge flap suction

Because of its sink effect, suction at the trailing edge may increase the circulation around an airfoil at constant angle of attack.

Griffin and Hickey (1956) studied boundary layer control by suction as applied to trailing-edge flaps mounted on a wing. The wing had a 45-deg sweep of the quarter-chord line of chords parallel to the plane of symmetry, an aspect ratio of 6, and taper ratio of 0.292. The airfoil section was constant across the span and streamwise thickness ratio amounted to 8.2 percent. The test model geometry is shown in Fig. 5.45. The flap hinge line was located on the lower surface at $0.75c'$ (c' is local chord measured perpendicular to the spanwise quarter-chord line). The flaps were deflected to 46, 55, 60, and 65-deg measured normal to the flap hinge line. The locations of the outboard ends of the flaps at the trailing edge corresponded to 0.50, 0.66, and 0.8 semispan.

Suction was provided by a centrifugal pump driven by a variable-speed electric motor. The boundary layer air was removed through the porous area, passed through wing ducts to the pump within the fuselage, and was expelled from an exhaust port beneath the fuselage. The configuration was tested without suction on the flap as well as with a varying flap suction flow coefficient C_{Qf} through a range of angles of attack from −2 to 20 deg. Here $C_{Qf} = Q_f/u_eA$, where Q_f is the volume rate of flap suction flow correlated to

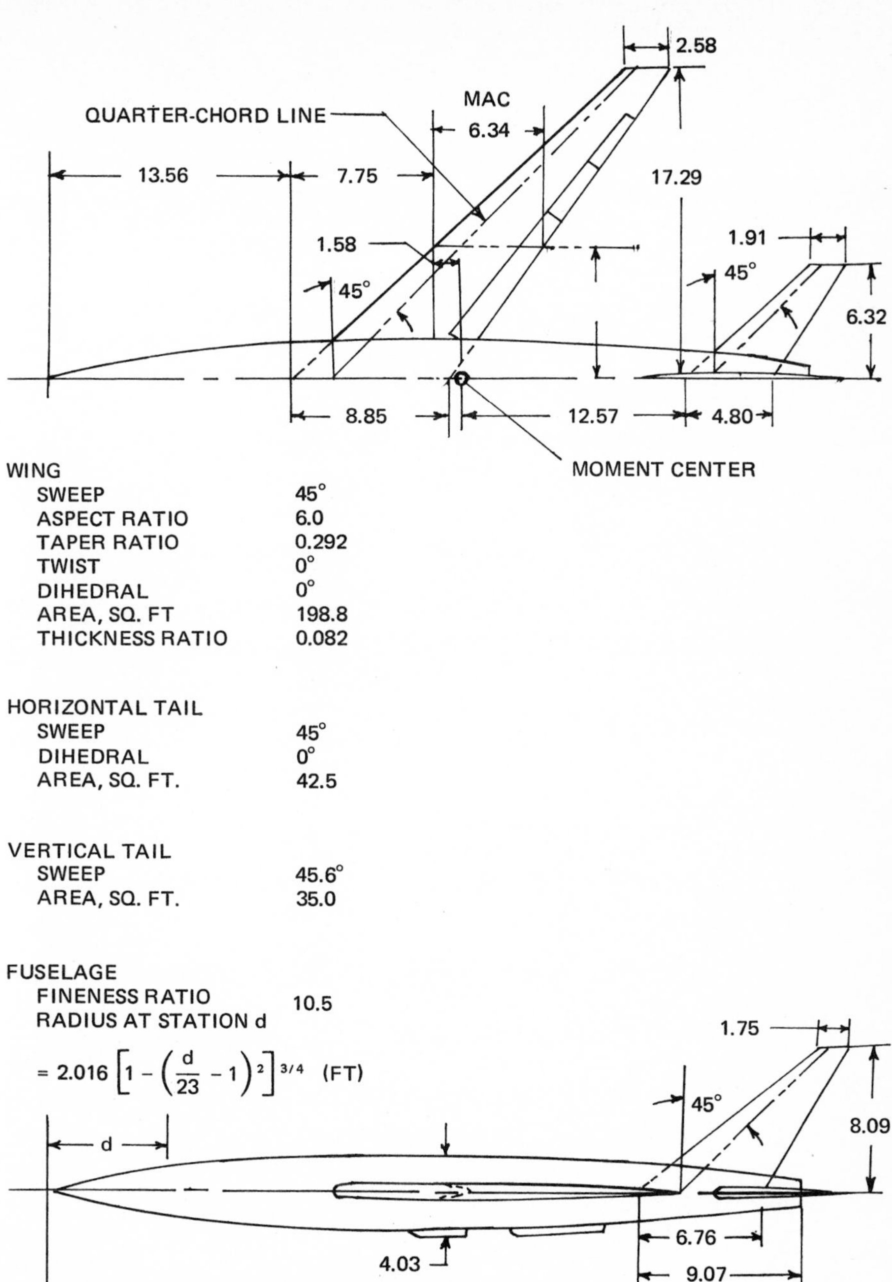

FIG. 5.45 Geometric characteristics of the model used for trailing edge flap suction tests (*from Griffin & Hickey, 1956*; unless otherwise noted, dimensions are in feet).

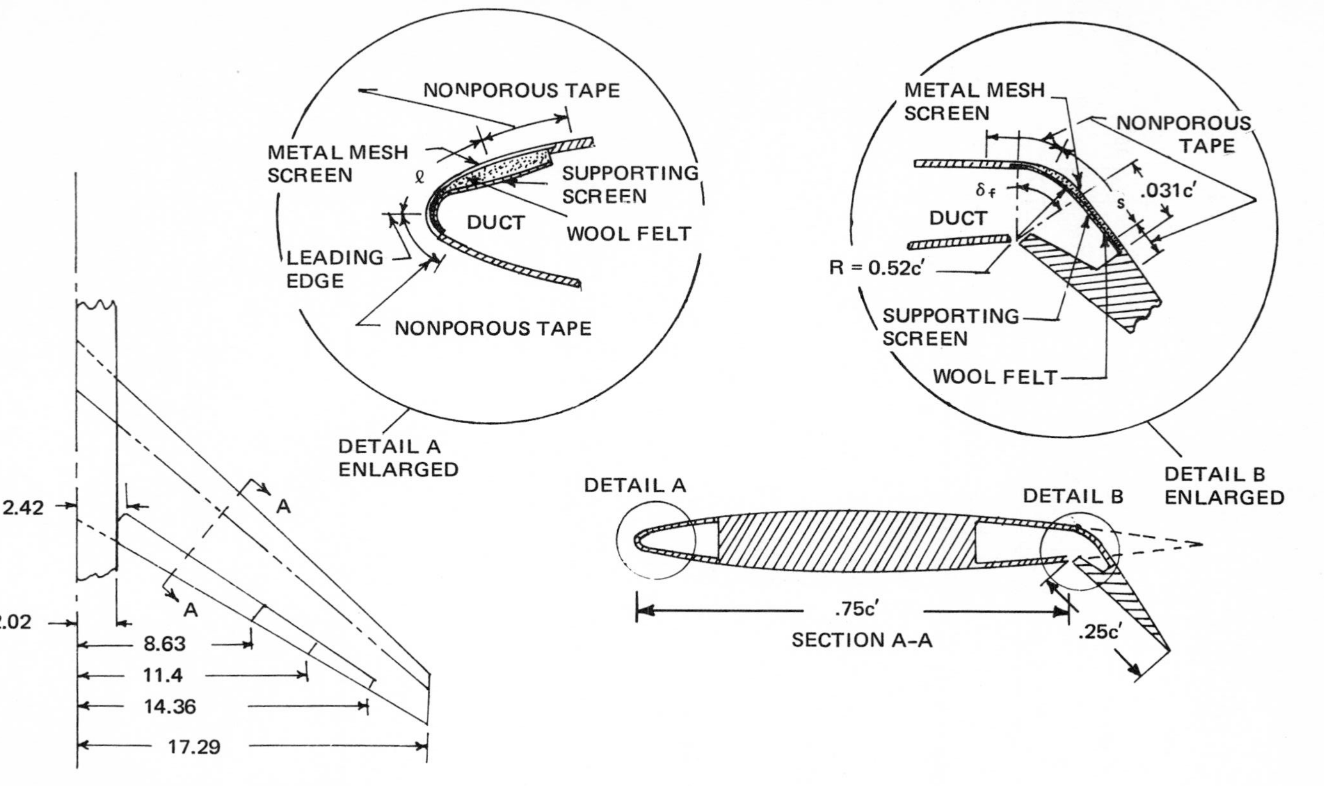

FIG. 5.46 Details of the flap and leading edge of the Griffin-Hickey model (all dimensions are in feet; *from Griffin & Hickey, 1956*).

standard atmosphere, u_e is the free-stream velocity, and A is the wing area. As mentioned previously, the critical suction flow coefficient $C_{Qf\text{crit}}$ is defined as the value of C_{Qf} for which the slope of ΔC_L versus C_{Qf} approaches zero. Thus, this critical position may be considered as the condition where suction is accomplished most economically. Griffin and Hickey (1956) used this $C_{Qf\text{crit}}$ to evaluate flap effectiveness at a constant angle of attack.

Details of the trailing-edge flap and the wing leading edge are presented in Fig. 5.46. The area-suction flaps attained the flap lift increment predicted by inviscid flow theory for small flap deflections and shorter flap spans tested. Suction did not entirely prevent separation at larger values of either the deflection or span area, and flap lift increments were somewhat lower than the analytical values.

Effect of location on the flap of porous material The effect of location of the porous material was investigated at both the rear and the leading edge of the flap. Theoretical and experimental flap lift increments and $C_{Qf\text{crit}}$ values are compared in Table 5.2 for a 55-deg flap deflection and various flap spans. Griffin and Hickey used the method of Cook et al. (1953) to obtain the estimated values.

The results presented in Table 5.2 and those in Fig. 5.47 are typical of what was found for all flap deflections.

On the basis of greatest lift increment and lowest suction requirement, the optimum position of the forward edge of the porous area was at the point of minimum external negative pressure. This result is in agreement with the theoretical findings of Cook et al. (1953).

Effect of suction on lift, drag, and pitching moment As demonstrated in Table 5.3, suction increases the lift that is available without suction.

At the highest available C_{Qf}, boundary-layer control was inadequate to prevent the occurrence of separation from about 0.50 semispan to the outboard end of the flap. In order to improve the flow condition over the outer end of the flap (and thereby increase the flap lift increment), two small

TABLE 5.2 Theoretical and experimental values of flap lift and critical suction for two chordwise porous areas

	$C_{Qf\text{crit}}$			ΔC_L		
		Experiment			Experiment	
Flap span	Esti-mated	$s/c'=0.030$	$s/c'=0.062$	Esti-mated	$s/c'=0.030$	$s/c'=0.062$
0.12–0.50 $b/2$	0.00050	0.0006	0.0010	0.755	0.70	0.715
0.12–0.66 $b/2$	0.00065	0.0007	0.0012	0.98	0.82	0.87
0.12–0.83 $b/2$	0.00085	0.0008	0.0016	1.18	0.905	0.985

Note: From Griffin and Hickey (1956). The symbol s refers to the chordwise extent of the porous opening on the flap measured along the surface in the plane perpendicular to the flap hinge line and c' is the local chord measured perpendicular to the quarter-chord line.

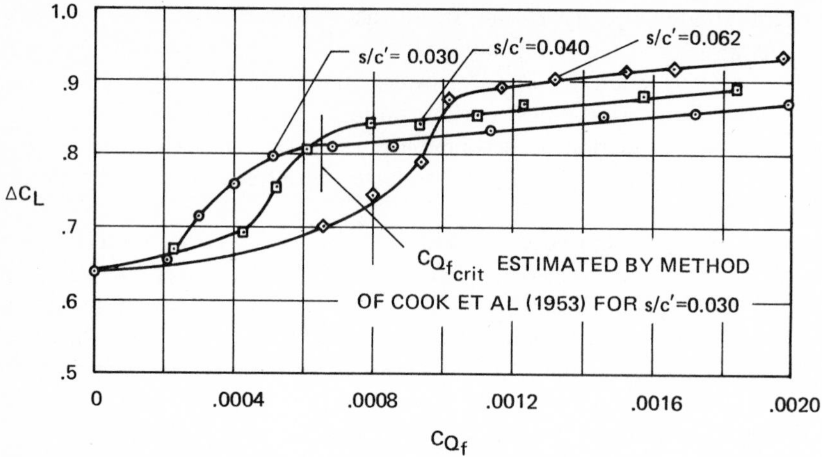

FIG. 5.47 Effect of suction flow coefficient on the flap lift coefficient increment for various chordwise extents of porous area (*from Griffin & Hickey, 1956*; $\alpha = 0.4$ deg, $\delta_f = 55$ deg, flap span from 0.12 to 0.66 $b/2$, wing leading edge sealed).

fences were installed on the flap upper surface at 0.50 and 0.58 semispan; these increased flap lift about 0.03 at $\alpha = 0$ deg. The improvement in flow over the outboard portion of the flap is shown in Fig. 5.48.

To determine the effect of boundary-layer control on the drag of the model with flaps deflected, drag coefficients of the same configuration were studied with and without suction at $\alpha = 0$ deg. The results showed that the drag was less for the larger flap spans when suction was applied. This indicated that the reduction of the drag due to flow separation was greater than the increase in induced drag caused by the higher lift. The magnitude of negative pitching moment coefficient was increased when suction was applied

TABLE 5.3 Effect of suction on lift

Flap span $b/2$	Flap deflection deg	ΔC_L	
		$C_{Qf} = 0$	$C_{Qf} = C_{Qf\,\text{crit}}$
0.12 to 0.50	46	0.475	0.625
	55	0.535	0.715
	60	0.550	0.750
	65	0.540	0.775
0.12 to 0.66	46	0.565	0.770
	55	0.625	0.870
	60	0.643	0.905
	65	0.635	0.965
0.12 to 0.83	46	0.642	0.875
	55	0.690	0.985

Source: Griffin and Hickey (1956).

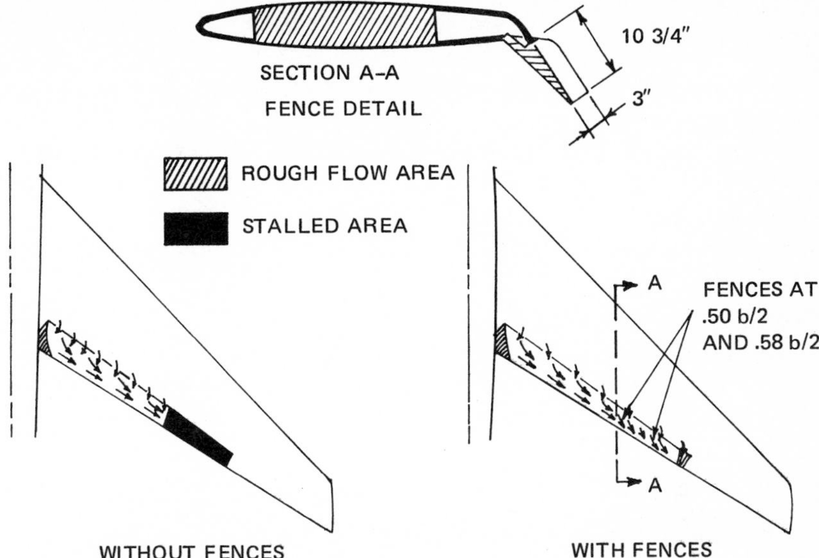

FIG. 5.48 Improved flow by installation of fences on the upper surface of the flap (*from Griffin & Hickey, 1956*; the tuft study was conducted for $\delta_f = 65$ deg and flap span from 0.12 to 0.66 $b/2$).

to the deflected flaps. However, at constant angle of attack, the incremental value of C_m/C_L with area suction flaps was slightly greater than for the same flaps without suction.

5.2.4.4 Porous material for suction surfaces

Danneberg et al. (1957) studied an NACA 0006 airfoil with area suction on the upper surface near the hinge line of a 30 percent chord plain trailing-edge flap. The distribution of the suction quantity taken into the porous region was controlled by porous metal sheets. The permeability of the sheets was regulated by the size and the spacings of the holes. The tests included a wide variety of hole sizes and perforation patterns with uniform and gradient arrangements of permeability.

Suitable material for the porous area used in area suction required two fundamental properties: (*a*) an outer surface which is not detrimental to the aerodynamic characteristics desired, and (*b*) a permeability which provides the proper control of the suction flow quantities. Other desirable properties of the material include strength and formability characteristics, ease of attachment, long service life (freedom from clogging by dirt or by corrosion), and ease of reproduction (especially of the flow resistance characteristics). A very promising practical material tested at NASA Ames Aeronautical Laboratory consists of a perforated metal sheet in which the permeability is controlled solely by the size and/or the spacing of the perforations. The control of permeability and its distribution was sufficient to meet the typical specification for boundary-layer control systems that use area suction. The strength and forming characteristics of perforated sheet material were superior to those of sintered powdered materials. The airflow through a perforated sheet is more concentrated than through a woven or sintered metal sheet because of the larger size and smaller number of openings. The concept of perforation is useful not only for the prevention of separation on a plain flap but also apparently for the elimination of separation in wide angle diffusers, elbows, and bluntly rounded trailing-edge sections.

5.2.5 Application of Suction and Bleeding to Prevent Separation of Compressible Flow

Bleeding and suction may be applied to control compressible flow separation upstream of a flare. The nature of separation can be laminar or turbulent, depending on Reynolds number. The extent of separation ahead of the flare may alter the stability derivatives of the vehicle, the drag caused by the flare, and the heat transfer to the flare in high speed flight. Several flares are often used to stabilize missiles or to control the drag of reentry vehicles. The influence of separation on their effectiveness is of practical interest.

An adequate boundary layer bleed at the start of a flare exposed to a supersonic air stream may be applied to prevent or reduce the extent of separation. It may also be used to remove the turbulent boundary layer ahead of the flare and to allow the start of a laminar boundary layer on the surface of the flare with a reduction in heat transfer. Although in contrast to ordinary suction, bleeding does not require auxiliary power, it is treated here because of its many similarities to suction.

5.2.5.1 Application of air bleed

Crawford (1961) studied experimentally, but rather qualitatively, the effect of boundary layer bleed at the start of a 30 deg half angle flare on the shape of the flow boundaries, the pressure distribution on the flare, and the heat transfer to the flare.

Two ogive cylinders of different diameters served as the forebodies (Fig. 5.49) and enabled the radial dimension of the flow bleed gap to be controlled. These forebodies could be positioned at different distances ahead of the flare to control the axial dimension of the flare bleed gap. Both forebodies had a Karman minimum-drag nose shape of fineness ratio 5 modified by placing a 10 deg half-angle cone at the tip, tangent to the nose. On Forebody I, the small ogive nose was followed by a 1.5-in. diameter cylinder five diameters in length. On Forebody II, the ogive-nose dimensions were increased by 20 percent and the length of the cylinder that followed was shortened correspondingly so that the overall length was 15.474 in. for both forebodies.

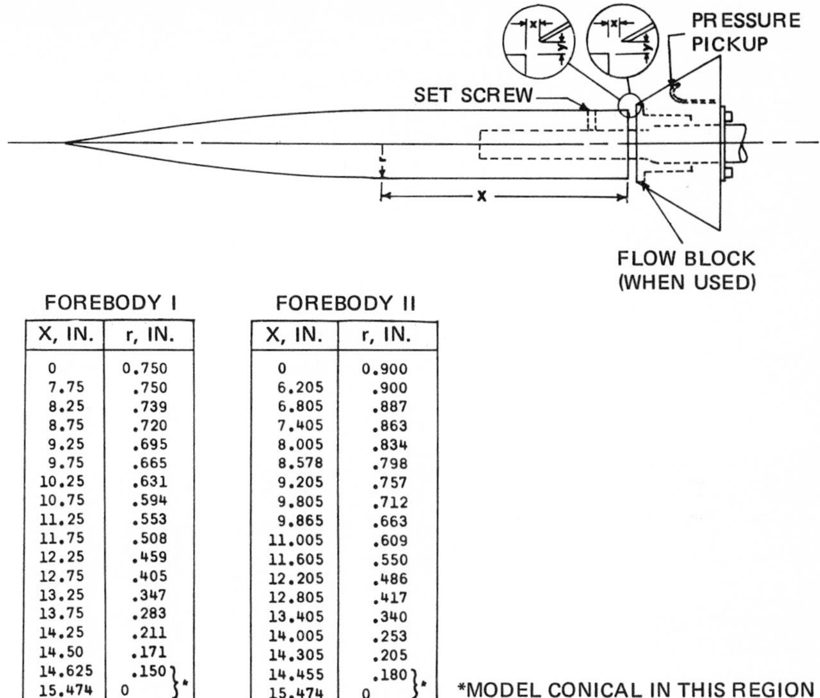

FOREBODY I

X, IN.	r, IN.
0	0.750
7.75	.750
8.25	.739
8.75	.720
9.25	.695
9.75	.665
10.25	.631
10.75	.594
11.25	.553
11.75	.508
12.25	.459
12.75	.405
13.25	.347
13.75	.283
14.25	.211
14.50	.171
14.625	.150 }*
15.474	0

FOREBODY II

X, IN.	r, IN.
0	0.900
6.205	.900
6.805	.887
7.405	.863
8.005	.834
8.578	.798
9.205	.757
9.805	.712
9.865	.663
11.005	.609
11.605	.550
12.205	.486
12.805	.417
13.405	.340
14.005	.253
14.305	.205
14.455	.180 }*
15.474	0

FIG. 5.49 Model for studying boundary layer bleed (*from Crawford, 1961*; asterisks indicate regions where the model was conical).

Three different flares (Fig. 5.50) were tested. All had a base diameter of 4.64 in. and were 0.10 in. thick. The tunnel stagnation pressure ranged from 5.2 to 41 atmospheres and stagnation temperatures from 1050 to 1190°R. The test Reynolds number (based on forebody length) ranged from 1×10^6 to 7.4×10^6 at a free-stream Mach number of 6.8. This range was sufficiently large to test the laminar, the transitional, and the turbulent separated boundary layers. The experimental data were compared with results of both two and three-dimensional analyses. Schlieren photography enabled observations of the effects of varying degrees of bleed on the flow boundaries in both radial and axial directions and in the leading edge of flare due to the blunting of the lip. The heat transfer and pressure distributions on the flare were correlated with the shape and nature of flow boundaries. The separation ahead of a 30 deg half-angle flare was reduced at increased Reynolds number when the separated boundary was transitional. The separation was almost eliminated when transition occurred on the body surface ahead of separation. The flow separated upstream of the flare when the pressure rise caused by the shock at the start of the flare exceeded the maximum allowable for an attached boundary layer. The size of the separation region varied from one in which no

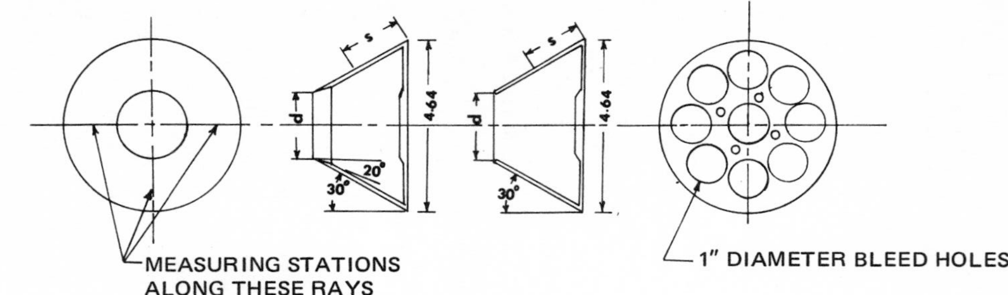

THERMOCOUPLE LOCATIONS

DISTANCE TO THERMOCOUPLE, S, IN ., FOR–		
SHARP FLARE; d = 1.80 IN.	BLUNT FLARE; d = 1.80 IN.	BLUNT FLARE; d = 2.10 IN.
0.31	0.31	0.31
.56	.56	.56
.81	.81	.81
1.06	1.06	1.06
1.31	1.31	1.31
1.56	1.56	1.56
1.81	1.81	1.81
2.06	2.06	2.06
2.31	2.31	
	2.56	
S = 2.84 IN.	S = 2.64 IN.	S = 2.34 IN.

PRESSURE-ORIFICE LOCATIONS

DISTANCE TO PRESSURE ORIFICE, S, IN., FOR–		
SHARP FLARE; d = 1.80 IN.	BLUNT FLARE; d = 1.80 IN.	BLUNT FLARE; d = 2.10 IN.
0.56	0.56	0.31
2.31	2.31	.81
		1.19
		1.56
		2.06

FIG. 5.50 Details of flares for the Crawford model (all dimensions are in inches unless otherwise indicated; *from Crawford, 1961*).

separation was observed upstream of the flare to one in which the boundary separated slightly ahead of the constant-diameter cylindrical segment of the body and attached at the trailing edge of the flare.

The schlieren movies showed a varying amount of unsteady flow in the separated regions. It was observed in some tests that a rapid small amplitude fluctuation of the position of the start of separation grew to a large amplitude fluctuation of the position of reattachment. In other tests, the separated region collapsed for an instant at random intervals of time. In tests wherein the free-stream Reynolds number was varied slowly, a tendency toward collapse of the separated region prevailed over the entire Reynolds number range. There is a strong possibility that the violent unsteadiness which led to complete collapse of the separated boundary was partly due to unsteadiness of the flow in the tunnel. This unsteadiness may have been coupled with the movement of transition on the separated boundary.

The schlieren movies also showed that the unsteady characteristics of the separated boundary were less pronounced with a flare gap. Bluntness of the flare lip influenced the flow boundaries. The pressure rise caused by the blunt lip was transmitted forward and thickened the boundary layer even though the bleed was sufficiently large to prevent any extensive separation. The larger bleeds were able to prevent extensive flow separation at the lowest test number (Reynolds number of 1×10^6 based on conditions ahead of the model and on the forebody length with a sharp leading edge). At that number, the separation was extensive and reattachment occurred at the trailing edge of the flare. At Reynolds numbers of about 1.9×10^6 and 1×10^6, the data showed the extensive effect of flare bleed gap on the flare drag. The use of a flare bleed can increase the value of $C_{DP}/\overline{St}$, where C_{DP} is the pressure drag coefficient of the flare and $\overline{St}$ is the average Stanton number on the surface of the flare, and allow the realization of high values of $C_{DP}/\overline{St}$ over an extended Reynolds number range, therefore increasing the effectiveness of the use of a flare as a drag brake. With the largest bleed, the observed drag coefficients were in agreement with the values for attached flow. The size of the largest radial bleed gap for the blunt lip was more than double that for the sharp lip. Separation could have been prevented in the case of sharp lip, had a larger bleed been tried.

5.2.5.2 Application of suction

For convenience, compressible flow control of separated flow upstream of a flare by suction is discussed separately for supersonic and hypersonic flow.

Control of supersonic flow separation Pate (1969) investigated experimentally the effects of continuous (distributed) suction on a transitional type of boundary layer at Mach numbers of 2.5, 3.0, and 3.5. The corresponding range of Reynolds numbers based on flare axial location was from 0.8×10^6 to 6.4×10^6. The model (Fig. 5.51) was a slotted, 20-cal tangent-ogive with flare angles of 7.5 and 15 deg. Continuous suction was applied through seven

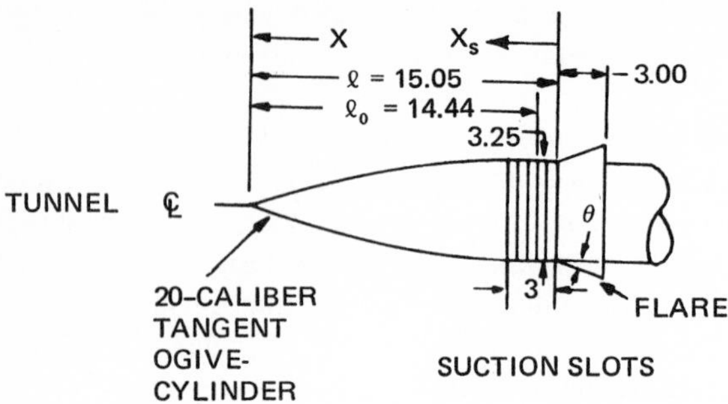

SUCTION SLOT DIMENSIONS

SLOT LOCATIONS X IN.	INITIAL SLOT WIDTH IN.	
12.0	0.006	
12.5	0.0065	
13.0	0.0070	
13.5	↓	
14.0		
14.5		
15.0	0.0070	
	MODIFIED SLOT WIDTH	
	TOP OF SLOT IN.	BOTTOM OF SLOT IN.
12.0	0.014	0.011
12.5	"	"
13.0	"	"
13.5	0.018	0.014
14.0	0.019	0.016
14.5	0.021	0.017
15.0	0.021	0.016

FIG. 5.51 Details of model for studying continuous suction on a transitional type of boundary layer (all dimensions are in inches; *from Pate, 1969*).

slots (equally spaced at 0.5-in. intervals) located in an area extending 3 in. upstream of the flare lip. In order to provide measurements of the suction mass flow, theoretical gas flow equations were used in conjunction with the metering box that recorded pressures and temperatures and the nozzle discharge coefficient.

The extent of separation decreased with increasing Reynolds numbers for the two flare angles. All flow separation data with no suction for these 15 and 7.5 deg flare configurations were classified as transitional in accordance with the investigations of Gray (1967) and Lewis et al. (1968). Because the maximum length of flow separation did not exceed more than about 3 in., the area to which suction was applied was always in and slightly upstream of the naturally separated flow region. At higher Reynolds numbers, transition occurred upstream of the flare and caused no separation. The separation that occurred at $\mathrm{Re}_l = 2.7 \times 10^6$ (based on the free-stream condition and a model reference length $l = 15.05$ in.) with 7.5 deg flare and at $\mathrm{Re}_l = 3.9 \times 10^6$ with a 15 deg flare was almost completely eliminated by suction.

The length of the region of separated flow with no suction (determined by schlieren photographs) is shown in Fig. 5.52 for various Mach numbers. Figure 5.53 indicates the effects of suction on the separated flow region at $M_\infty = 2.5$ for both flare angles over the range of Reynolds numbers tested.

The values of suction flow coefficient C_Q were obtained by dividing the removed mass flow through the suction chamber by $\rho_\infty u_\infty A$, where $A = 30$ in.^{2} For the models with 7.5 and 15 deg flares, the most favorable suction did not eliminate a small region of flow separation ($x_s \lesssim 0.3$ in.) although the extension of separated flow region decreased with an increase in C_Q. However, enlargement of the slot widths was not noticeably effective on the extent of flow separation.

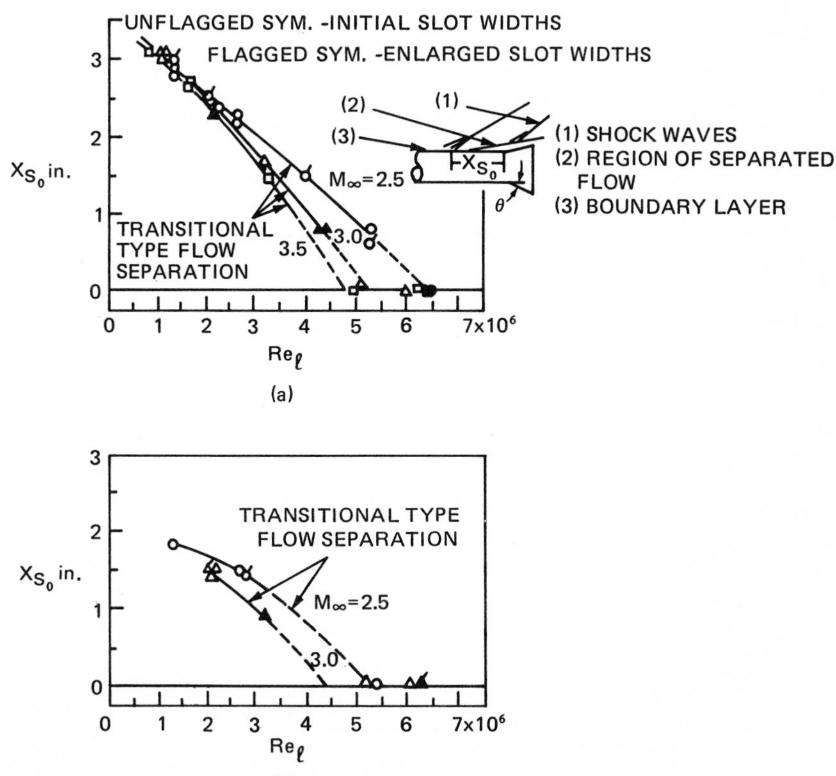

FIG. 5.52 Length of flow separation without suction for the Pate model at $M_\infty = 2.5$, 3.0 and 3.5: (*a*) flow angle of $\theta = 15$ deg; (*b*) flow angle of $\theta = 7.5$ deg (*from Pate, 1969*).

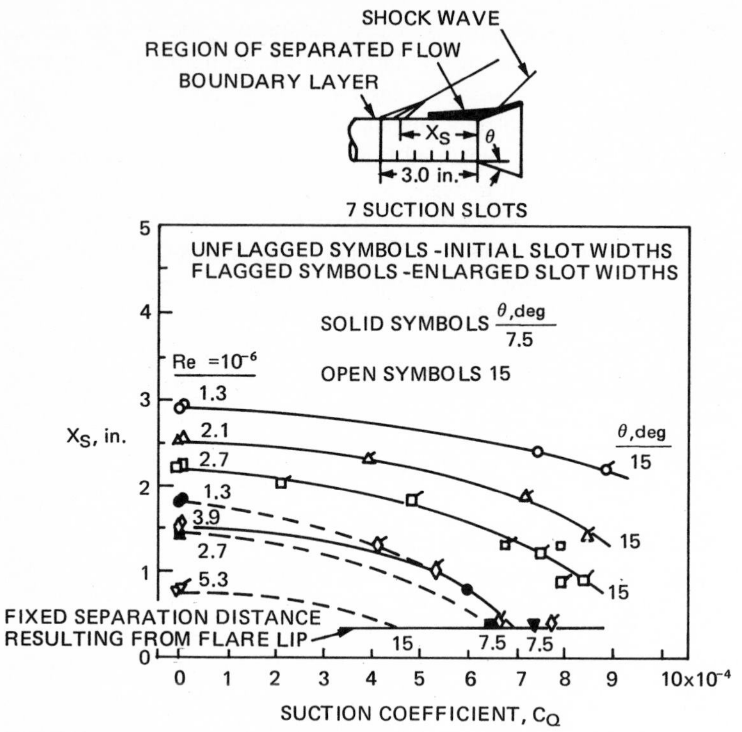

FIG. 5.53 Effects of localized, variable suction on the length of flow separation for Pate model at $M_\infty = 2.5$ (*from Pate, 1969*).

The mass flow rate in the boundary layer ($\dot{m}_{BL}$) was determined by $\dot{m}_{BL} = \int_0^\delta \rho u \, dA = 2\pi \int_0^\delta \rho u(r + y)dy$ where r is the radius of the body surface (1.625 in.) and y is the normal distance to the model surface. By defining the displacement thickness of the boundary layer on the body of revolution as $\delta^* = \int_0^\delta (1 - \rho u/\rho_e u_e)(1 + y/r)dy$, the mass flow rate becomes

$$\dot{m}_{BL} = 2\pi\rho_l u_l \left[\delta + \left(\frac{\delta^2}{2r}\right) - \delta^*\right] \tag{5.5}$$

where subscript l refers to local value outside the boundary layer at $x = 15.05$ in. fps.

The reduction in flow separation for a particular flare angle was a function of the amount of the boundary layer mass flow removed; see Fig. 5.54. The size of the separated flow region was very responsive to the application of continuous suction. A suction mass flow of approximately 10 percent of the total undisturbed mass flow in the boundary layer at the flare leading-edge station was sufficient to eliminate separation upstream of the 15 deg flare. Thus a continuous suction is an effective means of reducing the extensive areas of transitional-type flow separation that develop on an adiabatic wall body of revolution at relatively low Reynolds numbers. Caution should be exercised in applying the results of this investigation (adiabatic wall) to cold wall conditions because the relation between suction and flow separation could alter significantly.

Control of hypersonic flow separation Ball and Korkegi (1968) utilized a symmetrical wedge of 12-deg half-angle (Fig. 5.55) to determine experimentally the quantitative effect of suction on the laminar boundary layer separation in a compression corner at a nominal Mach number of 12 and free-stream Reynolds number per inch of 0.8×10^6. The model had an 8-in. span, a 3.5-in. chord, and its upper surface was extended by a 1.5-in. flap. The compression corner was formed by varying the angle of the flap from 0 to 20 deg with respect to the upper surface of the wedge. Suction was provided by means of a variable width gap between wedge and flap; this allowed controlled natural flow from the upper high pressure region of the corner to the relatively lower region of the base of the wedge as sketched in Fig. 5.56.

The trailing edge of the wedge and the leading edge of the flap were slightly rounded in order to achieve reasonably smooth flow across the gap; this could be viewed in cross section as a two-dimensional supersonic nozzle

SYM.	M_∞	Re_ℓ RANGE$\times 10^{-6}$	SUCTION RANGE
○	2.5	1.3 TO 3.9	0 TO MAXIMUM
△	3.0	1.1 TO 3.2	↓
□	3.5	1.6 TO 3.2	

$\dot{m}_{SUCTION}$ – TOTAL MASS FLOW REMOVED ($\dot{m} = C_m \rho_\infty u_\infty A$)

$\dot{m}_{BL}$ – THEORETICAL MASS FLOW IN THE BOUNDARY LAYER AT STATION 15.05 in. EQ. (5.5)

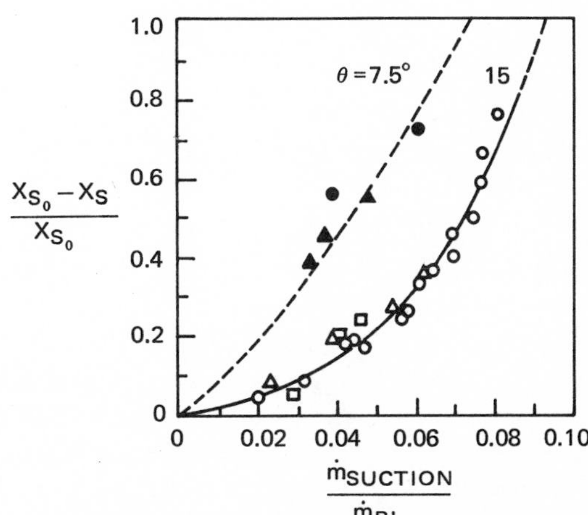

FIG. 5.54 Reduction in flow separation as a function of the amount of boundary layer mass flow removed (*from Pate, 1969*).

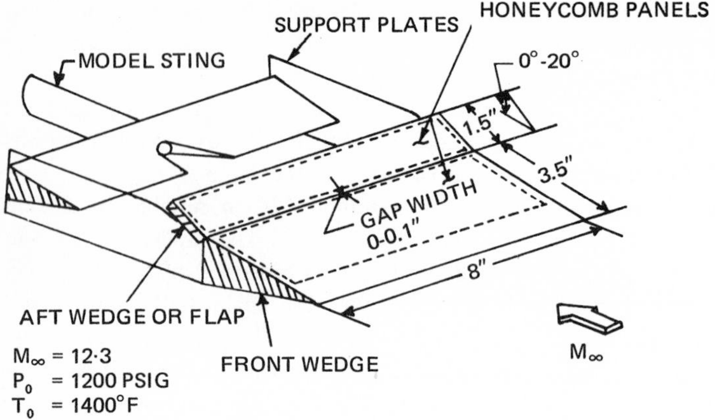

FIG. 5.55 Wind-tunnel model for determining separation in a compression corner (*from Ball & Korkegi, 1968*).

for all tests. Since the pressure ratio across the gap was expected to be larger than 2, the mass flow through the gap was computed under the assumption of one-dimensional sonic flow across the minimum section. For purposes of verification, a rig was designed to directly measure the mass flow through the gap. Since separated flows are sensitive to edge conditions, the spanwise pressure was determined with and without small triangular fences that just barely covered the region of separation and thus disturbed the outer flow as little as possible. These tests were carried out with the longest separated region corresponding to a 20-deg flap angle and closed gap because that configuration yielded the highest plateau pressure. The streamwise pressure distributions with and without suction are shown in Figs. 5.57 and 5.58 for flap angles of 5, 10, 15, and 20 deg with respect to the wedge surface.

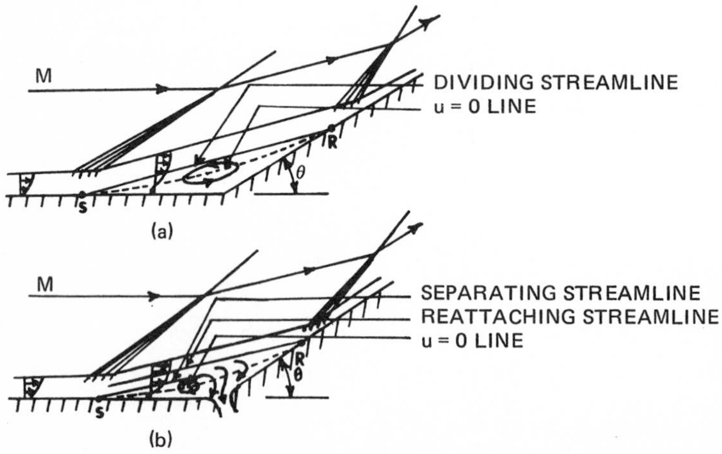

FIG. 5.56 Flow separation in a compression corner with and without suction: (*a*) no suction; (*b*) corner suction (*from Ball and Korkegi, 1968*).

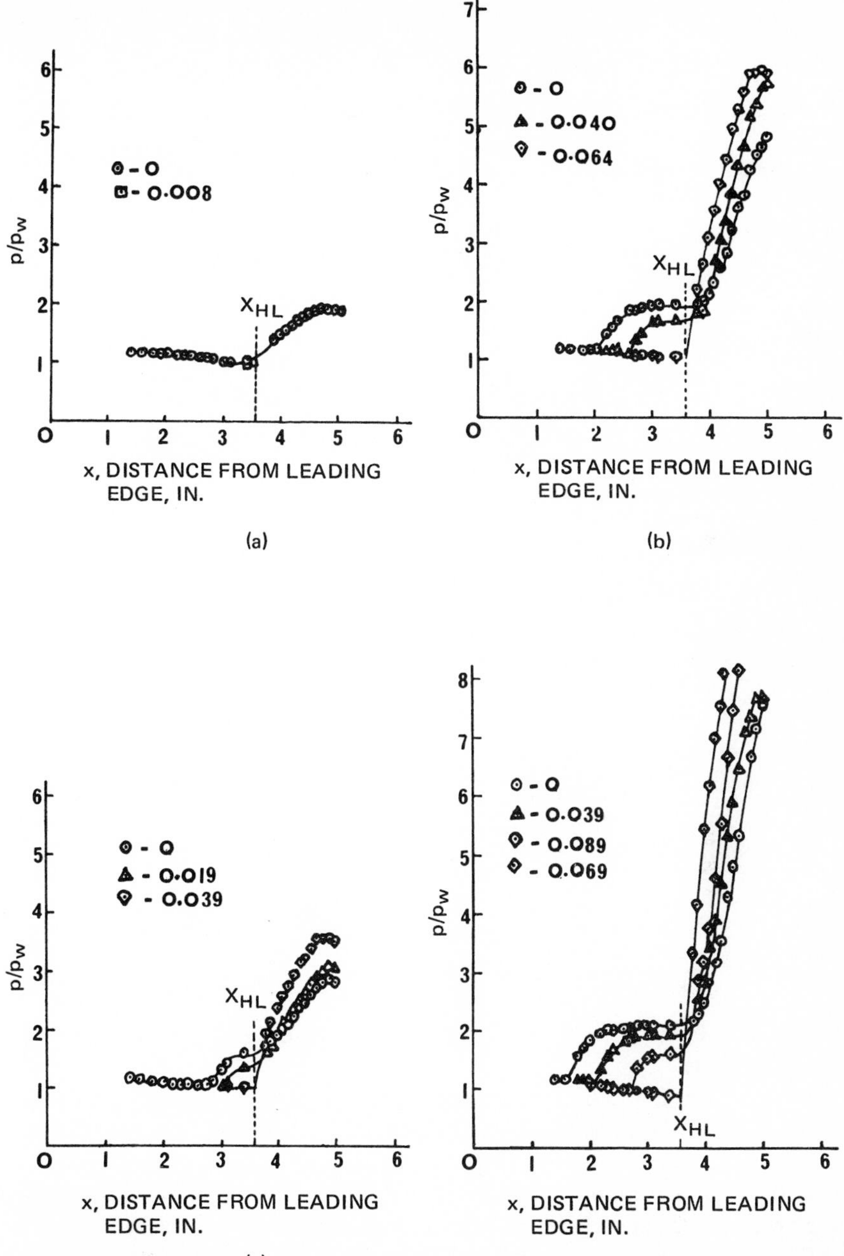

FIG. 5.57 Chordwise pressure distribution on a compression corner with suction: (*a*) flap deflection angle $\theta = 5$ deg; (*b*) flap deflection angle $\theta = 10$ deg; (*c*) flap deflection angle $\theta = 15$ deg; (*d*) flap deflection angle $\theta = 20$ deg ($M_\infty = 12.3$, Re_∞/in. $= 7.9 \times 10^4$, gap width θ is flap deflection angle, d^* in inches; *from Ball & Korkegi, 1968*).

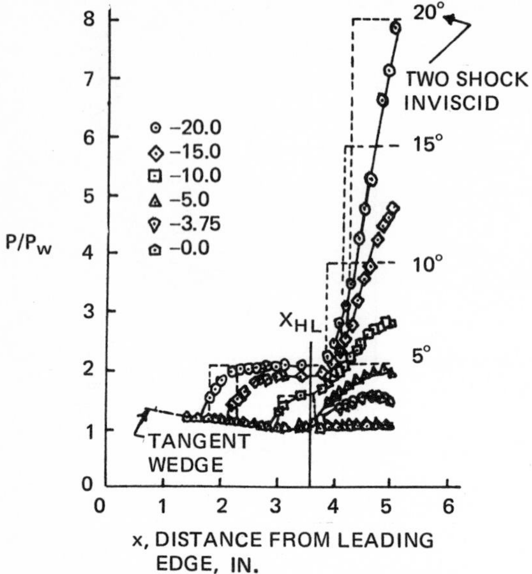

FIG. 5.58 Pressure distributions in a compression corner, no suction ($M_\infty = 12.3$, $Re_\infty/in. = 7.9 \times 10^4$, and $d^* = 0$; *from Ball & Korkegi, 1968*).

Note that suction is reflected by the gap width d^* whose relation to mass flow is given in Fig. 5.59. The symbol $\dot{m}$ refers to the mass flow rate in slug/ft^2-second and L_s refers to the length of suction gap in inches. The mass flow per unit span $\dot{m}_s/L_s$ is related to effective stagnation conditions p_{ES} and T_{ES} by

$$\frac{\dot{m}_s}{L_s} = Kp_{ES} \frac{d^*}{T_{ES}^{1/2}}$$

Here K is a function only of γ and the gas constant R, and T is given in degrees Rankine.

As shown in Fig. 5.60, the extent of separated flow region reduces and finally disappears with increasing gap width. The symbols x_{HL} and x_s refer to the chordwise surface distance from the model leading edge (in inches) to the hinge line and separation, respectively. The separation point is taken as the streamwise coordinate where the separation compression has reached one-half its final value, i.e., one-half the plateau pressure. The incipient separation is located at the point where no pressure rise is detected by the nearest pressure orifice upstream of the gap. The corresponding gap width is then labeled d^*. The gap widths required for incipient separation are shown in Fig. 5.60 and the reduction of plateau pressure due to mass suction (reflected by the mass ratio β) is shown in Fig. 5.61.

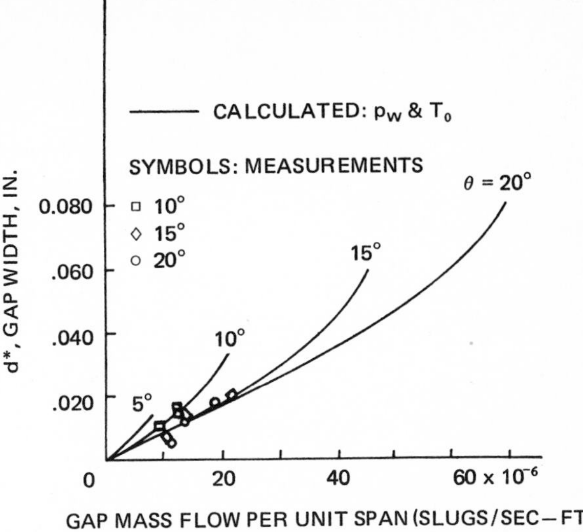

FIG. 5.59 Relation of gap width to mass flow ($M_\infty = 12.3$, $Re_\infty/in. = 7.3 \times 10^4$, and $T_w/T_0 = 0.56$; in $\dot{m}/L_s$, m is mass flow rate in slugs/foot² second and L is the length of the suction gap in inches; *from Ball and Korkegi, 1968*).

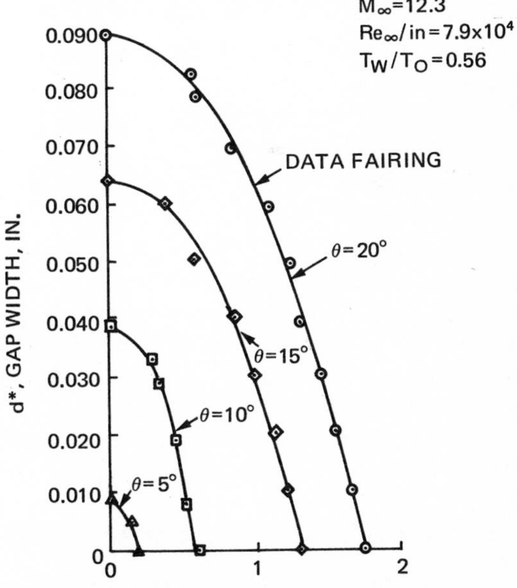

FIG. 5.60 Gap widths required for incipient separation at various flap angles (X_{HL} and X_s are chordwise surface distance from model leading edge (in inches) to hinge line and separation, respectively; *from Ball & Korkegi, 1968*).

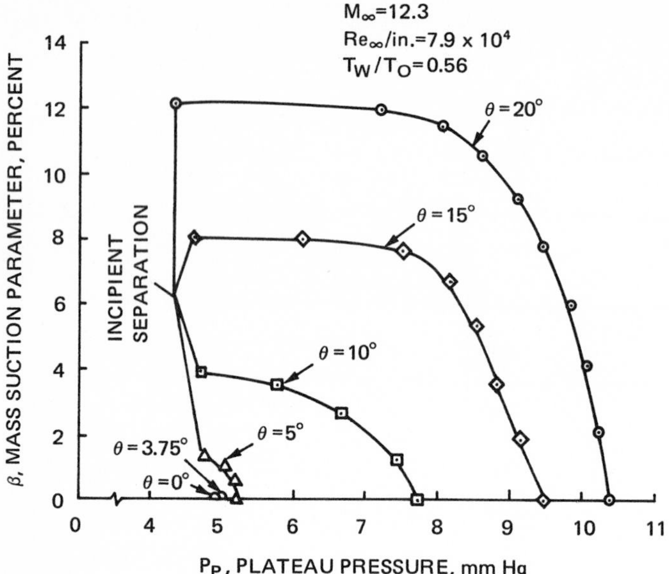

FIG. 5.61 Variation of the plateau pressure with mass suction (*from Ball & Korkegi, 1968*).

The mass-transfer ratio β is defined by $\beta = \dot{m}_s/\dot{m}_d$ and the mass defect per unit length is $\dot{m}_d/L = \rho_e\, u_e\, d^*$.

The mass suction required for incipient separation is small if expressed in terms of boundary-layer mass defect; β is approximately 1, 4, 8, and 12.5 percent for flap angles of 5, 10, 15, and 20 deg, respectively. Ball and Korkegi (1968) conjecture that for a large flap angle, the streamlines in the lower part of the boundary layer possess momentum well below that required to overcome the full corner compression, and therefore that removal of successive layers here does not significantly change the extent of separation. However, as more mass is sucked away, the remaining streamlines may possess higher momentum, and as the momentum of the reattaching streamline approaches that value required to negotiate the full corner compression, the extent of the separation zone becomes extremely responsive to mass removed by suction.

5.3 PREVENTION OF SEPARATION BY BLOWING

Blowing fluid tangentially along the wall is an effective technique for preventing separation. It may be applied not only for circulation control and aircraft stabilization but also for reduction of profile drag when located considerably rearward of maximum thickness.

Only the separation control aspect of tangential blowing is studied here. With increased cruising speeds, wings are designed thinner, and some sweep-back is necessary, together with efficient measures for avoiding excessive takeoff and landing distances. One suitable technique is to raise the maximum lift. Low speed lifting efficiency and stability characteristics can be thus improved to meet landing and takeoff requirements. When the wing angle of attack and the flap angles are large, the boundary layer is unable to negotiate the severe adverse pressure gradients on the upper surface of the wing nose and flap knee, and flow separation takes place at either or both of these places.

The blowing technique is particularly attractive because compressed air from the jet engine compressor may be used for blowing, and the high pressure air bleed can be bled to a choked blowing slot. The effectiveness of blowing depends primarily on the momentum of the air ejected from the injector pump and on the mixing conditions. Experiments have shown that flow separation over the trailing edge flap at least for flap angles up to 60 deg (streamwise) and hinge line sweeps up to 30 deg can provide sufficient blowing momentum to prevent practical engine-compressor bleeds. The lift increment due to the flaps is roughly doubled by blowing with flap angles between 45 and 60 deg. From the lift aspect, there seems little advantage in using flap angles much greater than 60 deg. Williams (1968) reports that the blowing requirements to prevent a rapid increase in separation with increased flap angle and the extra gain in lift can often be realized from "supercirculation" effects (jet extension of the flap chord) at, say, 60 deg flap angle.

5.3.1 Governing Parameters of Blowing

Blowing a thin jet into the neighborhood of the wall to increase the energy of fluid flow is an effective way to control flow separation. In overcoming the viscous effect and pressure rise along the flow path near the wall, fluid flow loses energy due to the momentum loss. The momentum of fluid flow is enhanced by blowing the jet tangentially from the wall; this then mixes with the shear layer flow, and a high energy level shear flow is thus obtainable. Therefore, the governing parameter of blowing may be based on the measure of the energy expended in the process of blowing. The most widely used governing parameter for the high pressure blowing system is the momentum coefficient C_μ defined by

$$C_\mu = \frac{mu_j}{q_\infty A}$$

where m is mass flow rate of the exiting jet in slug/second, u_j is the jet velocity in foot/second, q_∞ is the free-stream dynamic pressure in pound/foot^2, and A is a reference area such as total wing area in square feet. For a lower pressure system, $C_{\mu\text{net}} = C_\mu(1 - u_\infty/u_j)$ is a more suitable parameter; here u_∞ is free-stream velocity.

Because C_μ is also a similarity parameter, C_μ is an acceptable, adequate, and dependable parameter. The following other parameters are also used: flow coefficient C_Q, pressure coefficient C_p, and jet reaction coefficient C_j. However, these parameters are related to C_μ, as will be indicated later in this chapter.

Attinello (1961) found that for a given geometrical configuration, a given value of momentum coefficient C_μ resulted in similar lift, drag, and moment effects regardless of the particular combination of mass flow and jet velocity chosen to obtain this momentum coefficient. This finding permits a wide application of any experimental results or readily available data on pressure and volume flows and, furthermore, determines the practical configurations amenable to the application of blowing with structural, aerodynamic, and power plant limitations.

Another well-known flow coefficient C_Q is defined by $C_Q = Q/A\, u_\infty$ where Q is volume flow rate in foot^3/second and A is the reference area (usually wing area) in foot^2. When u_∞ is subsonic, C_μ is related to C_Q by

$$C_\mu = \frac{2C_Q^2}{h/c} = 2C_Q\sqrt{C_p}$$

because from the Bernoulli equation, $C_p = (u_j/u_\infty)^2$ and $u_j/u_\infty = C_Q/h/c$. In the above equation, h is the height of the slot or nozzle and c is the wing chord. The experimental determination of C_μ is rather difficult, but Poisson-Quinton (1956) suggests the following three ways:

1. Measure the jet thrust and take into consideration the under pressure caused by the jet stream on the airfoil.
2. Use C_Q and the scale of relative slot height h/c. (This procedure does not yield accurate results because the slot can be disfigured under the pressure and its effective cross-section is difficult to determine.)
3. Use the values of C_Q and C_p. (Error in the measurement may occur since the loss between the blowing chamber and the jet exit is not negligible.)

The effectiveness of blowing may also be formulated by the intensity of the blowing, denoted by $C_{\mu G}$; this is the value of C_μ necessary to reach the lift computed by the Glauert (1927) airfoil theory for unseparated potential flow. Another criterion is $C_{\mu R}$; this corresponds to the upper economic limit since only a slight increase in lift is achieved at a value of C_μ greater than $C_{\mu R}$. This point is to be found in the logarithmic expansion $\Delta C_L = f(C_\mu)$ at constant flap deflection. The theoretical lift coefficient of flap effectiveness is given by

$$C_{L_f} = \left(\frac{dC_L}{d\alpha}\right) \delta_f f\left(\frac{x_f}{c}\right)$$

where δ_f is the flap deflection angle, x_f is the effective chord length of the flap, and c is the total chord length of the airfoil. The values of function $f(x_f/c)$ are as follows:

x_f/c	20 percent	25 percent	30 percent
$f(x_f/c)$	0.55	0.61	0.66

As shown in Fig. 5.62, it is possible by blowing to reach and exceed the lift coefficient computed by the Glauert airfoil theory. The points labeled G form the logarithmic coordinate curve $\Delta C_L = f(C_\mu)$ at constant flap deflection.

According to classical airfoil theory, the momentum coefficient required to achieve the theoretical lift coefficient (without blowing) is designated as the critical momentum coefficient $C_{\mu\text{crit}}$. The value of $C_{\mu\text{crit}}$ may be roughly estimated from the graph shown in Fig. 5.63.

The value of $C_{\mu\text{crit}}$ depends primarily on wing aspect ratio, flap deflection, and the affected area ratio S_f/S_w; here S_f is the affected wing area and S_w is the total wing area. The affected wing area is the section of the wing ahead of the blowing slot plus the flap over which the jet sheet is directed. Momentum coefficient can also be defined on the basis of the "affected" area. This graph is valid for sweepback angles from 0 to 45 deg, flap chord ratios from 0.25 to 0.40, and angles of attack from 0 to 10 deg (but below the stall angle). Another coefficient is also used, namely C_{BLC}, the blowing jet boundary layer control parameter which is given by

$$\frac{C_{BLC}}{C_\mu} = 1 - \frac{u_e}{u_j}$$

where u_e is the local stream velocity over the nozzle. The concept of C_{BLC} was first introduced by Tolhurst (1958), and this equation is given by Kikuhara and Kasu (1968). The small value of u_j indicates the small value of C_{BLC} compared with C_μ, that is, the deterioration of the effectiveness of boundary layer control. From the practical point of view, Kikuhara and Kasu (1968) found that u_j should be more than ten times the free-stream velocity.

Attinello (1961) introduced the so-called excess momentum coefficient $C_{\mu_{\text{ex}}}$ defined by

$$C_{\mu_{\text{ex}}} = C_\mu - C_{\mu_{\text{crit}}}$$

This coefficient indicates the momentum coefficient in excess of that required to achieve theoretical flap effectiveness. The momentum coefficient is normally determined by the state of the gas leading to the nozzle exit. It is not easy to determine u_j during actual tests, particularly for the high velocity jets that result from compressor bleed air and ducted fan bypass air. However,

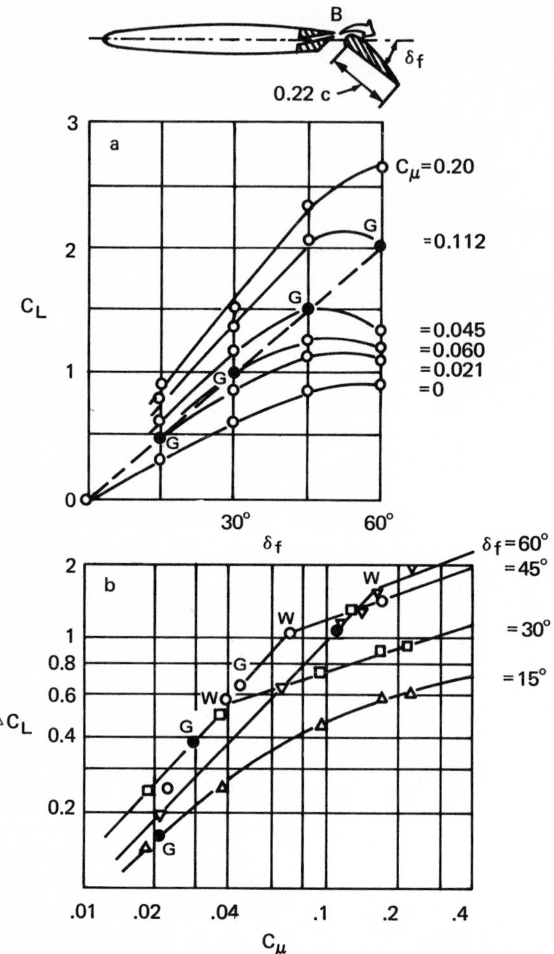

FIG. 5.62 Influence of flap deflection with blowing over a slotted flap (according to *Schwier, 1944; from Poisson-Quinton, 1956*). The sketches are for the NACA 0009 airfoil at a geometric angle of attack $\alpha_g = 0$. In the upper figure on the broken straight line, line *G* corresponds to the theoretical flap according to *Glauert, 1927.* The points *W* refer to the reattachment of flow on the flap.

in practice, u_j is computed by assuming that it is that velocity after isentropic expansion to standard sea-level conditions:

$$u_j = \left\{2gRT_{st}\left(\frac{\gamma}{\gamma-1}\right)\left[1-\left(\frac{p_\infty}{p_d}\right)^{\gamma-1/\gamma}\right]\right\}^{1/2}$$

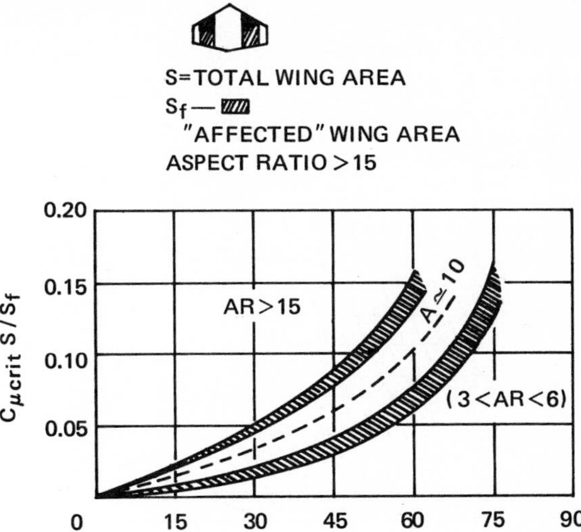

FIG. 5.63 Critical momentum coefficient as a function of flap deflection (*from Attinello, 1961*).

where R is the universal gas constant (53.3 ft/lb/deg R for air), T_{st} is the stagnation temperature in degrees Rankine in the duct and p_∞ and p_d are respectively the ambient free-stream static pressure and the duct stagnation pressure.

5.3.1.1 Effect of supply air pressure ratio

The value of C_μ is computed as a function of supply air pressure:

$$C_\mu = \left(\frac{w}{gqA}\right)(109.6\sqrt{T_{st}})\left[1 - \left(\frac{p_\infty}{p_d}\right)^{2/7}\right]^{1/2}$$

For testing at large values of C_μ, a jet reaction coefficient defined by $C_j = J/q_\infty A$ has been used in Great Britain. The symbol J is the total jet reaction in pound thrust at the nozzle exit. This coefficient is obtained in wind model tests by direct reading from the tunnel balance with the jet directed along the drag axis of the balance. For the convergent nozzle normally employed in blowing flap application and for supercritical pressure ratios, the measured jet reaction coefficient is only slightly smaller than the corresponding value of C_μ computed by isentropic flow. At pressure ratios of 5.1, the difference in values of C_j and C_μ is only 3 percent.

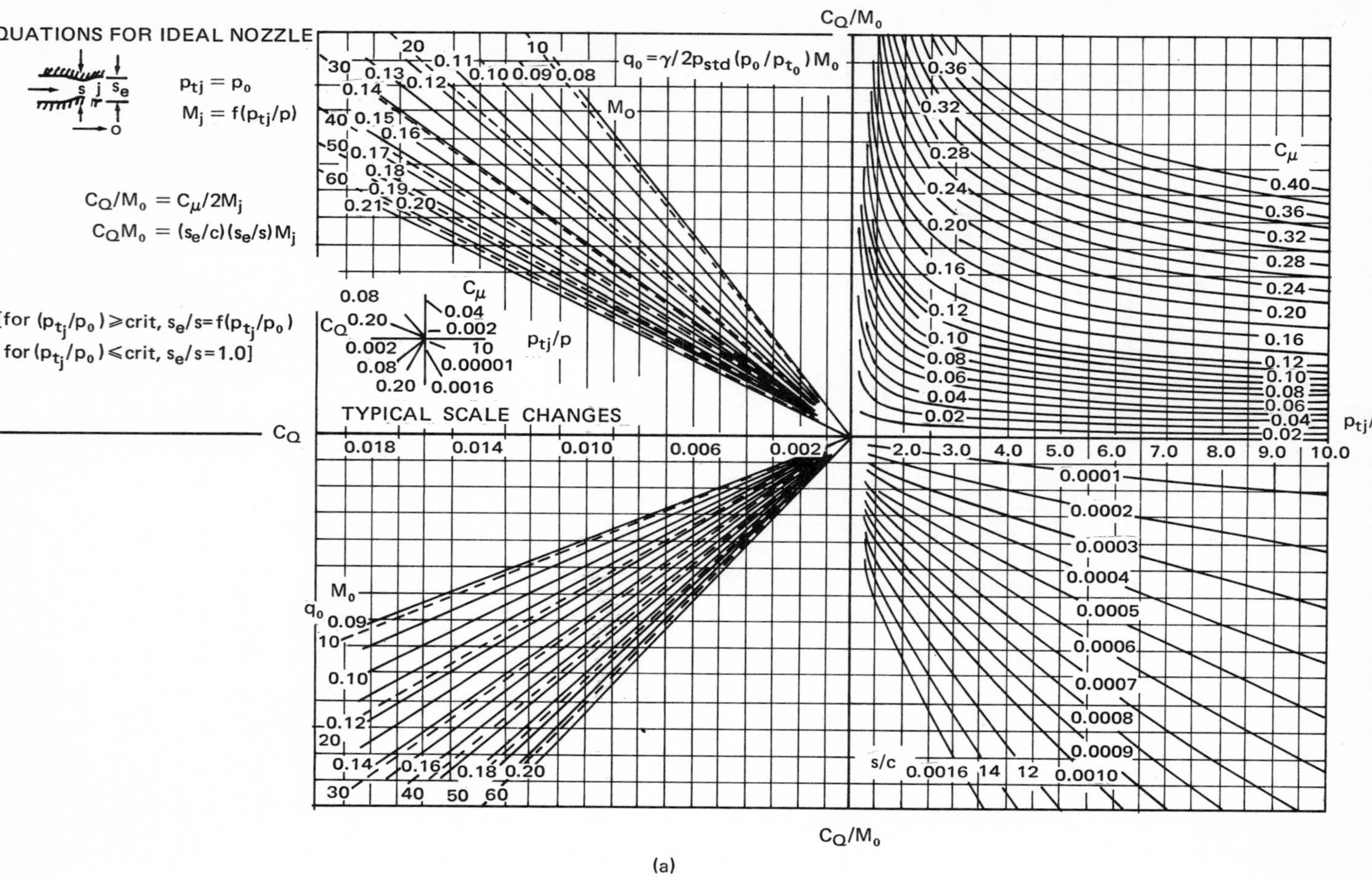
EQUATIONS FOR IDEAL NOZZLE
$p_{tj} = p_0$
$M_j = f(p_{tj}/p)$
$C_Q/M_0 = C_\mu/2M_j$
$C_Q M_0 = (s_e/c)(s_e/s)M_j$
[for $(p_{tj}/p_0) \geq$ crit, $s_e/s = f(p_{tj}/p_0)$
for $(p_{tj}/p_0) \leq$ crit, $s_e/s = 1.0$]
$q_0 = \gamma/2p_{std}(p_0/p_{t_0})M_0$
M_0
q_0
C_Q
C_Q/M_0
C_μ
p_{tj}/p
s/c
TYPICAL SCALE CHANGES
0.018
0.014
0.010
0.006
0.002
2.0
3.0
4.0
5.0
6.0
7.0
8.0
9.0
10.0
(a)

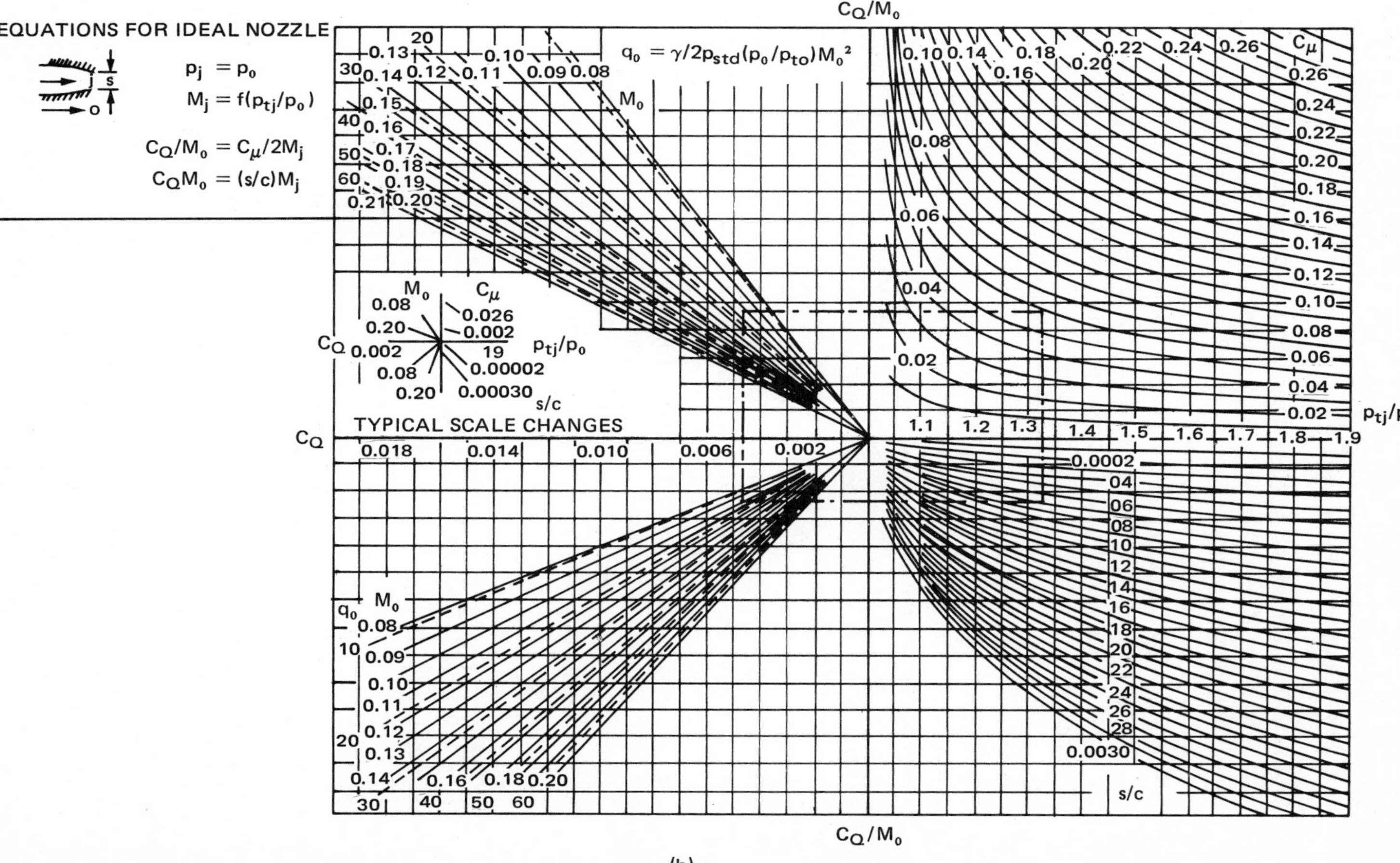

(b)

FIG. 5.64 Charts for correlating C_Q and C_μ values of a duct (*from Attinello, 1961*).

5.3.1.2 Effect of pressure ratio

Although an exiting jet becomes supersonic at high pressure ratios, this has little effect on lift for a given C_μ. Moreover, the exiting mass flow for supercritical pressure ratios is independent of small variations in the external pressure field along the wing span. However, at subcritical pressure ratios, variations in the external pressure field along the wing span do affect the local mass flows. If duct velocities are held to low values (less than $\frac{1}{6}$ of the jet velocity is a good rule-of-thumb value), spanwise variations in the jet velocity are obtainable by adjusting only the nozzle height for supercritical pressure ratios. This technique eliminates the need for elaborate internal guide vanes and throttles employed in low pressure systems.

The ideal nozzle configuration for pressure ratios greater than critical should be convergent-divergent and that for pressure ratios less than critical should be convergent. The graphic solution of equations shown in Fig. 5.64 correlates mass flow coefficients, free-stream Mach number, momentum coefficients, ratio of nozzle area to wing reference area (proportional to s/c for the two-dimensional case), and pressure ratio. In these figures, the subscript *cd* refers to ideal convergent-divergent nozzle.

The magnitude of the difference which may occur for pressure ratios above the critical may be obtained from the ratio of the jet momentum coefficient for a convergent nozzle to that for a convergent-divergent nozzle for isentropic flow; see Fig. 5.65. The ratio of momentum coefficients decreases with an increase in pressure until at a pressure ratio of 10 the ideally obtainable jet momentum coefficient $(C_\mu)^*$ is 0.93 of the C_μ value that could be ideally obtained with a convergent-divergent nozzle. The 2 to 3 percent

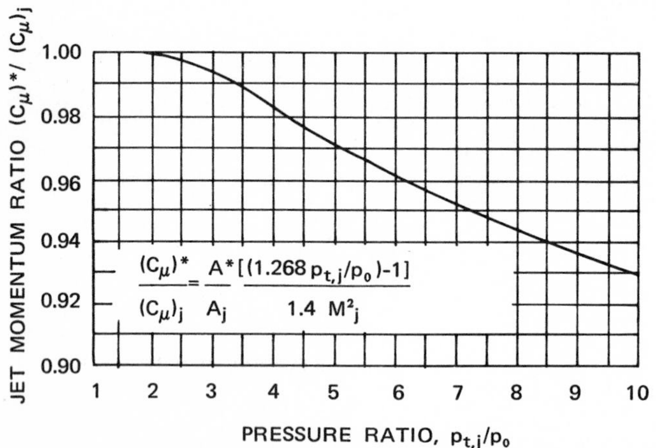

FIG. 5.65 Variation of jet momentum ratio with pressure ratio (*from Attinello, 1961*; the asterisk refers to the ideal convergent nozzle).

loss of jet momentum at pressure ratios on the order of 3 to 5 is never realizable in practice because of the tolerance that must be accepted in the nozzle fabrication and the range of pressure ratios over which the nozzle must operate. Practically, the convergent nozzle results in the best compromise of performance and fabrication. Since power is required for blowing, it is economically important to know the relationship between power expressed as $hp/\lambda s_w$ and the blowing similarity parameters C_Q and free-stream Mach number M_0; see the graph of Fig. 5.66; hp refers to horsepower, λ is the efficiency, and s_w is the wing area.

5.3.2 Physics of Blowing

A jet of higher velocity blown tangentially to the wall energizes the boundary layer flow; downstream of the trailing edge, the ejected jet acts as the flow trap and increases the circulation around the airfoil. Although blowing is not an effective technique for maintaining laminar flow, it may reduce form drag by preventing separation. Thus, blowing can also be effectively applied to a special laminarized airfoil. The Poisson-Quinton (1956) study of the physics of blowing in incompressible flow will now be discussed.

5.3.2.1 Effect of position of blowing slot

The effects of blowing on lift and pressure distributions as a function of angle of attack with uniform blow intensity are shown in Fig. 5.67 for various positions along the upper surface of a wing. This figure indicates that:

1. Blowing at the nose shifts the position of separation downstream and increases the maximum lift coefficient at increased critical angle of attack. However, this effect is small at small angles of attack.
2. Blowing at the knee of a movable airfoil nose prevents separation downstream of the maximum velocity position on the airfoil. The position at the nose must be selected so that the stagnation point lies near the nose.
3. Blowing over the flap may prevent separation. The effect of the jet resulting from a greater intensity of blowing is indicated downstream of the trailing edge. The angle of attack for the maximum lift coefficient decreases due to separation at the nose.
4. Blowing from the under side of the flap creates a jet flap which increases circulation.

Because the effect of the boundary layer is not significant, the lift increase obtainable by blowing may be evaluated analytically on the basis of potential theory.

Bamber (1931) showed that the optimum position of the blowing slot on the upper surface is beyond the midchord. The optimum for a 14.5-percent thick airfoil lies between the $\frac{1}{2}$ to $\frac{2}{3}$ chord from the leading edge. This

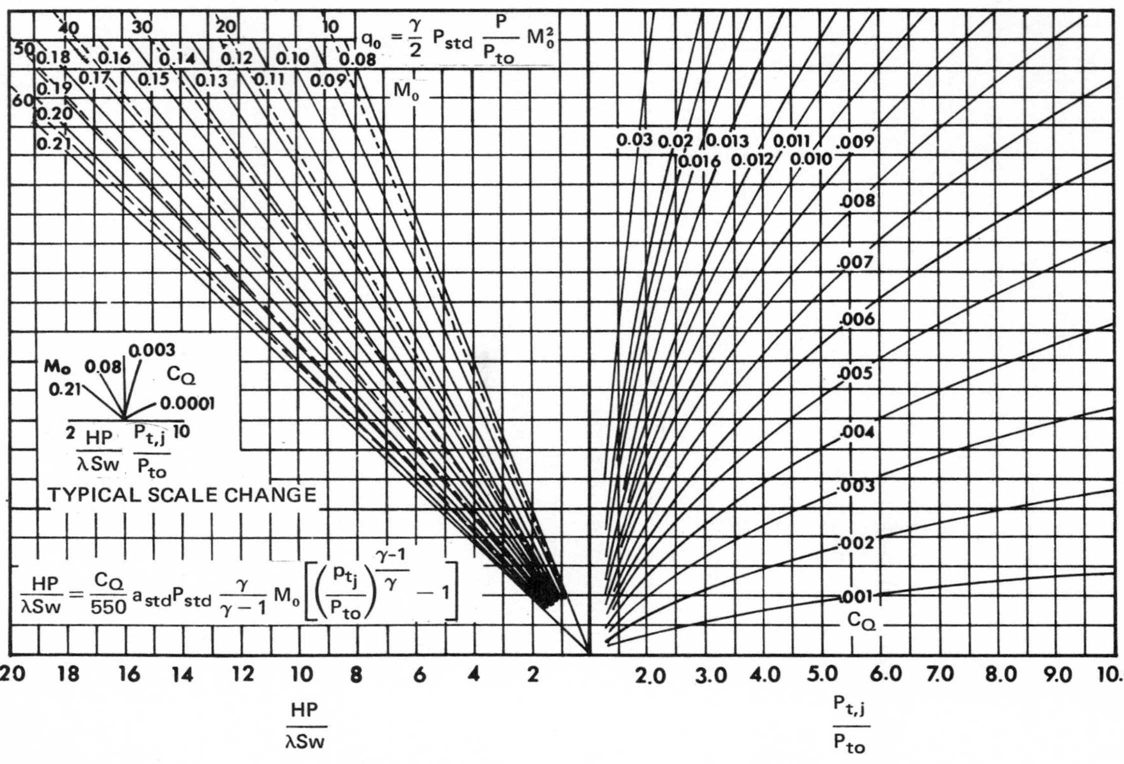

FIG. 5.66 Correlation of power and blowing similarity parameters (*from Attinello, 1961*).

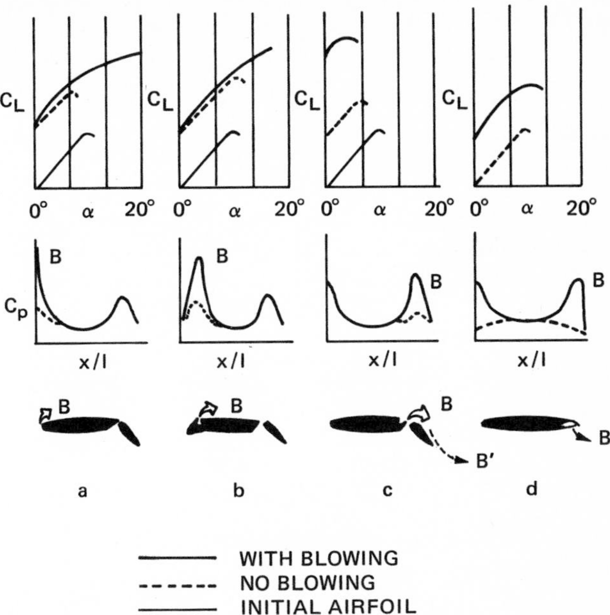

FIG. 5.67 Influence of blowing on boundary layer and circulation (*from Poisson-Quinton, 1956*; the upper series of sketches indicate lift coefficient as a function of angle attack and the lower series the pressure distributions on the upper side of the airfoil). B' is the position where increase of circulation is possible.

rearward position of the blowing slot yields the largest increase of lift at given values of C_Q and ratio of slot height with respect to chord length as a function of angle of attack. At small angles of attack, the rear slot resulted in the largest lift increment ΔC_L by circulation.

5.3.2.2 Effect of velocity of blowing jet

The initial thin jet of large velocity over a convex wall remains attached along the wall up to a certain distance due to the Coanda effect. The jet flow not only affects the velocity profile of the boundary layer but also the external flow due to the effect of induction. Figure 5.68 shows schematically the velocity profile over a flap downstream of the slot.

Since the velocity near the wall is increased because it is energized by the jet, it is apparent that separation can be prevented by blowing. The intensity of blowing is governed by the magnitude of two of the following three parameters: mass flow m, pressure in flowing chamber p', and the cross-section of the blowing slot sb' (s is the slot height and b' the slot width). If an adiabatic pressure decrease is assumed, then m and the pressure upstream and downstream of the slot may be used to determine the blowing velocity. In practice, if the jet velocity u_j does not exceed 150 m/sec, it may be

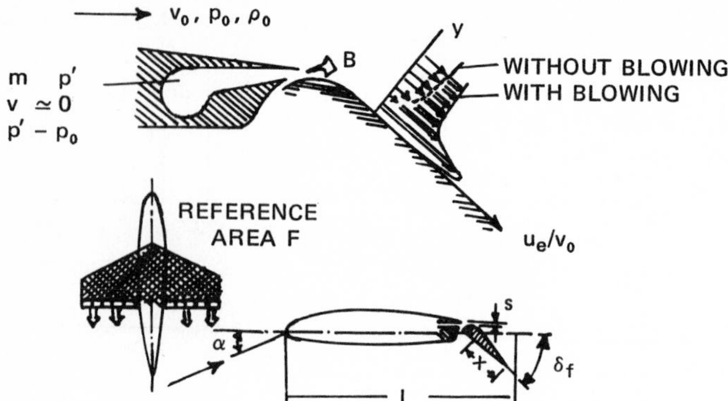

FIG. 5.68 Velocity profile over a flap downstream of the blowing slot (*from Poisson-Quinton, 1956*).

computed by using the Bernoulli equation. The pressure coefficient C_P is determined by

$$C_p = \frac{p' - p_\infty}{(\rho_\infty/2)u_\infty^2} = \left(\frac{u_j}{u_\infty}\right)^2$$

5.3.2.3 Momentum coefficient $C_{\mu A}$

A value of the momentum coefficient designated by $C_{\mu A}$ may be used as the criterion for the prevention or avoidance of separation by blowing. If $C_\mu < C_{\mu A}$, then separation may not be prevented although the effect of boundary layer control is quite effective. But if $C_\mu > C_{\mu A}$, then the boundary layer remains attached to the wall and lift may be increased further due to the additional supercirculation; however, the lift increment in this supercirculation is considerably less than in the boundary layer control region of $C_\mu < C_{\mu A}$.

Thomas (1962) studied this subject for a deflected flap. The solution of this problem entails a thorough investigation into the mechanism of the turbulent boundary layer with tangential blowing. However, since a rather lengthy presentation is necessary of the effect of blowing on the turbulent boundary layer, the details are given later (Sec. 5.3.3).

Thomas (1962) also studied the effect of a trailing edge flap on the flow field over an airfoil. He found that with an increase in angle of attack, a suction peak forms at the nose of the section and very soon leads to a separation of flow there. The behavior of this separated flow is very much influenced by the strength of the blowing jet that issues from in front of the trailing edge flap. With no blowing or with weak blowing, the flow remains detached along the whole of the upper surface of the sections as far as the trailing edge, resulting in an abrupt fall in lift after the maximum lift. On the other hand, with strong blowing over the flap, the flow that separated at the nose at large angle of attack becomes reattached after a certain length. This corresponds to a separation bubble. Although such local separation is not desirable, a further increase of lift may be achieved with an increase in angle of attack. In the region of boundary layer control, however, the flow completely breaks away shortly afterward because the collapse of the separation bubble causes an abrupt fall in lift. On the other hand, in the region of supercirculation, the separation bubble becomes even larger with increasing angle of attack, so that lift increases still further. Such flow phenomena may strongly depend on the shape of the section and on Reynolds number. Figure 5.69 demonstrates the effect of momentum coefficient on pressure distribution as calculated by the methods of Glauert (1927) and Spence (1958).

For the case $C_\mu \approx C_{\mu A}$, the flow is almost potential and predicted quite well by the Glauert airfoil theory, but in the region of supercirculation, the Spence theory gives more accurate predictions for smaller angles δ_F.

The momentum coefficients $C_{\mu A}$ necessary to just prevent separation are almost independent of angle of attack in the linear region of $C_L(\alpha)$ curves. It

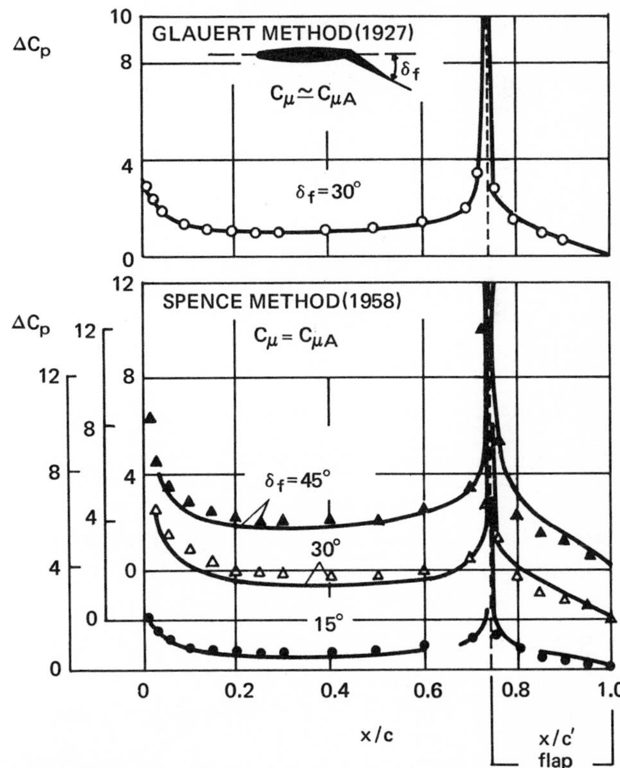

FIG. 5.69 Comparison of measured and calculated pressure distributions on an airfoil with a blown flap (*from Thomas, 1962*).

is clear that the lower the $C_{\mu A}$ values (i.e., the more favorable) the smaller the ratio of slot width w to wing chord c. The particularly low and hence economical $C_{\mu A}$ values may be found by interpolating the experimental results obtained at ONERA by Carriére et al. (1958):

$$C_{\mu A} = 0.015 \tan \delta_f$$

Thomas (1962) found that the separation of flow over the flap could be prevented by sufficiently strong blowing over the range of flap deflection investigated ($\delta_f = 0$ to 105 deg). Furthermore, the blowing momentum which leads to the largest gain in lift is such that it is just sufficient to prevent separation on the flap, and the lift is then the same as predicted by potential theory.

A comparison with various measurements shows that the effectiveness of a blowing jet depends not only on its momentum coefficients but also on the slot width. For the purpose of boundary layer control, it is much more profitable to generate a required momentum coefficient with the least possible mass flow and highest possible blowing velocity; therefore, the smallest possible slot width w/c should be used.

Thomas (1962) gives the best momentum coefficient necessary to prevent the separation over the airfoil as

$$C_{\mu A} = 2\,\frac{\delta_G}{c}\,\frac{1}{0.85(1 - \bar{u}_f/u_j)^2}$$

where δ_G is the nccessary momentum requirement of the boundary layer to keep the flow attached up to the trailing edge, $\bar{u}_f$ is the mean velocity on the flap, u_j is the velocity of the blowing jet, and the value of δ_G/c is evaluated by

$$\frac{\delta_G}{c} = \frac{0.037}{(U_\infty c/\nu)^{1.5}} \left(\frac{u_e}{u_\infty}\right)_{\text{te}}^{-3} \left[\int_{p_s}^{\text{te}} \left(\frac{\mu_e}{u_\infty}\right)^{3.5} d\left(\frac{s}{c}\right)\right]^{0.8}$$

Here u_e is the velocity on the section contour, s is the coordinate in section contour, p_s is the separation point, and te is the trailing edge.

5.3.3 Analysis of Incompressible Turbulent Wall Jet Involving Separation

Hubbartt and Bangert (1970) and Bangert (1971) analyzed incompressible turbulent wall jets in an adverse pressure gradient by dividing the wall jet into layers and using the integral momentum and mechanical energy equation. The analysis was made for two limiting cases: (*a*) the initial boundary layer was ignored (or considered rapidly consumed), and (*b*) it was assumed to be separated, forming a starting wake which persists downstream with the minimum velocity remaining zero. This uniquely specifies the allowable pressure gradient. It was shown that for a constant jet momentum, wall layer separation is relatively insensitive to the ratio of jet to free-stream velocity, to the jet and wake layer profile shape factors, and to the jet and wake layer dissipation rates. They computed the pressure rise and the corresponding separation distances for typical conditions and concluded that turbulent flow separation can be prevented by a blowing wall jet of sufficient jet momentum.

The distance and pressure rise to the separation point was found to be relatively insensitive to a reasonable range of dissipation rates in the jets and wake layers. Furthermore, entrainment of free-stream flow in the adverse pressure gradient increased the wall jet excess momentum and tended to offset the reduction caused by surface friction. Increasing the jet velocity at a constant jet momentum increased the jet excess momentum relative to the free stream, but this favorable effect was diminished by an increase in wall shear stress. Thus an optimum ratio of jet velocity at the slot can be found with respect to free-stream velocity at the slot (u_{js}/u_{es}) or at least a very small benefit may be obtained by increasing jet velocity at the slot.

The problem of the wall jet has been investigated by Glauert (1956), Bradshaw & Gee (1962), Harris (1965), Kruka and Eskinazi (1964), and Escudier and Nicoll (1966), but their studies were limited to the case of no (or mildly) adverse pressure gradients that have only a small influence on the flow downstream of the jet slot.

Thomas (1962) used empirical data and modeled the turbulent mixing process. Carriére and Eichelbrenner (1961) developed a wall jet analysis involving reattachment, but this analysis required empirical information. They determined the importance of velocity profiles with both a maximum and a minimum by characterizing the initial boundary layer effect as indicated in Fig. 5.70. High-velocity, narrow jets appear to require smaller momentum since only the inner part of the original boundary layer must be accelerated to prevent the separation; this was in agreement with the Thomas results.

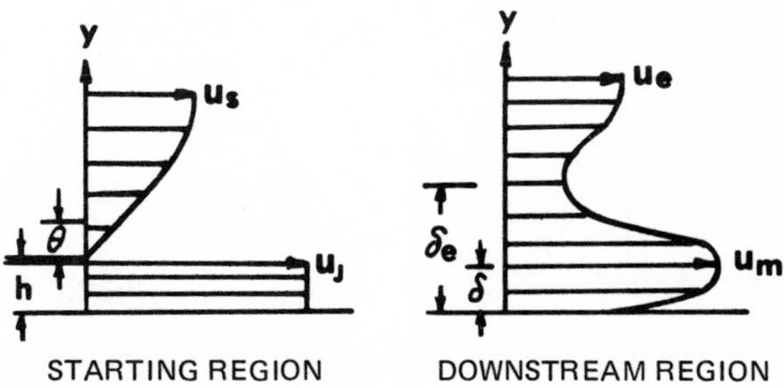

FIG. 5.70 Carriére-Eichelbrenner wall jet analyses involving reattachment (*from Hubbartt & Bangert, 1970*).

Gartshore and Newman (1969) developed a wall jet analysis that included prediction of separation, but it was restricted to a simple wall jet, i.e., with no velocity minimum, as shown in Fig. 5.71.

An estimate of the effect of the upstream boundary layer is obtainable by relating the actual velocity profile at the slot to an equivalent profile for a simple wall jet. The model developed by Hubbartt and Bangert (1970) for the wall jet-boundary layer mixing process is applicable for the general case, as indicated in Fig. 5.72. However, this model is limited to a simple wall jet (Fig. 5.72) and a situation in which the initial boundary layer is separated and forms a starting wake which persists downstream with zero minimum velocity.

A second limiting case considers the interaction of a given wall jet with a boundary layer that is at or on the verge of separation. The Hubbartt and Bangert (1970) analysis determines the limiting adverse external velocity distribution, distance of attached flow downstream of the slot, and pressure and pressure rise to separation. For convenience, their analysis is discussed separately for a simple wall jet and a wall jet with a wake region and zero minimum velocity.

5.3.3.1 Results of the analysis

Simple wall jet The profile of a simple wall jet is divided into a wall layer and a jet layer in the downstream region of the initial point of analysis. There the uniform jet core has been consumed by shear layers, as shown in Fig. 5.72.

The flow is modeled by the integral method, and incompressible flow with negligible curvature is assumed. The integral momentum and mechanical energy equations are used for each layer. The following simple power law is adopted for the wall level:

$$\frac{u}{u_m} = \left(\frac{y}{\delta}\right)^{(H-1)/2}$$

where u is the component of velocity parallel to the wall or along x; the subscript m refers to conditions at $y = \delta$, the maximum velocity location; y is the perpendicular distance from the wall; δ is the distance from the wall to the maximum velocity in the jet; and H is defined by $H = (1/\theta)\int_0^\delta [1 - (u/u_m)]\, dy$, where $\theta = \int_0^\delta (u/u_m)[1 - (u/u_m)]\, dy$.

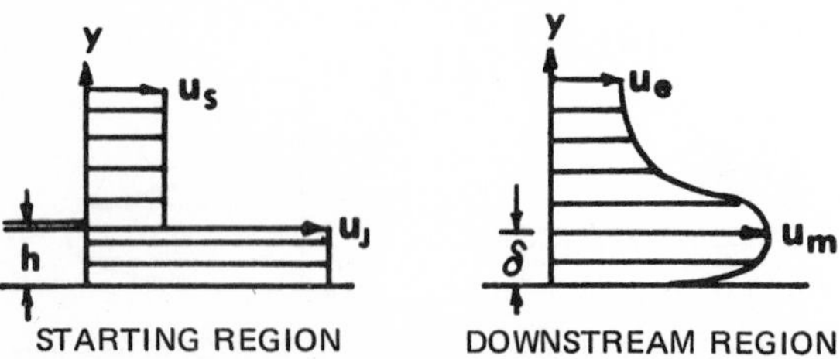

FIG. 5.71 Gartshore-Newman wall jet analyses including prediction of separation (*from Hubbartt & Bangert, 1970*).

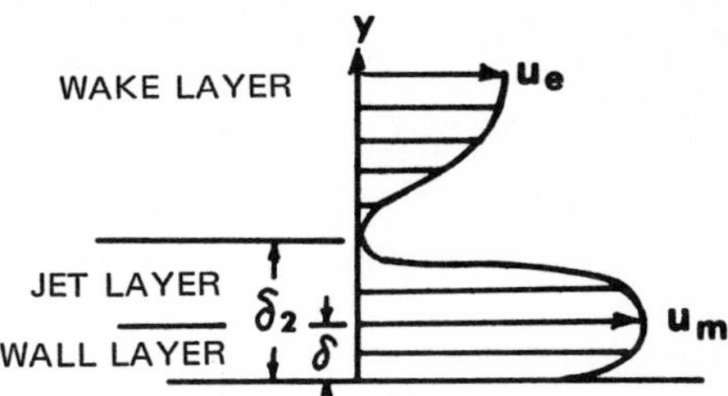

FIG. 5.72 Hubbartt-Bangert model for the wall jet-boundary layer mixing process (*from Hubbartt & Bangert, 1970*).

Similarity of the jet-layer velocity profile as expressed in the following form is also assumed:

$$\frac{u - u_e}{u_m - u_e} = f\left(\frac{y - \delta}{\theta_2}\right)$$

The symbol u_e refers to velocity at the outer edge of the boundary layer or wall jet, and θ_2 is defined by

$$\theta_2 = \int_\delta^\infty \frac{u - u_e}{u_m - u_e}\left(1 - \frac{u - u_e}{u_m - u_e}\right) dy \tag{5.6}$$

After utilization of the continuity equations and simplifying, we obtain

in the wall layer, the momentum equation:

$$\frac{d\theta}{dx} - \frac{2\theta}{(H-1)u_m}\frac{du_m}{dx} = -\theta H\left(\frac{H+1}{H-1}\right)\frac{u_e}{u_m^2}\frac{du_e}{dx} + \frac{c_{fw}}{2} - \frac{c_{fm}}{2} \tag{5.7}$$

in the wall layer, the energy equation:

$$\frac{d\theta}{dx} - \frac{\theta}{H(3H-1)}\frac{dH}{dx} - \frac{2\theta}{(H-1)u_m}\frac{du_m}{dx} = -\left(\frac{3H-1}{H-1}\right)\frac{u_e\theta}{u_m^2}\frac{du_e}{dx} + \frac{3H-1}{2H}(2dc_{f_w} - c_{fm}) \tag{5.8}$$

where

$$c_{fm} = -c_{f_w}\left(1 - \frac{u_e}{u_m}\right)^2$$

in the jet layer, the momentum equation:

$$A_{11}\frac{d\theta}{dx} + A_{12}\frac{dH}{dx} + A_{13}\frac{dI}{dx} + A_{14}\frac{du_m}{dx} = A_{15}\frac{du_e}{dx} - \frac{c_{fm}}{2} \tag{5.9}$$

in the jet layer, the energy equation

$$A_{21}\frac{d\theta}{dx} + A_{22}\frac{dH}{dx} + A_{23}\frac{dI}{dx} + A_{24}\frac{du_m}{dx} = A_{25}\frac{du_e}{dx} - \frac{c_{fm}}{2} - \left(\frac{u_\Delta}{u_m}\right)^3 \frac{F_3}{R_{T_2}} \tag{5.10}$$

where c_f is the shear stress coefficient given by $2\tau/\rho u_m{}^2$

$$u_\Delta = u_m - u_e$$

d is the dissipation integral given by

$$d = \int_0^\delta \frac{u}{u_m}\frac{\partial(\tau/\tau_w)}{\partial y}\,dy \tag{5.11}$$

(here τ is shear stress and the subscript w refers to wall), I refers to the jet layer thickness parameter given by

$$I = \int_\delta^\infty \left(\frac{u - u_e}{u_m - u_e}\right)^2 dy \tag{5.12}$$

and F_3 is the velocity profile shape factor of the jet given by

$$F_3 = \int_\delta^\infty \left(\frac{u - u_e}{u_m - u_e}\right)\left(1 - \frac{u - u_e}{u_m - u_e}\right) dy \Bigg/ \int_\delta^\infty \left[\frac{\partial\left(\dfrac{u - u_e}{u_m - u_e}\right)}{\partial y}\right]^2 dy \tag{5.13}$$

The symbol R_{T_2} refers to turbulent Reynolds number for the jet layer given by

$$\frac{1}{R_{T_2}} = \frac{\displaystyle\int_\delta^\infty \frac{\tau}{\rho(u_m - u_e)^2}\,\frac{\partial\left(\dfrac{u - u_e}{u_m - u_e}\right)}{\partial y/\theta_2}\,\frac{dy}{\theta_2}}{\displaystyle\int_\delta^\infty \left[\frac{\partial\left(\dfrac{u - u_e}{u_m - u_e}\right)}{\partial y/\theta_2}\right]^2 \frac{dy}{\theta_2}} \tag{5.14}$$

For $\tau = \rho\epsilon\,\partial u/\partial y$ and constant ϵ, this reduces to $R_{T_2} = (u_m - u_e)\,\theta_2/\epsilon$.

The coefficients of Eqs. (5.9) and (5.10) are

$$
\begin{aligned}
A_{11} &= \frac{2Hu_\Delta}{(H-1)u_m} \qquad A_{12} = -\frac{2\theta u_\Delta}{(H-1)^2 u_m} \\
A_{13} &= \frac{u_\Delta(u_e F_1 + u_\Delta)}{u_m^2} \qquad A_{14} = \frac{I}{u_m^2}(2u_\Delta + u_e F_1) + \frac{2\theta H u_\Delta}{(H-1)u_m^2} \\
A_{15} &= \frac{2I}{u_m^2}(u_m F_1 - 2u_e F_1 - u_\Delta) \qquad A_{21} = \frac{H}{H-1}\frac{u_\Delta}{u_m^2}(u_m + u_e) \\
A_{22} &= -\frac{\theta u_\Delta}{(H-1)^2 u_m^2}(u_m + u_e) \qquad A_{23} = \frac{u_\Delta}{2u_m^3}(u_\Delta^2 F_2 + 3u_e u_\Delta + 2u_e^2 F_1) \\
A_{24} &= \frac{I}{2u_m^3}(3F_2 u_\Delta^2 + 3u_e u_\Delta + u_e^2 F_1) + \frac{H\theta u_\Delta}{(H-1)u_m^3}(u_m + u_e) \\
A_{25} &= \frac{I}{2u_m^3}(3u_\Delta^2 - 6u_e u_\Delta - 3F_e u_\Delta^2 + 4u_e u_\Delta F_1 - 2u_e^2 F_1) \\
F_1 &= \frac{\displaystyle\int_\delta^\infty \left(\frac{u - u_e}{u_m - u_e}\right) dy}{\displaystyle\int_\delta^\infty \left(\frac{u - u_e}{u_m - u_e}\right)^2 dy} \\
F_2 &= \frac{\displaystyle\int_\delta^\infty \left(\frac{u - u_e}{u_m - u_e}\right)^3 dy}{\displaystyle\int_\delta^\infty \left(\frac{u - u_e}{u_m - u_e}\right)^2 dy}
\end{aligned}
\qquad (5.15)
$$

Equations (5.7)-(5.10) are for the wall layer momentum thickness θ and the ratio H of displacement thickness with respect to θ, the maximum u_m and the jet layer integral I, in terms of the jet layer profile shape factors F_1, F_2, and F_3, the friction factors c_{fm} and c_{fw}, the wall and jet layer dissipation integral d, and the turbulent Reynolds number R_{T_2}. The values of F_1, F_2, and F_3 are relatively insensitive to changes in the ratio u_{js}/u_e where u_{js} is jet velocity at slot. The wall friction factor c_{fw} (evaluated by using the results of Thompson, 1967, for incompressible turbulent boundary layer) is related to the friction factor c_{fm} (evaluated by the approach of Harris, 1965) given previously in Eq. (5.8*a*).

As mentioned previously, the dissipation integral d is evaluated by using the Ludwieg-Tillman skin friction formula so that

$$d = 0.0455 \times 10^{0.678H} \left(\frac{u_m \theta}{\nu}\right)^{0.101} \tag{5.16}$$

The Thompson (1967) method yields higher values of τ_w than does the Ludwieg-Tillman formula for commonly encountered wall jets with lower wall Reynolds numbers.

In the starting region where the jet core velocity is uniform, the jet layer and the wall layer develop independently. The analysis is essentially the same except that u_m is related uniquely to u_e by the Bernoulli equation $u_m(du_m/dx) = u_e(du_e/dx)$ and $c_{fm} = 0$.

The development of the outer free shear layer and the conventional boundary layer is followed downstream until these two layers merge and the initial conditions for the downstream region are specified. The only notable distinction between the analysis for the starting and downstream regions is that the wall friction factor is given by the simpler flat plate formula $c_{fw} = 0.0256\ (u_m\theta/\nu)^{-1/4}$ (Schlichting, 1968).

The turbulent boundary layer is assumed to develop along the entire starting length. Figure 5.73 compares typical experimental results with numerical values obtained by Eqs. (5.7)-(5.10) and based on $R_{T_2} = 18.2$, $F_1 = 1.5$, $F_2 = 0.8$, and $F_3 = 0.178$.

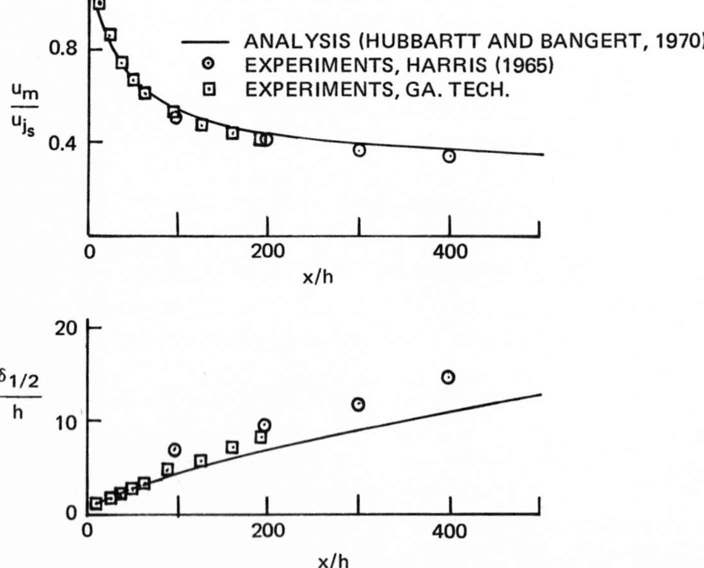

FIG. 5.73 Experimental and computed values for development of the turbulent boundary layer of a simple wall jet (*from Hubbartt & Bangert, 1970*).

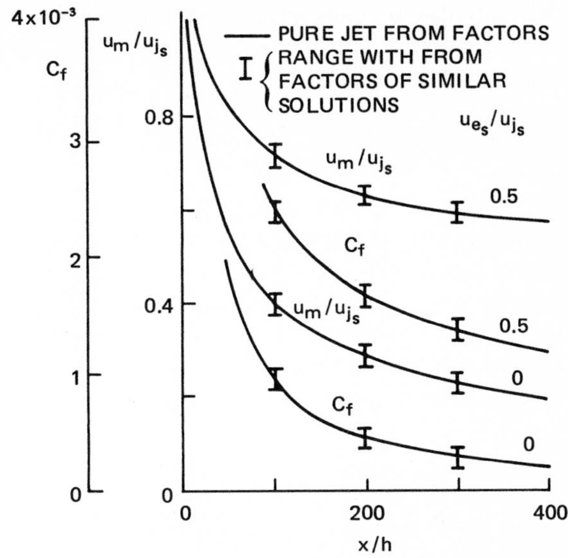

FIG. 5.74 Effect on wall jet development of varying jet layer form factors (*from Hubbartt & Bangert, 1970*).

The computed and experimental values of R_{T_2} are in good agreement. Values of F_1, F_2, and F_3 for asymmetric type jet flows with the eddy viscosity constant in the lateral direction were obtained by the classical Falkner-Skan laminar boundary layer equation, which is also applicable for the turbulent flow in this case. Solutions were obtained by using fourth-order Runge-Kutta numerical integration, starting at the maximum velocity point with specified values of the lateral velocity component. In the present case of the wall jet, this would represent entrainment of the jet layer into the wall layer. Solutions were obtained for values of u_m/u_e from 1.25 to 8 and lateral velocity components from zero (symmetric jets) to essentially the maximum value obtainable with the similar solutions (exceeding the values obtained in the wall jet analysis). These results indicated the following approximate ranges for the shape factors:

$$1.42 \leqslant F_1 \leqslant 1.58$$

$$0.776 \leqslant F_2 \leqslant 0.813$$

$$0.164 \leqslant F_3 \leqslant 0.187$$

In order to determine the effect on wall jet development, these parameters have been varied over this range for constant pressure wall jets. The solid line in Fig. 5.74 represents the pure jet profile and the bars indicate the range of values computed for the range in profile parameters. It is clear that the variations due to these parameters are small and within the accuracy of the analysis. Therefore, in the analysis that follows, the jet layer is represented by the pure jet shape factors.

5.3.3.2 Wall jet with wake region—Limiting case of zero minimum velocity

The model for this case is that shown already in Fig. 5.72. The only difference in the description of wall region and jet region here and for the case of simple wall jet, is in the definitions of jet layer parameters. Now u_e is replaced by u_2 and the upper limit of integration infinity ∞ is replaced by δ_2. The subscript 2 refers to the velocity maximum $y = \delta_2$. The following constant turbulent Reynolds number R_{T_3} is used to describe the turbulent stress and dissipation for the outer wakelike region:

$$\frac{1}{R_{T_3}} = \frac{\displaystyle\int_{\delta_2}^{\infty} \frac{\tau}{\rho(u_e - u_2)^2} \frac{\partial\left(\dfrac{u - u_2}{u_e - u_2}\right)}{\partial(y/\theta_3)} \, d\left(\frac{y}{\theta_3}\right)}{\displaystyle\int_{\delta_2}^{\infty} \left[\frac{\partial\left(\dfrac{u - u_2}{u_e - u_2}\right)}{\partial(y/\theta_3)}\right]^2 d\left(\frac{y}{\theta_3}\right)} \tag{5.17}$$

The nominal value of 15.6 selected for R_{T_3} is based on data from Townsend (1956) for a symmetrical wake. However, reasonable variations in this turbulent Reynolds number have only a relatively minor effect on the flow development. Again, the Falkner-Skan solutions for wake flow are used to obtain the velocity profiles and the shape factors. The nominal values are taken from the case of a symmetric wake ($v_2 = 0$) with $u_2 = 0$. This yields $H_3 = (1/\theta_3) \int_{\delta_2}^{\infty} (1 - u/u_e) dy$ or 4.028 and $He_3 = (1/\theta_3) \int_{\delta_2}^{\infty} (u/u_e)[(1 - u^2/u_2{}^2)] dy$ or 1.515; F_4 the velocity profile shape factor for the wake layer is given by $\theta_3 \int_{\delta_2}^{\infty} [(\partial/\partial y)(u/u_e)]^2 \, dy$ and amounts to 0.1565. The symbol θ_3 refers to

$$\theta_3 = \int_{\delta_2}^{\infty} \frac{u - u_2}{u_e - u_2} \left(1 - \frac{u - u_2}{u_e - u_2}\right) dy$$

Similar to the variation of R_{T_3}, a reasonable variation on values for these parameters also has only a minor effect.

To prevent separation of the original boundary layer by the action of a sufficient jet flow, it is assumed for this limiting case of zero minimum velocity that $u_2 = 0$ and that the wake region has a constant shape. Since $u_e(x)$ cannot be computed by this requirement, it is calculated as part of the solution.

The solution for this case yields the adverse pressure gradient, the length of the attached flow, and the pressure rise to separation that can be overcome by a boundary layer that is on the verge of separation and is mixing with a given wall jet. These quantities are determined by the given values of the initial velocity ratio u_{js}/u_{e_s} (u_{js} is jet velocity at the slot and u_{e_s} is the value of u_e at the slot), the ratio of the jet momentum to the boundary layer momentum deficit $u_{js}{}^2 h/u_{e_s}{}^2 \theta_s$ (here h is slot height), and the boundary layer Reynolds number Re.

The following integral equations are obtained by using the continuity equation and eliminating the y-velocity component.

momentum integral equation:

$$\frac{d\theta_3}{dx} + F_1 \frac{u_m}{u_e}\frac{dI}{dx} + \left(\frac{2H\theta}{H-1} + F_1 I\right)\frac{1}{u_e}\frac{du_m}{dx} + (2+H_3)\frac{\theta_3}{u_e}\frac{du_e}{dx} + \frac{2H}{H-1}\frac{u_m}{u_e}\frac{d\theta}{dx} - \frac{2\theta}{(H-1)^2}\frac{u_m}{u_e}\frac{dH}{dx} = 0 \quad (5.18)$$

For this case, the outer velocity of the jet is zero (corresponding to the zero minimum velocity) and

$$F_1 = \frac{\displaystyle\int_\delta^\infty \left(\frac{u}{u_m}\right) dy}{\displaystyle\int_\delta^\infty \left(\frac{u}{u_m}\right)^2 dy}$$

mechanical energy integral equation:

$$\left(\frac{H_3 - 1}{H_{e_3} - 1}\right) H_{e_3}\theta_3 \frac{du_e}{dx} + F_1 u_m \frac{dI}{dx} + \left(\frac{2H\theta}{H-1} + F_1 I\right)\frac{du_m}{dx} + \frac{2Hu_m}{H-1}\frac{d\theta}{dx} - \frac{2\theta u_m}{(H-1)^2}\frac{dH}{dx} = -\frac{2F_4 u_e}{(H_{e_3}-1)R_{T_3}} \quad (5.19)$$

Here F_4 is the velocity profile shape factor for the wake, defined (as before) by $F_4 = \theta_3 \int_{\delta_2}^{\infty} [(\partial/\partial y)(u/u_e)]^2\, dy$.

These equations are for $u_2 = 0$, $\tau_2 = 0$, and constant shape of the wake region (constant H_3 and H_{e_3}). Equations (5.7)-(5.10) and (5.18) and (5.19) are solved numerically by the Runge-Kutta method for the unknown θ, H, u_m, I, θ_3, and u_e. The position of separation of the wall jet is determined by $H = 2.5$. However, the obtainable results are relatively insensitive to the value of H at separation. For example, for $H = 2.0$ at separation, x_{sep} is reduced by about 4 percent, indicating a rapid increase in H near separation. A starting region calculation is used to proceed from the initial velocity profile (Fig. 5.70) to the fully developed profile (Fig. 5.72) a short distance downstream. The interaction of the upper part of the jet with the original boundary layer is analyzed in the same way as for the fully developed flow. The lower part of the jet develops along the wall in a manner similar to a turbulent boundary layer. The starting length (typically eight slot heights) is complete when the inviscid jet core is consumed by the shear layer and the boundary layer.

5.3.3.3 Results of the analysis

Simple wall jet The Hubbartt-Bangert (1970) analytical prediction of adverse pressure gradient leading to separation for simple wall jet flow with constant velocity gradient at $H = 2$ is presented in Fig. 5.75. This case represents a conservative estimate of the blowing required to prevent separation because for a more practical experimental velocity decay (i.e., a decreasing velocity gradient), there is a lower pressure gradient in the downstream region where the jet momentum available to overcome the adverse pressure gradient has diminished.

Although the jet Reynolds number Re_j is constant for a fixed ratio of the jet to free-stream velocity (u_{j_s}/u_{e_s}), it varies inversely with that ratio, Therefore, in effect, the jet momentum is held constant and is independent of the velocity ratio. The jet momentum coefficient $C_{jL} = (2u_{j_s}^2 h)/(u_{e_s}^2 \cdot L)$ is related to the length L over which the free-stream velocity decays from u_{e_s} to zero (i.e., $x = L$ at $u_e = 0$). If the flow separates ($x_{\mathrm{sep}} \leqslant L$), the momentum coefficient $C_{j\mathrm{sep}} = (2u_{j_s}^2 h)/(u_{e_s}^2\, x_{\mathrm{sep}})$ is related to the corresponding length of separation and, correspondingly, $C_{j\mathrm{sep}} \geqslant C_{jL}$. It is evident from these results that the pressure rise (or the distance to separation point) for a given jet momentum and velocity gradient is relatively insensitive to the velocity ratio within a range from 2 to 8. Apparently the increase in excess momentum (i.e., the jet flow rate times the difference between the jet and free-stream velocity) associated with the increasing velocity ratio is largely offset by the increased wall friction. For a given velocity ratio, the pressure rise to separation increases as L is increased. This is a reflection of the reduced length to separation and, hence, the diminished momentum decay of the wall jet. Limiting values of C_{jL} above which no separation occurs are obtained by extrapolation, leading to $\Delta p/q_s = 0$ in Fig. 5.78, where q_s, the dynamic pressure, is equal to $\frac{1}{2}\,\rho u_{e_s}^2$. The momentum deficit of an initial boundary layer can be approximately corrected by increasing the jet momentum on the basis of the Gartshore-Newman (1969) approach. They assume an instantaneous adjustment from the required jet to the final simple jet with no boundary layer and conservation of momentum and jet mass flow rate during the adjustment.

Wall jet with wake region For this limiting case of zero minimum velocity, a wall jet model is proposed such that the boundary layer upstream of the

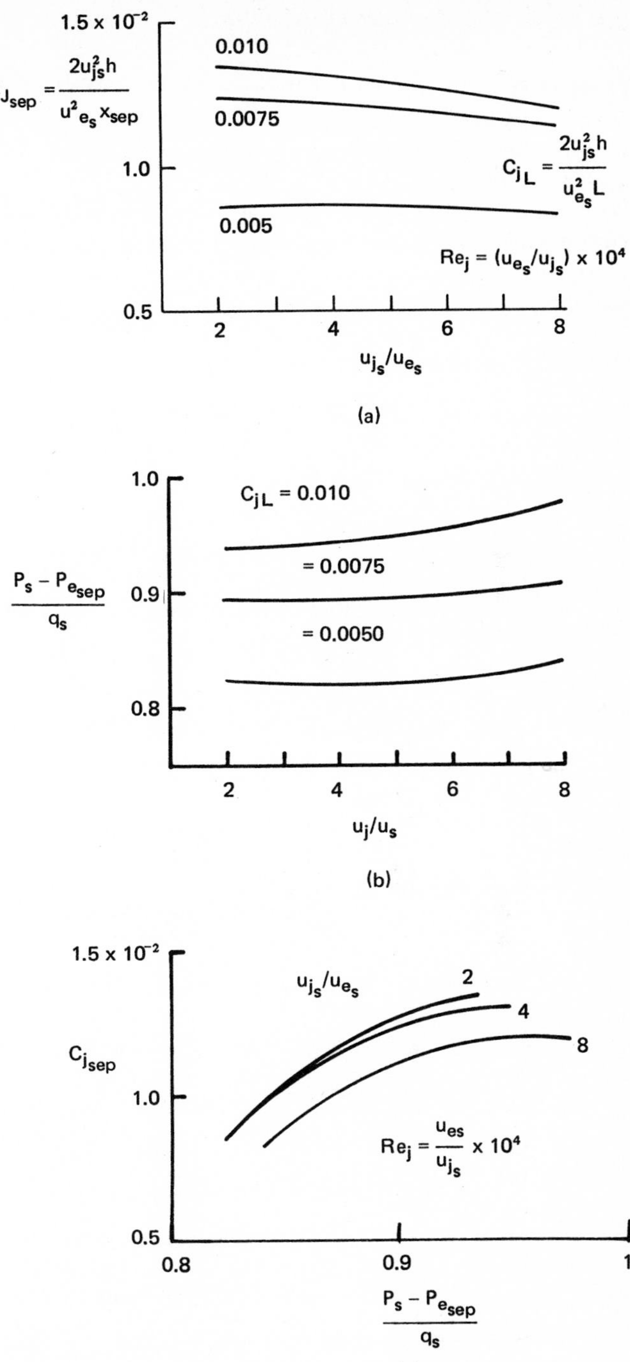

FIG. 5.75 Effect of velocity ratio on separation of a simple wall jet flow: (*a*) on separation length; (*b*) on pressure rise to separation; (*c*) on pressure rise to velocity ratio (*from Hubbartt & Bangert, 1970*).

slot is on the verge of separation, whereas downstream of the slot there is enough mixing with the jet to prevent reverse flow at the minimum velocity point. Primary results of the calculation were obtained from the limiting velocity distribution, the distance from the slot to separation of the wall layer, and the pressure rise from the slot to separation of the wall layer. The corresponding jet momentum and jet velocity are minimum values because under the condition of no boundary layer acceleration, flow reversal is just prevented. The calculations were carried out for the following nominal values:

$$R_{T_2} = 18.2 \quad F_1 = 1.437 \quad H_3 = 4.028$$

$$R_{T_3} = 15.6 \quad F_4 = 0.1565 \quad H_{e_3} = 1.515$$

Figure 5.76 shows the calculated external velocity distribution $u_e(x)$ for typical cases of different momentum ratio $u_j^2 h/u_{e_s}^2 \theta_s$; here θ_s is the value of θ_3 at the slot and velocity ratio u_{j_s}/u_{e_s}. These are the limiting velocity distributions for which the reverse flow is just prevented; they are qualitatively in agreement with those measured over the deflected flaps.

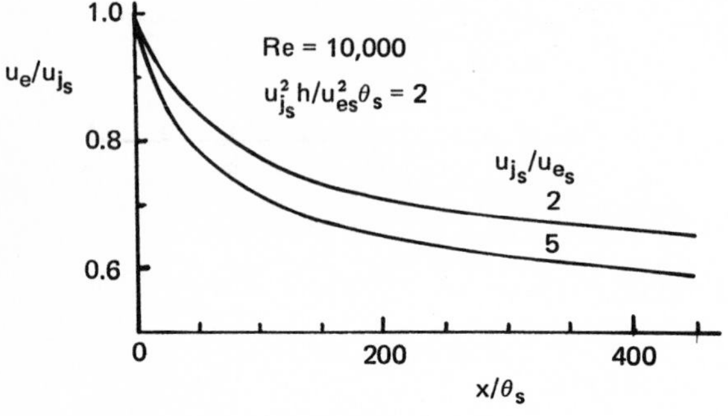

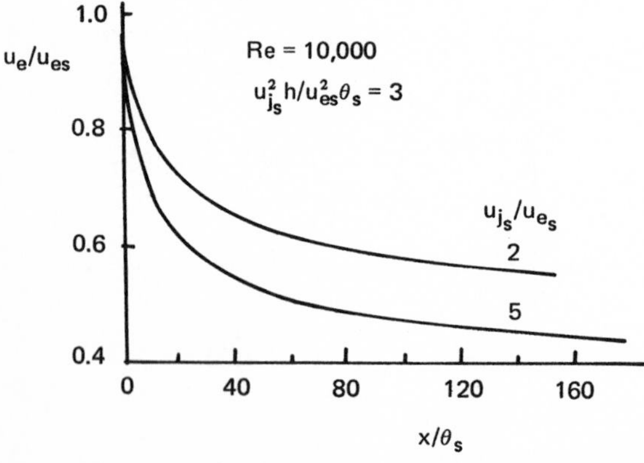

FIG. 5.76 Calculated external velocity distribution for typical cases of different momentum ratio (*from Hubbartt & Bangert, 1970*).

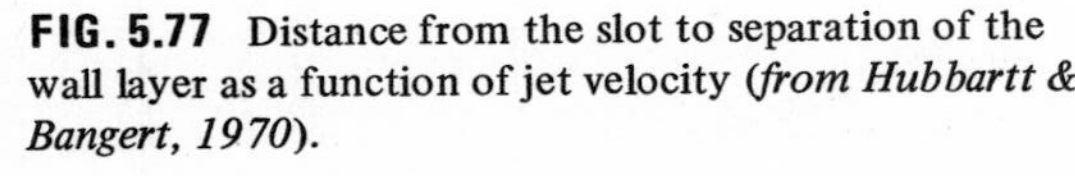
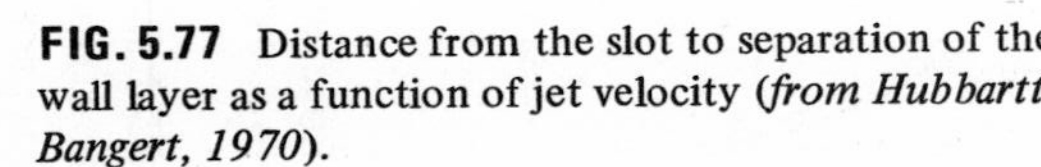

FIG. 5.77 Distance from the slot to separation of the wall layer as a function of jet velocity (*from Hubbartt & Bangert, 1970*).

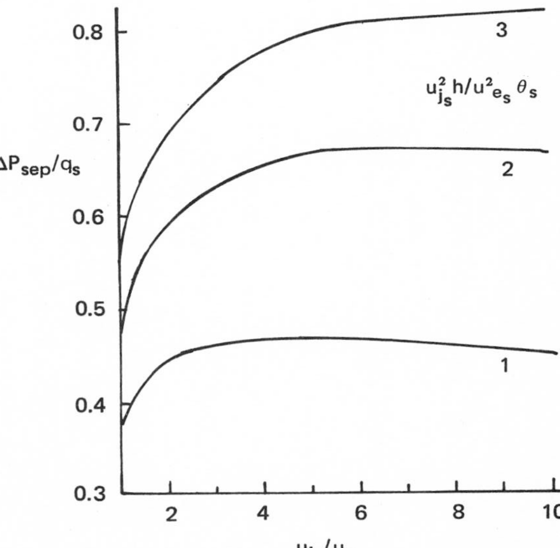

FIG. 5.78 Pressure rise to separation for the same conditions as Fig. 5.77 (*from Hubbartt & Bangert, 1970*).

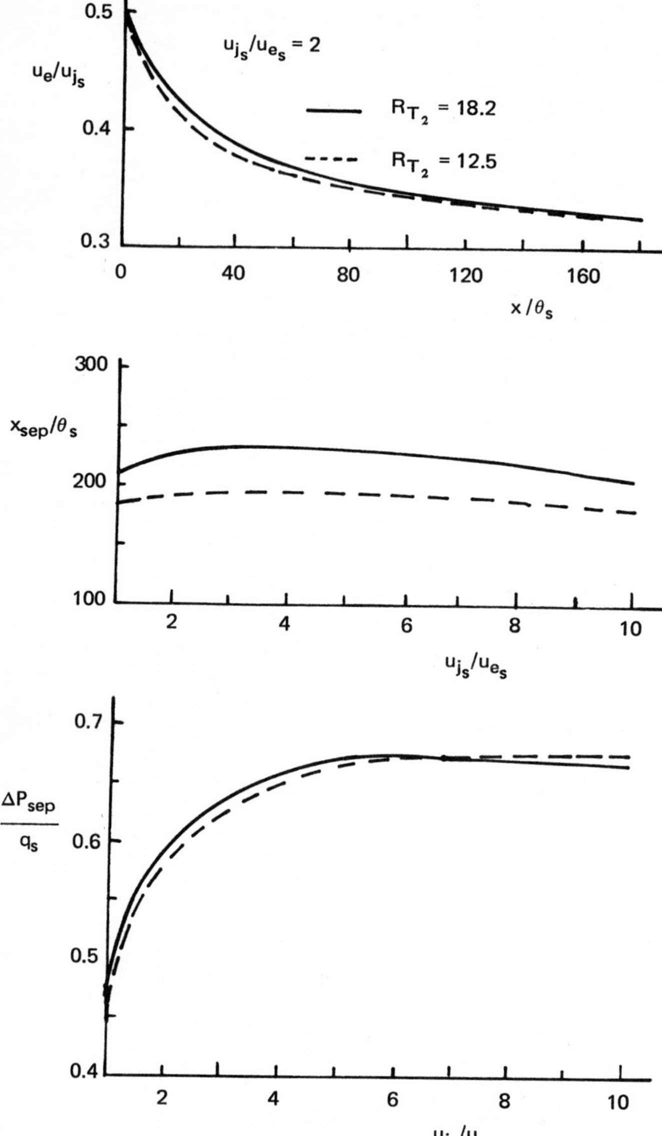

FIG. 5.79 Effect of turbulent Reynolds number R_{T_2} ($u_{js}^2 h/u_{es}^2 \theta_s =$ $2 \cdot q_s = \frac{1}{2} \rho u_{es}^2$; *from Hubbartt & Bangert, 1970*).

Figure 5.77 shows the distance from the slot to separation of the wall layer (x_{sep}/θ_s) as a function of jet velocity. For constant initial u_{es} and θ_s, each of the curves represents constant jet momentum. An increase in jet velocity and a decrease of slot height are indicated in moving from left to right on one of these curves. At a constant jet momentum, computations yield an optimum velocity ratio for maximum x_{sep}/θ_s; this is apparently caused by an interaction of two effects that result from increasing u_{js}: (*a*) the favorable effect on entrainment and (*b*) the adverse effect of more rapid jet decay caused by the increasing wall shear stress.

Figure 5.78 shows $\Delta p_{sep}/q_s$ for the same set of conditions with the same counteracting effects of increasing entrainment and increasing wall shear stress

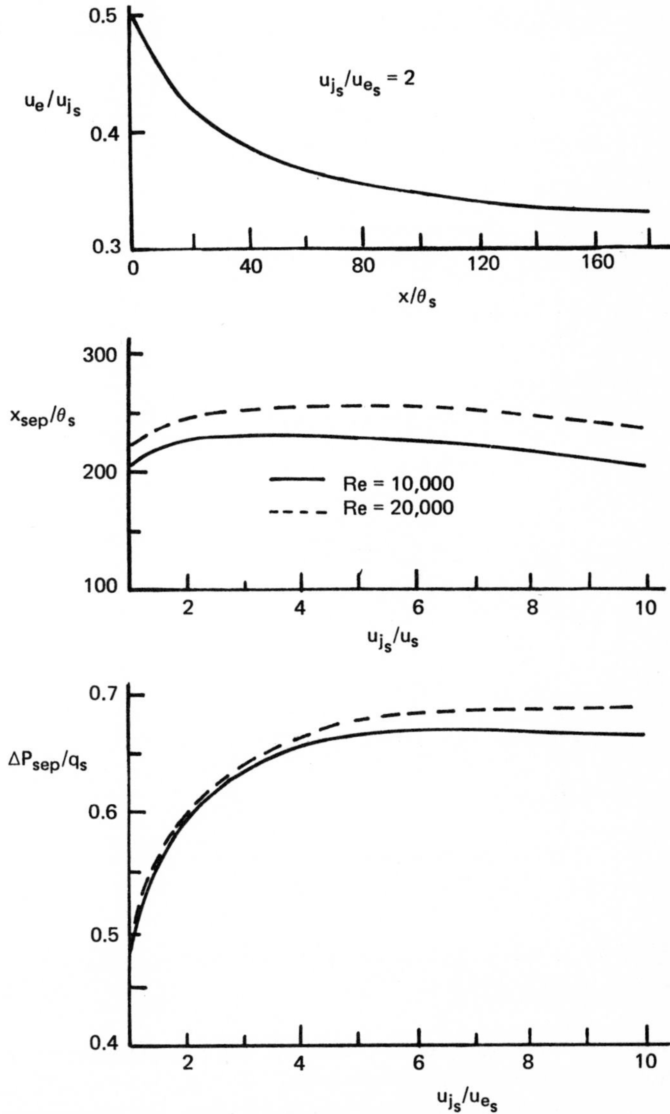

FIG. 5.80 Effect of turbulent stress and dissipation of Reynolds number R_{T_3} ($u_{js}^2/u_{es}^2\,\theta = 2$; *from Hubbartt & Bangert, 1970*).

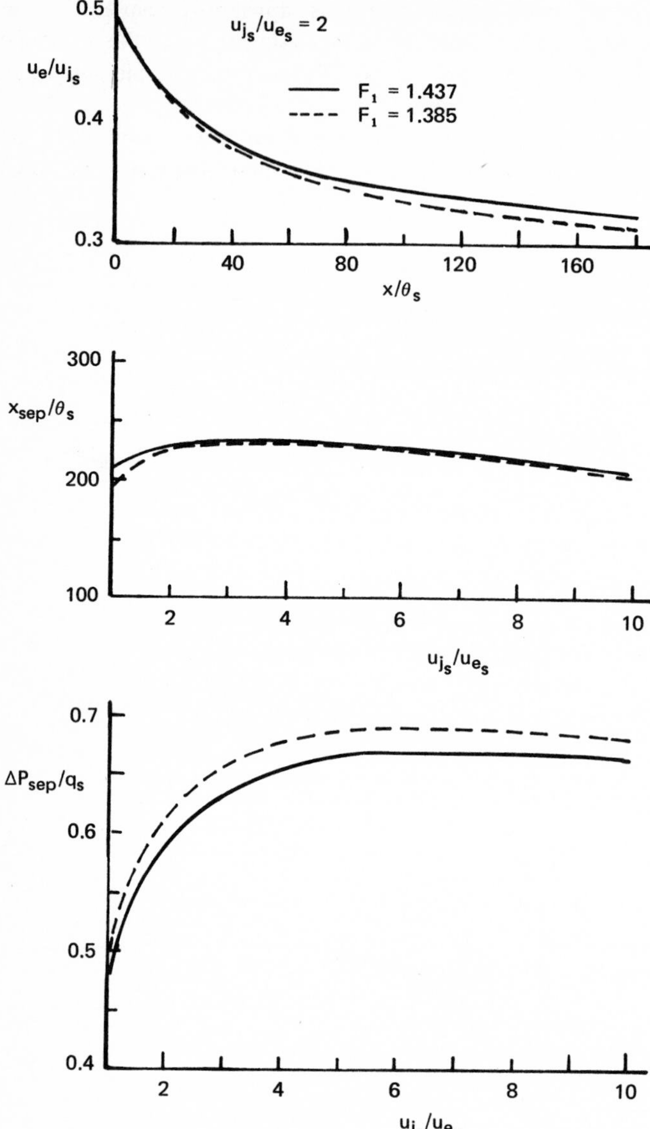

FIG. 5.81 Influence of changes in the jet layer shape ($u_{j_s}{}^2 h/u_{e_s}{}^2 \theta_s =$ 2; *from Hubbartt & Bangert, 1970*).

and with increasing u_{j_s} at constant $u_{j_s}h$. Again there appears to be an optimum velocity ratio.

Figure 5.79 indicates the effect of turbulent Reynolds number R_{T_2} on u_e/u_{j_s}, x_{sep}/θ_s, and $\Delta p_{sep}/q_s$ ($q_s = \frac{1}{2} \rho u_{e_s}{}^2$). Although the proper choice for R_{T_2} may be uncertain, note that the large change from 12.5 to 18.2 has a relatively minor effect on overall flow behavior. Figure 5.80 shows a similar finding for turbulent stress and dissipation R_{T_3}.

Figure 5.81 and 5.82 shows the influence of changes in the jet layer shape and the wake layer shape.

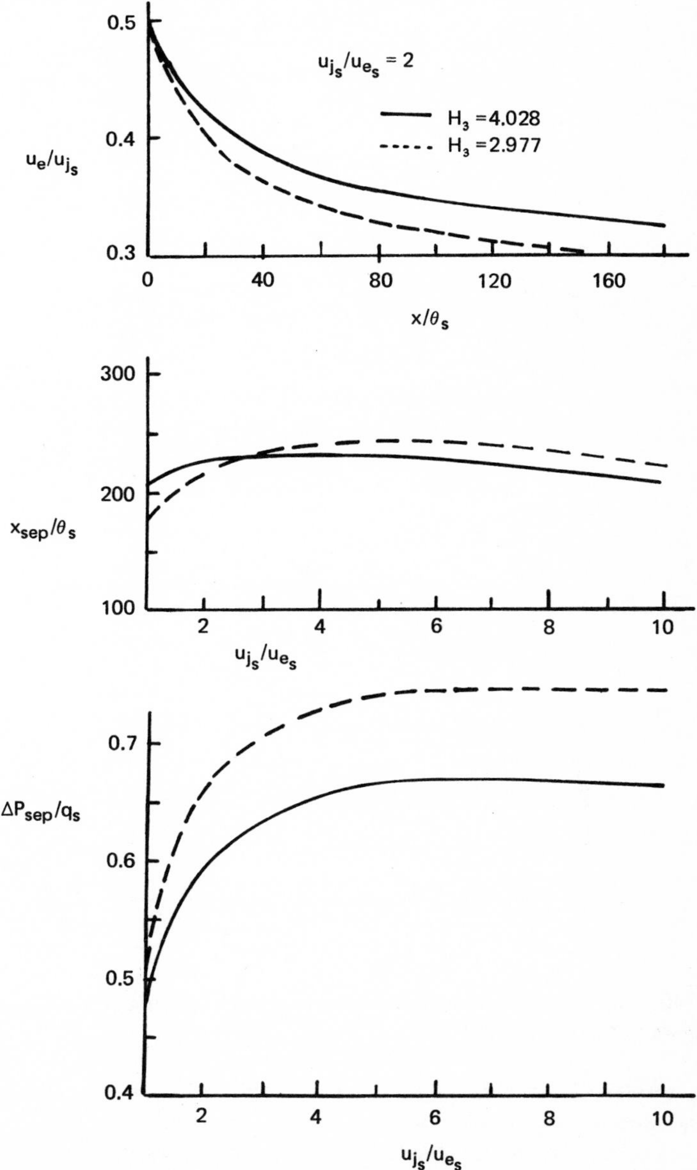

FIG. 5.82 Influence of changes in the wake shape ($u_{js}{}^2 h/u_{es}{}^2 \theta_s = 2$; *from Hubbartt & Bangert, 1970*).

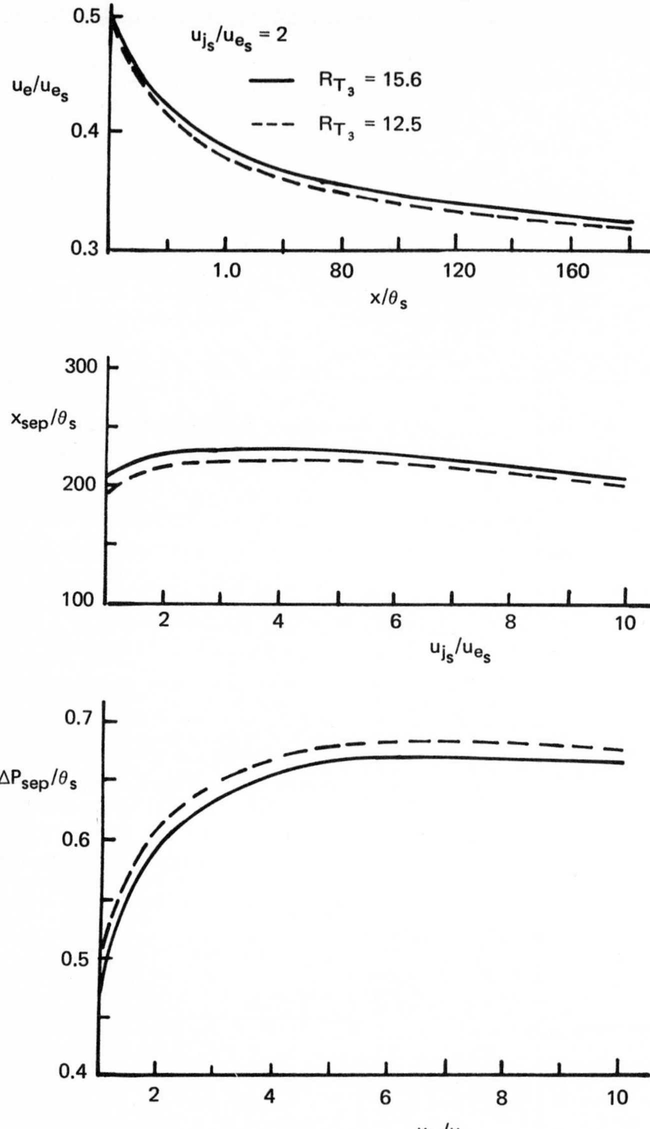

FIG. 5.83 Effect of Reynolds number on the wall jet ($u_{js}^2 h/u_{es}^2 \theta_S = 2$; *from Hubbartt & Bangert, 1970*).

When the velocity profile F_1 is changed from 1.437 to 1.385, a large range is covered for similar solutions, but results (Fig. 5.81) are only weakly dependent on the jet shape. Wall jet behavior is shown in Fig. 5.82 for two widely different values of H_3. The solution for $H_3 = 2.9777$ is about five times larger than the value calculated by Hubbartt and Bangert (1970), although the differences in $\Delta p_{sep}/q_s$ are less than 15 percent. This suggests that the value of $H_3 = 2.9777$ is unrealistically low.

Finally, Fig. 5.83 shows the influence of Reynolds number (defined by $Re = u_{e_s}\theta_s/\nu$) on the wall jet; note the relatively small change in performance for a variation from Re = 10,000 to 20,000.

5.3.4 Examples of Test Results Applicable for Practical Designs

5.3.4.1 Application to a full-scale aircraft

James and Maki (1957) carried out a wind-tunnel investigation of a full-scale aircraft in order to determine the lift effectiveness obtainable with trailing-edge blowing flaps in combination with leading-edge slats. The Reynolds number of 8.2×10^6 was based on a wing mean aerodynamic chord for a high-wing airplane with an aspect ratio 6.75 wing and approximately 36 deg of sweepback. Figure 5.84 presents three views of the test plane.

The investigators showed that at high angles of attack, leading-edge flow separation can be effectively controlled with a leading-edge slat and the effectiveness and the stability of trailing-edge flaps can be maintained. Maximum lift depends on the leading-edge configuration, the trailing-edge flap deflection angle, the blowing momentum, and flow and duct pressure coefficient. With a 55 deg trailing edge flap and a full-span simulated 24 deg slat, there was an increase in the maximum lift coefficient from 2.20 without boundary layer control to 2.54 with a momentum coefficient of 0.012 and a further increase to 2.69 with a momentum coefficient 0.032. For operation over a 50 ft obstacle, these results correspond to a reduction of 13 to 18 percent in takeoff distance and a reduction of 35 percent in landing distance.

Figure 5.85 indicates the increased lift coefficient due to flap deflection at low angles of attack and the momentum coefficients required for flow attachment on the flaps.

The losses in lift and marked increases in stability at angles of attack greater than 6 deg are attributed to inboard flow separation. The effect of increasing the momentum from $C_\mu = 0.012$ to 0.032 is to increase $C_{L\max}$ from 1.78 to 1.94 and to slightly increase the lift curve slope. It is reasoned that further increases in $C_{L\max}$ and maintainence of flap effectiveness to angles of attack greater than 6 deg may be obtained by eliminating the inboard flow separation through inboard slats.

Figure 5.86 indicates the characteristics of the airplane with trailing-edge flaps deflected 55 deg in combination with a simulated full-span slat deflected 24 deg.

With blowing off and a momentum coefficient $C_\mu = 0.012$, the value of $C_{L\max}$ increased from 2.20 to 2.54 and with $C_\mu = 0.032$, it increased to 2.69. Flap effectiveness and stability were maintained up to an angle of attack of about 14 deg.

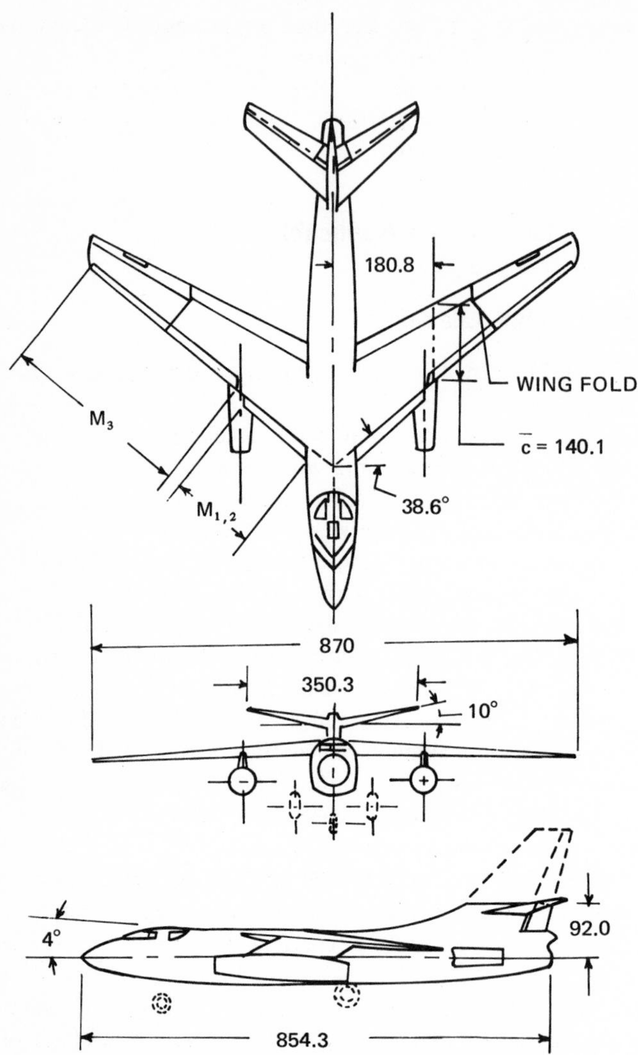

FIG. 5.84 Three views of airplane for the James-Maki tests (all dimensions in inches unless otherwise noted; *from James & Maki, 1957*).

5.3.4.2 Application to a STOL seaplane

Kikuhara and Kasu (1968) investigated blowing over the trailing-edge flaps and control surface of the Japanese STOL seaplane and successfully applied the design to a prototype seaplane with improved takeoff and landing characteristics on open oceans. The main feature of their system is the adoption of relatively low pressure blowing, which necessitates taking larger

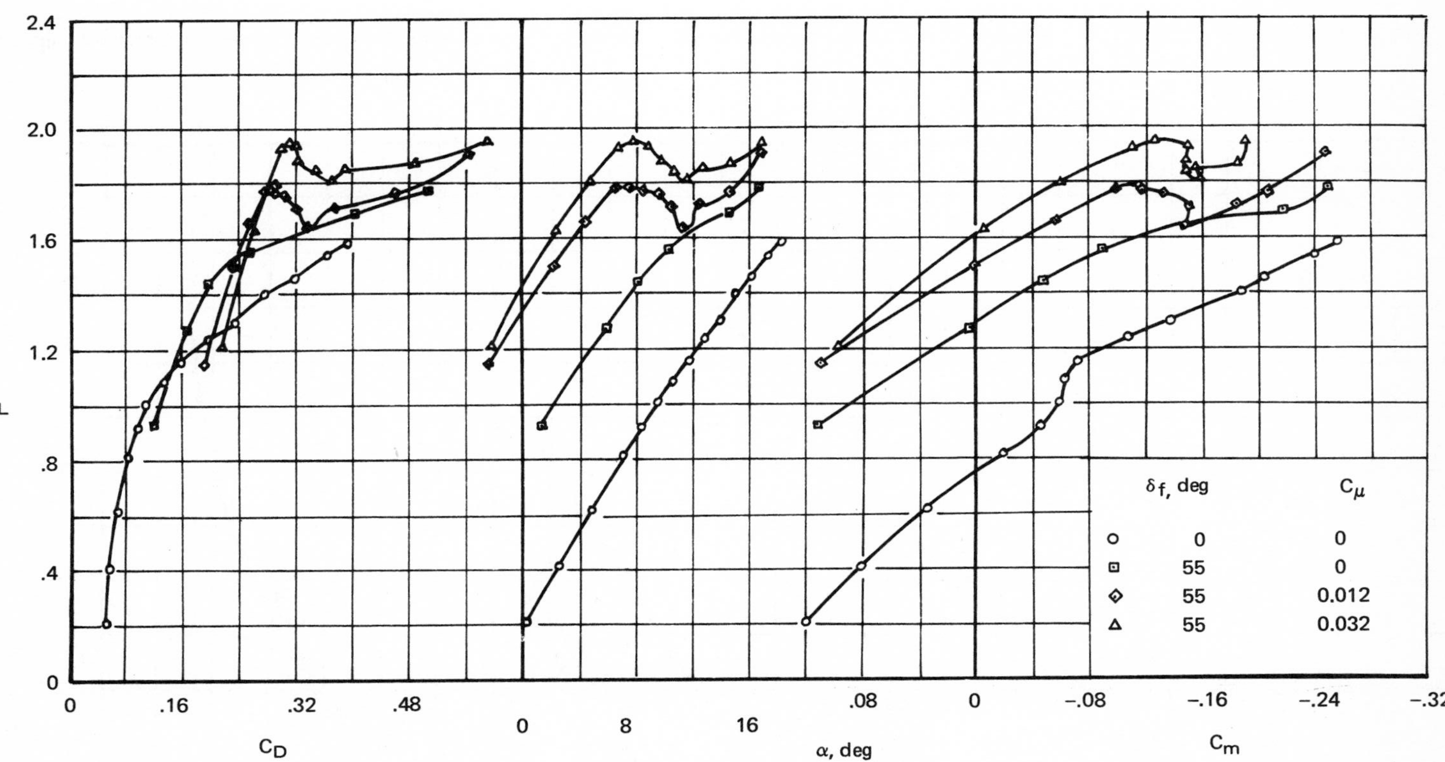

FIG. 5.85 Longitudinal characteristics of the basic configuration with and without blowing (normal airplane slat extended; *James & Maki, 1957*).

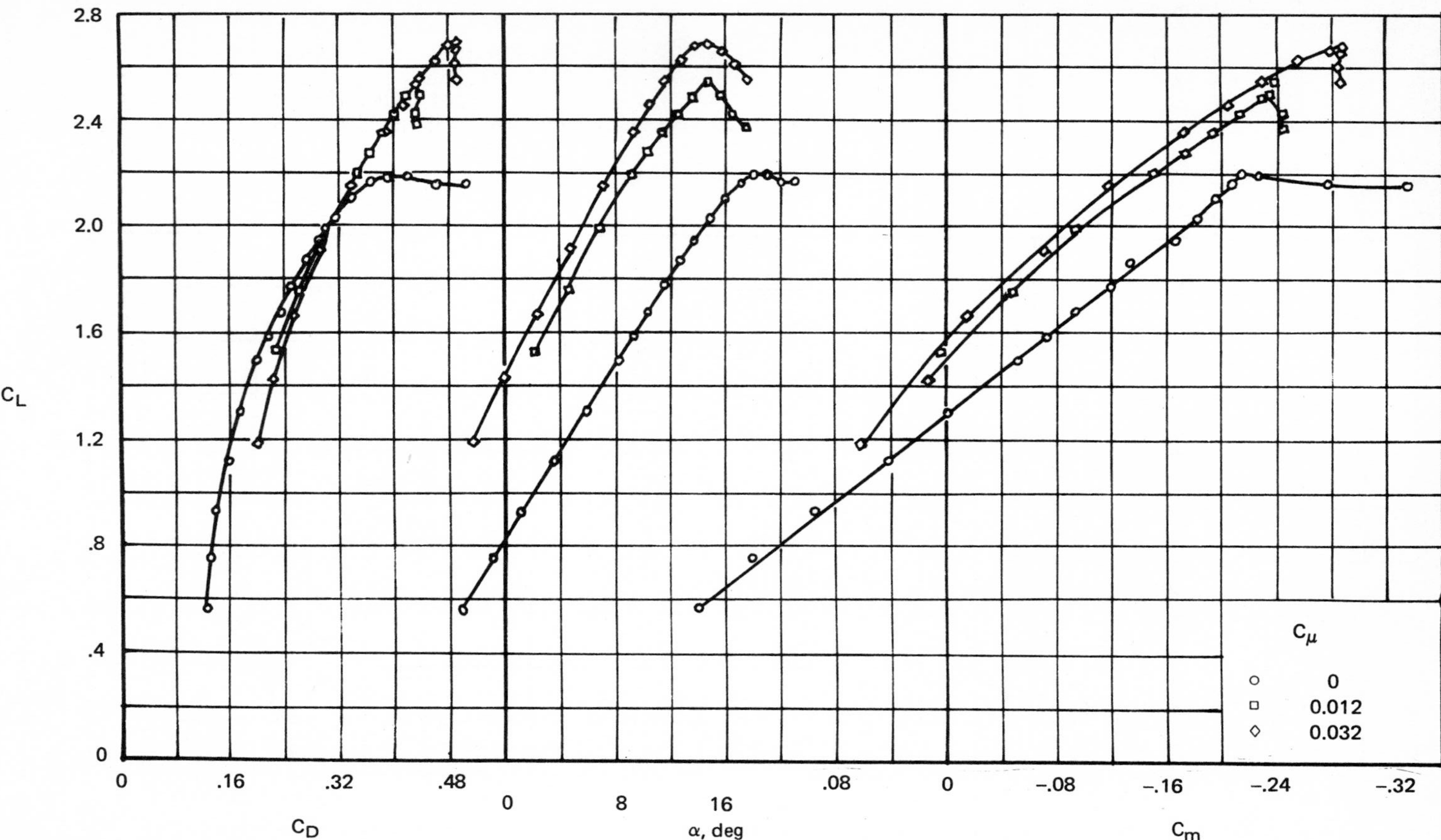

FIG. 5.86 Longitudinal characteristics of the airplane with trailing edge flaps deflected 55 deg in combination with a full-span simulated slat deflected 24 deg (the symbol M_2 designates an inboard slat and M_3 a modified normal slat with a removable glove; *from James & Maki, 1957*).

diameter air ducts into consideration. This has merit, since small horsepower is needed, and no special concern need be given to the thermal deformation of adjacent structures. Furthermore, the system has been designed in such a way that the largest possible value of C_μ may be obtained for a given horsepower but little increase is obtained at C_μ values greater than $C_{\mu A}$, the momentum coefficient required to prevent the separation over the flap. Here $C_{\mu A}$ corresponds to a minimum speed of 45 knots which is used for takeoff and landing. The investigators first estimated the minimum horsepower for $C_{\mu A}$ for various parameters and then they chose a minimum-diameter duct system for the selected engine. Next they tested two kinds of blowing, flap blowing (i.e., blowing from a nozzle located on the upper surface of the flap) and shroud blowing (i.e., blowing from a nozzle located in the wing shroud just ahead of the flap).

The general geometric dimensions of their model (Fig. 5.87) were as follows:

Wing span	1.5 m (5 ft)
Wing chord	0.25 m (10 in.)
Aspect ratio	6
Wing area	0.375 m^2 (4.04 ft^2)
Net wing area for blowing	0.34 m^2 (3.66 ft^2)
Wing section	NACA 23018
Nozzle throat width	0.25 mm (0.01 in.)
Ratio of nozzle throat width to wing chord	1/1000
Ratio of flap chord to wing chord	0.25

The blowing air was applied through a dummy fuselage from a specially devised blowing balance in the wing. The gap between the wing itself and the flap was adjustable from 1 to 4 mm. A gap of 1 mm on the model is equivalent to a gap of 20 mm for a wing with a 5-mm chord length. For the test of leading edge blowing, the leading edge of the wing could be replaced by three variations of nose droop as well as leading edge slat and a modified wing section similar to a NACA 75018.

Their comparison of the two types of blowing (Fig. 5.88) demonstrated the superiority of flap blowing at larger flap deflections. The lift coefficient and the blowing momentum coefficient shown there are local values based on the blowing area, which was about 90 percent of the gross wing area. At a flap angle of 40 deg, there was little apparent difference between the types, but at a flap angle of 60 deg, flap blowing had a slightly larger C_L at the same incidence and a 5-deg lower stalling incidence at the same values of $C_{L\max}$. For a flap deflection of 80 deg, flap blowing attained much higher values of C_L and $C_{L\max}$ and a stalling incidence about 8 deg lower.

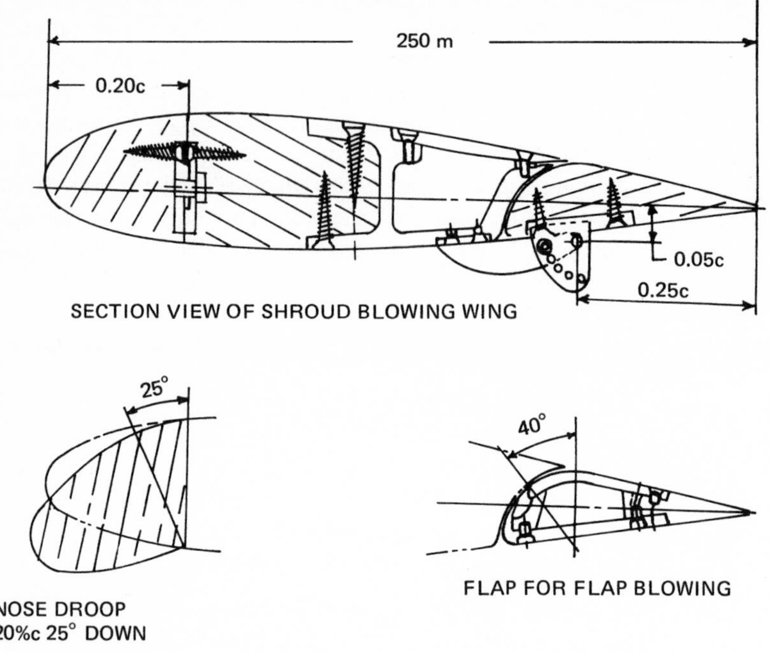

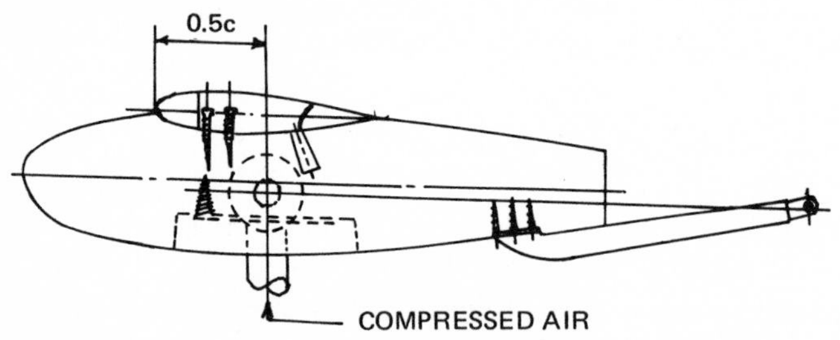

FIG. 5.87 Blowing model and nozzle, rectangular wing (*from Kikuhara & Kasu, 1968*).

The superiority of flap blowing is again demonstrated in Fig. 5.89, which shows the values of C_L obtained by the two types. It appears that the inferiority of shroud blowing may be due to the gap between the wing and the flap. Note in Fig. 5.90 the effect of the gap on stalling incidence. The increase in lift attained by shroud blowing was largely lost when the gap was increased, whereas with flap blowing, an increase in the gap tended to increase the stalling incidence. Flap blowing also gives a superior performance in the presence of a propeller slipstream; see Fig. 5.91.

In practice, the design of the nozzle throat width is an important factor because it is directly related to momentum coefficient C_μ by

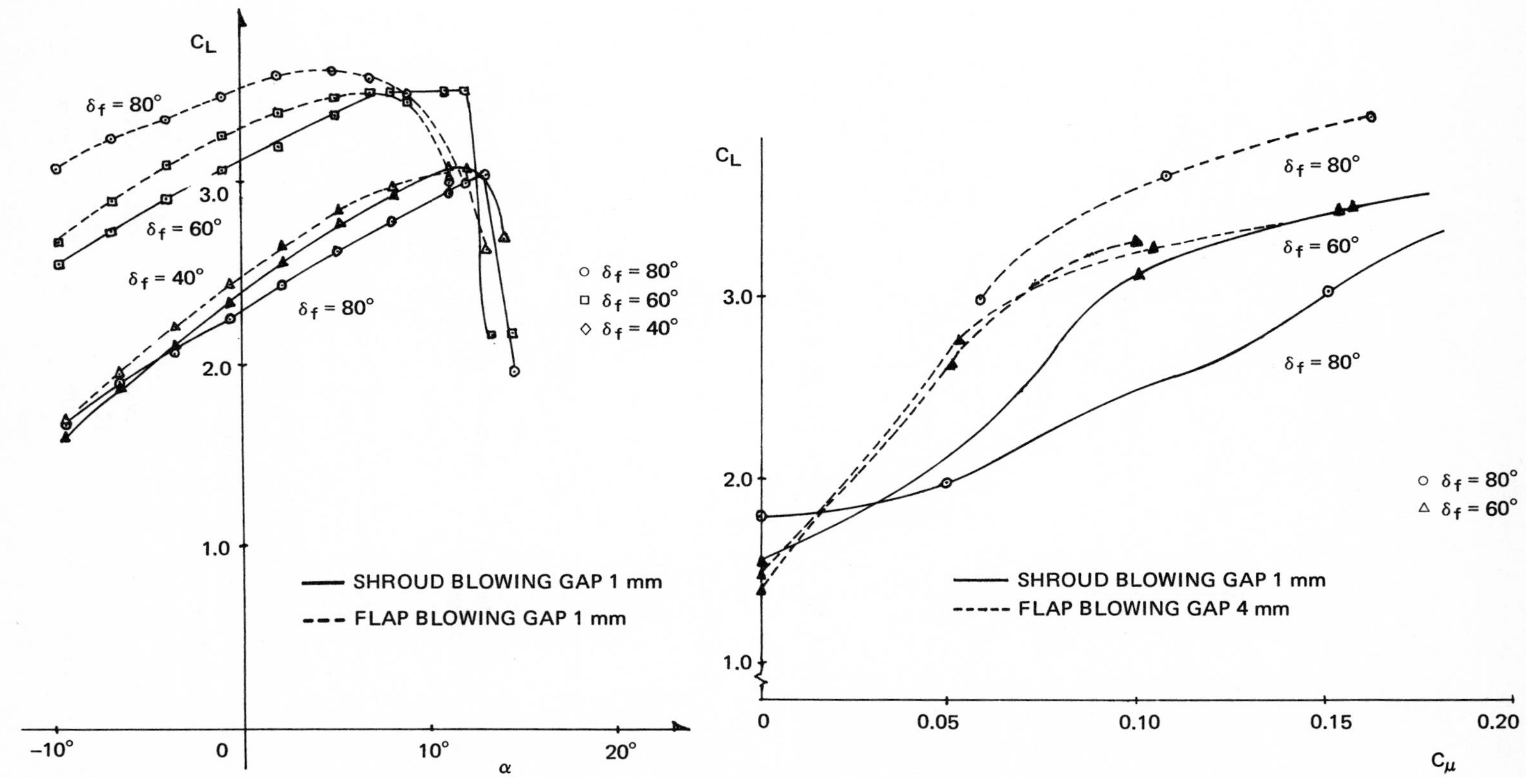

FIG. 5.88 Comparative effectiveness of the two types of blowing for various flap deflection angles (rectangular wing, $C_\mu = 0.10$; *from Kikuhara & Kasu, 1968*).

FIG. 5.89 Comparative effectiveness of the two types of blowing on lift at momentum coefficients below $C_{\mu A}$ (rectangular wing, $\alpha = 0$ deg; *from Kikuhara & Kasu, 1968*).

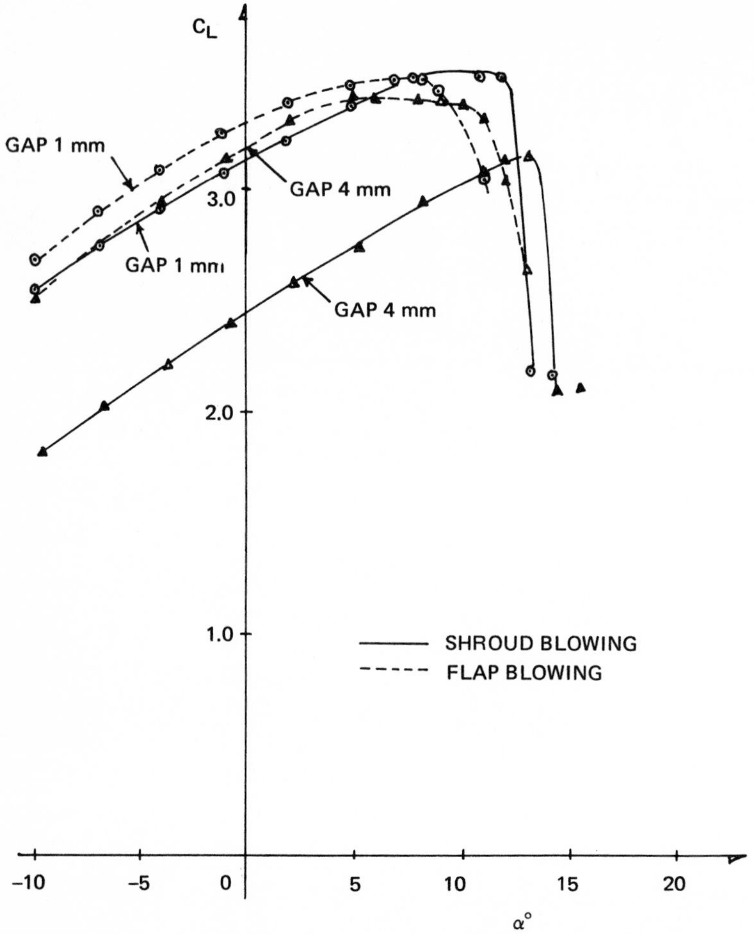

FIG. 5.90 Effect on stalling incidence of gap between wing and flap for the two types of blowing (rectangular wing, $C_\mu = 0.10$, $\delta_f = 60$ deg; *from Kikuhara & Kasu, 1968*).

$$C_\mu = \frac{m}{q}\frac{u_j}{s} = 2\left(\frac{A_t}{s}\right)\left(\frac{u_j^2}{u_\infty}\right) = 2k\left(\frac{W_t}{c}\right)\left(\frac{u_j^2}{u_\infty}\right)$$

where A_t is the area of the nozzle,
W_t is the nozzle throat width, and
k is a constant.

The thickness of the jet or the width of the nozzle throat should be thinner than the thickness of the boundary layer over the nozzle. Tolhurst (1958) found that if W_t/c is below 0.0012, C_μ is useful as a correlating parameter. Therefore, in consideration of manufacturing tolerance, the maximum value of

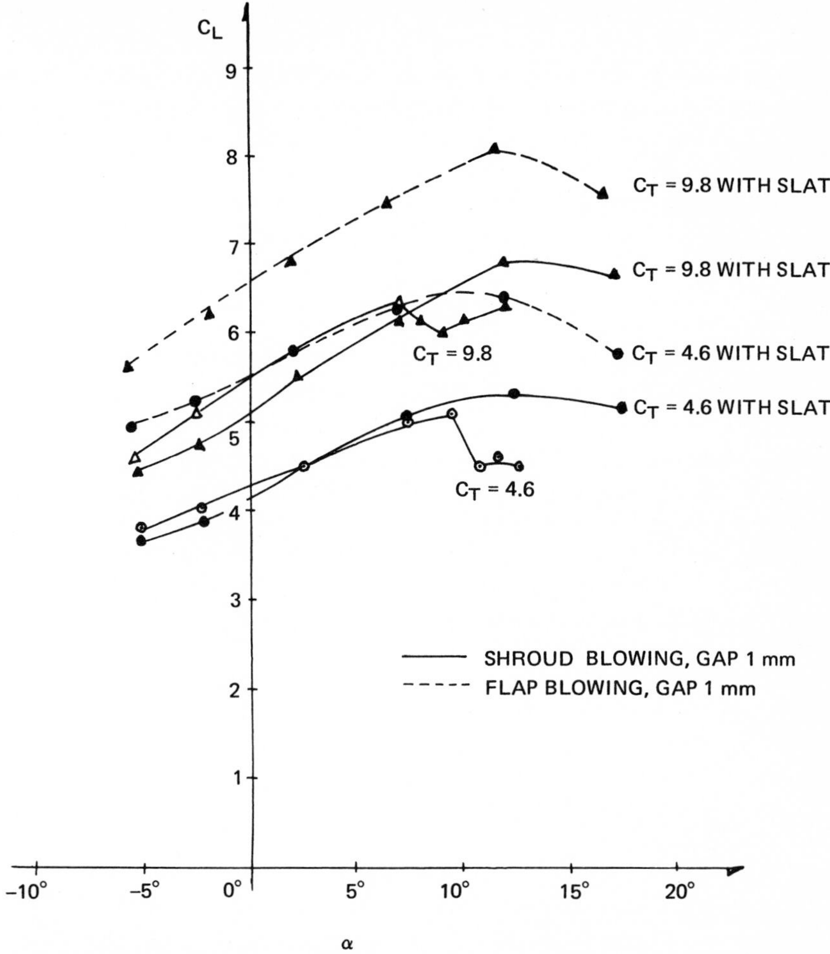

FIG. 5.91 Comparative effectiveness of the two types of blowing in the presence of a propeller slipstream (rectangular wing, $C_\mu = 0.2$, $V = 8.5$ m/sec (27.9 ft/sec), and $\delta_f = 80$ deg, C_T is propeller thrust coefficient based on disk area; *from Kikuhara & Kasu, 1968*).

W_t/c on the duct system was limited to 0.001. From the practical standpoint, it is desirable to have u_j more than ten times the free-stream velocity; this limits the selection of the value of W_t/c. Duct loss represents a third limitation on the nozzle slot area because a large air flow in the duct causes additional pressure losses, thus requiring high compressor pressure and more power.

5.3.5 Blowing into Supersonic Flow

Blowing of gas flow may affect the heat transfer and pressure distribution in the supersonic flow regime involving aerodynamic heating. With no mass injection, the heat transfer rate may be reduced by almost a factor of two if a laminar boundary layer is replaced by a laminar cavity flow. Thus this technique may be of interest for the protection of hypersonic vehicles against heat transfer. In the vicinity of cavity reattachment, local heat transfer coefficients are several times larger than the attached flow values, but these high rates decay to attached flow values downstream in approximately one cavity length. When integration is extended to include the downstream surface as well as the cavity, then the high heat transfer rates downstream of reattachment nullify almost all the reduction obtained in the separated flow region itself and laminar cavity flow offers no advantages for reducing overall heat transfer rate.

Nicoll (1965) conducted an experimental investigation on the effect of injecting helium into a hypersonic laminar cavity flow at free-stream Mach 11. He found that a mass injection approximately equal to 10 percent of the attached boundary layer flow rate was capable of removing the pressure peak at reattachment and of producing a flow with an almost constant pressure within and downstream of the cavity. Furthermore, such mass injection had a pronounced effect on the heat transfer rate. In the vicinity of reattachment, the rate was reduced by as much as a factor of six, and up to about one cavity length downstream of the reattachment shoulder it was reduced by as much as a factor of three.

Nicoll varied the mass injection rate and the height of the separation shoulder relative to the reattachment shoulder in order to determine the effects of those parameters on the distribution of the pressure and the heat transfer rate. The mass injection appeared to have no appreciable effect on transition Reynolds number. When the separation shoulder was raised relative to the reattachment shoulder, the effect on cavity floor pressure was about the same as the effect of mass injection; the two effects were virtually independent of one another. However, in the immediate vicinity of and downstream of the reattachment shoulder, flow was primarily controlled by the mass injection rate. Several of the configurations studied closely simulated the theoretical flow model of Chapman (1956), and integration of the local heat transfer coefficient for these configurations yielded values of the overall heat transfer to the cavity surfaces which were in excellent agreement with the Chapman theory.

The experimental study with helium injection conducted by Nicoll in 1965 utilized the model shown in Fig. 5.92. It was identical to one of the configurations he used in 1963, namely an annular cavity with a length-to-depth ratio of 5 mounted on the surface of a 20 deg total angle cone. He made only one geometric change in adapting the 1963 configuration for the mass injection studies, namely, the separation shoulder was raised above the basic cone surface by an amount of ϵ_s. Broadly speaking, this change provided clearance for the mass injected into the cavity to pass downstream. Two parameters, ϵ_s and mass injection rate m_i, were varied in the experiment.

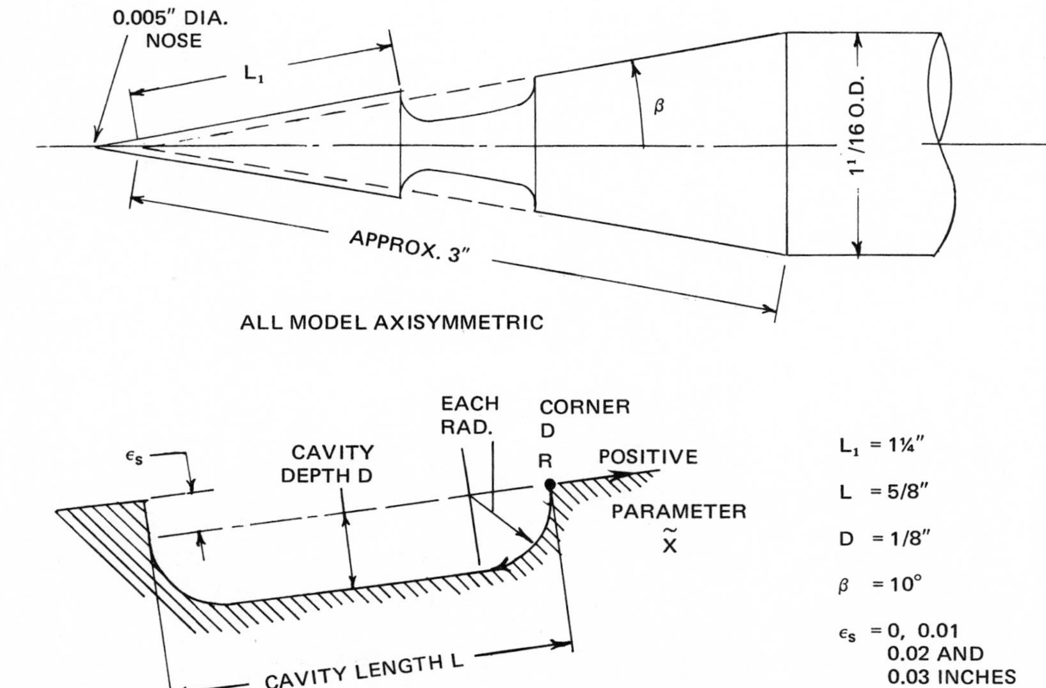

FIG. 5.92 Wind-tunnel geometry for cavity injection study (*from Nicoll, 1965*).

The following geometric and fluid-dynamic variables were held constant:

for injected helium:

$$\mathrm{Pr} = 0.688 \qquad \gamma = \frac{5}{3}$$

$$p_0 = 400 \text{ psia} \qquad T_0 = 535 \text{ deg } R$$

for the flow:

$$\mathrm{M}_\infty = 11 \qquad \mathrm{M}_{\text{cone}} = 6.46 \qquad \text{angle of attack} = 0 \text{ deg}$$

$$(\mathrm{Re}_{l_s})_{\epsilon_s=0} = 0.737 \times 10^6$$

$$0.82 < \frac{T_w}{T_0} < 1.08$$

The Reynolds number given is based on the conditions at the edge of the boundary layer on a pure cone of 20 deg total angle and the characteristic length $L_1 = 1\frac{1}{4}$ in. The model wall temperature was fairly close to the stagnation temperature, thus avoiding the normal hypersonic complication of variable fluid properties. The model nose diameter was chosen so that the bluntness effects were negligible. Two models, A and B, were used for the investigation. Model A refers to the cavity configuration with $\epsilon_s = 0$ and Model B had detachable noses with ϵ_s values of 0, 0.01, 0.02, and 0.03 in.

The "carpet plot" shown as Fig. 5.93 indicates the measured pressure at the midpoint of the cavity floor for a moderate range of both mass injection rate m_i and ϵ_s/L (L is cavity length). Each unit of the abscissa represents one μslug/sec in m_i or 0.016 in ϵ_c/L.

Note that the floor pressure rose with an increase in m_i, but dropped with an increase in ϵ_s/L. The network of Fig. 5.93 consists approximately of two sets of straight parallel lines, and a particular floor pressure can be determined by combining m_i and ϵ_s/L. The heat transfer rates were measured at three stations; Fig. 5.94 indicates those at $\bar{x}/L = 0.4$ ($\bar{x}$ is the wetted length along a generator of the axisymmetric cavity model with origin at the reattachment point). Data are presented for three values of ϵ_s/L and for mass injection rates up to 10 μslug/sec.

It is clearly seen that mass injection strongly reduces the downstream heat transfer rates, at least within one cavity length from the reattachment shoulder. For $m_i = 5$ μslug/sec, the reduction was about a factor of two. Increasing ϵ_s/L again appears to reduce the heat transfer rates, but the experimental inaccuracy allowed a less positive conclusion for downstream stations.

Experiments by Peake (1968) indicated that at a mainstream Mach 1.8 the wall jet blowing is capable of preventing turbulent flow separation of a

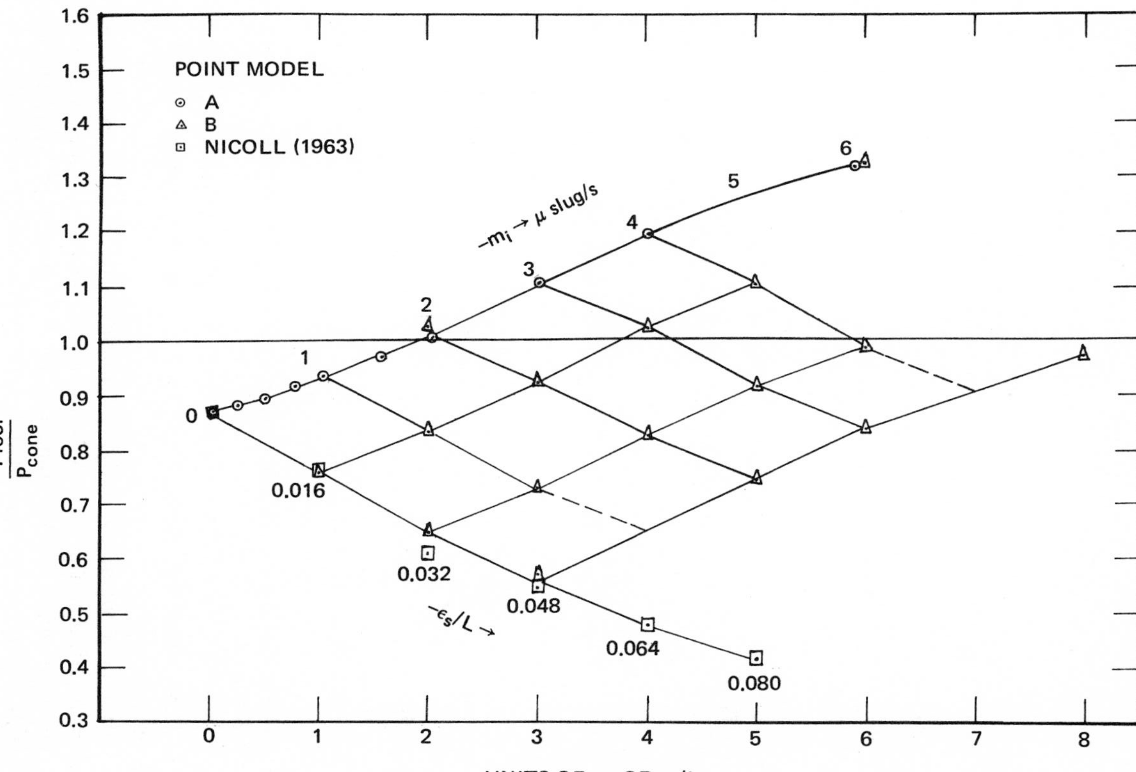

FIG. 5.93 Measured pressure at the midpoint of the cavity floor for a range of m_i and ϵ_s/L values (*from Nicoll, 1965*).

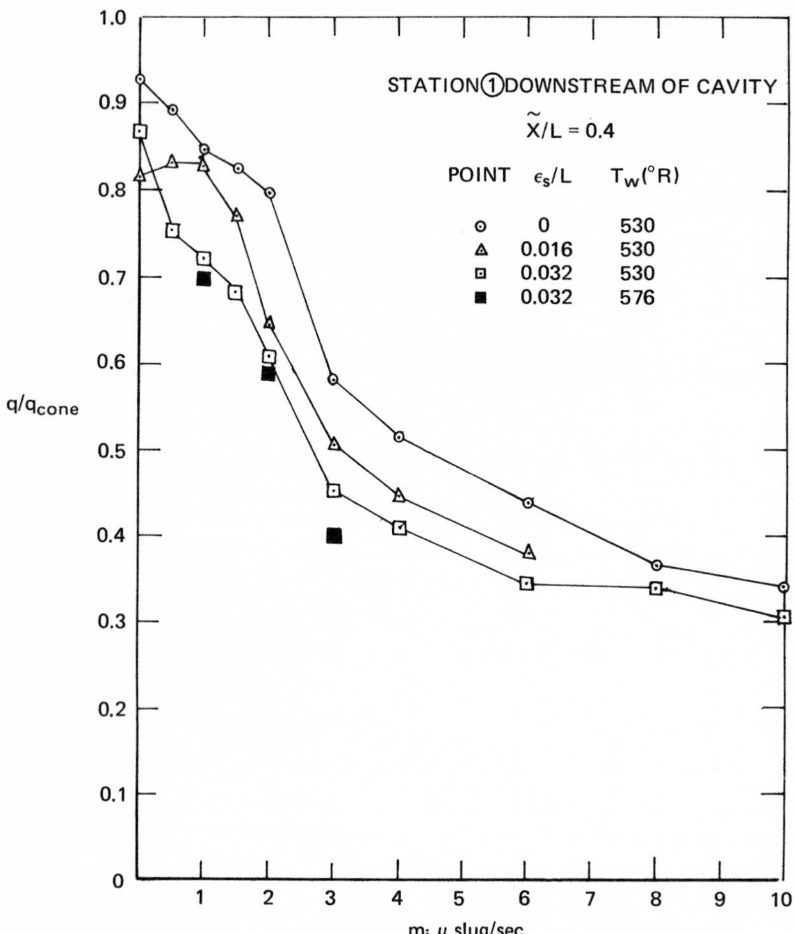

FIG. 5.94 Effect of mass injection on heat transfer rate at a downstream station (*from Nicoll, 1965*).

supersonic flow with large pressure rise upstream of the pressure rise at a Reynolds number of 4 million per foot.

The control of a supersonic turbulent boundary layer separation can be important in practical applications. In supersonic air intake, e.g., the boundary layer in the supersonic and transonic regions is subjected to steep adverse pressure gradients; these are generated by the intake shock system and can cause separation. The favored method of controlling such separation is to bleed away the low energy air before or at incipient separation. However, it may be difficult to design and position the bleed slots for ducting away the low energy air; in addition, the resultant loss of intake mass flow often means a significant loss of performance associated with the bleed flow itself.

5.4 SEPARATION CONTROL BY COMBINED SUCTION AND BLOWING

As already indicated, either suction or blowing can be effective in preventing separation. Each can result in a higher lift coefficient for an aircraft wing. Therefore, it appears possible to combine suction and blowing and attain optimum performance for an aircraft through boundary layer control; in this way, pumping power is not wasted but is usefully employed. Aircraft parameters such as the ratio of blown to the sucked area or relative blowing slot width are then involved in the optimization.

However, the technology for boundary layer control by blowing and suction is still in the process of development.

The idea of combined suction and blowing for controlling boundary layer was applied first by Arado Aircraft in Germany. Combined suction and blowing over the entire wing span was applied to two large BLC prototypes, the Arado 232 and Dornier 24, which were built between 1940 and 1945. Later ONERA applied chordwise combined suction and blowing. In the United States, a combined system enabled the XC-123D aircraft to achieve $C_{L\max} = 4.8$, and a number of single-engine lightweight Cessna aircraft built after World War II employed various combinations of pumps, blowers, and jet pumps. Samples of both spanwise and chordwise combined systems evolved by Wagner (1961) for Fairchild Aircraft are shown in Figs. 5.95 and 5.96, respectively.

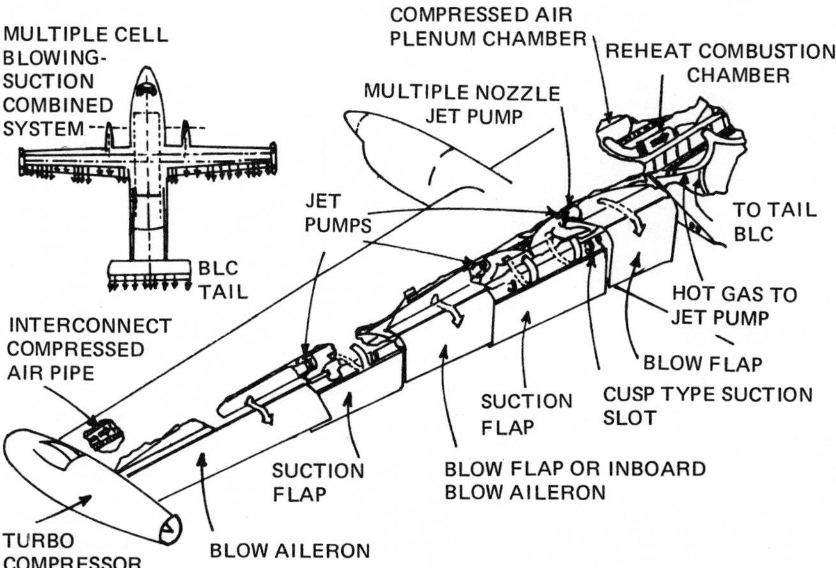

FIG. 5.95 Layout of spanwise system for combined suction and blowing (developed for Fairchild aircraft by *Wagner, 1961*).

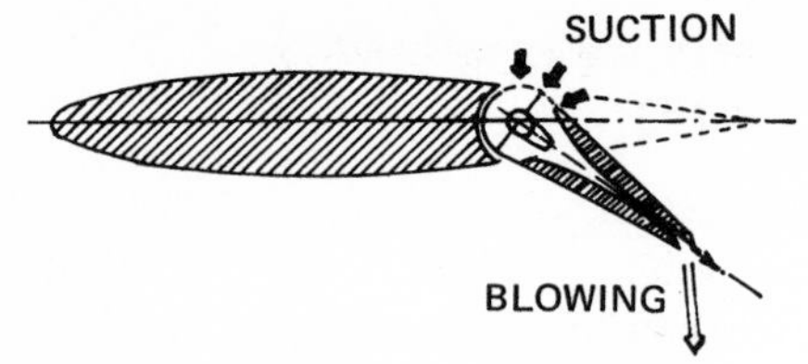

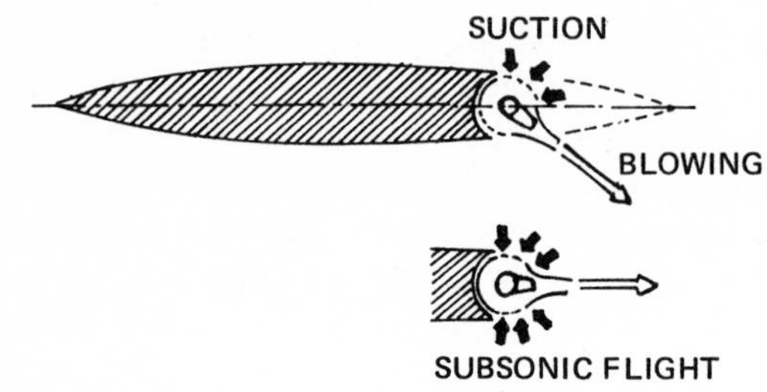

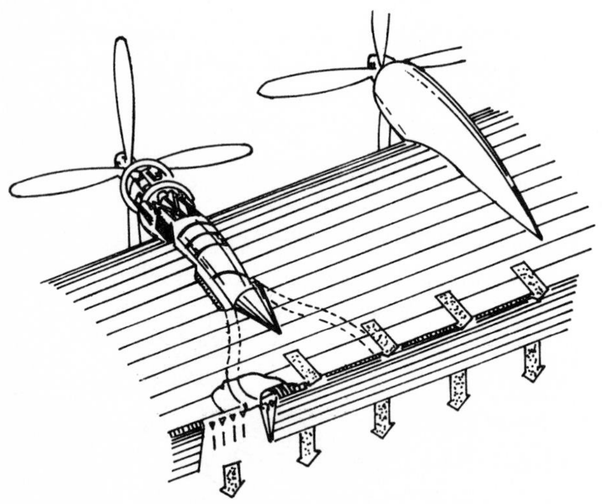

FIG. 5.96 Chordwise system for combined blowing suction airfoils designed by Wagner for practical applications (the use of a jet pump in the flap frees the airfoil from any spanwise duct; such airfoils can be developed as integrated lifting-thrusting systems; sources: *Malavard et al., 1956, and Wagner, 1961*).

Further development may well indicate the superiority of a chordwise arrangement over the spanwise system. Nevertheless, even the spanwise system of combined suction and blowing is far superior to other BLC systems, at least for the type of subsonic transport under consideration.

Simple power calculations show that the power consumption is larger for either a suction system or a blowing system than for a combined system. In the combined system, the air that is sucked in at one position of the span or the chord has to be blown out at another position. Matching of the quantities of air to be sucked in and blown out can be achieved by proper selection of blowing slot width and the ratio of sucked to blown wing area. In the simple case where the air volume flux is half the flux of an all-suction or an all-blown system, the dynamic pressure in the same available duct cross section is reduced to one-fourth. Half the airflow at one quarter of the dynamic duct pressure may result in an 87.5 percent saving in duct pump power.

Duct losses increase with the duct dynamic pressure; therefore it is possible to further reduce these losses by $(1/N)^2$ where the BLC air is divided into two or more spanwise portions in a multiple-cell system; here N is the number of cells. The actual power required is even smaller; the reduction in function to $(1/N)^{2.2}$ achieved for several designs was mainly due to shorter ducts. In an optimized system (total pump power is minimum), the duct losses cause a major portion of the total pump power required. The Wagner (1961) presentation contains additional details on required pumping power.

Interconnected pumps are required for the safe operation of spanwise combined blowing and suction. Pumps independent of the propulsion engines enhance safety and provide improved takeoff and landing. On the other hand, chordwise combined systems blow out and suck in at different chordwise stations or on the upper and lower side of the airfoil. They can be classified as multiple-cell arrangements with an "infinitely" large number of cells. Such systems have no duct losses due to spanwise flow or to turning from the chordwise to the spanwise direction. The internal losses include inlet (suction slot) losses, chordwise duct losses, and exit (blowing slot) losses. The internal losses are very small when slots are well designed and the passage from the suction to the blowing slot is kept very short.

Another arrangement of a chordwise combined system was that applied by Breguet Vultur to flaps at two bends; see Fig. 5.97. This chordwise system applies suction at the trailing edge in connection with blowing over a plain flap. As illustrated by the smoke tunnel flow visualization in Fig. 5.98, the effect of combined suction and blowing for this system is to shift the forward

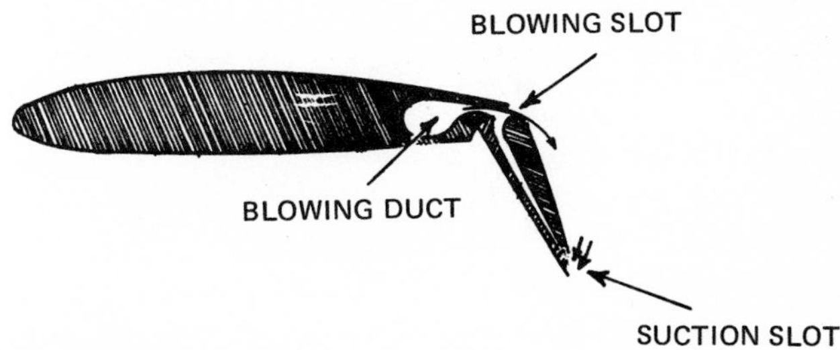

FIG. 5.97 Arrangement of a chordwise combined system ("recirculation" flap) proposed by H. B. Helmbold for Breguet Vultur; trailing edge suction is induced by conventional blowing *(from Wagner, 1961)*.

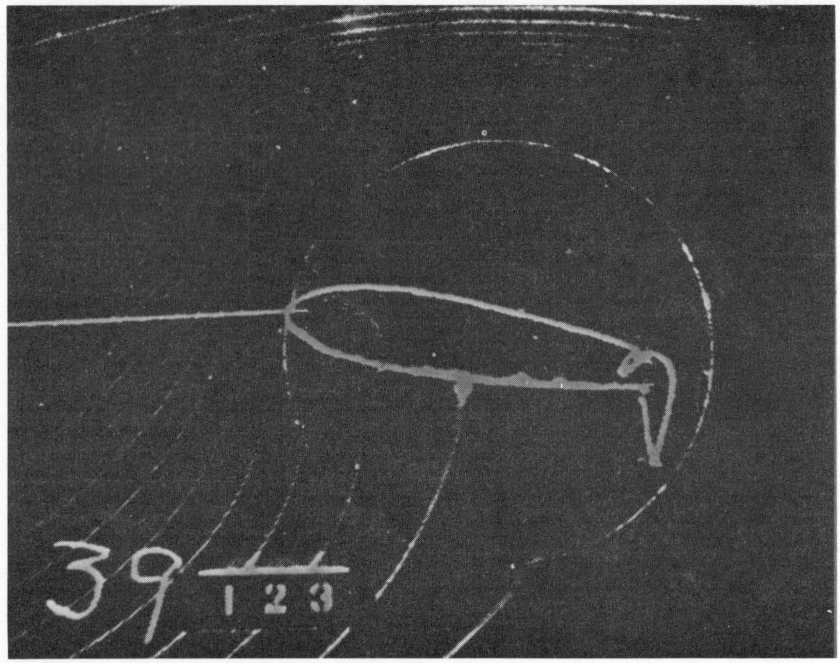

FIG. 5.98 Two-dimensional smoke-tunnel visualization of the "recirculating" flap airfoil (note the location of the forward stagnation point far downstream on the lower side of the airfoil; test performed by Fairchild Aircraft; *from Wagner, 1961*).

stagnation point far downstream on the lower side of the airfoil. Furthermore, this configuration produces higher lift than the conventional blowing wing at the same values of C_μ and the effect of trailing edge suction helps in attaching the jet to the flap. The "recirculating" suction air appears to reduce the friction losses over the flap.

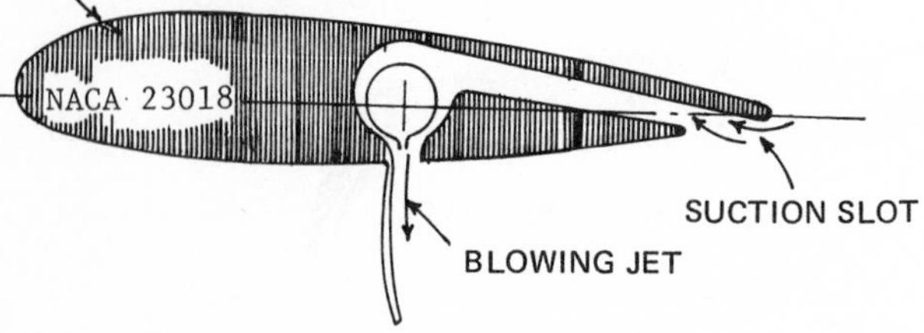

FIG. 5.99 Basic configuration for the TESABOF airfoil chordwise combined system (the basic airfoil is the NACA 23018; trailing edge suction is induced by jet and blowing is over a highly deflected midchord split flap; *from Wagner, 1961*).

FIG. 5.100 Smoke-tunnel test of TESABOF model (note high upwash and downwash angles even at this low angle of attack $\alpha = -20$ deg; the test was performed by Fairchild Aircraft Company; *from Wagner, 1961*).

The unorthodox chordwise combined system shown in Fig. 5.99 was provided by Wagner (1967) for the so-called "TESABOF" airfoil developed by Professor D. Hazen at Princeton. A smoke flow visualization (Fig. 5.100) indicates high upwash and downwash angles even at a negative angle of attack $\alpha = -20$ deg. By producing ΔC_L linearly with C_μ, the combined system for the TESABOF airfoil demonstrated decisive advantages for very high lift over the conventional blowing wing where there is a substantial shift of the center of pressure and where ΔC_L increases as a function of C_μ. Higher efficiency rotary pumps are not easily adaptable for the combined system because the ducts require circular cross-sections (these reduce the duct area compared to other shapes) and cause large diffuser losses. Furthermore, rotary pumps require reliable interconnecting systems. On the other hand, a jet pump has the advantage of maximum flexibility, cross-section adaptability, and small weight. Moreover, all that is needed in order to secure even operation on both wing panels is interconnecting pipes for the primary jet flow. These advantages offset the low efficiency of the jet pump. With efficiencies of $\eta > 0.2$ (diffuser loss included) the jet pump is able to compete with rotary pumps in the BLC installation of combined systems.

Poisson-Quinton (1956) presented some details on the application of combined suction and blowing to various wings with flaps. Strahle (1967) analyzed two-dimensional adiabatic laminar separated regions in supersonic flow that may be caused by a forward-facing step or a control jet exhausting from the side of a flight vehicle. An order of magnitude analysis indicates that blowing or suction at the wall in the separated region could be highly effective in altering the force and moment characteristics of the separated pressure field that acts on the wall. Thus, theoretically, several schemes offer potential for controlling the trim and stepped thrust of flight vehicles with a small expenditure of control fluid.

NOMENCLATURE

A — Reference area

$A_1, A_{11}, A_{12}, A_{13}, A_{14}, A_{21}, A_{22}, A_{23}, A_{24}$, and A_{25} — Coefficient

A_t — Area of nozzle

a — Streamwise suction distance along the surface

a_1, a_2 — Coefficients

b — Span of wing

C_{BLC} — Blowing jet boundary-layer control parameter

C_D — Drag coefficient

C_{Dp} — Drag coefficient of pump

C_{Dpi} — Ideal value of C_{Dp}

C_{Dpk} — Porous resistivity drag coefficient

C_j — Jet reaction coefficient

C_{jL} — Jet momentum coefficient

$C_{j\mathrm{sep}}$ — Jet momentum coefficient to prevent separation

C_L — Lift coefficient

C_m — Pitching moment coefficient

C_p — Pressure coefficient

C_{p_s} — Suction energy coefficient, suction power coefficient, or coefficient of pumping energy

C_{p_t} — Total energy loss coefficient

$C_{p_{tm}}$ — Minimum energy loss coefficient

C_{p_w} — Boundary-layer kinetic loss coefficient

C_Q — Flow coefficient of suction or blowing

C_Q' — Suction flow coefficient per unit span

C_{Qf} — Suction flow coefficient of flap

C_{Q_s} — Suction flow coefficient of individual chamber

C_{Q_t} — Total suction flow coefficient

C_μ — Momentum coefficient

$C_{\mu A}$	Value of momentum coefficient to prevent separation by blowing
$C_{\mu \mathrm{ex}}$	Excess momentum coefficient
$C_{\mu \mathrm{G}}$	Value of C_μ necessary to reach the theoretical value computed by Glauert
$C_{\mu \mathrm{R}}$	Value of C_μ corresponding to upper economic limit
c	Chord length
c'	Local chord measured perpendicular to quarter-chord line
c_{f_m}	$c_{f_m} = c_{f_w}(1 - u_e/u_m)^2$
c_{f_w}	Friction coefficient at wall
d	Dissipation integral, Eq. (5.11)
F_1, F_2, F_3	Velocity profile parameter for jet region
F_4	Velocity profile shape factor for wake $(1/\theta_3) \int_{\delta_2}^{\infty} [\partial/\partial_y(u/u_e)]^2 dy$
g	Similarity scale factor
H	Shape factor of boundary layer; also $H = (1/\theta) \int_0^\infty [1 - u/u_m)dy$
$\bar{H}$	ϵ/θ
H_{e_3}	$(1/\theta_3) \int_{\delta_2}^{\infty} (u/u_e)(1 - u^2/u_e^2)dy$
H_0	Total head of free stream
H_3	$(1/\theta_3) \int_{\delta_2}^{\infty} (1 - u/u_e)dy$
h	Slot height
I	Jet layer thickness parameter, Eq. (5.12)
J	Total jet reaction
K	Function of γ relating to $\dot{m}_s$
k	Constant
L	Cavity length
$\dot{m}$	Mass flow rate
$\dot{m}_{\mathrm{BL}}$	Mass flow rate in boundary layer
$\dot{m}_d$	Mass defect per unit length
m_i	Mass injection rate
$\dot{m}_s$	Mass flow rate related to effective stagnation condition
N	Number of cells
P	Power required for unit span
P_s	Energy required to bring suction air to ambient velocity and static pressure
Q	Total suction quantity
Q_s	Suction mass flow of individual chamber
R	Universal gas constant
R_{T_2}	Turbulent Reynolds number for jet layer, Eq. (5.14)
R_{T_3}	Turbulent stress and dissipation, Eq. (5.17)
r	Radius
S_f	Affected wing area
S_w	Wing area
s	Streamwise distance of the suction zone
T_{st}	Stagnation temperature

t	Thickness of wall
u	Cylinder peripheral velocity
u_{e_s}	Free-stream velocity at slot
u_{j_s}	Jet velocity at slot
u_m	Velocity component parallel to wall at $y = \delta$
$\bar{u}_F$	Mean velocity on flap
u_2	Maximum velocity of component u at $y = \delta_2$
u_j	Jet velocity
v_0	Velocity of suction or blowing, or tunnel speed
v_0^*	Dimensionless suction velocity, defined by $v_0^* = (v_0/u_\infty)\sqrt{\mathrm{Re}_c}$
W_t	Nozzle throat
w	Velocity on the section contour
x_{HL}	Chordwise surface distance from model leading edge to hinge line
x_s	Chordwise surface distance from model leading edge to separation
$\bar{x}$	Wetted length along a generator of the axisymmetric cavity model with origin at the reattachment point
β	Separation angle or mass ratio, $\beta = \dot{m}_s/\dot{m}_d$
Γ	Circulation
δ_f	Flap deflection angle
δ_G	Momentum required of boundary layer necessary to keep the flow attached to the trailing edge
ϵ	Energy loss thickness
ϵ_s	Difference of height upstream and downstream edge of cavity
η_a	Efficiency of the propulsive system of aircraft
η_p	Efficiency of pump
θ	Momentum thickness of boundary layer or $\theta = \int_0^\delta (u/u_m)(1 - u/u_m)\,dy$
θ^*	Boundary layer kinetic energy thickness
θ_2	Quantity defined by Eq. (5.6)
θ_3	$\theta_3 = \int_{\delta_2}^{\infty}(u - u_2/u_e - u_2)[1 - (u - u_2/u_e - u_2)]\,dy$
κ	Resistance coefficient of material
ξ	x/c
Ω	Quantity defined by Eq. (5.2*a*)

REFERENCES

Alvarez-Calderon, A. A. (1964). "Rotating cylinder flaps of V/STOL Aircraft," *Aircraft Eng.*, Vol. 36, No. 10, pp. 304–309.

Attinello, J. S. (1961). "Design and engineering features of flap blowing installations," in *Boundary-layer and flow control*, Vol. 1, G. V. Lachmann, ed., Pergamon Press, New York.

Ball, K. O. and R. H. Korkegi (1968). An investigation of the effect of suction on hypersonic laminar boundary-layer separation, *J. AIAA*, Vol. 6, No. 2, pp. 239–243.

Bamber, M. J. (1931). "Wind tunnel tests on airfoil boundary layer control using a backward opening slot," NACA Report 385.

Bangert, L. H. (1971). "The turbulent wall jet with an initial boundary layer," AIAA Paper No. 71-612, 4th Fluid and Plasma Dynamics Conference, 21–23 June, Palo Alto, California.

Bradshaw, P. and M. T. Gee (1962). "Turbulent wall jets with and without an external stream," ARC R&M 3252.

Bradshaw, P. et al. (1967). "Calculations of boundary layer development using the turbulent energy equation," *J. Fluid Mech.*, Vol. 28, pp. 593–616.

Brady, V. G. and G. R. Ludwig (1963). "Research on unsteady stall of axial flow compressors," Cornell Aeronautical Laboratory Report AM-1762-S-4.

Carriére, P. and E. A. Eichelbrenner (1961). "Theory of flow reattachment by a tangential jet discharging against a strong adverse pressure gradient," in *Boundary-layer and flow control*, Vol. I, G. V. Lachmann, ed., Pergamon Press, New York.

Carriére, P. et al. (1958). "Contribution théorique et experimentale à l'étude du contrôle de la couche limite par soufflage," (translated as "Theoretical and experimental contribution to the study of boundary layer control by blowing") *Advances in Aeronautical Sciences*, Proc. First Internat. Cong. Aeron. Sci., Madrid, 1958, Vol. 2, Pergamon Press, New York, 1959.

Chapman, D. R. (1956). "A theoretical analysis of heat transfer in regions of separated flow," NACA TN 3792.

Cook, W. L. et al. (1953). "The use of area suction for the purpose of improving trailing-edge flap effectiveness on a 35° swept-back wing," NACA RM A53E06.

Crawford, D. H. (1961). "The effect of air bleed on the heat transfer and pressure distribution on 30° conical flares at a Mach number of 6.8," NASA TM X-439.

Danneberg, R. E. et al. (1957). "Perforated sheets as the porous material for a suction-flap application," NACA TN 4038.

Escudier, M. P. and W. B. Nicoll (1966). "The entrainment function in turbulent boundary-layer and wall-jet calculations," *J. Fluid Mech.*, Vol. 25, p. 337.

Flettner, A. (1924). "Die Anwendung der Erkenntnisse der Aerodynamik zum Windantrieb von Schiffen," *Jb. Schiffbautech. Ges.*, Vol. 25, No. 222.

Gartshore, I. S. and B. G. Newman (1969). "The turbulent wall jet in an arbitrary pressure gradient," *Aeronaut. Quart.*, Vol. 20, p. 25.

Glauert, H. (1927). "Theoretical relationship for an airfoil with hinged flap," ARC R&M 1095.

Glauert, M. B. (1956). "The wall jet," *J. Fluid Mech.*, Vol. 1, p. 625.

Glauert, M. B. (1959). "The flow past a rapidly rotating circular cylinder," *Proc. Roy. Soc., Ser. A*, Vol. 242, p. 108.

Görtler, H. and H. Witting (1957). "Zu den Tani'schen Grenzschichten," *Österr. Ingr.-Arch.*, Vol. 9, p. 111.

Gray, J. D. (1967). "Investigation of the effect of flare and ramp angle on the upstream influence of laminar and transitional reattaching flows at Mach 3 to 7," AEDC-TR-66-190 (AD 645 840), Arnold Engineering Development Center.

Gregory, N. et al. (1950). "Wind-tunnel tests on the prevention of boundary-layer separation by distributed suction at the rear of a thick airfoil," (NPL 153), ARC R&M 2788.

Griffin, R. N., Jr. and D. H. Hickey (1956). "Investigation of the use of area suction to increase the effectiveness of trailing-edge flaps of various spans on a wing of 45° sweepback and aspect ratio 6," NACA R&M A56B27.

Gross, L. W. (1968). "Investigation of the behavior of boundary layer suction in decelerated flows, experiments and comparison with theory," Northrop Corporate Laboratories Tech. Report AFFDL-TR-68-117, for Air Force Flight Dynamics Laboratory, Air Force Systems Command, Wright-Patterson Air Force Base, Ohio.

Harris, G. L. (1965). "The turbulent wall jet on plane and curved surfaces beneath an external stream," von Karman Institute for Fluid Dynamics TN 27.

Head, M. R. (1961). "Approximate methods of calculating the two-dimensional laminar boundary layer with suction," in *Boundary-layer and flow control*, Vol. 2, G. V. Lachmann, ed., Pergamon Press, New York.

Holstein, H. (1943). "Ähnliche Laminare Reibungsschichten an durchlässigen Wänden," UM 3050.

Howarth, L. (1938). "On the solution of the laminar boundary layer equations," *Proc. Roy. Soc.*, Ser. A, Vol. 167, p. 547.

Hubbartt, J. E. and L. H. Bangert (1970). "Turbulent boundary layer control by a wall jet," AIAA Paper 70-107, 8th Aerospace Sciences Meeting, 19–21 Jan, New York.

James, H. A. and R. L. Maki (1957). "Wind-tunnel tests of the static longitudinal characteristics at low speed of a swept-wing airplane with blowing flaps and leading-edge slats," NACA RM A57D11.

Kelly, M. W. and W. H. Tolhurst, Jr. (1954). "The use of area suction to increase the effectiveness of trailing-edge flap on a triangular wing of aspect ratio 2," NACA RM A54A25.

Kikuhara, S. and M. Kasu (1968). "Design of blowing type BLC system on the Japanese STOL seaplane," Interna. Cong. Subsonic Aeron., NY Acad. Sci., pp. 397–424.

Kruka, V. and S. Eskinazi (1964). "The wall-jet in a moving stream," *J. Fluid Mech.*, Vol. 20, p. 555.

Lewis, J. E. et al. (1968). "Experimental investigation of supersonic laminar two-dimensional boundary-layer separation in a compression corner with and without cooling," *J. AIAA*, Vol. 6, No. 1, pp. 7–14.

Ludwieg, H. and W. Tillmann (1949). "Untersuchungen über die Wandschubspannung in turbulenten Reibungsschichten," *Ingr.-Arch.*, Vol. 17, p. 288.

Ludwig, G. R. (1964). "An experimental investigation of laminar separation from a moving wall," Second Aerospace Sciences Meeting, New York, AIAA Preprint 64-6.

Malavard, L. et al. (1956). "Theoretical and experimental investigation of circulation control," Princeton University, Translation Report 358.

Mangler, W. (1944). "Laminare Grenzschicht mit Absaugen und Ausblasen," UM 3087.

Maus, J. R. (1968). "Calculation of turbulent boundary layers in decelerated flows with suction using Bradshaw-Ferriss method," Appendix 2 to Gross, L. W. (1968), "Investigation of the behavior of boundary layer suction in decelerated flows: Vol. 1. Experiments and comparison with theory," Northrop Corporate Laboratories Tech. Report AFFDL-TR-68-117.

Mickley, H. S. and R. S. Davis (1957). "Momentum transfer for flow over a flat plate with blowing," NACA TN 4017.

Mickley, H. S. et al. (1954). "Heat, mass and momentum transfer for flow over a flat plate with blowing and suction," NACA TN 3208.

Nicoll, K. M. (1963). "An experimental investigation of laminar hypersonic cavity flows," Ph.D. Thesis, Princeton University.

Nicoll, K. M. (1965). "Mass injection in a hypersonic cavity flow," Aerospace Research Laboratories Report ARL 65-90.

Pankhurst, R. C. and N. Gregory (1952). "Power requirements for distributed suction for increasing maximum lift," ARC C.P. 82.

Pankhurst, R. C. et al. (1948). "Wind-tunnel tests of the stalling properties of an eight-percent thick symmetrical section with nose suction through a porous surface," ARC R&M 2666.

Pate, S. R. (1969). "Experiments on control of supersonic transitional flow separation using distributed suction," *J. AIAA*, Vol. 7, No. 5, pp. 847–851.

Peake, D. J. (1966). "The use of air injection to prevent separation of the turbulent boundary layer in supersonic flow," Aeronautical Research Council CP No. 890.

Pechau, W. (1958). "Ein Näherungsverfahren zur Berechnung der ebenen und der rotationssymmetrischen turbulenten Grenzschicht mit beliebiger Absaugung und Ausblasung," Bericht 58/13, Inst. Strömmungsmechanik, Techn. Hochschule Braunschweig; also Jahrbuch 1958, Wiss. Ges. Luftfahrt.

Poisson-Quinton, P. (1956). "Einige physikalische Betrachtungen über das Ausblasen an Tragflügeln," Jahrb d. W. G. L. Fridr. Vieweg & Sohn, Braunschweig & Akademie-Verlag, Berlin, pp. 29–51.

Prandtl. L. (1925). " Magnuseffeckt und Windkraftschiff," *Naturwissenschaften*, Vol. 13, No. 93.

Rheinboldt, W. (1955). "Zur Berechnung stationärer Grenzschichten bei kontinuierlicher Absaug mit unstetig veränderlicher Absaugegeschwindigkeit," Inaugural Dissertation, University of Freiberg (1955); *J. Rat. Mech. Anal.* 5, 539 (1956).

Rotta, J. (1952). "Schubspannungsverteilung und Energiedissipation bei turbulenten Grenzschichten," *Ingr.-Arch.*, Vol. 20, p. 195.

Schlichting, H. (1948). "Ein Näherungsverfahren zur Berechnung der laminaren Reibungsschicht mit Absaugung," *Ingr.-Arch.*, Vol. 16, No. 201 (translated as "An approximate method for the calculation of the laminar boundary layer with suction for bodies of arbitrary shape," NACA TM 1216).

Schlichting, H. (1958). "Some recent developments in boundary layer control," First Inter. Cong. Aeron. Sci., Madrid.

Schlichting, H. (1968). *Boundary layer theory*, Sixth Ed. McGraw-Hill Book Company, Inc., New York.

Schlichting, H. and K. Bussmann (1943). "Exakte Lösungen für die laminare Grenzschicht mit Absaugung und Ausblasen," *Dtsch. Akad. Luftfahrtforsch*, 7B, p. 25.

Schlichting, H. and W. Pechau (1959). "Auftriebserhöhung von Tragflügeln durch kontinuierlich Verteilte Absaugung," *Zeitschrift für Flugwissenschaften*, Vol. 7, p. 113.

Schwier, W. (1944). "Versuche zur Auftriebssteigerung eines Pfeilflügels durch Ausblasen von Luft," Zentrale f. wiss. Beichtswesen Berlin-Adlershof, UM 3114.

Smith, S. H. (1963). "Research on unsteady stall of axial flow compressors," Cornell Aeronautical Laboratory Report AM-1762-S-4.

Spence, D. A. (1958). "The lift on a thin airfoil with a jet-augmented flap," *Aeronaut. Quart.*, Vol. 9, pp. 287–299.

Strahle, W. C. (1967). "Theoretical studies on the effects of blowing and suction in laminar separated regions," AIAA Paper 67-192, Fifth Aerospace Sciences Meeting, 23–26 Jan, New York.

Tani, I. (1949). "On the solution of the laminar boundary layer equations," *J. Phys. Soc. Japan*, Vol. 4, pp. 149–154.

Tani, I. (1954). "On the approximate solution of the laminar boundary layer equations," *J. Aerospace Sci.*, Vol. 21, No. 7, pp. 487–504.

Thomas, F. (1962). "Untersuchungen über die Erhöhung des Auftriebes von Tragflügeln mittels Grenzschichtbeeinflussung durch Ausblasen," *Z. Flugwiss*, Vol. 10, No. 2, pp. 46–65. (Translated as "Investigations into increasing the lift of wings by boundary layer control through blowing," Royal Aircraft Establishment Library Translation 1267, Nov, 1967.)

Thompson, B. G. J. (1965). "The calculation of the shape-factor development in incompressible turbulent boundary layers with or without transpiration," AGARDograph 97, Part 1.

Thompson, B. G. J. (1967). "A new two-parameter family of mean velocity profiles for incompressible turbulent boundary layers on smooth walls," ARC R&M 3463.

Thwaites, B. (1946). "Investigations into the effect of continuous suction on laminar boundary layer flow under adverse pressure gradients," ARC R&M 2514.

Thwaites, B. (1949). "The development of laminar boundary layers under conditions of continuous suction," unpublished A.R.C. Report 12699.

Thwaites, B. (1960). *Incompressible aerodynamics*, Oxford at the Clarendon Press.

Tolhurst, W. H. (1958). "Full-scale wind-tunnel tests of a 35° sweptback-wing airplane with blowing from the shroud ahead of the trailing-edge flaps," NACA TN 4283.

Townsend, A. A. (1956). "The structure of turbulent shear flow," Cambridge University Press.

Truckenbrodt, E. (1952). "Ein Quadraturverfahren zur Berechnung der laminaren und turbulenten Reibungsschicht bei ebener und rotationssymmetrischer Strömung," *Ingr.-Arch.*, Vol. 20, p. 211. (Translated as "A method of quadrature for calculation of the laminar and turbulent boundary layer in case of plane and rotationally symmetrical flow," NACA TM 1379, 1955.)

Truckenbrodt, E. (1955). "A simple approximate method for calculating the laminar boundary layer," Institut für Strömmungsmechanik, Technische Hochschole, Braunschweig Report 55/6a.1.

Truckenbrodt, E. (1956). "Ein einfaches Näherunges-Verfahren zum Berechnen der laminaren Reibungsschicht mit Absaugung," *Forsch. Ing. - Wesen*, Vol. 22, No. 147; also *Z. Angew, Math. Mech.*, p. 49, Sonderheft (1956).

Vidal, R. J. (1959). "Research on rotating stall in axial flow compressors: Part 3–Experiments on laminar separation from a moving wall," Wright-Patterson Air Development Center, WADC TR-59-75.

Wagner, F. G. (1961). "Design and engineering features for flap suction and combined blowing and suction," in *Boundary layer and flow control,* Vol. 1, G. V. Lachmann, ed., Pergamon Press, New York.

Williams, J. (1968). "British research on boundary-layer control for high lift by blowing," *Z. Flugwiss,* Vol. 6, No. 5.

Wuest, W. (1958). "Messungen an Absaugegrenzschichten," Bericht Aerodyn. Versuchsanst, Göttingen 58/A/31 and 32.

Wuest, W. (1961a). "Theory of boundary suction to prevent separation," in *Boundary layer and flow control*, Vol. 1, G. V. Lachmann, ed., Pergamon Press, New York.

Wuest, W. (1961b). "Survey of calculation methods of laminar boundary layers with suction in incompressible flow," in *Boundary layer and flow control*, Vol. 2, G. V. Lachmann, ed., Pergamon Press, New York.

SUPPLEMENTARY REFERENCES

Alzner, E. and V. Zakkay (1971). "Turbulent boundary-layer shock interaction with and without injection," *AIAA J.*, Vol. 9, No. 8, p. 1769.

Anderson, G. F., V. S. Murthy, and S. P. Sutera (1969). "Laminar boundary-layer control by combined blowing and suction in the presence of surface roughness," *J. Hydronaut.*, Vol. 3, No. 3, pp. 145–151.

Ball, K. O. W. (1970). "Further results on the effects of suction on boundary layer separation," *AIAA J.*, Vol. 8, No. 2, p. 374.

Brown, S. L. and R. D. Murphy (1974). "Design and test of ejector thrust augmentation configurations," AGARD-CP-143, pp. (19-1)–(19-12).

Clauser, F. H. (1954). "Turbulent boundary layers in adverse pressure gradients," *J. Aeronaut. Sci.*, Vol. 21, No. 2, pp. 91–108.

Collins, D. J. et al. (1970). "Near wake of a hypersonic blunt body with mass addition," *AIAA J.*, Vol. 8, No. 5, p. 833.

Durando, N. A. (1971). "Vortices induced in a jet by a subsonic flow," *AIAA J.*, Vol. 9, No. 2, p. 325.

Ericsson, L. E. and R. A. Guenther (1973). "Dynamic instability caused by forebody blowing," *AIAA J.*, Vol. 11, No. 1, p. 23.

Fong, M. C. (1971). "An analysis of plum-induced boundary layer separation," *J. Spacecraft and Rockets*, Vol. 8, p. 1107.

Fukusako, S. et al. (1970). "Laminar wall jet with blowing or suction," *J. Spacecraft and Rockets*, Vol. 7, p. 91.

Granville, P. S. (1968). "Iterative method for the laminar boundary layer with pressure gradient and suction or blowing," Rept. 2946, U.S. Naval Ship Research and Development Center, Washington, D.C.

Griffin, O. M. and S. E. Ramberg (1974). "The vortex-street wakes of vibrating cylinders," *J. Fluid Mech.*, Vol. 66, No. 3, pp. 553–576.

Haering, G. W. (1972). "Boundary-layer separation: Effect of low-speed wall jets," *J. Aircraft*, Vol. 9, p. 751.

Johnson, W. G. and G. E. Kardas (1974). "A wind tunnel investigation of the wake near the trailing edge of a deflected externally blown flap," NASA-TM-X-3079, L-9665, NASA Langley Research Center, Langley Station, Virginia.

Jones, W. P. and J. A. Moore (1972). "Flow in the wake of cascade of oscillating airfoils," *AIAA J.*, Vol. 10, No. 12, p. 1600.

Karashima, K. (1970). "An integral analysis of heat transfer downstream of a rearward-facing step with small coolant injection," NASA-TN-D-5970, National Aeronautics and Space Administration, Washington, D.C.

Kassoy, D. R. (1971). "On laminar boundary-layer blow-off: Part 2," *J. Fluid Mech.*, Vol. 48, p. 209.

Kelly, M. W. et al. (1958). "Blowing-type boundary layer control as applied to the trailing edge flaps of a 35° swept-wing airplane," NACA Rept. 1369.

Lewis, J. E. et al. (1971). "Experimental investigation of the effects of base mass addition on the near wake of a slender body," *AIAA J.*, Vol. 9, No. 8, p. 1506.

Liu, T. M. and P. R. Nachtsheim (1973). "Numerical stability of boundary layer with massive blowing," *AIAA J.*, Vol. 11, No. 8, p. 1197.

Liu, T. M. and P. R. Nachtsheim (1973). "Shooting method for solution of boundary layer with massive blowing," *AIAA J.*, Vol. 11, No. 11, p. 1548.

Malikapar, G. I. (1972). "The effect of injection and curvature of walls on flow separation," Foreign Technology Div., U.S. Air Force Systems Command, Wright-Patterson AFB, Ohio. (Transl. into English from *Uch. Zap. Tsent. Aerogidrodinamicheskii Inst.* (USSR), 1970, Vol. 1, pp. 60–68.)

McGregor, I. (1971). "Some applications of boundary layer control by blowing to air inlet for V/STOL aircraft," AGARD-CP-91-71, pp. (12-1)–(12-13).

Miles, J. W. (1970). "The oseenlet as a model for separated flow in a rotating viscous liquid," *J. Fluid Mech.*, Vol. 42, p. 207.

Miles, J. W. (1971). "Boundary-layer separation on a sphere in a rotating flow," *J. Fluid Mech.*, Vol. 45, p. 513.

Nachtsheim, P. R. and M. J. Green (1971). "Numerical solution of boundary-layer flows with massive blowing," *AIAA J.*, Vol. 9, No. 3, p. 533.

Narain, P. S. and M. S. Uberoi (1974). "The swirling turbulent plume," *Trans. ASME, J. Appl. Mech.*, Vol. 41, No. E2, p. 337.

Pennell, W. T. et al. (1972). "Laminarization of turbulent pipe flow by fluid injection," *J. Fluid Mech.*, Vol. 52, p. 451.

Polak, A. (1971). "The prediction of turbulent boundary-layer separation influenced by blowing," NOL-TR-71-125, U.S. Naval Ordnance Lab., White Oak, Md.

Polak, A. (1972). "Prediction of turbulent boundary-layer separation influenced by blowing," *AIAA J.*, Vol. 10, No. 4, p. 555.

Polhamus, E. C. (1971). "Prediction of vortex-lift characteristics by a leading-edge suction analogy," *J. Aircraft*, Vol. 8, p. 193.

Poppleton, E. D. (1971). "Effect of air injection into the core of a trailing vortex," *J. Aircraft*, Vol. 8, p. 672.

Raithby, G. D. and D. C. Knudsen (1974). "Hydrodynamic development in a duct with suction and blowing," *Trans. ASME. J. Appl. Mech.*, Vol. 41, No. F4, p. 896.

Seebaugh, W. R. and M. E. Childs (1970). "Influence of suction shock wave-turbulent boundary layer interactions for two-dimensional and axially symmetric flows," NASA-CR-1639, National Aeronautics and Space Administration, Washington, D.C.

Snedecker, R. S. (1972). "Effect of air injection on the torque produced by a trailing vortex," *J. Aircraft*, Vol. 9, p. 682.

Spaid, F. W. and J. C. Frishett (1972). "Incipient separation of a supersonic, turbulent boundary layer including effects of heat transfer," *AIAA J.*, Vol. 10, No. 7, p. 915.

Suslov, A. I. (1974). "On the effect of injection on boundary layer separation," *J. Appl. Math. and Mech.*, Vol. 38, No. 1, p. 148.

Telionis, D. P. (1972). "Heat transfer at reattachment of a compressible flow over a backward facing step with a suction slot," *AIAA J.*, Vol. 10, No. 8, p. 1108.

Telionis, D. P. and M. J. Werle (1973). "Boundary-layer separation from downstream moving boundaries," *Trans. ASME J. Appl. Mech.*, Vol. 95, pp. 369–374.

Tennant, J. S. (1973). "A subsonic diffuser with moving walls for boundary layer control," *AIAA J.*, Vol. 11, No. 2, p. 240.

Thompson, E. R. and W. T. Snyder (1968). "Drag reduction of a non-Newtonian fluid by fluid injection at the wall," *J. Hydronaut.*, Vol. 2, No. 4, pp. 177–180.

Valentine, D. T. et al. (1970). "Turbulent axisymmetric near-wake at Mach four with base injection," *AIAA J.*, Vol. 8, No. 12, p. 2279.

Wickens, R. H. (1974). "The spanwise lift distribution and trailing vortex wake downwind of an externally blown jet-flap," AGARD-CP-143, pp. (5-1)–(5-23).

Wu, J. (1973). "Injection of drag-reducing polymers into a turbulent boundary layer," *J. Hydronaut.*, Vol. 7, p. 129.

chapter 6

PROVOCATION OF SEPARATION

INTRODUCTORY REMARKS

Because flow separation is generally considered in connection with energy loss, the aim is usually to suppress it. Therefore, control by provocation may be construed as unconventional, especially if the resultant separation is not the primary goal of control but its by-product. An example is that caused by the interaction of a jet with a main stream. Nevertheless, this rather unconventional type should be studied in order to provide the necessary means to achieve the high performance and high efficiency needed.

For convenience, the subject is treated in terms of

1. The change in body shape to produce a free streamline flap and trapped vortices.
2. The change in shock shape produced by a spike mounted upstream of a blunt body.
3. The force induced by interaction of the jet with the main stream flow.

Investigations on this type of separation have been few and so control techniques are less well explored than for other aspects of the phenomenon.

6.1 CHANGE IN BODY SHAPE

When a free streamline is produced along the outer edge of the separated region, a pseudo-body surface may be created. Vortices are formed within the separated region, but the extent of this significance for the formation of the free streamline may differ depending on the technique applied. Hence, separate discussions are presented for the free streamline flap and for trapped vortices.

6.1.1 Free Streamline Flap

An aircraft cruising at supersonic speeds requires a thin wing and highly curved or even sharp leading edges for good performance. However, at low speeds, i.e., landing and takeoff, the performance of such a wing will be poor because separation at the leading edge leads to reduced values of maximum lift, large drag, poor controllability, and buffeting. It is already known that a thicker wing section with curved leading edge is more suitable for good performance at low speeds. Hence, one possible way to utilize a thin solid wing more effectively at low speeds is to aerodynamically change it to a thick wing. This can be accomplished by provoking separation at the leading edge and using boundary layer control to ensure that the reattachment is maintained; in this way, the separating free streamline will hopefully form a thick pseudo-body shape, as already shown in Chap. 1 (Fig. 1.23).

6.1.1.1 The Hurley potential flow analysis

Hurley (1961) carried out fundamental and applied studies of the free streamline flap. The Hurley model consists of two flat plates (of lengths l_1 and l_2) hinged together at their rear at an angle τ. The lower plate is at an incidence α_D to a uniform stream velocity u_∞. The free streamline must meet each of the plates tangentially; otherwise there would be an infinite velocity at the junction. It is assumed that the pressure (and thus the fluid velocity) is nearly constant along the free streamline and that the circulation is fixed by satisfying the Kutta-Joukowski condition at the trailing edge. The analysis requires consideration of the z-plane (Fig. 6.1), the Q-plane (Fig. 6.2), and the ζ-plane (Fig. 6.3).

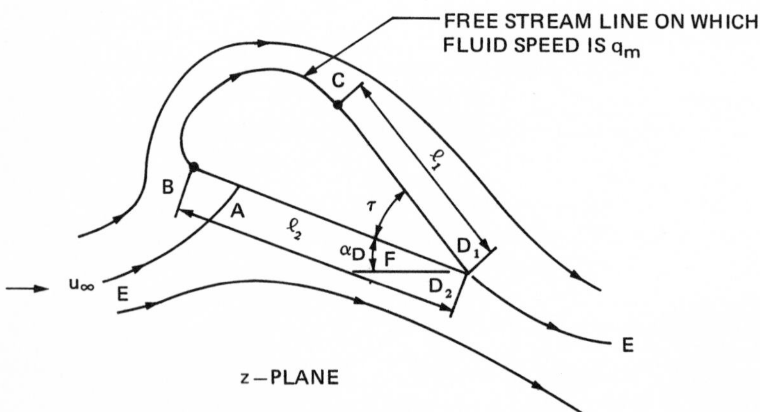

FIG. 6.1 Model used in potential flow calculations (*from Hurley, 1961*).

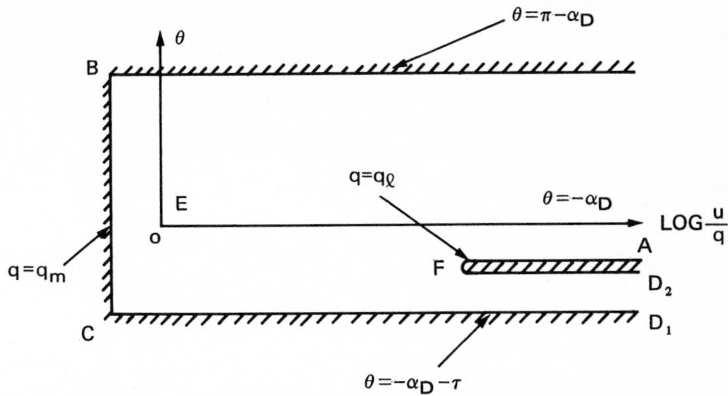

FIG. 6.2 Q-plane for the Hurley analysis (*from Hurley, 1961*).

The trace of the airfoil in the Q-plane is

$$
\begin{aligned}
Q &= \ln\left(-u_\infty \frac{dz}{dw}\right) \\
&= \ln \frac{u_\infty}{q} + i\theta
\end{aligned}
\tag{6.1}
$$

where w is the complex velocity potential $w = \phi + i\psi$,
q is the fluid velocity,
θ is the direction of flow.
The velocity components u and v are given by

$$
u = -\frac{\partial\phi}{\partial x} \quad \text{and} \quad v = -\frac{\partial\phi}{\partial y} \tag{6.2}
$$

FIG. 6.3 ζ-plane for the Hurley analysis (*from Hurley, 1961*).

Since the trailing edge angle of the airfoil is not zero, there will be stagnation points D_1 and D_2 on either side, mapped by

$$
Q = +\infty - i(\alpha_D + \tau) \quad \text{and} \quad Q = +\infty - i\alpha_D
$$

where τ is the trailing edge angle and α_D is the angle between the lower surface and the free-stream direction. At A and at the trailing edge, q is zero;

therefore a local maximum value of q_l is reached at some point between them. That point is designated F. The point at infinity in the z-plane maps to the origin of the Q-plane, and the region exterior to the airfoil maps to the unhatched region of the Q-plane shown in Fig. 6.2. The following Schwartz-Christoffel transformation is used to map the unhatched portion of the Q-plane on the upper half of the ζ-plane:

$$\frac{dQ}{d\zeta} = K_1(\zeta + 1)^{-1/2}(\zeta - 1)^{-1/2}(\zeta - d)^{-1}(\zeta - f) \tag{6.3}$$

Here points A, B, and C of the Q-plane have been mapped to $\zeta = \infty, -1$, and $+1$, respectively, and K_1 is a constant. The points to which D, E, and F map are designated by $\zeta = d$, ζ_1, and f, respectively. By integrating Eq. (6.3)

$$Q = K_1 \ln [\zeta \pm (\zeta^2 - 1)^{1/2}] + \frac{K_1(d-f)}{(d^2 - 1)^{1/2}}$$

$$\ln\left[\frac{1}{\zeta - d}(1 - d\zeta \pm (d^2 - 1)^{1/2}(\zeta^2 - 1)^{1/2}\right] + K_2 \tag{6.4}$$

where K_2 is an arbitrary constant.

In Eq. (6.4), the positive sign is taken in each place where there is an alternative, $(\zeta^2 - 1)^{1/2}$ is taken to behave as ζ for ζ large, and each of the logarithms is taken to have its principal value. It can then be shown that point-to-point correspondences between the Q- and the ζ-planes (illustrated in Figs. 6.2 and 6.3) are satisfied if

$$K_1 = 1 \tag{6.5}$$

$$K_2 = \ln \frac{u_\infty}{q_m} - i(\alpha_D + \tau) \tag{6.6}$$

where q_m is the theoretical value of q on the free streamline when $\alpha = \alpha_D$. Also

$$\frac{d-f}{(d^2 - 1)^{1/2}} = -\frac{\tau}{\pi} \tag{6.7}$$

$$\ln \frac{q_m}{q_z} = \ln [f + (f^2 - 1)^{1/2}] - \frac{\tau}{\pi} \ln \frac{1}{f-d} [df - 1 - (d^2 - 1)^{1/2}(f^2 - 1)^{1/2}] \tag{6.8}$$

where q_z is the velocity in the z-plane and

$$\frac{q_m}{u_\infty} \exp i(\alpha_D + \tau) = \frac{\zeta_1 + (\zeta_1{}^2 - 1)^{1/2}}{\{[1/(\zeta_1 - d)] - d[1 - d\zeta_1 + (d^2 - 1)^{1/2}(\zeta_1{}^2 - 1)^{1/2}]\}^{\tau/\pi}} \tag{6.9}$$

From Eqs. (6.1) and (6.4)-(6.7),

$$\frac{d\zeta}{dw} = -\frac{\exp\,[-i(\alpha_D + \tau)]\,[\zeta + (\zeta^2 - 1)^{1/2}]}{q_m \{[1/(\zeta - d)]\,[1 - d\zeta + (d^2 - 1)^{1/2}(\zeta^2 - 1)^{1/2}]\}^{\tau/\pi}} \tag{6.10}$$

Figure 6.4 shows the mapping of the exterior of the airfoil in the z-plane onto the exterior of the unit circle in the z_1-plane. The stagnation points at A and D are located symmetrically in the z_1-plane and the complex velocity potential is

$$w(z_1) = -u_\infty z_1 - \frac{u_\infty}{z_1} - \frac{iK}{2\pi} \ln z_1 \tag{6.11}$$

where K is the circulation given by

$$K = 4\pi u_\infty \sin\beta \tag{6.12}$$

The angle β is as indicated in Fig. 6.4. The exterior of the unit circle in the z_1-plane is mapped onto the interior of the unit circle in the z_2-plane by

$$z_1 = \frac{1}{z_2} \tag{6.13}$$

and then onto the upper half of the ζ-plane by the transformation

$$z_2 = \lambda \frac{\zeta - \bar{\zeta}_1}{\zeta - \zeta_1} \tag{6.14}$$

where $|\lambda| = 1$. When z_2 is eliminated,

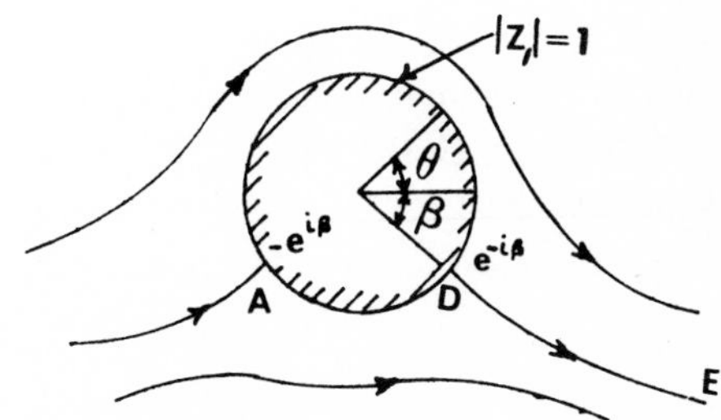

FIG. 6.4 The z_1-plane (*from Hurley, 1961*).

$$z_1 = \frac{1}{\lambda} \frac{\zeta - \zeta_1}{\zeta - \bar{\zeta}_1} \tag{6.15}$$

The stagnation points at $z_1 = -e^{i\beta}$ and $z_1 = e^{-i\beta}$ must map to $\zeta = \infty$ and $\zeta = d$; thus Eq. (6.15) becomes

$$\lambda = -e^{-i\beta} \tag{6.16}$$

and

$$d = \zeta_1 + \eta_1 \tan \beta \tag{6.17}$$

where $\zeta_1 = \xi_1 + i\eta_1$ and d is as indicated in Fig. 6.3.

It follows that from Eqs. (6.11), (6.15), and (6.16)

$$w(\zeta) = u_\infty e^{i\beta} \frac{\zeta - \bar{\zeta}_1}{\zeta - \zeta_1} + u_\infty e^{-i\beta} \frac{\zeta - \zeta_1}{\zeta - \bar{\zeta}_1} - \frac{iK}{2\pi} \ln \left(- e^{-i\beta} \frac{\zeta - \bar{\zeta}_1}{\zeta - \zeta_1} \right) \tag{6.18}$$

Differentiation of Eq. (6.18) yields

$$\frac{dw}{d\zeta} = - \frac{2iu_\infty \eta_1 e^{i\beta}}{(\zeta - \zeta_1)^2} + \frac{2iu_\infty \eta_1 e^{-i\beta}}{(\zeta - \bar{\zeta}_1)^2} - \frac{iK}{2\pi} \left(\frac{1}{\zeta - \bar{\zeta}_1} - \frac{1}{\zeta - \zeta_1} \right) \tag{6.19}$$

From Eq. (6.12), by substituting K,

$$\frac{dw}{d\zeta} = - \frac{8u_\infty \eta_1^2 (d - \xi) \cos \beta}{[(\xi - \xi_1)^2 + \eta_1^2]^2} \tag{6.20}$$

for $\zeta = \xi$, i.e., for ζ on the real axis.

The shape of the free streamline airfoil is obtained from

$$z = \int \frac{dz}{d\zeta} d\zeta = \int \frac{dz}{dw} \frac{dw}{d\zeta} d\zeta$$

Since dz/dw and $dw/d\zeta$ are given by Eqs. (6.10) and (6.20), the integration may be carried out along the real ζ-axis.

From Eqs. (6.10) and (6.20), $dz/d\zeta \sim K_3/\zeta^2$ for large values of ζ. Here K_3 is a constant which is determined for the condition of airfoil close up, expressed by

$$\oint_C \frac{dz}{d\zeta} d\zeta = 0 \tag{6.21}$$

The path of integration C is the real axis, indented above $\zeta = \pm 1$ and d, and a large semicircle lying in the upper half of the ζ-plane. The only singularity of $dz/d\zeta$ within C is at $\zeta = \zeta_1$, so that the closure condition is that of the residue of $dz/d\zeta$ at $\zeta = \zeta_1$, namely, zero. Thus, from Eqs. (6.10) and (6.19)

$$\frac{1}{(\zeta_1^2 - 1)^{1/2}} - \frac{(\tau/\pi)(d^2 - 1)^{1/2}}{(\zeta_1 - d)(\zeta_1^2 - 1)^{1/2}} = -\frac{i(\xi_1 - d)}{\eta_1(\zeta_1 - d)} \tag{6.22}$$

6.1.1.2 Calculation procedure

Hurley (1961) used the values of dz/dw and $dw/d\zeta$ (given by Eqs. (6.10) and (6.20)) to calculate the shape of a particular "airfoil" from Eq. (6.21). Similar to his preliminary work, it was necessary to obtain a set of parameters which satisfy Eqs. (6.7)-(6.9), (6.17), and (6.22). Since two of these equations, (6.9) and (6.22), are complex, there are seven relationships to be satisfied by the nine parameters f, q_l/u_∞, q_m/u_∞, ξ_1, η_1, d, τ, α_D, and β, resulting in a family of solutions given by two parameters.

Equation (6.22) has one and only one acceptable root for $\zeta_1 = \xi_1 + i\eta_1$. It is given by

$$\xi_1 = \frac{-fd^2 + f^3 + 2f - f(f-d)[(d+f)^2 - 4]^{1/2}}{2(2f^2 + 1 - 2df)} \tag{6.23}$$

$$\eta_1 = \frac{\xi_1^{1/2}(d - \xi_1)}{(f - \xi_1)^{1/2}} \tag{6.24}$$

where $f = d + (\tau/\pi)(d^2 - 1)^{1/2}$.

The various relationships can be satisfied by the following procedure:

1. Select values for d that lie in the range $(1, \infty)$; see Fig. 6.3 and τ.
2. Solve Eq. (6.23) for ξ_1.
3. Solve Eq. (6.24) for η_1.
4. Solve Eq. (6.9) by q_m/u_∞ and α_D.
5. Solve Eqs. (6.17), (6.7), and (6.8) in turn for β, f, and q_l, respectively.

After parameters have been set up by the following this procedure, the shape of the airfoil may be calculated from Eq. (6.21) and the velocities on the surface from Eq. (6.10). Hurley (1961) carried out the numerical calculation to determine values of C_L and a family of airfoils for various values of τ and d. The results are shown in Table 6.1 and Fig. 6.5.

Figure 6.5 indicates that for a particular relative geometry of the plates (given values of l_1/l_2 and τ), there is a solution for only one particular value of α_D. It also shows that large values of C_L may be obtained although the

TABLE 6.1 Main features and parameter values of various airfoils

τ°	d	q_m	$\alpha^\circ D$	l_1	l_2	l_1/l_2	ξ_1	η_1	β°	C_{LD}
10	1.0	1.000	0	0	4.00	0	1.0000	0	0	0
	1.1	1.711	7.28	2.32	3.59	0.645	1.0814	0.0922	11.41	1.38
	1.3	2.292	13.18	2.72	3.49	0.778	1.2226	0.2434	17.63	2.18
	1.6	2.941	19.65	3.00	3.48	0.862	1.3875	0.4714	24.26	2.97
	2.0	3.625	26.26	3.00	3.39	0.886	1.5383	0.7666	31.06	3.82
	5.0	5.792	54.10	3.24	3.39	0.957	1.3988	2.1641	59.00	6.35
	∞	6.727	85.00	3.34	3.34	1.000	0	2.9594	90.00	7.52
30	1.0	1.000	0	0	4.00	0	1.0000	0	0	0
	1.1	1.891	8.87	1.32	3.38	0.392	1.0425	0.1605	19.72	2.51
	1.3	2.442	17.20	1.68	3.16	0.531	1.0905	0.3709	29.46	3.91
	1.6	2.953	25.66	1.96	3.01	0.649	1.1012	0.6225	38.70	5.22
	2.0	3.382	33.67	2.11	2.93	0.722	1.0563	0.8737	47.20	6.29
	5.0	4.248	56.86	2.29	2.73	0.839	0.5962	1.4883	71.33	8.72
	∞	4.464	75.00	2.60	2.60	1.000	0	1.6638	90.00	9.67
60	1.0	1.000	0	0	4.00	0	1.0000	0	0	0
	1.1	2.058	7.79	0.61	3.12	0.197	0.9831	0.2233	27.64	3.76
	1.3	2.558	16.92	1.02	2.80	0.363	0.9361	0.4399	39.60	5.72
	1.6	2.925	24.88	1.28	2.60	0.493	0.8472	0.6408	49.59	7.36
	2.0	3.180	31.70	1.47	2.47	0.594	0.7343	0.7989	57.74	8.60
	5.0	3.579	48.49	1.81	2.18	0.829	0.3341	1.0745	77.03	11.23
	∞	3.659	60.00	2.00	2.00	1.000	0	1.1340	90.00	12.67
90	1.0	1.000	0	0	4.00	0	1.0000	0	0	0
	1.1	2.196	7.01	0.42	2.95	0.144	0.9254	0.2643	33.44	4.69
	1.3	2.621	14.07	0.75	2.59	0.288	0.8189	0.4598	46.30	7.01
	1.6	2.895	20.38	0.98	2.36	0.413	0.6925	0.6101	56.08	8.85
	2.0	3.065	25.51	1.14	2.21	0.517	0.5700	0.7125	63.51	10.18
	5.0	3.301	37.30	1.47	1.87	0.786	0.2380	0.8651	79.70	13.22
	∞	3.342	45.00	1.68	1.68	1.000	0	0.8944	90.00	14.96
120	1.0	1.000	0	0	4.00	0	1.0000	0	0	0
	1.1	2.290	5.00	0.34	2.87	0.117	0.8712	0.2921	38.06	5.40
	1.3	2.659	10.00	0.62	2.47	0.252	0.7274	0.4601	51.21	7.93
	1.6	2.870	14.30	0.83	2.22	0.371	0.5880	0.5714	60.55	9.86
	2.0	2.992	17.69	0.98	2.04	0.481	0.4695	0.6400	67.31	11.37
	5.0	3.148	25.20	1.33	1.70	0.781	0.1875	0.7332	81.34	14.61
	∞	3.175	30.00	1.50	1.50	1.000	0	0.7499	90.00	16.76
150	1.0	1.000	0	0	4.00	0	1.0000	0	0	0
	1.1	2.361	2.64	0.30	2.82	0.108	0.8211	0.3109	41.89	5.95
	1.3	2.682	5.23	0.58	2.40	0.240	0.6544	0.4515	55.03	8.58
	1.6	2.850	7.40	0.77	2.15	0.356	0.5125	0.5336	63.86	10.49
	2.0	2.941	9.07	0.98	1.98	0.495	0.4012	0.5806	70.04	11.93
	5.0	3.052	12.71	1.25	1.62	0.767	0.1561	0.6405	82.47	15.38
	∞	3.083	15.00	1.42	1.42	1.000	0	0.6507	90.00	17.70
180	1.0	1.000	0	0	0	0	1.0000	0	0	0
	1.1	2.418	0	–	–	–	0.7753	0.3231	45.14	–
	1.3	2.698	0	–	–	–	0.5947	0.4389	58.11	–
	1.6	2.833	0	–	–	–	0.4551	0.4992	66.44	–
	2.0	2.904	0	–	–	–	0.3515	0.5316	72.13	–
	5.0	2.987	0	–	–	–	0.1344	0.5708	83.31	–
	∞	3.000	0	–	–	1.000	0	0.5773	90.00	–

Source: Hurley (1961).
– not calculated.

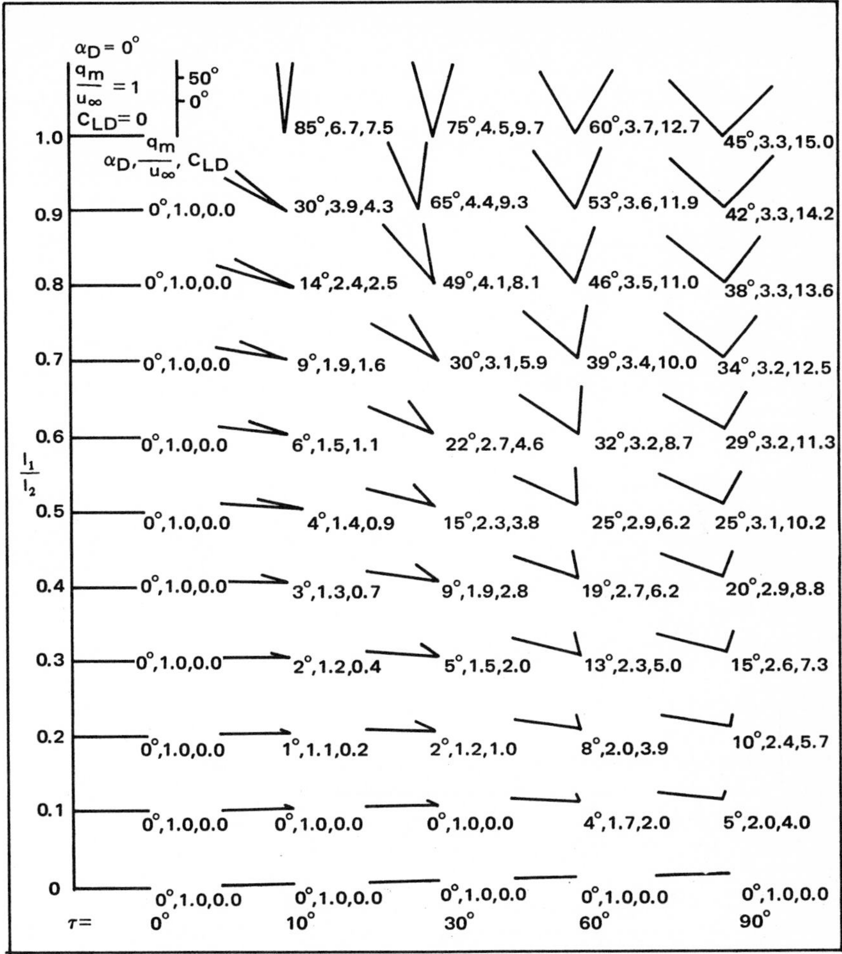

FIG. 6.5 Family of "airfoils" (*from Hurley, 1961*).

corresponding plate configurations appear to be impractical. The airfoil shape and velocity distribution are given in Figs. 6.6 and 6.7 for $l_1/l_2 = 0.6$, $\tau = 30$ deg, $\alpha_D = 21.8$ deg, $q_m/u_\infty = 2.73$, and $C_{LD} = 4.63$ (C_{LD} is the theoretical value of C_L when $\alpha = \alpha_D$).

As shown in Fig. 6.7, the flow on the upper plate may separate because of the large adverse pressure gradient; therefore it appears necessary to apply a preventive technique. As the most likely way to establish the desired flow, Hurley (1961) proposed that the upper plate be fitted with a rounded leading edge that incorporated a single blowing slot. He theorized that the high velocity air discharged from the slot (affected by entrainment of ambient air) would cause reattachment of the flow at the leading edge of the upper plate and prevent separation there.

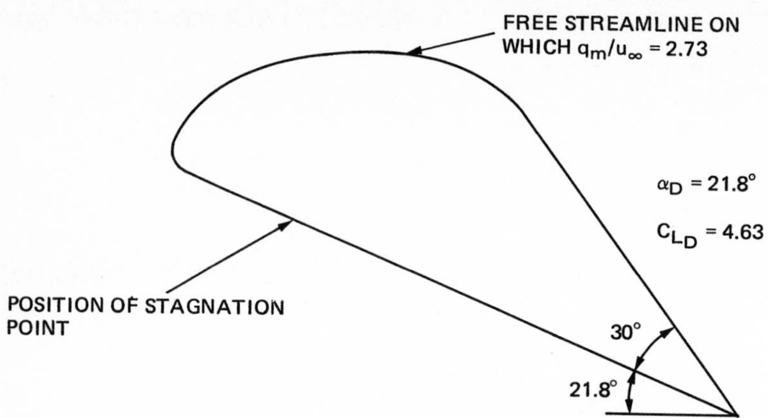

FIG. 6.6 Potential solution when $l_1/l_2 = 0.6$ and $\tau = 30$ deg (*from Hurley, 1961*).

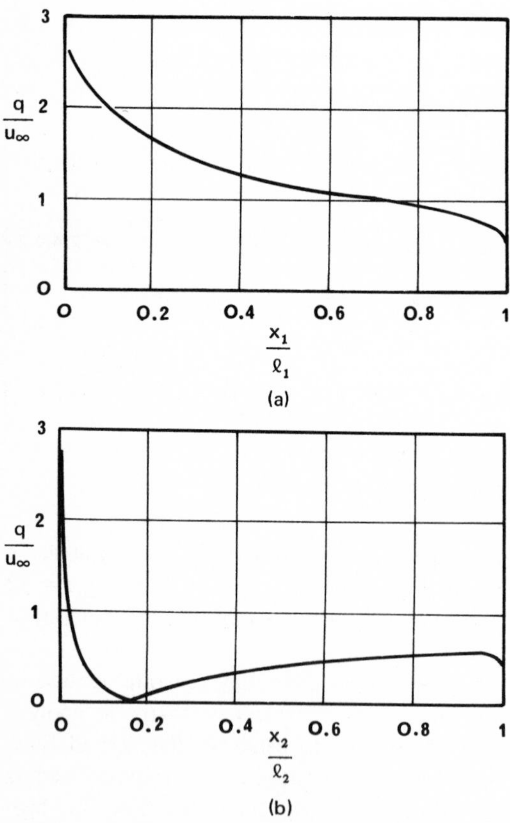

FIG. 6.7 Velocity on the plates when $l_1/l_2 = 0.6$ and $\tau = 30$ deg: (*a*) upper plate; (*b*) lower plate (*from Hurley, 1961*).

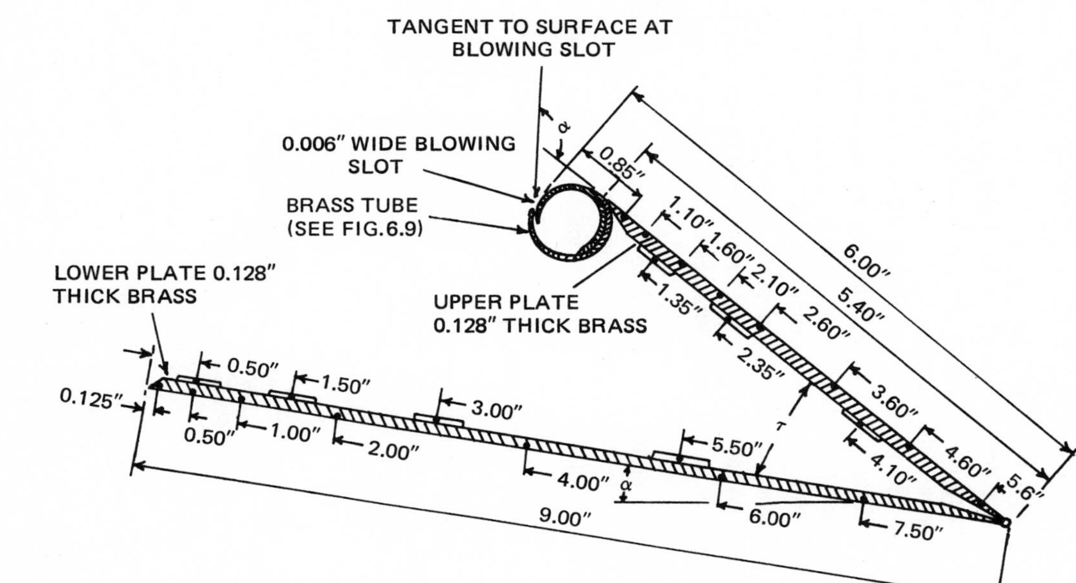

FIG. 6.8 Model dimensions (*from Hurley, 1961*).

6.1.1.3 Experiment and control technique

To determine the accuracy of predictions given by his proposed technique, Hurley (1961) tested the two-dimensional model shown in Fig. 6.8 for various values of α and C_μ at $\tau = 30$ deg, a slot angle $\gamma = 130$ deg, wind speed of 50 ft/sec, and a Reynolds number of 2.43×10^5 based on the length of the lower plate l_2. Tests were run in a 36- by 20-in. tunnel; Fig. 6.9 presents details of the blowing tube and slot designs. The front of the lower plate was bevelled at 45 deg to form a sharp leading edge, and two plates were hinged along the 20-in. dimension so that τ could be changed. The hinge was sealed to prevent leakage. The mass flow and the total pressure of the air discharged from the slot were measured, and the values of C_μ (the momentum coefficient of the blown air) were calculated empirically.

Measured and computed pressure distributions on the lower plate and on the outer and inner surface of the upper plate are compared in Figs. 6.10, 6.11, and 6.12 for various values of C_μ. Note from these figures that for each value of C_μ, the pressures on the inner surfaces of the two plates are nearly uniform and that they become more negative as C_μ is increased. The predicted and measured pressures along the free streamline were approximately equal for $C_\mu = 0.37$. The desired leading edge separation at the lower plate and reattachment to the front of the tube was accomplished for each value of C_μ. However, for low values of C_μ, separation took place once again only a short distance downstream of the reattachment point. As C_μ was increased, the point of separation moved downstream along the outer surface of the upper plate until flow attached completely to the upper surface at $C_\mu = 0.37$; this was confirmed by the measured pressures on the outer surface of the upper plate (Fig. 6.11).

In a flow visualization study using tufts, Hurley found that in the separated region between the two plates, the circulating motion of vortices was fairly

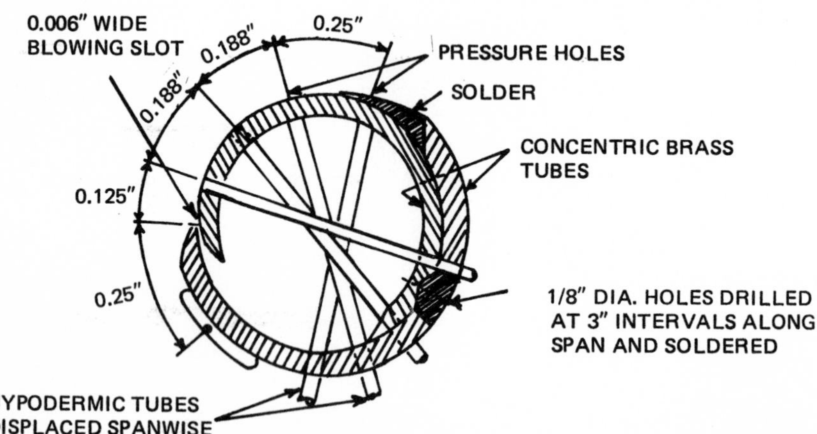

FIG. 6.9 Blowing tube and slot (*from Hurley, 1961*).

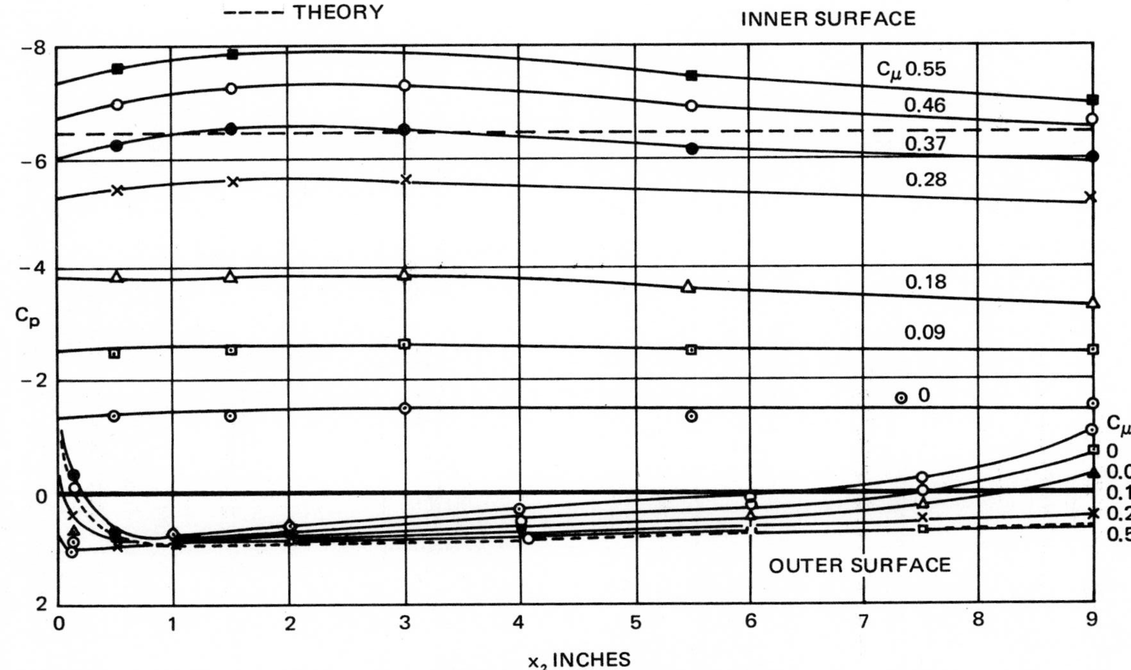

FIG. 6.10 Pressure distributions on lower plate (for $\alpha = 22$ deg, $\gamma = 130$ deg, and $\tau = 30$ deg; *from Hurley, 1961*).

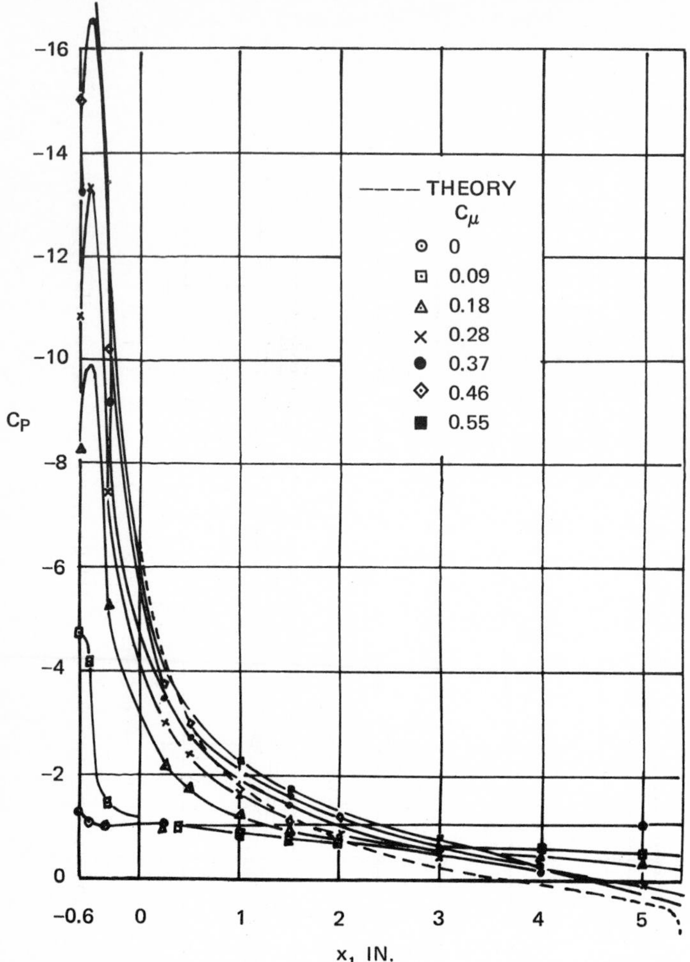

FIG. 6.11 Pressure distributions on outer surface of upper plate (for $\alpha =$ 22 deg, $\gamma = 130$ deg, and $\tau = 30$ deg; *from Hurley, 1961*).

low. As envisioned by the analysis, the flow left the leading edge of the lower plate tangentially to the lower plate except for the smallest values of C_μ. The flow outside a certain line joining the leading edge of the lower plate to the tube appeared to be perfectly steady; inside this line, the flow was turbulent and the velocity decreased to low values at or a short distance inside the line. As C_μ was increased, the line became progressively more convex; this behavior is consistent with the observation that the pressure between the plates is more negative (Figs. 6.10 and 6.12).

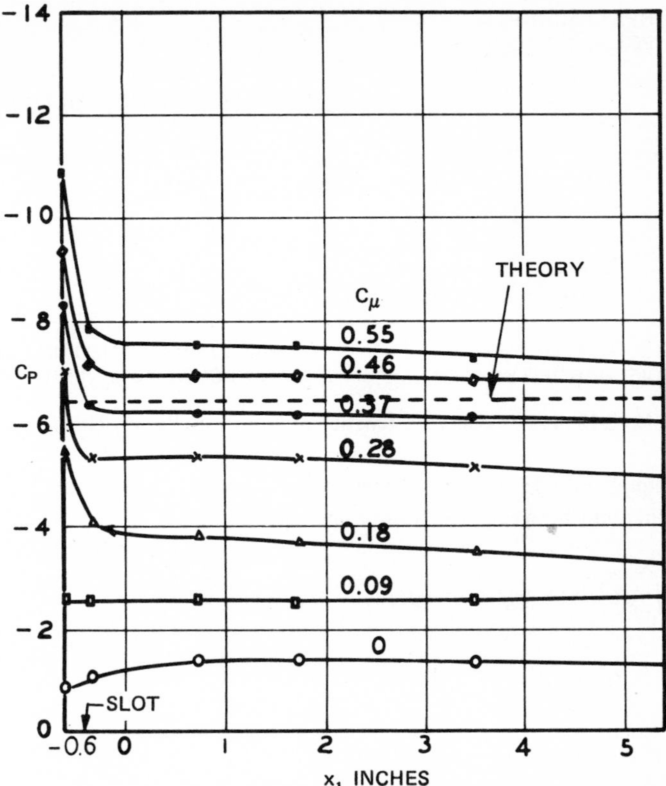

FIG. 6.12 Pressure distributions on inner surface of upper plate (for $\alpha = 22$ deg, $\gamma = 130$ deg, and $\tau = 30$ deg; *from Hurley, 1961*).

6.1.1.4 Evaluation of $C_{\mu\text{crit}}$ and its relation to C_L and C_D

As previously indicated from the successful performance of the free streamline attachment to the upper surface, a necessary minimum amount of blowing $C_{\mu\text{crit}}$ has to be produced. For the test conditions described previously, $C_{\mu\text{crit}} = 0.37$. The prediction of $C_{\mu\text{crit}}$ may be made by using mixing length theory. Hurley (1961) evaluated $C_{\mu\text{crit}}$ based on a model (Fig. 6.13) which is first related to $C_{\mu R}$. The symbol $C_{\mu R}$ refers to the value of C_μ needed to restore the average velocity in the boundary layer to u_e^*, the velocity outside the layer.

It is assumed that the reattaching boundary layer of momentum thickness θ_R mixes at pressure C_p^* with the high velocity air discharged from the slot. Here C_p^* is the pressure coefficient on the free streamline whose velocity is u_e^*. Furthermore, the velocity in the boundary layer is assumed to be uniform and

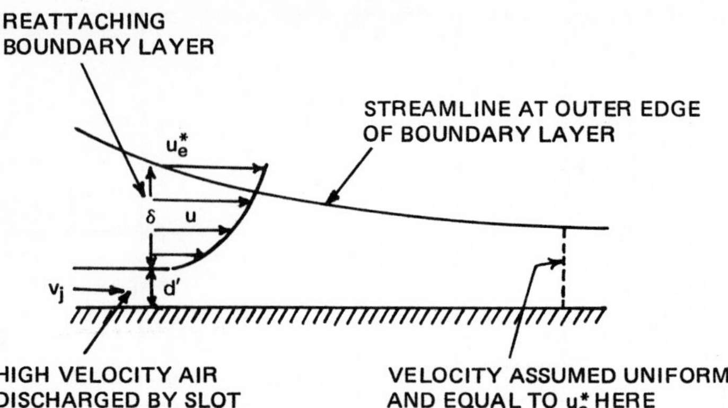

FIG. 6.13 Flow model for estimation of $C_{\mu R}$ (*from Hurley, 1961*).

equal to u_e^* at some downstream station. Then, applying the momentum theorem between this station and upstream,

$$\int_0^\delta \rho u^2 dz + \rho v_j{}^2 d' = u_e^*(\rho v_j d' + \int_0^\delta \rho u\, dz)$$

where v_j is the jet slot velocity, so that

$$\int_0^\delta u(u_e^* - u)\, dz = v_j{}^2 d' - u_e^* v_j d'$$

Dividing by $u_e{}^2 l_2$, for $v_j \gg u_e^*$, we obtain

$$\frac{u_e^{*2}}{u_e^2}\frac{\theta_R}{l_2} = \frac{v_j{}^2 d'}{u_e{}^2 l_2} = \frac{1}{2} C_{\mu R}$$

or

$$C_{\mu R} = \frac{2u_e^{*2}\theta_R}{u_e^2 l_2}$$

$$= 2(1 - C_p^*)\frac{\theta_R}{l_2}$$

$$= 0.024(1 - C_p^*)\frac{L}{l_2}$$

because $\theta_R = 0.012L$ by the Tollmien analysis of mixing, where L is the length of the free streamline. Taking $L/l_2 = 0.677$, then

$$C_{\mu R} = 0.016(1 - C_p^*) \tag{6.25}$$

In order to correlate the values of $C_{\mu R}$ and $C_{\mu \text{crit}}$, the measured values of $C_{\mu \text{crit}}$ are compared in Fig. 6.14 with those computed by Eq. (6.25).

Note from this figure that values of $C_{\mu \text{crit}}$ computed by Eq. (6.25) vary between one-half and one-third of the measured values for $C_{\mu \text{crit}}$. Therefore, $C_{\mu \text{crit}}$ may be evaluated approximately by

$$C_{\mu_{\text{crit}}} = 2.5 C_{\mu_R} = 0.06(1 - C_p^*) \frac{L}{l_2} \tag{6.26}$$

The validity of this equation was checked for $\tau = 45$ deg and $\alpha = 29$ deg. From the potential flow solution, $q_m/u_e = 3.1$. Then $1 - C_p^* = (3.1)^2 = 9.6$. If L/l_2 is assumed to be 3/2 times its value when $\tau = 30$ deg gives $L/l_2 = 1.0$, then from Eq. (6.26), $C_{\mu \text{crit}} = 0.58$. The value of $C_{\mu \text{crit}}$ is useful to correlate C_L with respect to C_μ as shown in Fig. 6.15. The predicted value of C_L at $\alpha = 22$ deg is also shown in Fig. 6.15 and indicates the good agreement with the values when $C_\mu = C_{\mu \text{crit}}$. The variation of C_D, the form drag coefficient, with respect to C_μ is shown in Fig. 6.16 with α as a parameter.

For large values of α, the values of C_D decrease with an increase in C_μ and for small values of α and large values of C_μ, the values of C_D become negative. This is due to the thrust produced by the sheet of high velocity air that leaves the trailing edge of the airfoil. Therefore, for the evaluation of drag, it is more convenient to consider the total drag given by ($C_{D_t} = C_D + C_\mu$) associated with the wing flap arrangement except for skin friction drag as shown in Fig. 6.17.

The value of C_{D_t} includes the loss of thrust involved when air is discharged out of the blowing slot. The values of the pitching moment coefficient C_m

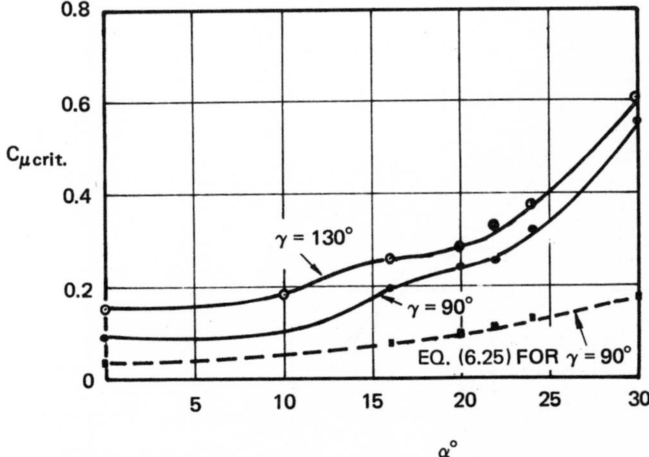

FIG. 6.14 Values of C_μ needed for complete attachment of flow to the outer surface of the plate (*from Hurley, 1961*).

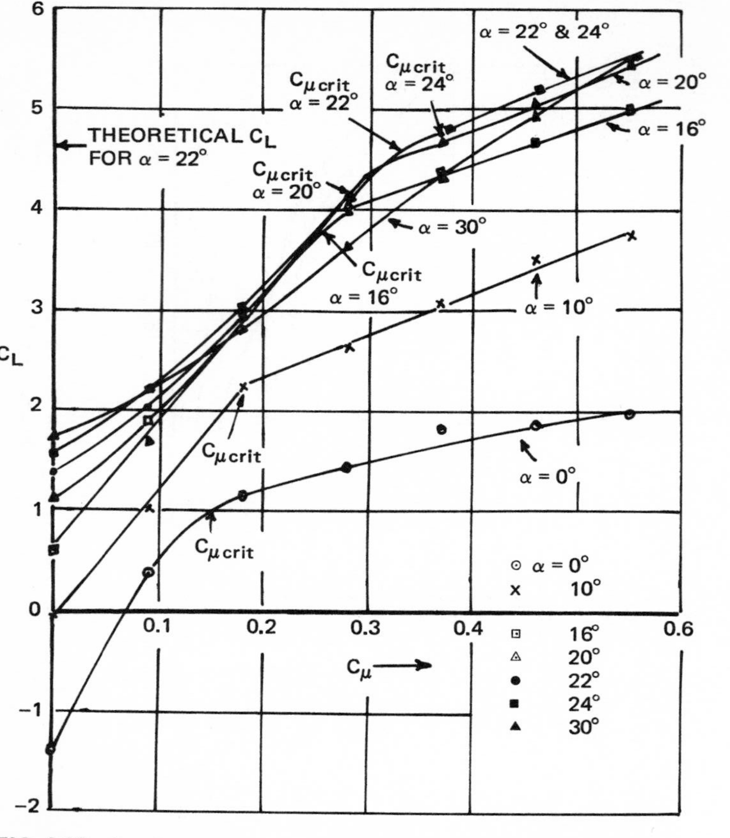

FIG. 6.15 C_L–C_μ curves for $\gamma = 130$ deg and $\tau = 30$ deg (*from Hurley, 1961*).

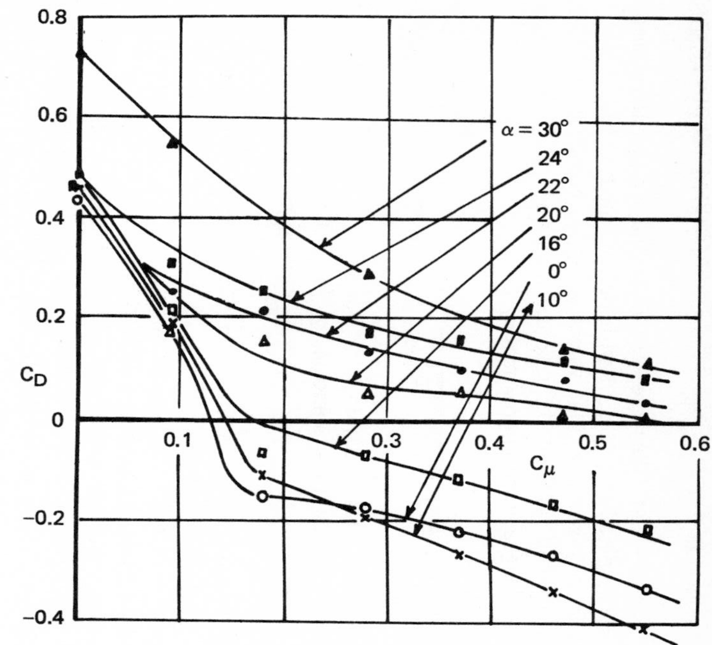

FIG. 6.16 C_D–C_μ curves for $\gamma = 130$ deg and $\tau = 30$ deg (*from Hurley, 1961*).

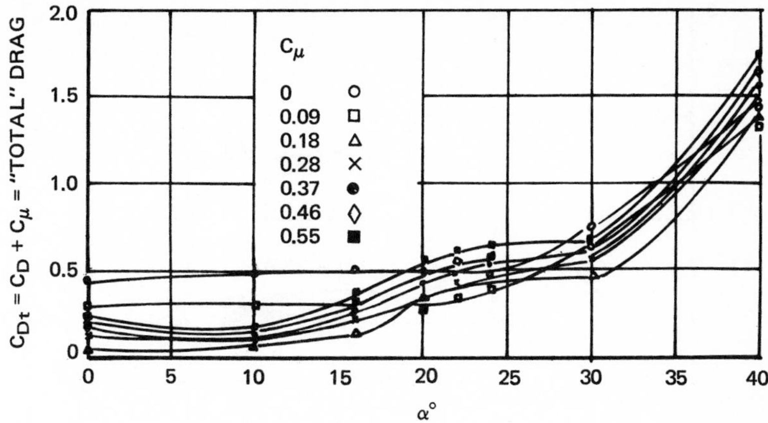

FIG. 6.17 C_{D_t}–α curves for $\gamma = 130$ deg and $\tau = 30$ deg *(from Hurley, 1961).*

about the quarter-chord point of lower plate are shown in Fig. 6.18; note that the absolute value increases as values of C_μ increase. At higher angles of attack, the absolute values of C_m increase with increasing values of C_μ except in the region of small C_μ. Finally, values of C_L obtained by the Hurley design are compared in Fig. 6.19 with a jet flap and blowing over a trailing edge flap at $\alpha = 10$ and 24 deg, $\gamma = 90$ deg, and $\tau = 30$ deg; note the superiority of the

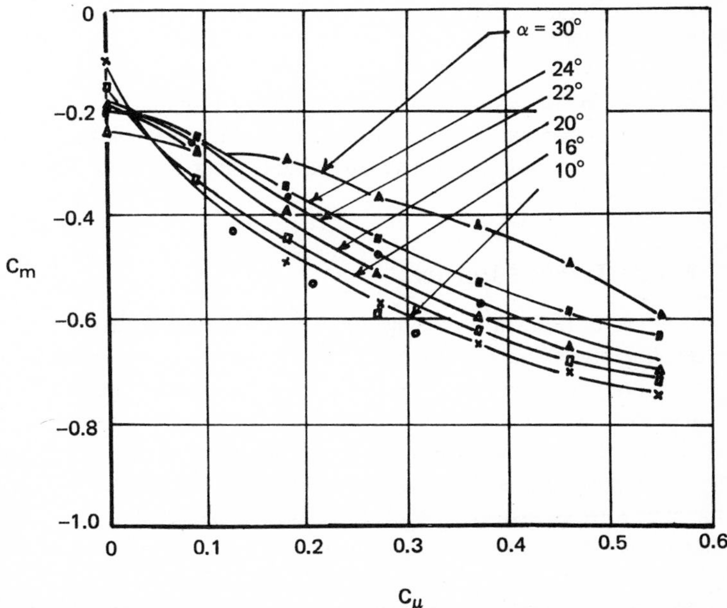

FIG. 6.18 C_m–C_μ curves for $\gamma = 130$ deg and $\tau = 30$ deg *(from Hurley, 1961).*

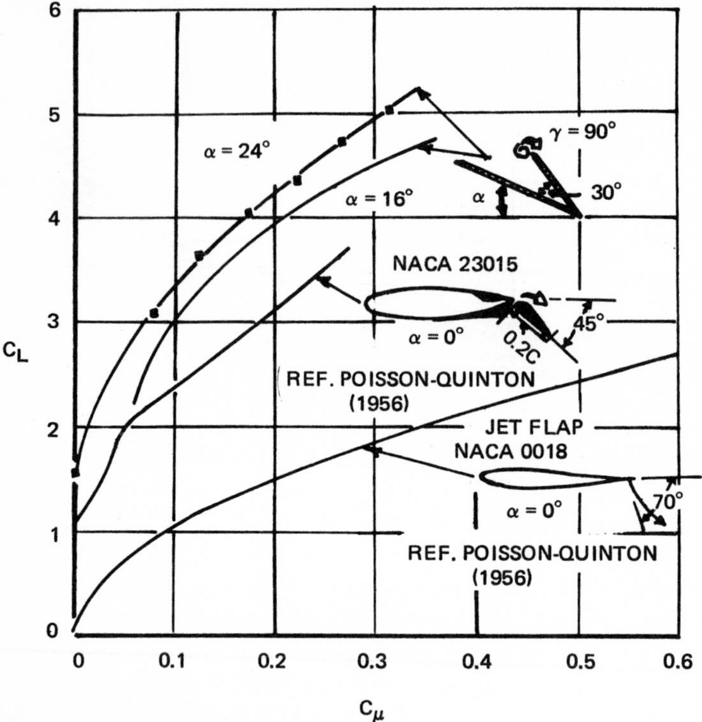

FIG. 6.19 Comparison of C_L values with those obtained by other methods (*from Hurley, 1961*).

Hurley design. However, it should be pointed out that the comparison would be less favorable had it been made for equal values of the incidence; such comparison was not made because suitable data for these flap arrangements were not available.

6.1.2 Trapped Vortices

The main flow in a passage involving separation may be controlled by trapped vortices. Such a control technique implies the change of an existing steady potential flow in a desirable way by creating standing vortices that change the contour of the flow area. The standing vortices may be understood as those that occur when the flow that contains the vortices is a steady one for an observer at rest with respect to the center of the vortices. In nature the standing vortices that may be observed behind snow cornices at high mountain ridges may be considered as natural, self-forming, automatic flow control devices.

An analysis by Ringleb (1961) leads to the design of a cusp surface configuration for a diffuser. First consider the snow cornice effect which

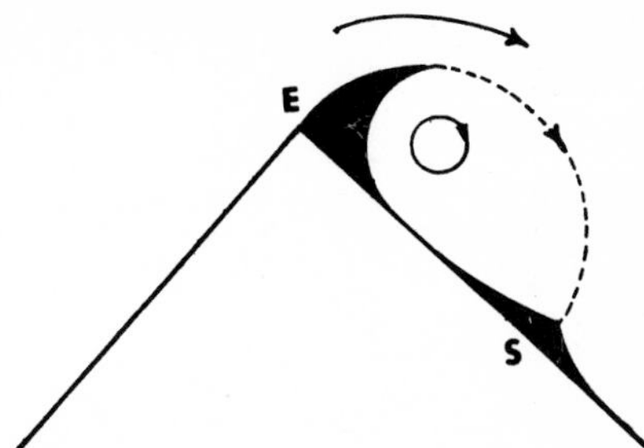

FIG. 6.20 A snow cornice (*from Ringleb, 1961*).

occurs because of the action of wind and snow. When wind passes over the mountain edge, it creates a vortex at the downstream side of the edge as shown schematically in Fig. 6.20. The vortex is within a region enclosed by a streamline, leaving the edge E, which ends on the downstream slope in a stagnation point S and the wall. Snow is deposited at point S because of the slow flow velocity; above S, snow is carried upstream toward the leading edge E where it freezes, forming a cusp-like cornice. The noted geologist and alpinist W. Welzenbach (1930) studied the shapes of cornices; one of his sketches of a double cornice is shown in Fig. 6.21.

The snow cornice is a natural flow-control device with an effect similar to that which produces the circulation around an airfoil with a sharp leading edge. With no cornice and vortex, the flow will have its maximum velocity over the leading edge; downstream, the velocity will decrease rapidly and flow separation will result near the trailing edge. However, the formation of the cornice with its ideal knife-edged cusp causes a change of the boundary of the flow area, creating a standing vortex in the flow and preventing the downstream flow separation and the formation of a long trail. Furthermore, it should be noted that the cusp prevents a sudden decrease of the velocity because a streamline may leave the contour tangentially with no disruption of velocity.

Another case of flow control by means of a standing vortex may be observed behind a dam protruding into a river; see the sketch in Fig. 6.22. For this case a streamline leaves the end point E of the dam and terminates at a stagnation point on the river bank. Sand is deposited near the stagnation point S where the flow velocity is small, thus contributing to the protection

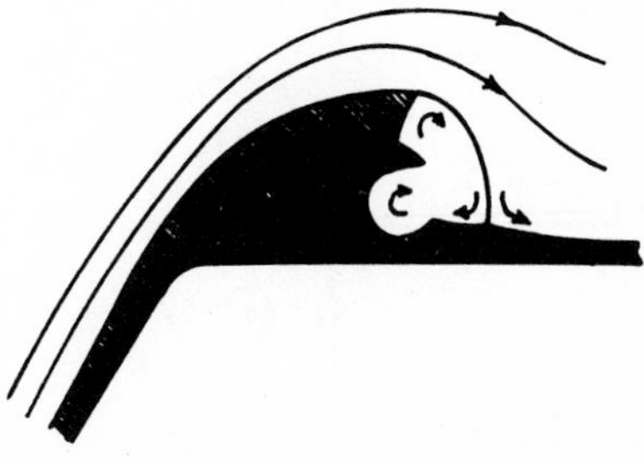

FIG. 6.21 A double snow cornice (*after W. Welzenbach; from Ringleb, 1961*).

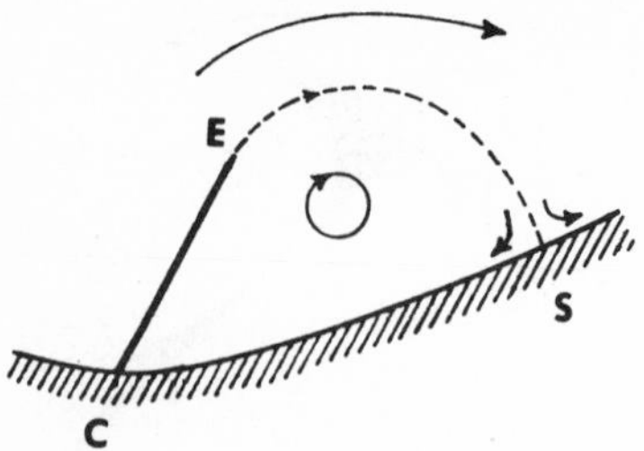

FIG. 6.22 Standing vortex behind a dam protruding into a river (*from Ringleb, 1961*).

of the bank. A standing vortex is then situated within the cavity of the cornice, causing a deep water hole in the center.

The stability of motion of the trapped vortex is important for any practical application. The fact that a vortex forms immediately downstream of a rearward facing step and within a rectangular cut indicates that the vortices are fairly stable. For the surface shape of snow cornice shown in Fig. 6.20, it appeared that in order to create a steady separation, the cusp should be formed so that a proper velocity distribution is provided along the wall and no separation takes place from positions other than the cusp.

6.1.2.1 Analysis of flow over snow cornices

The Ringleb technique of studying cornice flow by mapping leads to the practical surface configuration involving trapped vortices. The mapping function (see Fig. 6.23) used is

$$z = \zeta + \frac{C}{\zeta - \zeta_1} = f(\zeta)$$

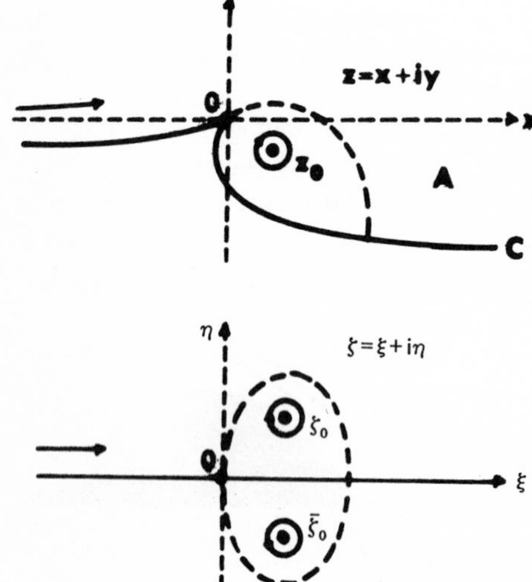

FIG. 6.23 Conformal map of the flow area (*from Ringleb, 1961*).

Here $z = x + iy$, $\zeta = \xi + i\eta$, and ζ_1 is an arbitrary point in the lower ζ-half plane. The constant C is determined by the conditions of $f'(0) = 0$ and $f''(0) \neq 0$. Then any angle at $\zeta = 0$ is doubled. Thus, the boundary of flow area A which is bounded by a simple piecewise analytical curve in the z-plane has a cusp corresponding to $\zeta = 0$. The condition $f'(0) = 0$ is satisfied if $C = \zeta_1{}^2$. Therefore, the mapping function that results is

$$z = \zeta + \frac{\zeta_1{}^2}{\zeta - \zeta_1} \tag{6.27}$$

This transformation can be performed similarly to the Joukowski case as indicated in Fig. 6.24.

Point P is the image of ζ by inversion in a circle whose center ζ_1 passes through 0. Point Q is the image of P with respect to the straight line $\zeta_1 0$, and point z is the fourth corner of the parallelogram with corner points of ζ, Q, and ζ_1. The angle of γ is formed between the straight line $\zeta_1 0$ and the ξ-axis.

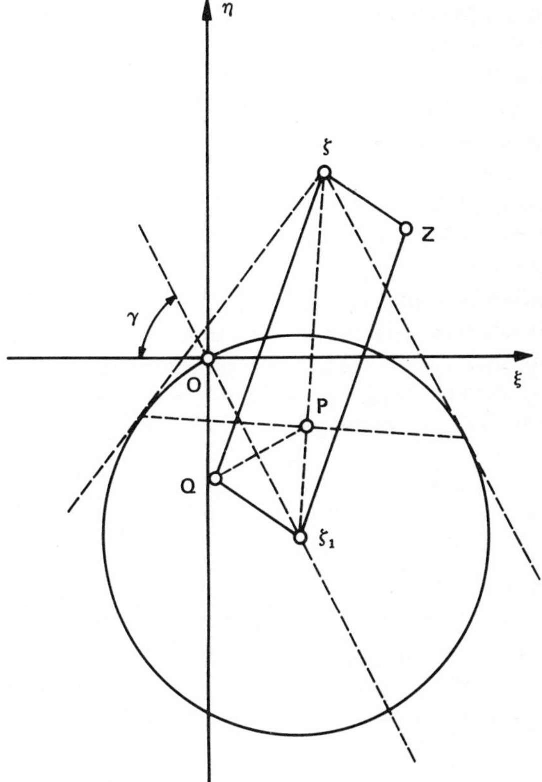

FIG. 6.24 Construction of the conformal map $z = \zeta + [\zeta_1{}^2/(\zeta - \zeta_1)]$ (*from Ringleb, 1961*).

The function $z = f(\zeta)$ given by Eq. (6.27) is a single-valued function with respect to ζ. The flow with a standing vortex in area A is obtained by mapping the flow within the ζ-plane with a standing vortex at ζ_0 given by

$$\frac{w}{u_\infty} = \zeta + 2\eta_0 i \ln \frac{\zeta - \zeta_0}{\zeta - \bar{\zeta}_0}$$

Here w is the complex potential flow function $w = \phi + i\psi$, $\bar{\zeta}_0 = \xi_0 - i\eta_0$, and the subscript 0 refers to the vortex position.

If $z = f(\zeta)$ is a single-valued function of ζ, the equilibrium condition of the vortex at z_0 is satisfied by only the following conditions

$$C_\Gamma = 2\eta_0 \quad \text{and} \quad C_Q = 0 \tag{6.27a}$$

where $C_\Gamma = \Gamma/2\pi u_\infty$, Γ is vortex strength, $C_Q = Q/2\pi u_\infty$, and $2Q$ is sink strength. It is assumed that a sink is situated at a point $\zeta = a$ on the real ξ-axis. The stagnation condition given by

$$1 - \frac{2\eta_0}{\xi_0^2 + \eta_0^2} C_\Gamma + \frac{2}{a} C_Q = 0 \tag{6.28}$$

and the equilibrium condition expressed by Eq. (6.27*a*) leads to the condition

$$\eta_0 = \frac{1}{\sqrt{3}} \xi_0 \tag{6.29}$$

for the coordinates of the center of the standing vortex in the ζ-plane. It is situated on the straight line through the origin at an angle of 30 deg with the positive ξ-axis. Figure 6.25 shows that area A corresponds to $\gamma = 90$ deg with a possible situation of a standing vortex at z_0 and the corresponding stagnation streamline.

A somewhat different shape may be obtained by using the following function:

$$z = \zeta + \zeta_1 \ln (\zeta - \zeta_1) \tag{6.30}$$

where ζ_1 represents an arbitrary point in the lower ζ-half plane. For this case, the boundary of the area A has a cusp corresponding to $\zeta = 0$, and the function $z = f(\zeta)$ is multivalued. If we set $\tau = \phi(\zeta) = ln(\zeta - \zeta_1)$, then $z = \zeta_1(1 + \tau) + \exp(\tau)$ and $\zeta = \zeta_1 + \exp(\tau)$. For a multivalued function of ζ, the necessary and sufficient condition that the vortex at z_0 is a standing one is

$$\frac{\phi''(\zeta_0)}{\phi'(\zeta_0)} = i\left(\frac{1}{\eta_0} - \frac{2}{C_\Gamma} + \frac{4C_Q}{C_\Gamma} \frac{1}{\zeta_0 - a}\right) \tag{6.31}$$

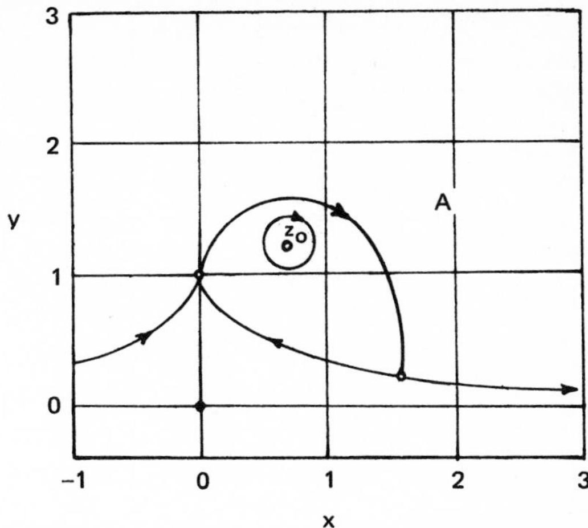

FIG. 6.25 Snow cornice flow when position of standing vortex is not uniquely determined (*from Ringleb, 1961*).

where $\tau = \phi(\zeta)$ is the function which conformally transforms the Riemann surface of the function $z = f(\zeta)$ covering the ζ-plane into the single-sheeted τ-plane whereby the branch points of the ζ-plane are excepted.

From Eq. (6.31) relating to the existence of a standing vortex,

$$-\frac{1}{\zeta_0 - \zeta_1} = i\left(\frac{1}{\eta_0} - \frac{2}{C_\Gamma}\right)$$

or

$$\xi_1 = \xi_0 \qquad \frac{1}{\eta_0 - \eta_1} = \frac{1}{\eta_0} - \frac{2}{C_\Gamma} \tag{6.32}$$

From the stagnation point condition given by Eq. (6.28),

$$C_\Gamma = \frac{\xi_0^2 + \eta_0^2}{2\eta_0}$$

By introducing these results into Eq. (6.32),

$$\xi_1 = \xi_0 \qquad \frac{\eta_1}{\eta_0} = \frac{4}{3 - (\xi_1/\eta_0)^2} \tag{6.33}$$

These equations provide a relation between the position of the standing vortex ζ_0 and the singularity ζ_1 of the mapping function; an example is shown in

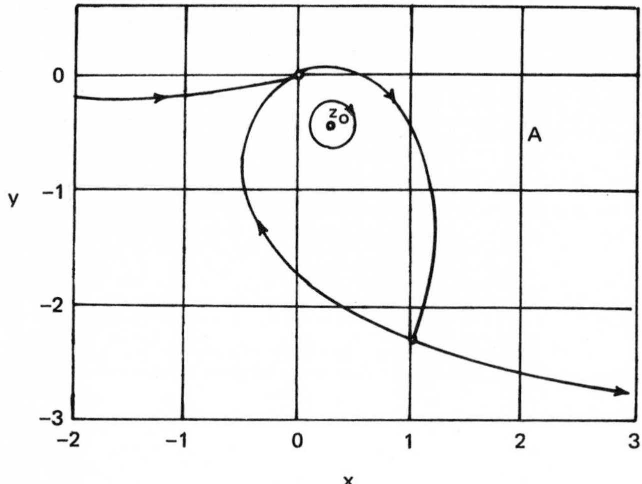

FIG. 6.26 Snow cornice flow when position of standing vortex is uniquely determined (*from Ringleb, 1961*).

Fig. 6.26. These two examples indicate that two different possibilities of flow phenomena exist over a cusp with a standing vortex. In the first case, Eq. (6.27), since the position of the standing vortex is not uniquely determined, the center of the vortex is still free to change its position on a curve, determined by Eq. (6.29), and thereby retains the characteristic of the center of a standing vortex. In the second case, however, Eq. (6.30), the position of the standing vortex is uniquely determined by the condition given by Eq. (6.33). The stability condition of a standing vortex is $[\eta_0 = (1/\sqrt{3})\xi_0]$. The physical interpretation is that a vortex whose center is initially at $\xi_1 = \xi_0$ and η_1 will move into the equilibrium position ξ_0 and $\eta_0 = (1/\sqrt{3})\xi_0$ with an increase in time; see Fig. 6.27.

If a vortex is initially in equilibrium and if the stagnation point condition is satisfied at any time during a disturbance, then that vortex will eventually return to the equilibrium position. If it temporarily moves parallel to the ξ-axis, that will be only during the period of the disturbance. Thus, the equilibrium is a stable one. Ringleb (1961) states that the same is true for all flows obtained by mapping within the ζ-plane into the z-plane by means of a single-valued function $z = f(\zeta)$, provided the Kutta-Joukowski condition is satisfied at any moment during the disturbance. In reality, however, this stability consideration based on potential flow theory and the assumption that a sharp edge with the angle 2π enforces the Kutta-Joukowski condition are not satisfied. Ringleb (1961) listed several supplementary arrangements for stabilization of a vortex for which an equilibrium position exists:

1. Application of suction downstream of the vortex.
2. Application of vanes located along parts of one or several streamlines.

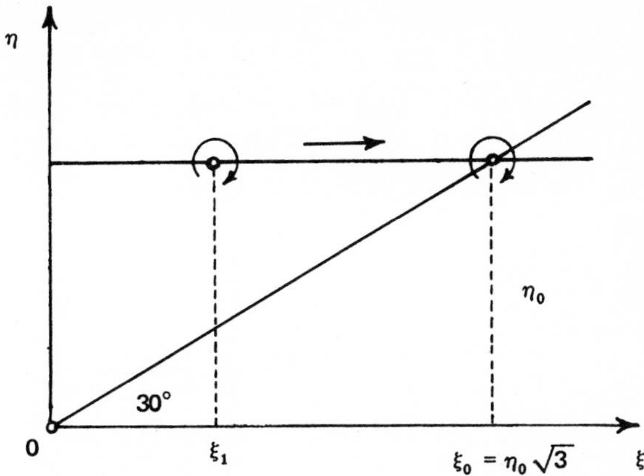

FIG. 6.27 Stability of standing vortex (*from Ringleb, 1961*).

3. Application of weak suction in the direction of the vortex centerline through the parallel walls that enclose the two-dimensional flow (boundary layer suction).

Figure 6.28 illustrates the vortex profile of applied suction. The profile has a suction slot near the trailing edge and a cusp cavity in order to trap a vortex on its upper side. It was not based on analysis, but rather was designed in such a way as to create an improved flow into the suction slot by guiding it over the trapped vortex and to save the power required for suction at the trailing edge. Since the measured pressure distributions on this profile are interesting, the two most distinct examples are shown in Figs. 6.29 and 6.30. It should be noted first that the lift-coefficient $C_L = 0.096$ at $\alpha = 0$ deg, obtained with a suction coefficient of $C_Q = 0.0307$, increased suddenly to

FIG. 6.28 Suction vortex profile (*from Ringleb, 1961*).

$C_L = 1.275$ at the slightly higher value of $C_Q = 0.0364$. This is attributable to the formation of the standing vortex and the considerably higher lift that resulted. The coefficients of lift and drag are related to C_Γ and C_Q by

$$C_L = 2C_\Gamma \quad \text{and} \quad C_D = 2C_Q$$

The lift is affected by the standing vortex because of the relation of C_Γ with C_{Γ_0} by the stagnation point and the equilibrium conditions. This paradox is explained by Ringleb (1961) as an effect similar to a change in the profile

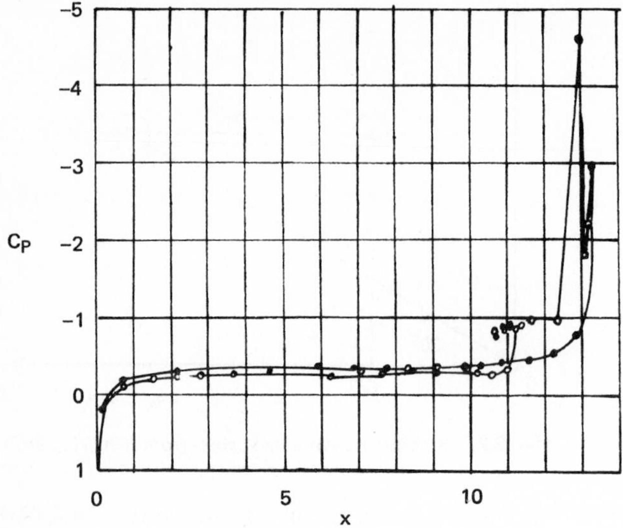

FIG. 6.29 Pressure distribution of suction vortex profile $\alpha = 0$ deg, $C_L = 0.096$, and $C_Q = 0.0307$ (*from Ringleb, 1961*).

shape. If the stagnation streamline that encloses the standing vortex is assumed to freeze, then a different profile may result. The direction of zero lift changes; consequently the lift also changes. A more favorable result is reported by Smith (1956) with a somewhat different design.

Mandl (1959) provides another example of suction applied for the stability of trapped vortex at a split flap; see the sketch in Fig. 6.31. Mandl

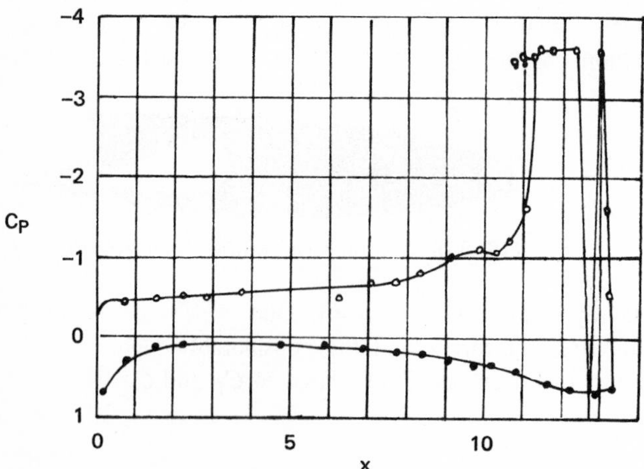

FIG. 6.30 Pressure distribution of suction vortex profile $\alpha = 0$ deg, $C_L = 1.275$, and $C_Q = 0.0364$ (*from Ringleb, 1961*).

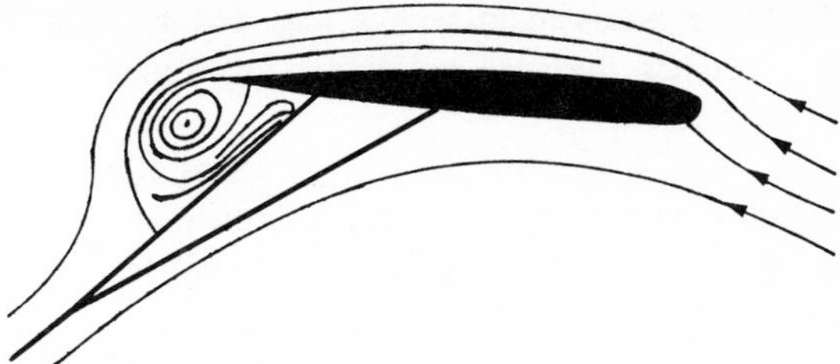

FIG. 6.31 Suction split flap with standing vortex (*after P. Mandl, 1959; from Ringleb, 1961*).

investigated this problem theoretically. The flow behavior obtained in his flow visualization study (Fig. 6.31) demonstrated that the Kutta-Joukowski condition is satisfied by streamlines leaving the trailing edge of the profile and of the split flap tangentially.

6.1.2.2 Cusp diffuser with standing vortices

Ringleb successfully designed a two-dimensional cusp diffuser which provided a smooth flow and good diffuser efficiency, as evidenced by the cusp diffuser in the Princeton wind tunnel. Thus he contributed to the solution for the control of separation along the diffuser wall, as shown in Fig. 6.32. No artificial stabilization of vortices was applied in this design; instead, a proper mapping of the outside of the circle $|\zeta| > 1$ was made. The flow outside the circle $|\zeta| > 1$ (corresponding to the flow within the diffuser containing two standing symmetrical vortices within the cusp cavities) had a source at $\zeta = -1$, a sink at $\zeta = +1$, and two symmetrical vortices at ζ_0 and $\bar{\zeta}_0$. Equilibrium conditions for the vortices and stagnation point conditions for the cusp points were derived. Lippisch used a similar axisymmetrical diffuser in a smoke tunnel.

The following items should be taken into account for the design: Cusps should not be bent toward the wind tunnel wall because such a design does not work. The angle of the cusp tangent with the general flow direction

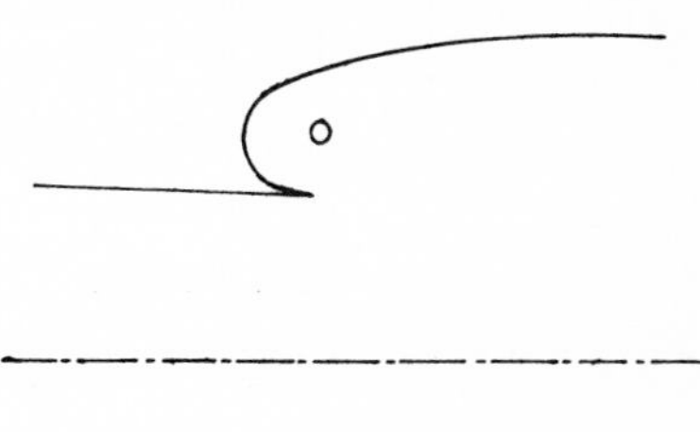

FIG. 6.32 Cusp diffuser (*from Ringleb, 1961*).

should be arranged in such a way that no flow separation occurs before the flow reaches the cusp point; this angle plays an important role in stabilizing the vortices.

For the Princeton diffuser, the cusps were slightly bent toward the tunnel center. Figure 6.33 gives curves of constant C_p obtained by measurement. The actual location of the center of the vortex was slightly different from that computed.

6.1.2.3 Discussion

Frey and Vasuki (1964) recently tested subsonic cusp diffusers and observed the flow development. The flow visualization with aluminum powder on the water surface was useful for observing the flow phenomena and the decay process of the starting vortices to the final status of the flow.

Items (*a*) and (*b*) of Fig. 6.34 show the limited stages of the starting vortices and clearly indicate the layers of discontinuity. Note in item (*c*) that the layers of discontinuity became wavy and swirled into the jet boundaries. The starting vortex center moved downstream and the elongated starting vortices–items (*d*)-(*g*) of Fig. 6.34–did not extend their rotary components into the cusps. The initial stage of the disintegration of the starting vortices is shown in item (*g*).

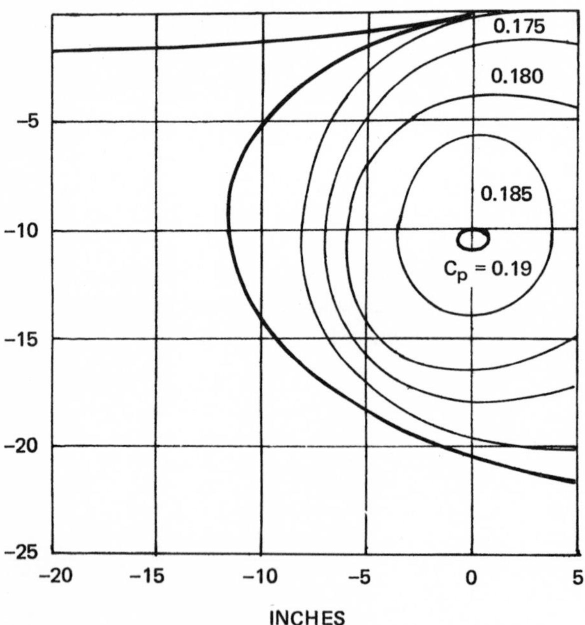

FIG. 6.33 Pressure distribution measurement of a standing vortex within a cusp diffuser (cusp diffuser such as shown in Fig. 6.32; *from Ringleb, 1961*).

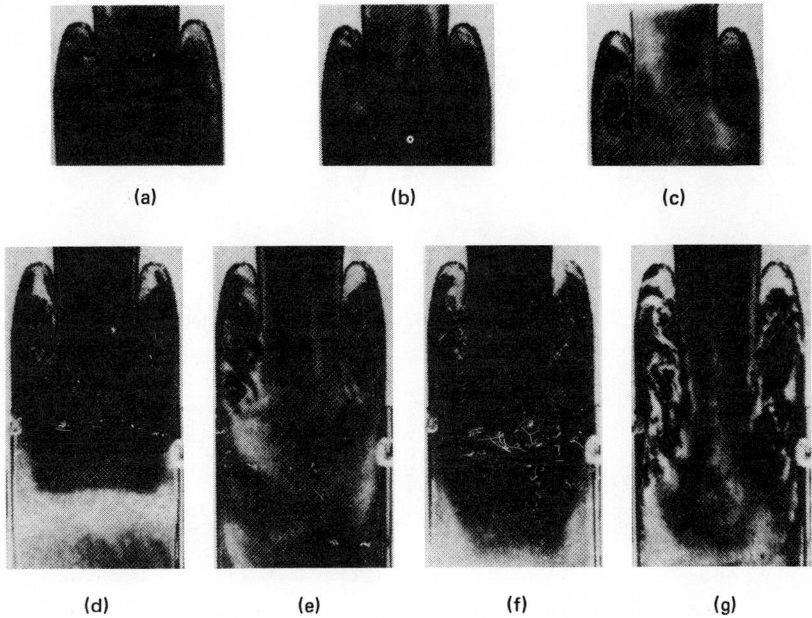

FIG. 6.34 Starting phase of flow in a cusp diffuser (*from Frey & Vasuki, 1964*).

Figure 6.35 illustrates time-dependent development of the first stage of the final status of flow, unmistakably showing the disintegrated and moving vortices. The boundary had no appreciable effect on the jet flow. Ringleb (1964) made the following comment on the experiment of Frey and Vasuki (1964):

The diffuser shape with cusp tested is one which never worked perfectly in completely two-dimensional arrangement, although it has been used successfully as part of a three-dimensional set-up where the flow has been guided away in a perpendicular direction behind the cusp-cavities.

Ringleb applied Eq. (6.30) to calculate the velocity distribution (Fig. 6.36) for the cusp diffuser used by Frey and Vasuki to obtain the photographs shown in Figs. 6.34 and 6.35. He also calculated the velocity distribution for another cusp diffuser as shown in Fig. 6.37. Where the two sets of velocity distributions are compared, it is not surprising that a standing vortex was not obtained for the cusp of Fig. 6.36 but was obtained for the cusp shown in Fig. 6.37. This is apparently due to the marked differences in the velocity distribution and the fact that the line of vortex equilibrium positions of Fig. 6.37 is only a single point in Fig. 6.36.

The dashed line in Fig. 6.37 indicates the locus of equilibrium positions for the vortex. The cusp diffuser of Fig. 6.36 has only the one equilibrium point

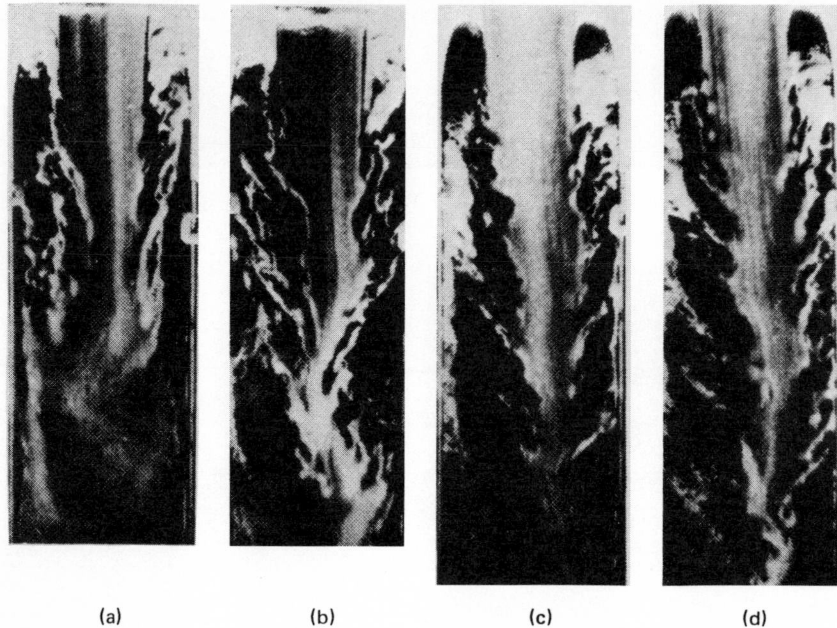

FIG. 6.35 Final phase of flow in a cusp diffuser (*from Frey & Vasuki, 1964*).

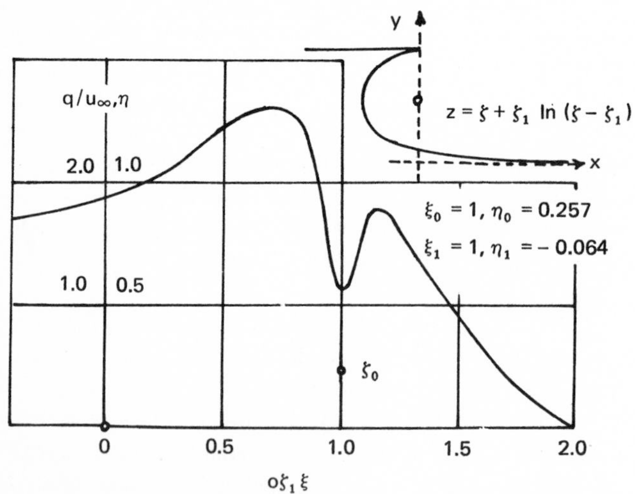

FIG. 6.36 Computed velocity distribution for the cusp diffuser (used by Frey & Vasuki, 1964, for the flow shown in Figs. 6.34 and 6.35; *from Ringleb, 1964*).

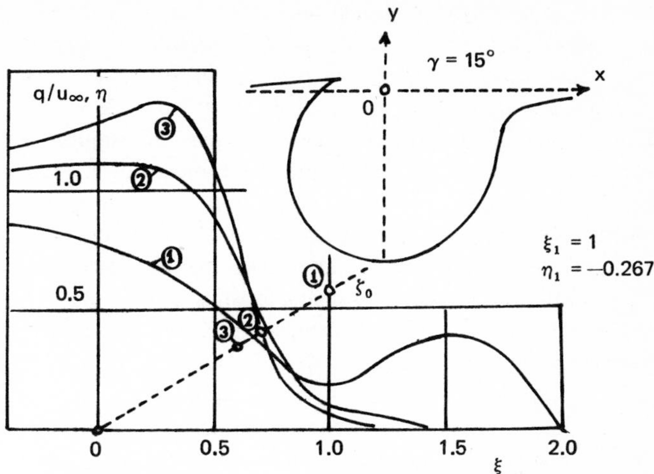

FIG. 6.37 Computed wall velocity distributions for a cusp diffuser (*from Ringleb, 1964*).

for the vortex, denoted by the small circle and subscript 0. The three velocity distributions of Fig. 6.37 correspond to those three stable vortex positions that are marked in the figure; the symbol q refers to the velocity at any point z in the flow area corresponding to ζ by $z = f(\zeta)$ and is given by

$$\frac{q}{u_\infty} = \left| \frac{(\zeta - 2\xi_0)(\zeta - \zeta_1)^2}{(\zeta - 2\zeta_1)(\zeta - \zeta_0)(\zeta - \bar{\zeta}_0)} \right|$$

Ringleb (1964) considered the formation of mushroom vortices in the Frey-Vasuki photographs rather remarkable. The regular series of vortices is of particular interest for the explanation of the driving mechanism of a standing vortex.

6.2 CHANGE IN SHOCK SHAPE

As already discussed in Chaps. 1 and 3, and shown in Fig. 1.20, a spike protruding from a blunt body exposed to a supersonic flow may change the shock shape and because of the streamwise pressure rise may provoke laminar or turbulent flow separation along the spike length, indicated in Fig. 1.21. The details of the flow phenomena and characteristics of spiked blunt body are presented in Chap. 9 of *Separation of Flow* (Chang, 1970). A spike protruding from a blunt body at supersonic speeds can reduce drag, increase lift, and correspondingly change the pitching moment. Thus, a spike can be an effective control device. A high-speed airborne vehicle can utilize the spike to make best use of the thrust of the propulsive system. For example, a boost-glider that travels in the atmosphere at nearly constant Mach number can considerably reduce the needed thrust and the rate of heat transfer. A blunt body with a given heat capacity may accomplish reentry at a higher Mach number, accompanied by increased impact of a warhead; this makes the use of countermeasures more difficult. Since a spike reduces the base pressure and probably the rate of heat transfer in the base, the design problems of the structure could be eased. A nose spike may be used for takeoff and other flight conditions. Since flow characteristics can be controlled by varying the length and diameter of the protruding probe, the thin straight probe is a convenient and simple control device compared to variable geometry skirts. There is, however, a limitation on its application. An unfavorable effect of flow oscillation on the probe may cause aerodynamic disturbance. Let us first consider the flow regimes on spiked blunt body.

6.2.1 Flow Phenomena over Spiked Body

6.2.1.1 Oscillating flow

Wood (1961) used a model of a spiked body of revolution involving separation (Fig. 6.38) and observed five distinct flow regimes for zero angle of attack at $M_\infty = 10$ (Fig. 6.39). Several spikes of different length were used for

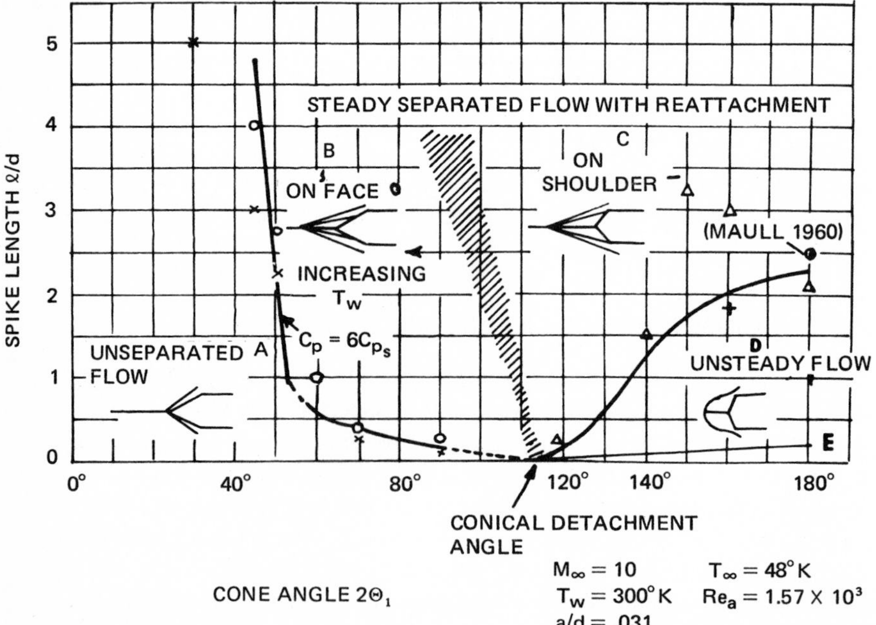

FIG. 6.38 Separated flow on a spiked model (*from Wood, 1961*).

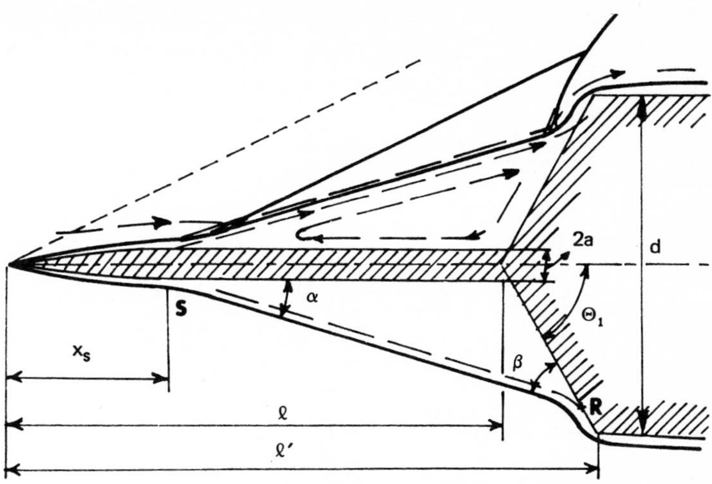

FIG. 6.39 Flow regions for spiked cones (*from Wood, 1961*).

this photographic study, and their apex angles were also varied. The flow was laminar and the shape of the separated region was conical except for two very small regions near the points of separation and reattachment.

All boundaries of the different flow regimes were centered on the conical separation angle. The oscillating flow occurred at a cone angle larger than the conical detachment angle and while fully attached flow was observed at cone angles smaller than this. It appears that the classification of flow regions that Wood (1961) made by using the cone model may also be applied for the flow region of a spiked blunt body whose shape differs from the cone. Mair (1952) found that at $M_\infty = 1.96$ and $Re = 1.3 \times 10^5$/cm, the relation between conical shock wave and the angle of separation was independent of the shape of the blunt body; see Fig. 6.40. It is seen from this figure that for nearly steady conical flow with separation at points on the spike, the angle of the conical separated region increased as the angle of the conical shock became larger. Analytically, this relation can be predicted by using the Kopal table (1947), although predicted values are generally higher than measured values.

Each model configuration in Fig. 6.38 is represented by a single point. The unseparated flow is found in Region A and the steady flow with reattachment in Regions B and C, although the boundary between Regions B and C is ill-defined. Region D is for unsteady flow. The fifth and last type of flow, Region E, is relatively unimportant and makes no difference to the flow. This type of flow occurred when the spike failed to pierce the detached flow wave. In Region A, for cones of less than about 30 deg total apex angle, the pressure rise was not large enough to cause separation; therefore, no separation is expected for a slender body. In Region B the shear layer reattached on the conical face in a steady flow condition, enclosing a fixed mass of flow in the separated region. Flow occurred in Region C as the angle

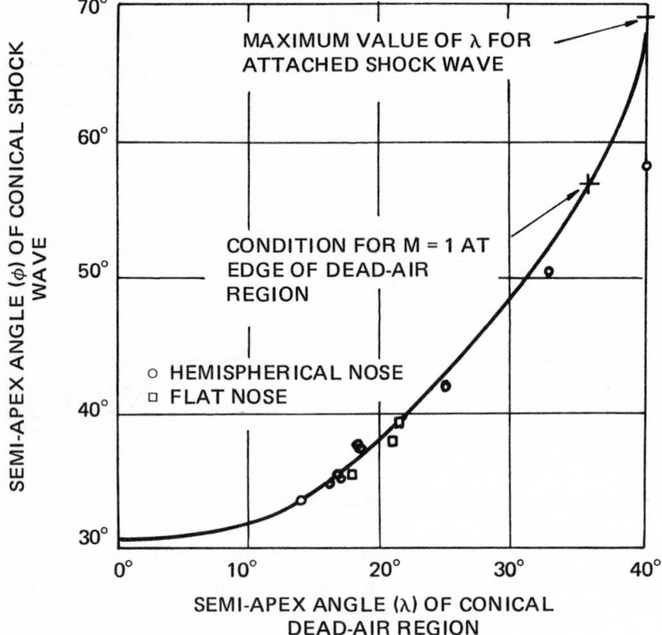

FIG. 6.40 Relation between conical shock-wave angle and conical angle of separated region (*from Mair, 1952*).

of the body surface increased and the reattachment point moved nearer to the shoulder to maintain the required pressure rise. The flow oscillated in Region D. Since this unsteady oscillating flow may cause the aerodynamic disturbance, it is of interest to study more details of this type of flow.

6.2.1.2 Oscillating flow—Mair and Maull

The hypothesis for oscillating flow presented by Mair (1952) and Maull (1960) is based on the mass balance. Photographs of the starting phase of the cycle of oscillating flow of a spiked axial symmetric body—item (*a*) in Fig. 6.41—indicate that the end of the separated conical region near the nose was almost below the shoulder. For this case, the shock must be detached in order to turn the flow just outside the separated region up to the nose of the cylinder. Now consider the equilibrium of fluid flow. The pressure rise across this shock and the area behind the shock, through which fluid may be reversed into the separated region, is too great to establish an equilibrium with the fluid mass salvaged from the separated region upstream of the shock. Hence, fluid flowing into the separated region down the nose is in a condition of nonequilibrium. This condition enlarges the separated region and the nearly normal shock wave formed at the tip of the spike grows and replaces the conical shock. As more fluid is fed into the separated region, the shock wave

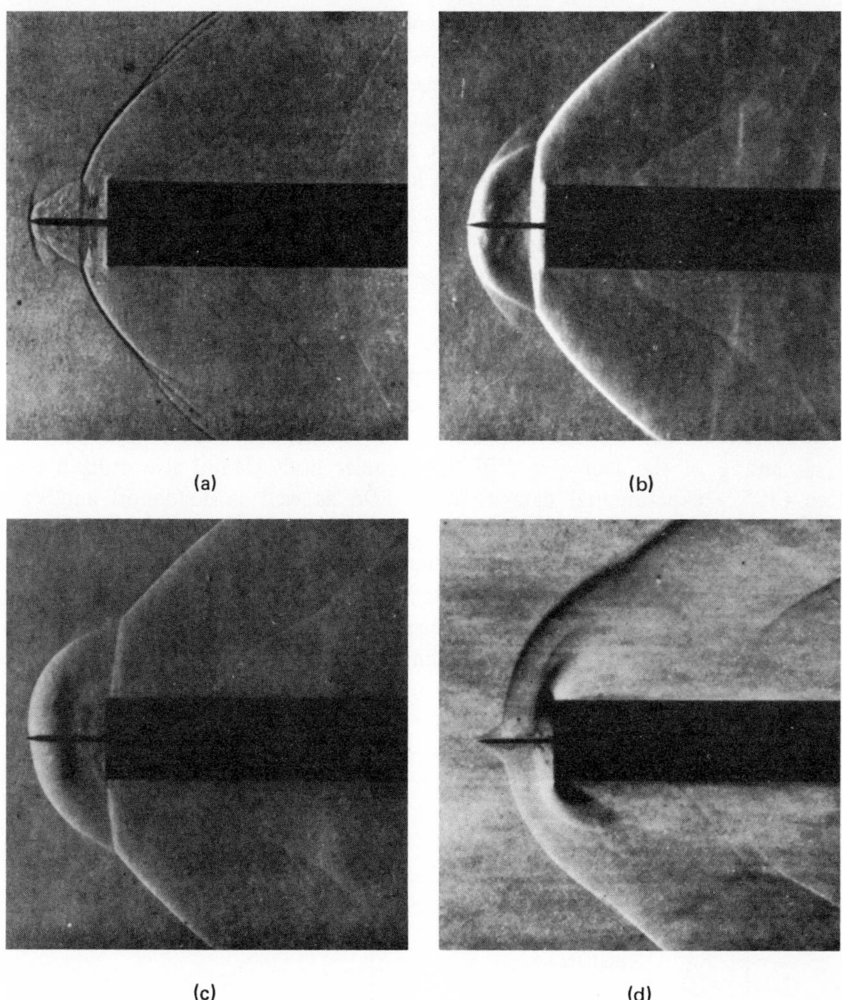

FIG. 6.41 Effect of a thin spike protruding from a blunt-nosed axially symmetric body (*from Mair, 1952*).

grows and the interaction with the bow shock of the body moves outward toward the shoulder of the cylinder as shown in item (*b*) of Fig. 6.41. Consequently, the shape of the separated region changes from a conical form to a blunt one, and the reattachment region moves up to the face of the blunt body. With this broadening of the separated region, there is a decrease in the angle through which the external flow must turn to pass around the nose; the pressure ratio across the shock that turns it also decreases. Therefore, the feeding of fluid into the separated region ceases. At this stage the fluid is fed from the region out past the shoulder, and this results in the collapse of the separated region, a movement of the shock wave from the tip toward the body (item [c] in Fig. 6.41), and a weakening of the shock. (Compare the sharpness of the shock wave for items [b] and [c] of Fig. 6.41.) As this shock moves downstream along the probe, separation again occurs at the probe tip, shaping a conical shock wave as seen in item (d) of Fig. 6.41. When the excess fluid in the separated region has escaped and the strong shock wave that originated at the tip of the probe has become the bow shock wave of the blunt body, the cycle starts again.

The oscillation depends on the nose shape, the length of probe, and the Mach number. Figures 6.42 and 6.43 show the oscillating regions obtained experimentally by Maull (1960). These were based on M_∞ and 6.8 and $Re = 0.17 \times 10^6$/in. with spikes whose lengths varied in the range of $0.25 < l/d \leqslant 2.5$. The basic configuration was a plain flat-nosed cylinder with rounded nose-shoulders. The respective radii r of the four models used were $\frac{1}{8}$, $\frac{1}{4}$, $\frac{3}{8}$, and $\frac{1}{2}$ of the diameter d of the cylinder body. Maull also utilized the Mair (1952) experimental data at $M_\infty = 1.96$ as well as Bogdonoff and Vas (1959) data at $M_\infty = 14$.

It is clear from these two figures that the oscillating range increased with an increase in Mach number. Although oscillation occurred for the flat-nosed cylinder, it was not observed for a hemisphere nose with $r/d = 0.5$.

It is interesting to note that oscillation was also suppressed for the following two cases:

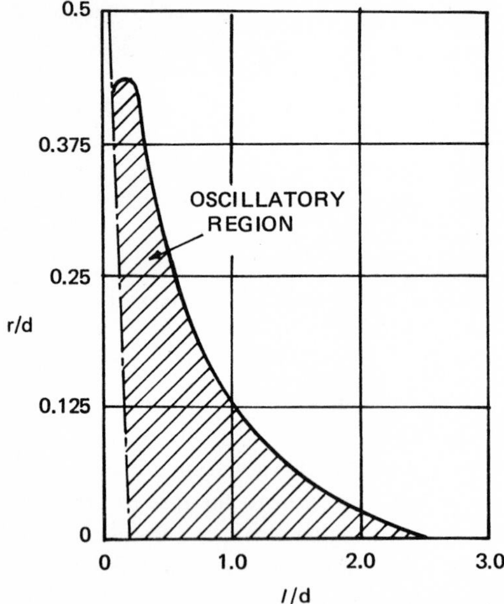

FIG. 6.42 Effect of shoulder radius on the oscillating region (*from Maull, 1960*).

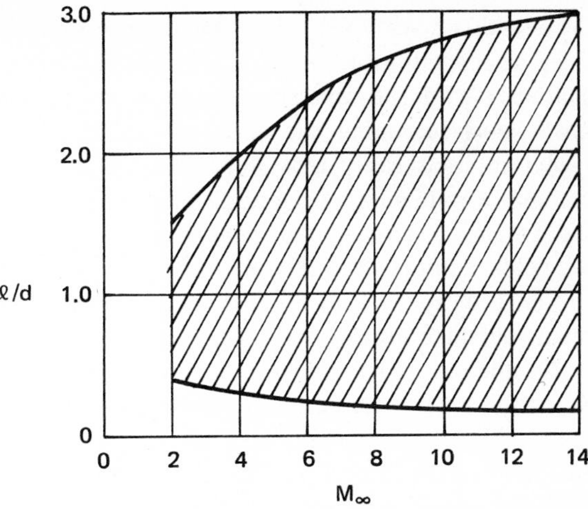

FIG. 6.43 Oscillatory range versus Mach number for flat-faced cylinder $(r/d) = 0$ (*from Maull, 1960*).

1. When the spike was extended beyond the critical length at which the drag became a minimum, the separation became unstable and oscillated rapidly up and down the spike. Such oscillation persisted as long as the boundary layer on the spike remained laminar. However, the oscillation stopped as soon as the boundary layer became turbulent before the separated region was reached.
2. Album (1961) observed that a frequency of 3000Hz at zero angle of attack dropped to roughly 2800Hz at a 5.5 deg angle of attack and that oscillation ceased at an angle of attack of 13.5 deg.

The oscillating flow differentiated two-dimensional flow from axial symmetric flow. For example, although for the two-dimensional flow case, the change from one type of separation to another was accompanied by oscillation, no unsteady flow was observed for the axially symmetric flow. The position of separation with respect to the spike length depends on the length of the spike; length is one of the governing parameters for flow control of a spiked *blunt* body. But the position of the shock wave is close to the surface of a spiked *slender* body, and there is no appreciable alternative of the shock wave (and consequently no significant reduction of drag).

A change in the flow characteristics can be achieved by varying the length of the spike. If the spike is short, "tip separation" occurs. If the spike becomes longer, the separation point moves downstream toward the shoulder and remains there through a range of lengths. For this case Hahn (1966) identified the onset of separation at the origin of Mach wave emanating from the spike surface. This kind of separation is called *retarded separation* to differentiate from *tip separation.* At zero angle of attack with an increase in the length of the spike, the flow on the spike becomes turbulent and the angle of separation decreases, consequently decreasing the drag. This decrease continues until the separation point leaves the shoulder to move along the spike, thus increasing the angle of separation and consequently the drag. Therefore, it is easily sensed that there is a spike length for which the drag becomes a minimum as the angle of separation varies. Such a length is called the *critical length of the minimum drag.* It is principally a function of Reynolds number, but is also influenced by Mach number, angle of attack, and ratio of spike diameter to nose diameter of blunt body. Hahn found that when the spike exceeded its critical length, which was approximately equal to twice the diameter of the hemispherical nose at $M_\infty = 1.61$-1.81 and $Re_d = 0.3 \times 10^6$, the separation moved suddenly downstream along the spike and could not be maintained at the shoulder. The critical length for minimum drag was obtained, for example, with the longest conical tipped spike for which separation occurred on the shoulder.

Further, it is conceivable that there exists a spike length at which the jump of the separation point from the tip of the spike to a position of retarded separation, or vice versa, takes place. This is called *critical length for the jump of separation point.* For this critical length, both kinds of flow separation (tip and retarded) occur in a random fashion, and such flow is called *flow multiplicity.* This critical length increases slightly when Mach number is increased but is reduced when spike length increases for a constant spike diameter. When the diameter of relatively slender spikes is increased, the critical length increases, but when the angle of attack is increased, the critical length for jump of separation reduces, as was observed by Album (1961). The point of separation may not be fixed for a longer spike length, but it remains in a retarded position. This type of fluctuation of the separation point is accompanied by a wavy shock pattern as well as by a *flexing* separated region. If the separated layer is bent in a continuous circle, this kind of separated layer is called *flexion.* This type of separated layer does not oscillate or pulsate. At hypersonic speeds, the tip shock becomes alternately convex and concave.

6.2.2 Drag of a Spiked Body at Zero Angle of Attack

The drag of spiked bodies at zero angle of attack has been investigated in considerably more detail than have lift or pitching moment at nonzero angle of attack. For the evaluation of drag, the pressure distribution on the blunt body and on the base should be known. Two examples of measured pressure distribution on the blunt body at low and high Mach numbers are presented.

Figure 6.44 shows the pressure distribution on a hemisphere with a spike on which the occurrence of retarded separation formed a conical separation angle λ of 16.7 deg measured by Mair (1952). Note that for small values of the hemispherical angle θ measured from the stagnation point, the measured

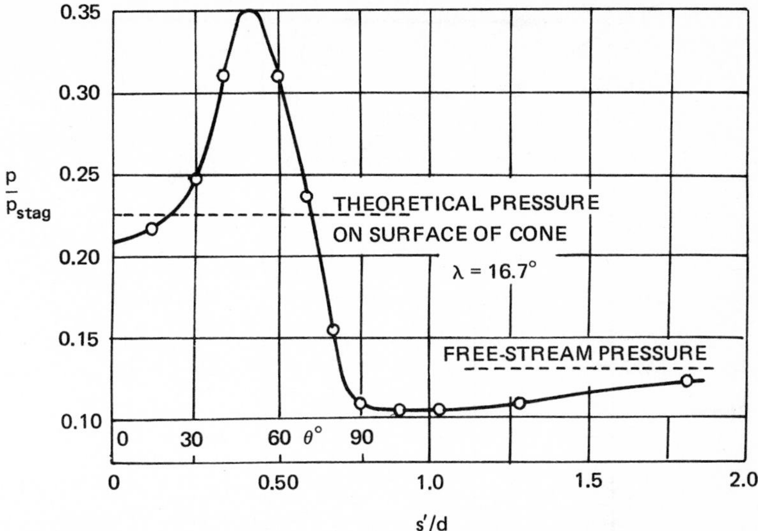

FIG. 6.44 Pressure distribution on hemispherical-nosed body with probe (l/d = 1.44) (P_{stag} is stagnation pressure in free stream, l is the length and d the diameter of the body of revolution, s' is the distance along the surface of the body measured downstream from the center of the probe, θ is the hemispherical angle, and λ is the angle of separation; *from Mair, 1952*).

pressures on the hemispherical nose face were slightly below the predicted values but that the effect of compression near the shoulder of the body became important at $\theta > 30$ deg and caused a rapid rise in pressure. The maximum pressure of $p/p_{stag} = 0.35$ was reached at $\theta \sim 50$ deg. After this point, the pressure fell rapidly to a minimum value, just behind the shoulder. There was a considerable rise in pressure between the shock wave and the body surface.

Figure 6.45 shows the pressure distribution on the nose of a flat-nosed cylinder at $M_\infty = 12.7$-14.0 as measured by Bogdonoff and Vas (1959). For these high Mach number cases, the conical angle of separation was reduced to $6\frac{1}{4}$ to 5 deg for spike lengths from 4 to 8 body diameters.

The results of the pressure distribution on the nose given by C_p/C_{max} at hypersonic speeds clearly demonstrate the pressure differences incurred by attaching a spike. Here

$$C_p = \frac{p - p_\infty}{(\gamma/2)p_\infty M_\infty^2}$$

$$C_{pmax} = \frac{p_t - p_\infty}{(\gamma/2)p_\infty M_\infty^2}$$

where p_t is the total pressure behind the normal shock,

s is the distance along the surface measured from the stagnation point,

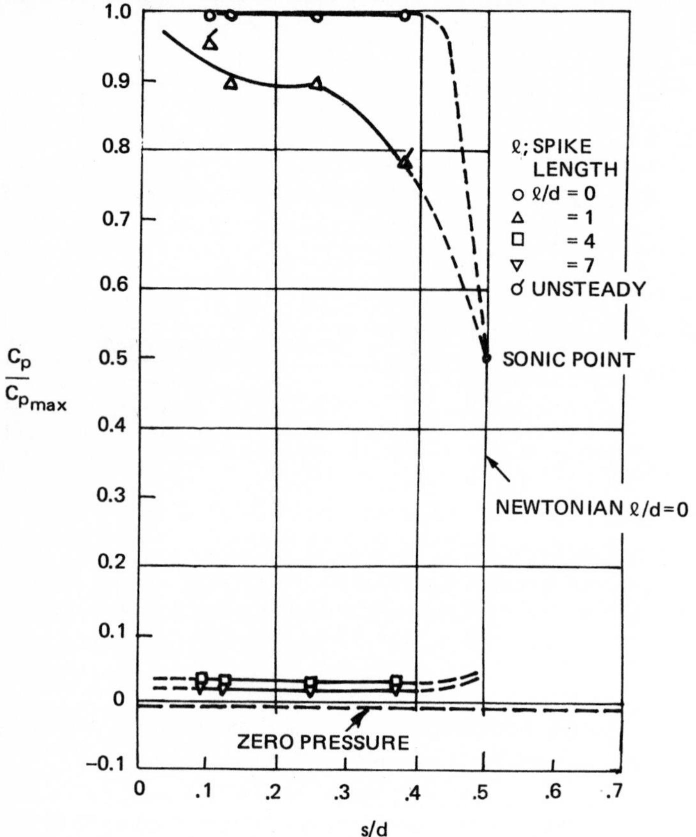

FIG. 6.45 Summary of the pressure coefficient ratio on the nose of a flat-nosed cylinder for various l/d (*from Bogdonoff & Vas, 1959*).

l is the spike length, and
d is the body diameter.

For the unspiked case, the measured pressure distribution is close to that predicted by Newtonian theory. For the spiked case, a sudden change of pressure distribution occurs at a certain spike length with a flat-nosed cylinder at $l/d > 4$ (the pressure drops rapidly between $l/d = 1$–1.5), but at $l/d > 6$ a change in spike length does not influence the pressure significantly.

Consider now the drag of a spiked body at zero angle of attack as a function of spike length at low Mach numbers. We relate it to the pressure distribution on the blunt body nose by using the measurements of Mair (1952), already shown in Fig. 6.44, and those by Daniels and Yoshihara (1954) at $M_\infty = 1.75$–2.5 and $Re = 4.15 \times 10^6/\text{ft}$ as illustrated in Fig. 6.46. Mair did not measure drag as a function of spike length, and Daniels and Yoshihara did not evaluate pressure distribution on the hemisphere nose.

Because the free-stream conditions and the shapes of the hemispherical nose were the same for both experiments, the results can be utilized to relate drag to pressure distribution on the blunt nose. The sameness of the free-stream condition is evidenced by the fact that the measured value of $C_D = 0.36$–defined by $C_D = D/[(\frac{1}{2}\rho_\infty u_\infty^2)(\pi d^2/4)] = 0.36$ where drag D is evaluated by neglecting the skin friction and assuming that the pressure downstream of the body is equal to the static pressure of the undisturbed stream–for $l/d = 1.44$ and $M_\infty = 1.96$ was in close agreement with Daniels-Yoshihara data (Fig. 6.46) at the same conditions.

On the other hand, Daniels and Yoshihara defined C_D by

$$C_D = \frac{D}{qA} + \frac{p_B - p_\infty}{q}$$

where q is the free-stream dynamic pressure,
D is the total drag measured by the balance,
A is the cross-sectional area of the body,
p_B is the base pressure, and
p_∞ is the static pressure of the ambient free stream.

Note that at a certain length of spike, the drag first levels off and then decreases to reach a minimum value. There is then an abrupt and considerable increase in C_D, and with a further increase of spike length, C_D remains constant. The sudden change of drag is attributed to the occurrence of

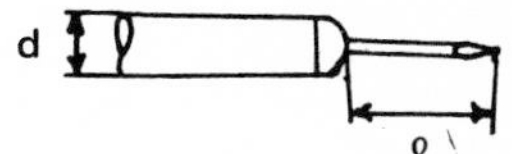

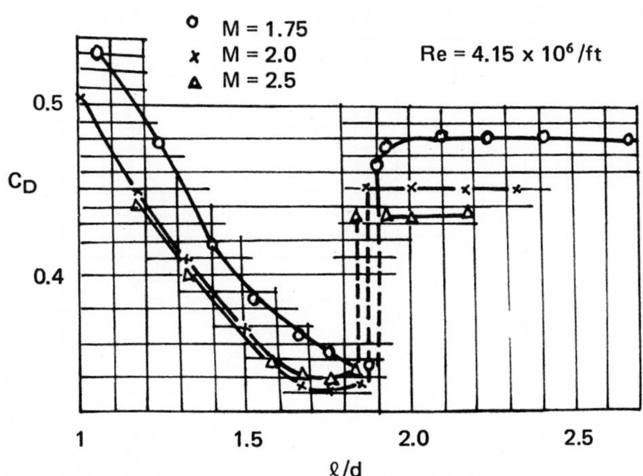

FIG. 6.46 Drag of a spiked hemispherical nose (*from Daniels & Yoshihara, 1954*).

transition. This was also observed by Crawford (1959) at $M_\infty = 6.8$ and by Hahn (1966) at $M_\infty = 3.3$. Chapman, et al. (1957) and Hahn (1966) determined the transition point of separated flow by optical observations at the starting point of the wavy boundary, where many Mach waves are generated. Since such a wavy boundary is unsteady, an average value is taken for location of transition point.

The behavior of the drag for a spiked flat-nosed cylinder at a low Mach number is similar to that of a spiked hemispherical nose body as shown in Fig. 6.47. Figure 6.48 shows the measured drag data of Bogdonoff and Vas (1959) for the spiked flat-nosed cylinder at high Mach number $M_\infty = 12.7$–14.0. The drag coefficient C_D presented here is evaluated by integrating the measured

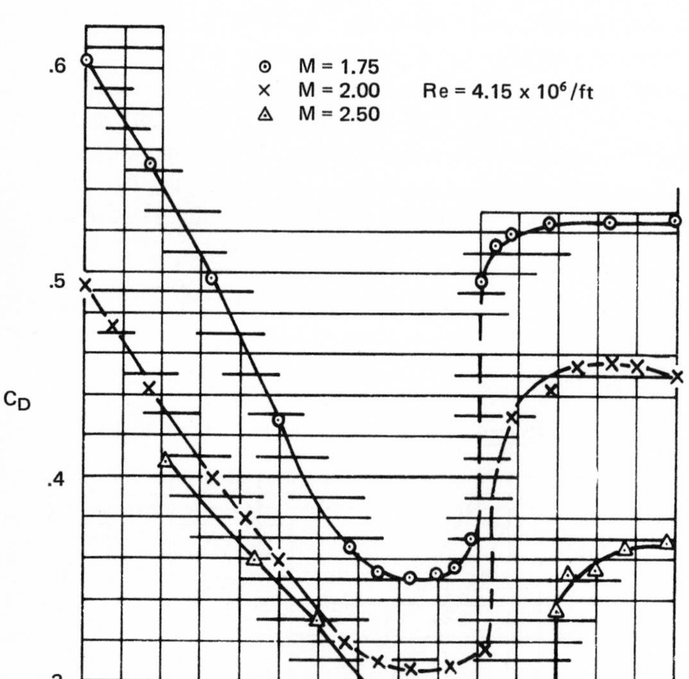

FIG. 6.47 Drag of a spiked flat-nosed cylinder (*from Daniels & Yoshihara, 1954*).

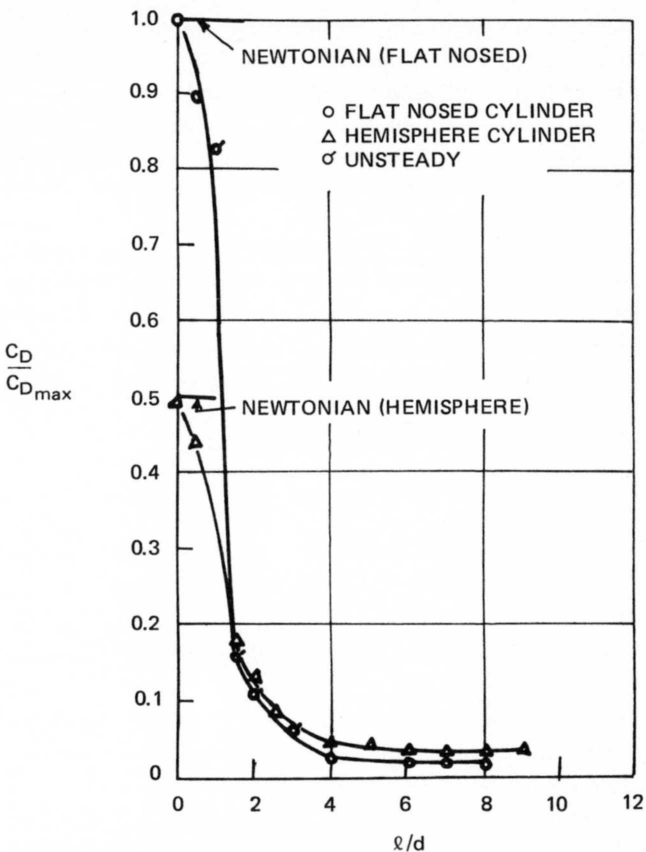

FIG. 6.48 Drag coefficient for spikes of various lengths (*from Bogdonoff & Vas, 1959*).

pressure on the face of the blunt body. A maximum drag coefficient $C_{D_{max}}$ (which is used as a reference value) is obtained by the pressure behind a normal shock and the nose area. As this figure indicates, $C_D/C_{D_{max}}$ approaches unity for a flat-nosed cylinder if a spike is absent, but this ratio reaches only 0.5 for the hemispherical cylinder. The values of $C_D/C_{D_{max}}$ drop to 0.02-0.03 with a spike of $l/d > 4$ on a flat-nosed cylinder. The sudden increase of drag shown in Fig. 6.47 did not occur in this experiment, probably because of a purely laminar boundary layer involving no transition.

The available experiment is utilized at zero angle of attack to explore the feasibility of jet injection to control separation. Stability is affected by the injection of an air jet from the spike tip or from the base of the tip and the reattachment region of separated flow. Although a mass injection from the spike tip has a strong destabilizing effect, the effect of an injection from the spike base and reattachment region can be either slightly stabilizing or destabilizing depending on the flow condition. This is evidenced from the experimental investigation of Hahn (1966) at $M_\infty = 3.3$ involving transition on the spike protruding from a hemispherical nose.

A forward injection of air from the spike tip causes disturbing perturbation and an increase in the mass-flow rate. The disturbing perturbation raises the initial level of turbulence and triggers transition and downstream shifting of the separation point. However, once the separated flow becomes turbulent following a small injection, then the separation distance remains constant at a higher rate of injection. The increase in mass-flow rate tends to cause an inflated volume of separated flow and upstream of the separation point of initially laminar separation.

At the tested sonic condition of injection orifice, injection from the base of the spike moved the separation point forward and did not radically alter the shape of the separated flow up to about

$$C_{\dot{m}} = \frac{\dot{m}}{\rho_\infty u_\infty \pi \left(\frac{d}{2}\right)^2} = 0.03$$

(here $\dot{m}$ is mass flow rate and d is the diameter of the hemispherical-cylinder body) for which the separation was shifted to the spike tip. If the injection was such that pure turbulent separation changed into laminar separation, then the separation point oscillated back and forth. However, the location of the reattachment point remained nearly the same as without base injection unless injection was at an extremely high rate.

Consider now the case of combined injection from the tip and base of the spike. If the tip is more influential, then the separation moves downstream. But if the base injection becomes dominant, the separation moves upstream, even ahead of the tip.

Injection from the reattachment region generally keeps the separated flow and the shape of the separated region stable up to $C_{\dot{m}} = 0.585$ or $p_{0j}/p_0 = 3.43$ (here p_0 is the total pressure of the free stream and p_{0j} is the stagnation pressure of the injected air). However, a serious oscillation was observed at a rate of injection higher than this value. The effect of injection of the reattachment region is similar to that of increasing step height in a two-dimensional flow over a forward facing step.

6.2.3 Drag, Lift, and Pitching Moment of a Spiked Body at Angle of Attack

Coefficients of drag, lift, and moment in aircraft flight at angles of attack should be known in order to permit practical application of the behavior of a spiked body. Unfortunately, investigations that involve complex nonsymmetrical flow characteristics are rather scarce compared to the availability of data for zero angle of attack, at which no lift and pitching moment are produced.

As indicated in Chap. 3, Hunt (1958) varied spike length in his studies of the flow phenomena of a spiked body at angle of attack (see Fig. 3.7). Since the dominant features of flow with different spike lengths have been described in Chap. 3, no further presentation is made here.

Album (1961) conducted his experiments—the most extensive to date—in the range of $M_\infty = 2\text{-}3$, $Re = 0.3\text{-}0.9 \times 10^6$ (based on the body diameter), and angles of attack from −2 to 10 deg with a blunt afterbody of fineness ratio 3. He varied the degree of nose bluntness by increasing the area of the flat-faced portion of the nose (denoted by A_f) as various percentages of the maximum cross-sectional area. Measurements were made at values of 0, 4, 50, and 80 percent for the hemispherical flat-nose and at 100 percent for the cylindrical nose. The 0.2 body diameter of the basic spike cylinder was common to all spikes.

Hunt (1958) also measured the coefficients of drag and lift by varying the ratio of cylindrical spike diameter to afterbody diameter in the range of 0.2, 0.133, and 0.067.

6.2.3.1 Drag

The Hunt (1958) measurements of drag were made at $M_\infty = 1.8$ and $Re_d = 310{,}000$ with a spike whose tip semiangle was 10 deg as already shown in Fig. 3.8. Note from this figure that the spike greatly reduced the drag at zero angle of attack, but that the reduction was less at nonzero angle of attack. Furthermore, drag was sensitive not only to angle of attack but also to the length of the spike. The value of C_D in Fig. 3.8 is the corrected one based on the measured value and assuming that the free static pressure acted uniformly over the base of the model.

6.2.3.2 Lift

The experiment by Album (1961) showed that body flatness had a considerable effect on the lift force if $A_f > 50$. The value of C_L may be increased by more than 50 percent if A_f is increased from zero to 100. For the experimental conditions given in the preceding paragraph, the Hunt measured values of C_L were corrected by accounting for the lift increment on a model mounting elbow and a component of the base pressure force. Hunt found that the lift reached its maximum value at the critical length of the minimum drag; see Fig. 3.9. Beyond this length, values of C_L remained almost constant, independently of spike lengths. However, with an increase in angle of attack, the values of C_L also increased.

6.2.3.3 Pitching moment

Album (1961) used $M_\infty = 2.5$, $Re_d = 0.60 \times 10^6$, and a flat tip spike to measure the pitching moment coefficient $C_{mX=0.4}$ at a single point $X = 0.4$; here X is the nondimensional distance in body lengths from the model nose to a point along the body longitudinal axis. If the value of $C_{mX=0.4}$ was positive, the moment tended to increase the angle of attack; thus the body was unstable. Figure 6.49 illustrates the complete qualitative presentation of the effect of A_f on $C_{mX=0.4}$. Note that the basic bodies were unstable compared to the spiked bodies. Depending on its shape, a body by the addition of a flat-tip spike can be either greatly stabilized or destabilized. The spike tends to destabilize a body with $A_f = 4$ and 50 but to considerably stabilize one with $A_f = 80$ and 100. A decrease in Mach number tends to decrease the stability, hence there is a possibility that the spike can be used as an effective stabilizing device only above a certain Mach number. At present, no measurements of chordwise force component or of moments of yaw and roll with spiked blunt body are available.

Hartley (1957) has studied the applicability of a solid spike to a supersonic wing. He carried out an experimental investigation of a 12-percent thick

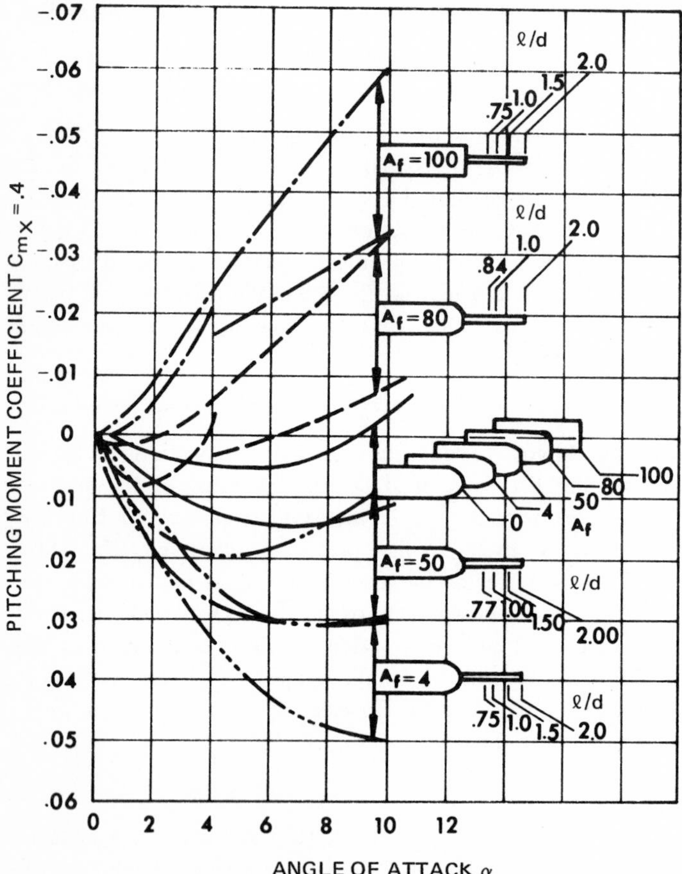

FIG. 6.49 Boundaries of pitching moment coefficient (about $X = 0.4$) for various bodies and flat-tip spikes of different lengths ($M_\infty = 2.5$, $Re_d = 0.60 \times 10^6$; *from Album, 1961*).

TABLE 6.2 Effect of spike on wing

Characteristics	Section NACA 0012 with Spikes	NACA 65-006 without Spikes
$C_{D\min}$ ($M_\infty = 1.88$)	0.079	0.031
$C_{L\max}$ ($M_\infty < 1$)	0.85	0.65
$C_{L\max}/C_{D\min}$	10.8	21.0

Source: Hartley (1957).

section (NACA 0012) wing panel by placing a number of leading-edge spikes and testing at Mach numbers of 1.56 to 1.88 and Reynolds numbers of 2.25 to 3.02×10^6. In order to compare the aerodynamic characteristics of this thick spiked wing with a high-speed wing section, he also tested a thinner 6-percent section (NACA 65-006). Hartley found that the solid leading-edge spikes caused the flow to become three-dimensional and not axisymmetric; without spikes, the flow around the leading edge of a wing was two-dimensional except at the tip. However, the mechanics of separation whereby flow on the spike formed a cone-shaped separated region and thus reduced the drag were essentially the same as those of the axisymmetrical case mentioned previously. The spikes were less effective in reducing drag for the thinner wing. When spikes with lengths ten times the radius of the leading edge were used with a spike spacing of 5 percent of the chord, a 20-percent reduction was obtained compared to the drag of the plain wing. However, this reduction was not sufficient for the spike configuration to compare favorably with the plain thin wing on the basis of drag and ratio of $C_{L\max}/C_{D\min}$, as shown in Table 6.2. However, it may be possible to achieve aerodynamic characteristics for a thick wing that are superior to those of the thin wing by providing proper combinations of wing, spikes, and high lift devices.

Instead of a solid spike, an air spike or an aerodynamic spike might be used for the control of flow and vehicle performance since air spikes can be created by blowing the air jet from the centerline of a blunt body at supersonic speeds. Like the solid spike, an air spike also forms a conelike flow region upstream of the nose. Brindle and Malia (1963) showed experimentally that the aerodynamic cone created by an air spike also reduces drag at low angles of attack. At higher angles of attack, however, this cone nearly disappears, reducing drag slightly and causing no appreciable changes in lift and pitching moment compared to those of the body shape with no spike; furthermore, surface heat transfer increases due to turbulent mixing caused by air blowing.

6.3 INDUCED FORCE BY SEPARATION

As mentioned in Chap. 1, a transverse gas jet issuing from the wall into a supersonic stream forms an effective spiked blunt body involving separation and shock formation upstream of the jet port. Since pressure in the separated flow region rises rapidly downstream from the shock, a side force is induced on the wall. This phenomenon of side force generation occurs not only in a supersonic main stream but also in a subsonic one because the issuing jet presents an obstacle to the mainstream and causes upstream separation. In view of the separation, such a flow phenomenon is similar to the case of flow which causes a pressure rise upstream of a blunt body such as a cylinder (Bernstein & Brunk, 1955) and to the case of supersonic flow over a forward facing step as evidenced by Spaid and Zukoski (1968), Sterrett and Barber (1966), and other ivestigators. Less attention has been devoted to the reattaching flow region downstream of the jet port, and no similarity to flows over other bodies has been established. The influence of negative pressure reattaching flow downstream of the jet is rather small. Letko (1963) found that the elimination of areas of negative pressure on a flat plate at a main stream Mach number of 4.5 resulted in an increase in effectiveness up to approximately 12 percent for both sonic and supersonic jets. More research on this flow region is desirable. The available studies on the flow phenomenon of a gas jet issuing from a wall perpendicular to the main stream refer to a turbulent boundary layer at subsonic and supersonic speeds.

A transverse gas jet issuing into a subsonic mainstream may be used to produce lift from the jet reaction for V/STOL aircraft. This lifting jet may be attained from engines set vertically in the fuselage or from fans buried in the wing, with the jet issuing from the bottom surface of the aircraft. At supersonic speeds, the mainstream transverse jet interaction may be used to achieve thrust vector control for rocket motors and jet flaps. Thrust vector control is attainable when the secondary jet issues into the divergent part of the thrust nozzle; the interaction produces an asymmetric force field that results in a greater side thrust than produced by the secondary jet alone. The expansion waves from the jet exhaust reflect as compression waves which coalesce to form the intercepting shock. The integrated effect of the compression waves causes a strong adverse axial pressure gradient and separation downstream of the jet. Thus, in addition to the usual jet reaction, local high pressures associated with an induced shock wave "amplify" the jet reaction.

Walker et al. (1963) used transducers to measure the side forces generated by secondary gas injection in a 15 deg conical rocket exhaust nozzle. They obtained an amplification factor $I_s/I_s^* \approx 3$ with a mainstream propellent Mach number at injection 2.34 (I_s is the measured effective specific impulse of the injectant and I_s^* is the specific impulse of the injectant for sonic flow into a vacuum). For a particular orifice size, the amplification factor was maximum at or near the transition from sonic to subsonic injection. Injection was normal to the axis of the nozzle through a single circular orifice in the diverging portion of the nozzle. A variety of ambient temperature gaseous injectants (H_2, He, He + Ar, N_2, CO_2, and Ar) and orifice diameters were carefully studied while the injectant flow rate was varied. The nominal motor chamber was 400 psia with a propellent exhaust temperature of 1845°R. The maximum pressure of injectant was 500–1000 psia with temperature of 70°F. For a fixed pressure ratio across the orifice, performance increased with decreasing orifice size. Essentially no chemical reaction occurred in the nozzle between the injectant H_2 and the propellent.

The hot catalytically decomposed H_2O_2 thrust vector control methods have the advantage of minimizing the exposure of moving potentially reactive parts to propellant exhaust products. It has been demonstrated that transverse jets can generate control forces comparable to those developed by aerodynamic control surfaces at fairly high deflection angles.

The gas jet issuing from a location close to the base of a space vehicle and directed normal to the body axis can be used to create a force that controls the flight path or altitude of the vehicle. Such control is attractive because it provides a lightweight system that is capable of fast response at any altitude and that is not susceptible to the usual severe heating problems. The interaction force appears to be independent of the mainstream Mach number, boundary layer condition (laminar or turbulent), angle of attack, and forebody length. The ratio of interaction force to jet force is found to be inversely proportional to the square root of the product of the ratio of jet stagnation to free-stream pressure and the ratio of jet diameter to body diameter. Vinson et al. (1959) used a sting balance to measure the interaction force, i.e., the normal force due to the discharge of a side jet into a supersonic mainstream and into a vacuum. Comparison of these two cases gives the interaction force.

6.3.1 Physics of Jet Interaction with Free-stream and Boundary Layer

When jet issues normally from a port, it interacts with the mainstream and the jet flow is bent downstream by the mainstream flow. Such an interaction phenomenon is quite complex and is not yet thoroughly understood. Nevertheless, an attempt is made here to describe this interaction phenomenon in such a way as to lead to the similarity of the flow over a forward facing step.

Consider first the jet flow issuing from a port into an under-expanded, unmoving fluid medium. This study is useful because there is a similarity between a jet discharging into a quiescent medium and a jet discharging into cross flow, as pointed out recently by Billig et al. (1970).

6.3.1.1 Structure of jets issuing into still air

Considerable effort has been made to gain an insight into the structure of a jet plume. Adamson and Nicholls (1959) studied the structure of jets from highly underexpanded nozzles into still air. Figure 6.50 indicates that as the gas issues from the port, it goes through an expansion fan and expands to the ambient pressure at the separating jet boundary. The condition of constant

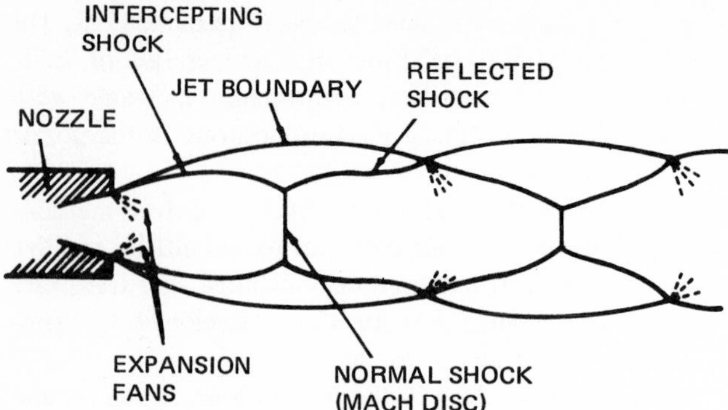

FIG. 6.50 Jet structure behind a highly underexpanded nozzle (*from Adamson & Nicolls, 1959*).

pressure along the separating jet boundary causes this boundary to be bent back toward the axis of the flow. As the jet flow changes its direction along this boundary, some of the compression waves formed at the intersection of expansion waves with the jet boundary are sent back into the flow; these waves coalesce to form the intercepting shock or barrel shock. For slightly underexpanded nozzles, these intercepting shocks meet at the axis and form the familiar diamond shock configuration. With an increase of the pressure ratio across the nozzle, these intercepting shocks become stronger; they no longer meet at the axis but are connected with a normal shock known as a Riemann wave or a Mach disk at the triple point. In both cases, reflected shocks are formed which intersect with the jet boundary, reflect as expansion waves, and the whole process is repeated. The repetition is continued until viscous effects predominate and the jet structure is no longer observed.

The downstream location of the first Mach disk depends essentially on the static pressure ratio and the exit Mach number. Since viscous effects appear to be insignificant, an isentropic quasi-one-dimensional assumption may be made for flow up to and beyond the shock. The shock configuration that occurs in the underexpanded jet is similar, although stretched out; thus the location of the Mach disk may be predicted without having to calculate the whole shock structure. The free jet can then be considered as an extension of the actual nozzle that is bounded by a fictitious surface, with an area distribution corresponding to the actual pressure distribution of the center or core flow up to the first Mach disk. The fictitious boundary extends only to the axial position of the shock wave; thus as the total-to-atmospheric-pressure ratio changes, the length of the extension changes.

Adamson and Nicholls (1959) presented a method for calculating the position of the first Mach disk in the jet behind a highly underexpanded nozzle. In the calculation for a sonic orifice, the axial pressure distribution on the centerline of the flow downstream of an orifice (calculated by the method of characteristics) is used to define a fictitious nozzle extension, and the shock is then assumed to exist at the point where atmospheric pressure would be attained downstream of the shock, i.e., the shock is assumed to exist at the end of the fictitious nozzle extension. Physical arguments may be employed to extend this calculation to nozzles with supersonic exit Mach numbers. The results compare favorably with experimental data.

Abbett (1970) recently presented a theory that predicts when and where a Mach disk forms in an underexpanded exhaust plume. It involves dividing the flow field qualitatively in two subregions, a quasi-one-dimensional core flow and an outer flow. Location of the Mach disk is quantitatively determined by the requirement that the subsonic core flow must pass smoothly through a throatlike region, thereby becoming supersonic. The results of experiments by Abbett (1970) are shown in Fig. 6.51 together with the triple-point predictions given by his theory and by five others, namely, Love et al. (1959), Abdelhamid and Dosanjh (1969), Boyer et al. (1963), Eastman and Radtke (1963), and Adamson and Nicholls (1959).

Note that because of the Mach disk curvature, predicted values for X_{TP}—the location of the triple point—are slightly upstream of the corresponding position of normal shock X_{ns}. Neglecting possible inaccuracies in

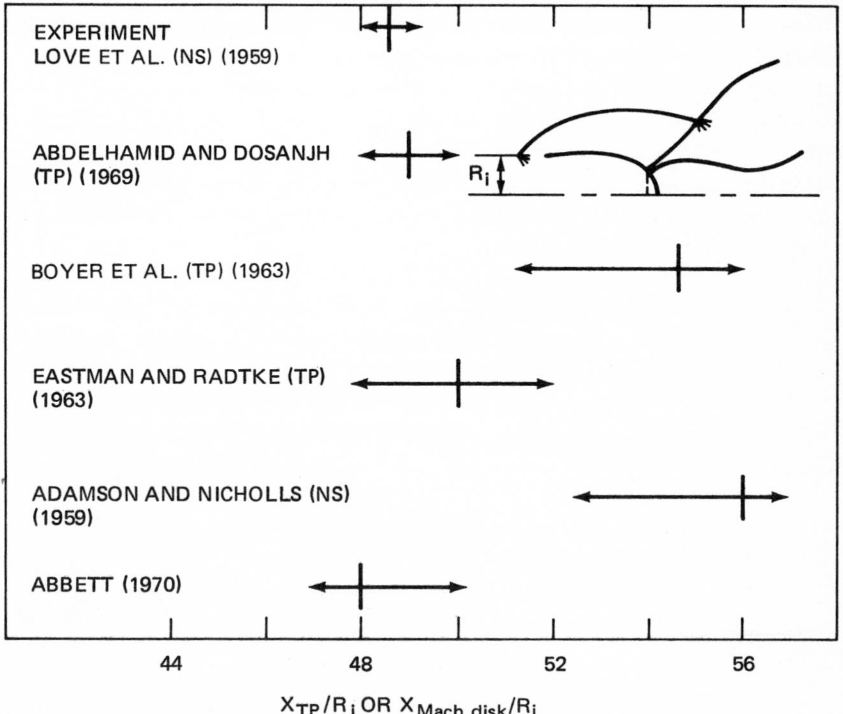

FIG. 6.51 Comparison of theoretical and experimental triple point locations (*from Abbett, 1970*).

the computation, the maximum difference between the various theories and hypotheses is about 20 percent. The Abbett theory for prediction of X_{TP} is accurate within 1 to 2 percent of the experimental Mach disk coordinate measured from jet axis $X_{\text{Mach disk}}/R_j = (4.8\text{–}4.9)$ where R_j is the radius of the jet exhaust. However, the Abbett theory is obviously inadequate in cases where viscous effects are important upstream of the throat.

Boyer et al. (1964) assumed that the Mach disk is a normal shock at the triple point, and Abdelhamid and Dosanjh (1969) proposed a model for Mach disk location in which it is assumed that the reflected shock remains linear out to its intersection with the inviscid jet boundary. The Mach disk is predicted to occur when the integral continuity equation is satisfied over a control surface located just downstream of the Mach disk and the reflected shock.

Peters and Phares (1970) considered the turbulent mixing along the plume boundary based on the model sketched in Fig. 6.52 and developed a method for computing the interior plume structure with a coupled computation of the turbulent mixing layer at the boundary. In addition, they developed a new model for predicting the Mach disk position by assuming that the Mach disk is uniquely related to the reaccelerating transonic flow downstream of the disk. When mixing at the boundary is included, the compression waves originate from the inner edge of the shear layer rather than from the corresponding inviscid plume boundary. Because the inner edge of the shear layer is closer to the centerline than the inviscid plume boundary, the boundary shock wave becomes stronger and displaced toward the centerline compared to the shock wave in the corresponding inviscid plume.

Computed results for an underexpanded rocket nozzle are shown in Fig. 6.53 to illustrate the influence of boundary mixing on the plume structure. The symbols r and x respectively denote the radial distance measured from the centerline and the axial distance from the nozzle exit. The subscript n refers to conditions at the nozzle exit. Note that mixing causes the boundary shock wave to be significantly displaced relative to the inviscid plume shock

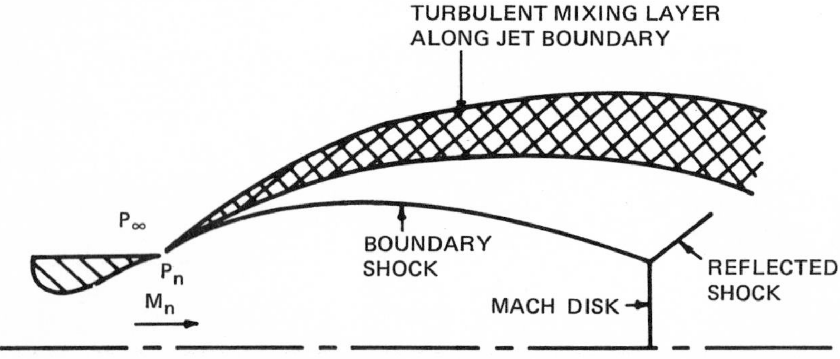

FIG. 6.52 Supersonic plume structure (*from Peters & Phares, 1970*).

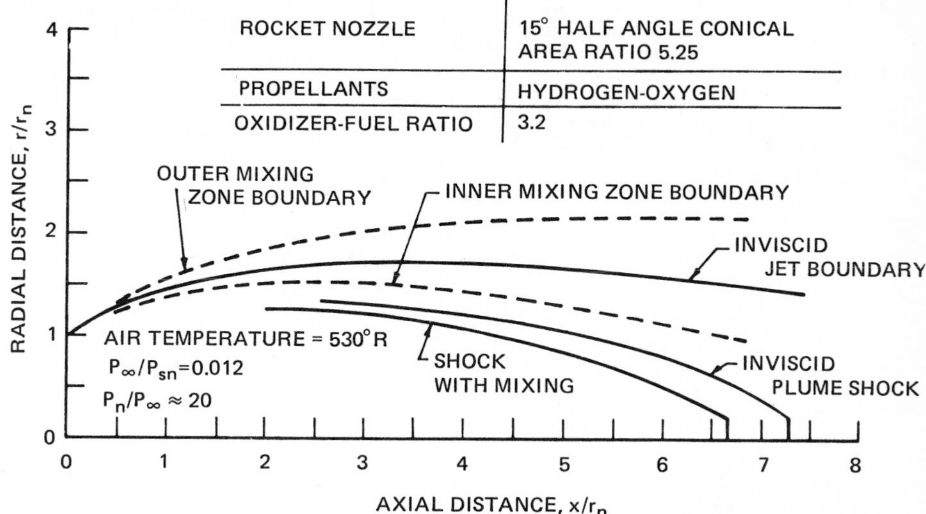

FIG. 6.53 Computed structure of rocket plume exhausting into still air (*from Peters & Phares, 1970*).

wave. When mixing is included, the location of the Mach disk is predicted to occur approximately 9 percent upstream of the Mach disk position for the inviscid plume.

The computed axial distance to the Mach disk with and without mixing is given in Fig. 6.54 as a function of p_n/p_∞ for $M_n = 2.0$. Peters and Phares (1970) found that for an inviscid plume analysis, the model of Eastman and Radtke (1963) gave the best overall correlation of experimental Mach disk position. When boundary mixing is included in the computation, one of the Mach disk models considered was generally superior. This finding is for all M_n considered.

Crist et al. (1966) carried out an experimental study of a highly underexpanded sonic jet in a wide range of stagnation pressures from 15,000 psia to 100 μ in. Hg at a temperature of 4200K. They found that Mach disk location was insensitive to such parameters as the ratio of specific heats γ, condensation, nozzle lip geometry, and absolute pressure level. Mach disk location varied as the square root of the overall pressure ratio up to about 300,000 psia. Diameters of Mach disk, jet boundary, and intercepting shock increased with a decrease in γ, increased with condensation, and decreased at high stagnation densities where intermolecular forces become important. At the high-pressure ratios, the ratio of Mach disk diameter to Mach disk distance was found to be constant for a given gas.

6.3.1.2 Interaction of jets issuing into an air stream

Most investigations of the interaction of jets issuing into an air stream have been concerned with a supersonic main stream. However, a subsonic

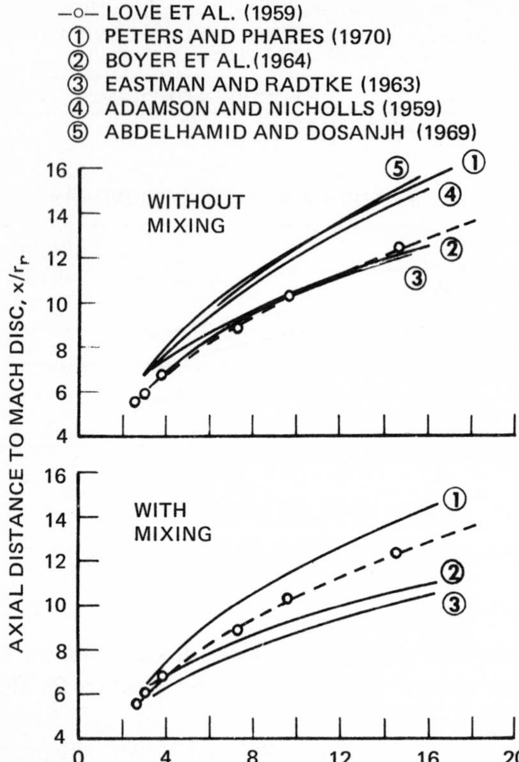

FIG. 6.54 Axial position of Mach disk for $M_n = 2.0$ *(from Peters & Phares, 1970).*

main stream may be useful to correlate the jet interaction problem with flow over the step.

Interaction of jets into a subsonic air stream The jet produces a region of positive pressure upstream of the jet and a larger region of negative pressures laterally and downstream of the jet. Melbourne (1960), Spreemann (1961), Otis (1962), Cubbison et al. (1961), and Janos (1961) indicated that a negative lift is induced on a wing when a jet issues from the bottom surface of the wing or fuselage in free-stream flow. Model tests by Spreemann (1961) and Otis (1962) show that large nose-up pitching moments occur for VTOL aircraft as a result of jet and free-stream interaction.

Vogler (1963) measured the pressure distribution on the surface of a flat plate induced by a round cold air jet into a subsonic air stream at jet velocities of 204, 510, and 1020 ft/sec and by a rod. Free-stream velocities from 82 to 408 ft/sec were used as representative of transition flight speeds of VTOL aircraft. He also used a cylindrical wooden rod extending from within the round nozzle to the tunnel floor in order to determine any similarity of the flow around the air jet to that of flow around a cylinder.

The test results at a free-stream velocity of 300 ft/sec are shown as C_p contour lines in Figs. 6.55 and 6.56, with the jet and with the rod, respectively. The major effect of the jet on the pressure coefficients in the area directly downstream of the jet occurred within 5 or 6 in. or jet diameters of the jet; outward and rearward ($\beta = 120$ and 150 deg) of the jet exit, the affected area may extend from 8 to 10 jet diameters, producing a swept lobate form for the contours. (β is the angle, in degrees, between the orifice rays and the longitudinal centerline of the plate; it is measured counter-clockwise from the upstream part of the centerline.) This shape results from the deflection of the free stream by the jet and from an expansion and flattening of the jet normal to a free stream. Upstream of the jet, the shape is fairly uniform but downstream especially on the centerline, the shape of the contours varies with jet and tunnel velocities. This variation in shape may have been caused by vortices rolling backward and upward off the edges of the downstream curving jet. The short-dashed curves on Fig. 6.55 show the pressure coefficient variation with distance from the jet center on the longitudinal centerline ($\beta = 0$ and 180 deg).

Although the lift loads on the plate were not integrated from pressures, the lift would be negative on the plate region affected by the jet, as evidenced by the pressure-coefficient contours and as reported by Spreemann (1961) and Otis (1962) for the VTOL investigation. The data also indicate that the positive pressures ahead of and the negative pressures behind the jet would produce large nose-up pitching moments on the plate similar to those shown by Spreemann (1961) and Otis (1962). Swept-wing models would suffer more than unswept-wing models because more of the swept wing would be in the strong negative pressure field.

Figure 6.56 shows the pressure-coefficient contour and pressure-coefficient variation along the centerline of the large plate with a rod. The variation in free-stream velocity had little effect on the magnitude of the pressure coefficients or the contour shape for the plate with the rod. For the highest free-stream velocity of 408 ft/sec (corresponding to a Reynolds number of 191,000 based on the rod diameter), the pressure coefficients were slightly more negative over most of the plate than at the lower velocities. Since there was no expansion jet, the lobate form of the contours was less pronounced than shown with the jet.

It is evident from Figs. 6.55 and 6.56 that the rod and jet produce positive pressure fields of about equal area on the plate just upstream of the jet location. The contours of negative pressure coefficient caused by the round jet generally enclosed a larger area and were located farther downstream than those of the rod. Upstream of jet and rod, there was a close similarity between the pressure field produced by a cold round jet into the subsonic

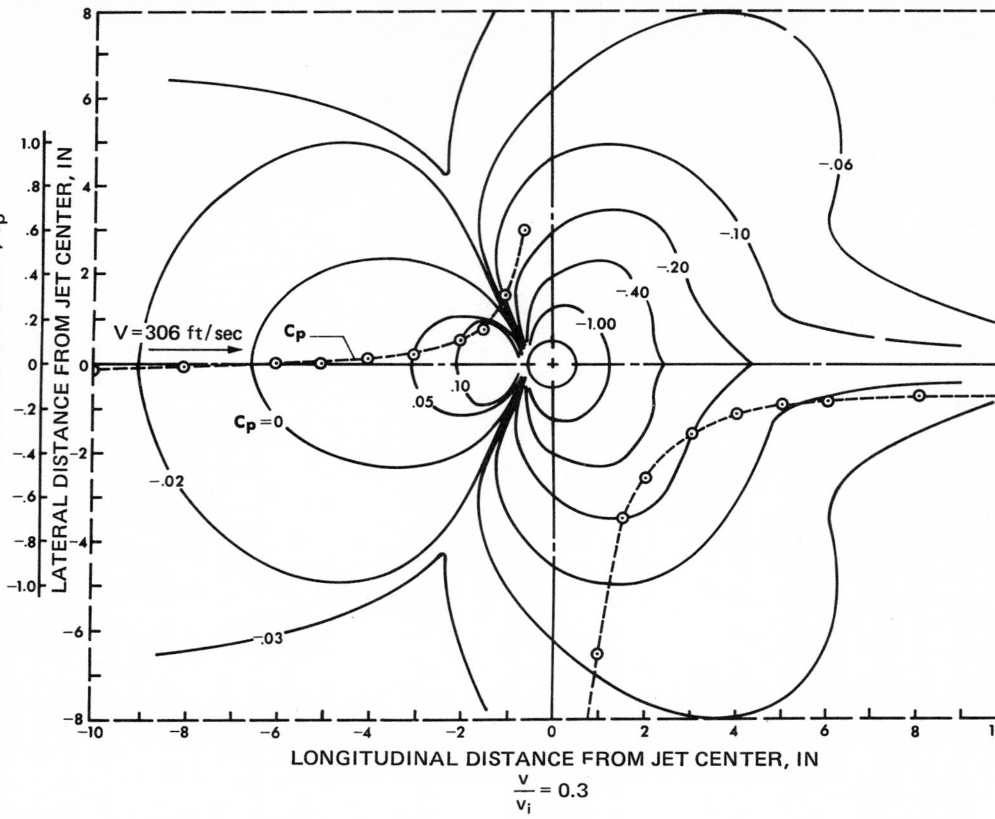

FIG. 6.55 Pressure-coefficient contours on the large plate with the round jet (also pressure-coefficient variation along the longitudinal center-line of the plate surface, $v_j = 1{,}020$ ft/sec; *from Vogler, 1963*).

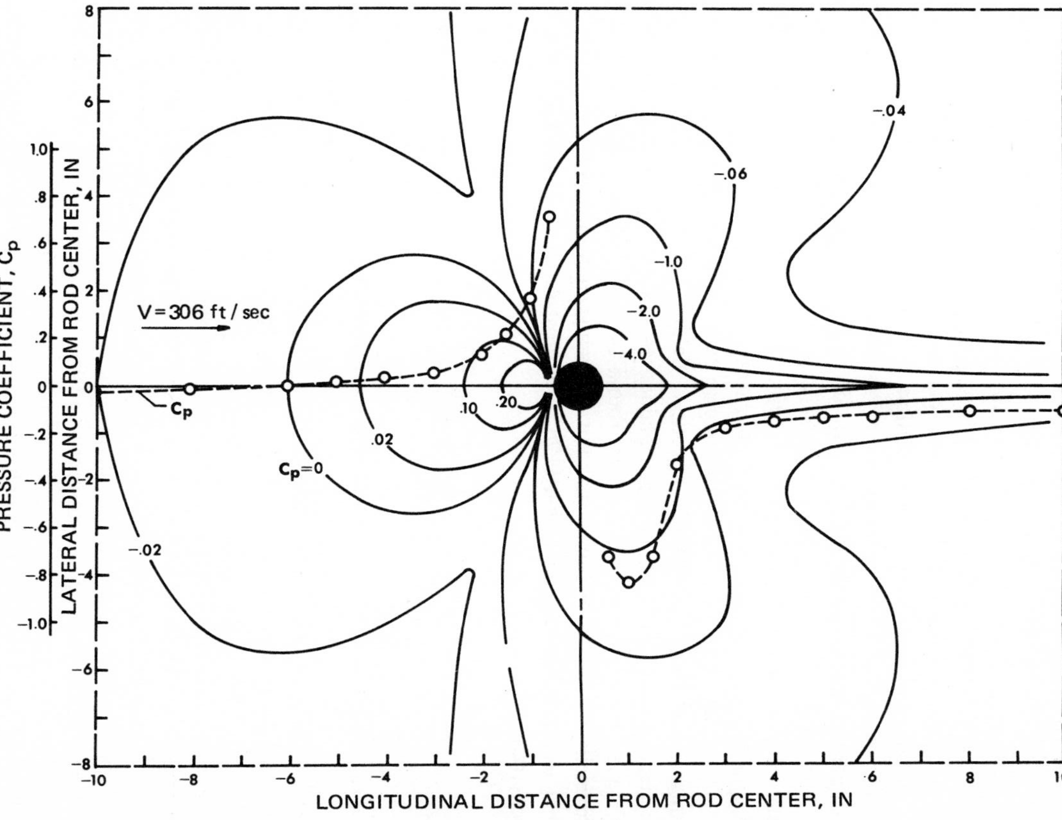

FIG. 6.56 Pressure-coefficient contours on the large plate with the rod (rod extended from plate to tunnel floor, and there was pressure-coefficient variation along the longitudinal centerline of the plate surface; *from Vogler, 1963*).

mainstream and a rod but downstream, differences existed between the two flow fields.

Interaction of jets into a supersonic air stream The penetration of a single gas jet directed normally from a flat plate to an air stream has been investigated experimentally by Povinelli et al. (1970), who determined the penetration distance by the concentration of the injectant. The orifice of the jet was located upstream, close to the center of the plate. The vertical penetration distance of a helium jet was determined at a location where helium concentration amounted to 1 percent because it was difficult to fix the location of zero helium concentration.

The range of the experiments by Povinelli et al. (1970) covered an air mainstream Mach 2, with total pressure of 13.54 psia and total temperature of 627°R, and injection of Mach numbers of 1, 2.4, 2.7, 3.5, and 4.0. The Reynolds number at the injection location was 5.35×10^5 and the turbulent boundary layer thickness was computed to be 0.061 in.

Jet penetration was found to increase with injection Mach number. For equal mass flows, supersonic injection yielded as much as a 25-percent increase in penetration over that obtained with sonic injection. Much of the observed increase in penetration was found to occur between Mach 1 and 2.5. Povinelli et al. (1970) found that the following equation, which is similar to that used by Vranos and Nolan (1965), is most suitable for the prediction of penetration:

$$\frac{y}{d_e} = 1.92\left(\frac{q_j}{q_a}\right)^{0.350}\left(\frac{\mathrm{M}_a}{\mathrm{M}_j}\right)^{0.094}\left(\frac{x}{d_e} + 0.5\right)^{0.277} \tag{6.34}$$

The symbols q, d, M, and x refer to momentum, nozzle diameter, Mach number, and downstream distance from nozzle centerline, respectively. The subscripts e, a, and j refer to exit, free stream, and jet, respectively. Figure 6.57 compares measured penetration with that predicted by Eq. (6.34).

In wind-tunnel experiments by Ruggeri et al. (1950), jet penetration was measured from circular, square, and elliptical orifices. The circular orifices had diameters of 0.375, 0.500, and 0.625 in., and the squares and ellipses had areas equivalent to the 0.375 and 0.625-in. diameter circles. The elliptical orifices were mounted with the major axis parallel to the air stream and the square orifices with two edges parallel to the air stream. Air velocities of 160, 275, and 380 ft/sec were used in the tunnel, and pressure ratios ranged from 1.15 to 3.2 at a jet total temperature of about 400°F. At low air velocities and jet pressures, penetration was best for the square orifices and the elliptical orifices that had an axis ratio of 4.1. At higher values of air velocity and jet total pressures, the best penetration was achieved by the square orifices.

Billig et al. (1970) investigated the more complex feature of a jet issuing into a supersonic air stream compared to the quiescent medium. The new unified model they introduced was in excellent agreement with measured flow

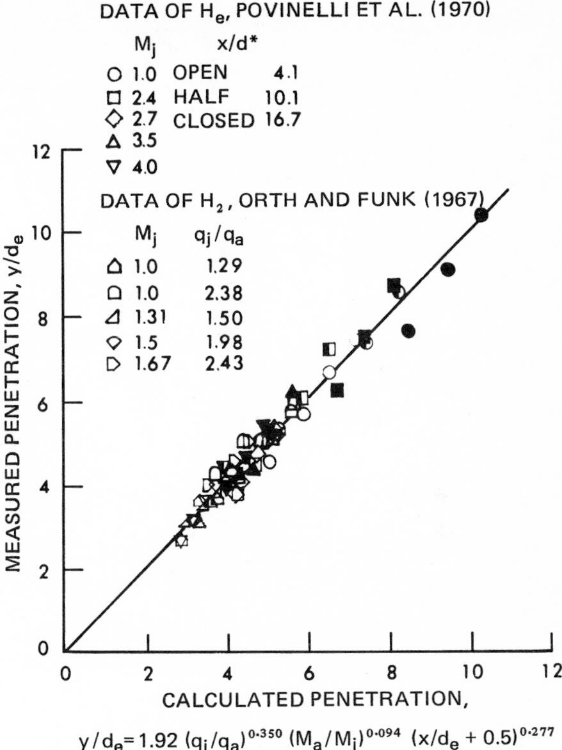

FIG. 6.57 Comparison of measured penetration with that predicted by Eq. (6.34) (*from Povinelli et al., 1970*).

field properties. The development of this model rests on a similarity between a jet issuing into a quiescent medium and a jet issuing into a cross flow. The suitable definition of an "effective back pressure" enabled a correlation between the normal distance to the center of the Mach disk and the ratio of injection pressure to effective back pressure. Then the complete trajectory of the injectant and the boundaries of penetration and the jet were investigated. The generalized model used by Billig et al. (1970) is shown in Fig. 6.58, and the model for calculating the jet centerline trajectory is sketched in Fig. 6.59.

In order to obtain the average properties of the flow just downstream of the Mach disk, Billig et al. (1970) used the following empirical equations to compute the Mach disk location and the cross-sectional area distribution:

$$\frac{y_1}{d_j^*} = M_j^{1/4}\left(\frac{p_j^*}{p_{e_b}}\right)^{1/2} \tag{6.35}$$

$$\left(\frac{A}{A_j}\right)^{1/2} = \frac{r}{r_j} = 1 + 1.45 \ln\left(\frac{p_j}{p_{e_b}}\right)\left[1 - \exp - 0.322\left(\frac{s}{d_j}\right)\right] \tag{6.36}$$

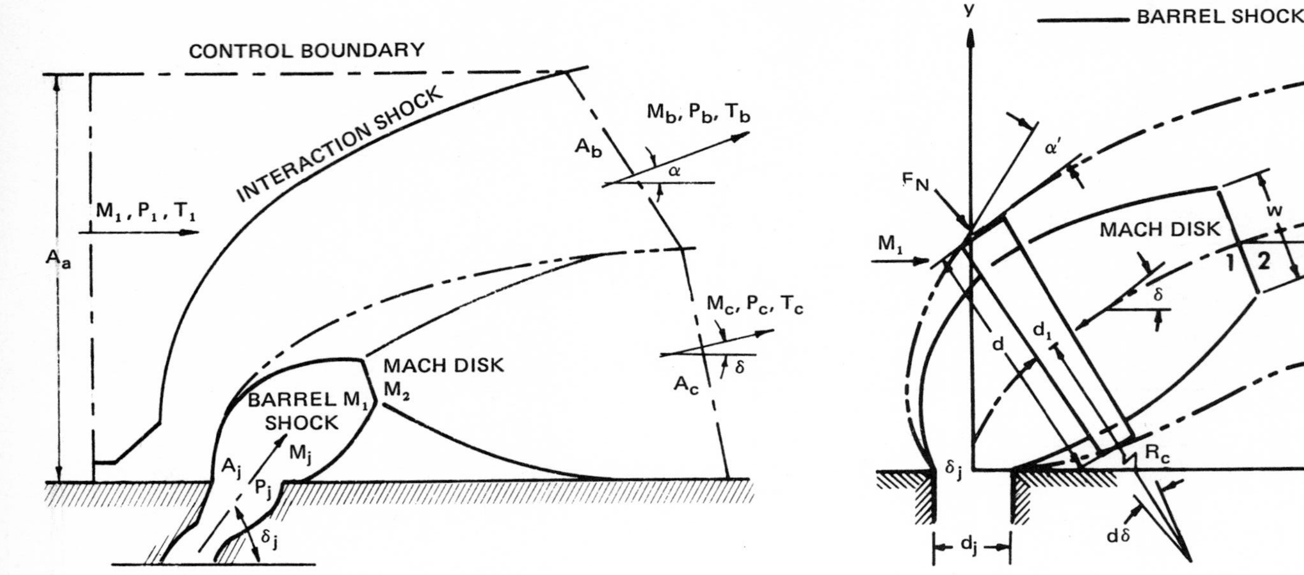

FIG. 6.58 Generalized model configuration (*from Billig et al., 1970*).

FIG. 6.59 Model for calculating jet centerline trajectory (*from Billig et al., 1970*).

Here d_j^* is the diameter of the jet orifice at the sonic point, M_j is the Mach number of jet conditions at the point of injection, p_j^* is the pressure at the sonic point in the jet orifice, and p_{e_b}—the effective back pressure—is related to the pitot pressure by $\frac{2}{3}\,p_{t_a}$; the subscript a refers to the undisturbed free stream upstream of the control volume (see Fig. 6.58), and s is the distance along the jet trajectory. In the region ahead of the Mach disk, the ratio of s to y is given by $s_1/y_1 \approx 1 + 0.14\,(M_j\,ln\,p_j/p_{e_b})^{-1}$; here subscript 1 refers to the region ahead of the Mach disk. The normalized Mach disk distance y_1/d_j^* for the jet issuing into the cross flow is plotted against $M_j^{1/2}(p_j/p_b)$ in Fig. 6.60 and compared with available experimental data; p_b is the pressure of the free-stream conditions at the downstream edge of control volume C (see Fig. 6.58).

The agreement between experimental and theoretical results was good for $1.0 \leqslant M_j \leqslant 2.2$ and $1.9 \leqslant M_a \leqslant 4.5$.

The width of the Mach disk W is computed by

$$\frac{W}{d_j} = 0.4\left(\frac{p_j}{p_\infty} - 2\right) - 0.154\left(\frac{p_j}{p_\infty} - 3\right)^{1.12} \tag{6.37}$$

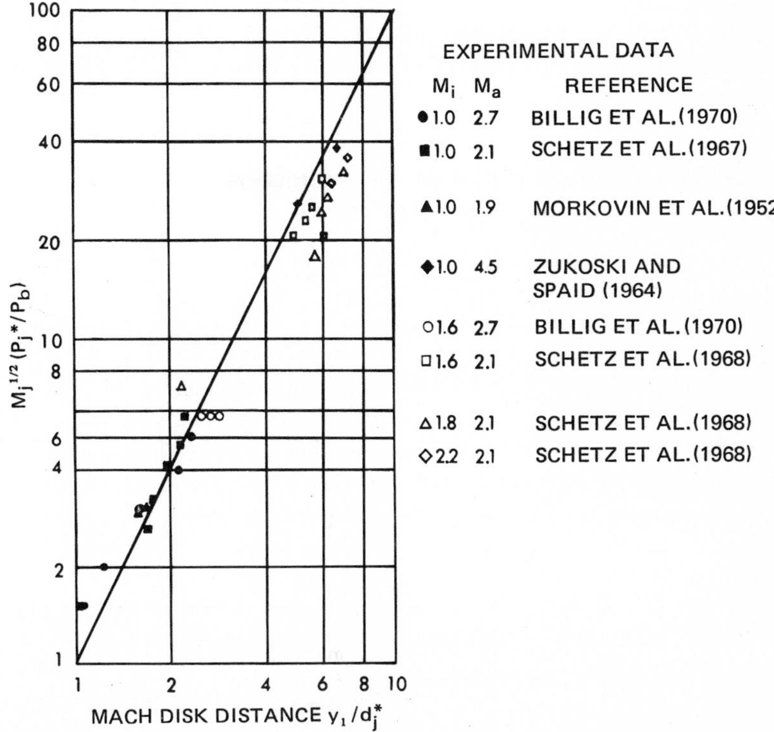

FIG. 6.60 Normalized Mach disk distance for jet issuing into cross flow (Mach disk distance versus the product of the square root of injection Mach number times the injection throat pressure ratio for sonic and supersonic injection into a supersonic cross flow; *from Billig et al., 1970*).

and the maximum coordinate of the Mach disk $\tilde{y}$ is given by

$$\frac{\tilde{y}}{d_j^*} \cong M_j^{1/4}\left(\frac{p_j}{p_{e_b}}\right)^{1/2} + \left[0.4\left(\frac{p_j}{p_{e_b}} - 2\right) - 0.154\left(\frac{p_j}{p_{e_b}} - 3\right)^{1.12}\right] \cdot \left\{\cos\left[\tan^{-1}\left(0.4M_j \ln \frac{p_j^*}{p_{e_b}}\right)\right]\right\} \tag{6.38}$$

Equation (6.37) was derived by Owen and Thornhill (1952) for a jet issuing into a quiescent medium, but Billig et al. (1970) found that the same equation also holds for $1 \leqslant M_j \leqslant 3$ and $0 \leqslant p/p_\infty \leqslant 54$. The shape of the leading shock involving interaction of jet issued normal to the mainstream is of considerable interest in obtaining the total side force. Narashima et al. (1967) observed that turbulent separation always occurred for a main air stream of $M_\infty = 1.64$ to 3.08 even with the lowest injection pressure and that such separation caused a distinct "separation" shock. The most striking feature was that at a given Mach number, variation of jet total pressure hardly affected the separation shock inclination to the free stream but moved only its foot. It was also found from the experiment of two-dimensional jet injection that the shock shape may be predicted by a slight modification of the blast analogy.

6.3.2 Analytical Model of Jet Interaction with a Supersonic Mainstream

Spaid and Zukoski (1968) set up an analytical model for the complex interaction of a jet issuing into a supersonic mainstream, and applied the conservation of momentum to a control volume at the jet nozzle exit. In addition, they conducted a series of flat-plate experiments with normal sonic jets at external flow Mach numbers of 2.61, 3.50, and 4.54. The situation chosen for study was the sonic injection of a gaseous jet into an external flow that is uniform and rectilinear outside of a turbulent boundary layer. The analysis was two-dimensional and experiments were conducted under conditions as nearly two-dimensional as possible. This analytic model utilized the experimental findings of the two-dimensional turbulent separation upstream of a step.

The sketch of the flow field and pressure distribution used by Spaid and Zukoski (1968) (Fig. 6.61) was based on available experimental data from Sterrett and Barber (1966), Hawk and Amick (1965 and 1967), Maurer (1966), Romeo (1963), Spaid (1964), Mitchell (1964), Romeo and Sterrett (1961), and Heyser and Maurer (1964). In this example, the jet is underexpanded, and the effective obstruction to the external flow that is produced by the jet is significantly larger than the undisturbed boundary-layer thickness.

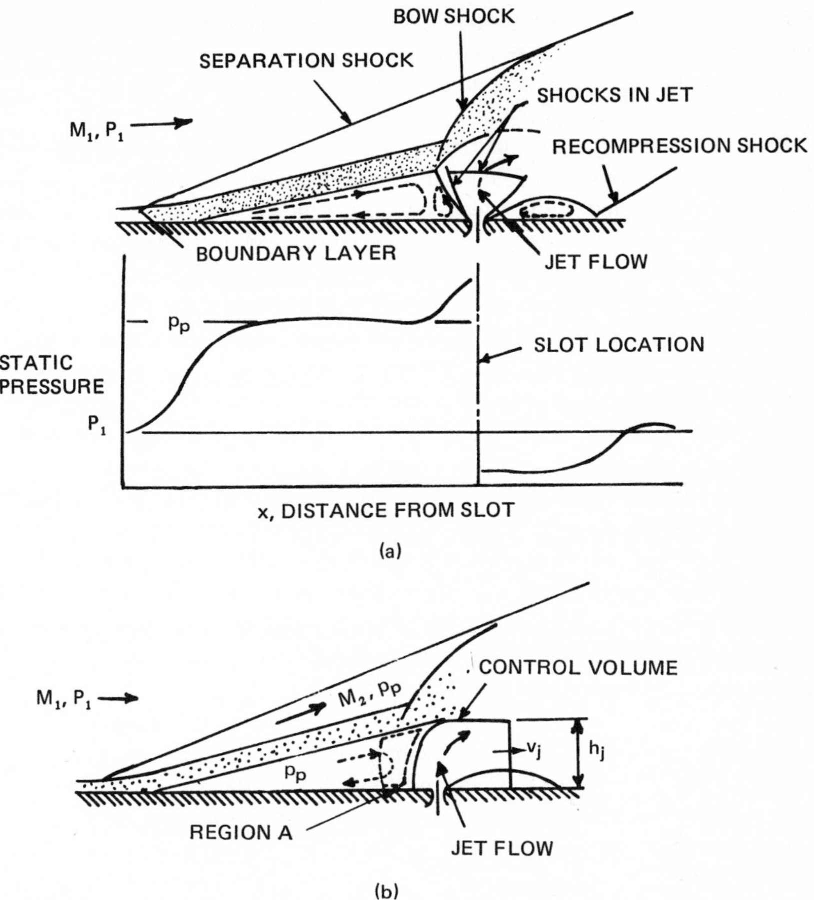

FIG. 6.61 Flow field and pressure distribution for issuance of jet into a supersonic mainstream: (*a*) flow field and pressure distribution; (*b*) simplified flow field used in calculation (*from Spaid & Zukoski, 1968*).

The pressure distribution of Fig. 6.61 is the same as described by Zukoski (1967) for the separated flow region upstream of a forward facing step. The results of the pressure distribution were found to be almost independent of Reynolds number, Mach number, and ratio of step height to boundary-layer thickness for the following conditions: (*a*) a turbulent boundary layer, (*b*) a Mach number range between 2 and 6, and (*c*) a boundary-layer thickness less than the step height. The plateau pressure p_p was found to depend only on the static pressure p_1 and Mach number M_1; here subscript 1 refers to ambient upstream stream of separation and is roughly correlated by

$$\frac{p_p}{p_1} = \left(1 + \frac{M_1}{2}\right)$$

The static pressure distribution obtained downstream of the jet showed a region of static pressure less than p_1 immediately downstream of the nozzle exit. The value of this static pressure was affected only by the external flow Mach number and the molecular weight of the injectant gas.

Because of the complex nature of a jet-interaction flow field, Spaid and Zukoski (1968) constructed a highly simplified analytic model of the flow field which included only those features believed to have a first order influence on the structure of the interaction and restricted only to the separation upstream of the jet. A simple body shape was selected to represent the dividing streamline between injectant and external flow. Momentum was used to calculate a characteristic dimension of an equivalent body to enable the interaction force to be estimated, and the calculation was partially based on the experimental information obtained in studies of flows over forward-facing steps.

A simplified sketch of the flow field is shown in the lower part of Fig. 6.61. The control volume selected consisted of the upstream interface between the jet and the external flow, the exit plane of the nozzle and a portion of the wall downstream, and a plane which is normal to the wall and to the primary flow direction and which passes through the downstream separated region. It was assumed that no mixing takes place between the jet and the primary flows along the boundaries of the control volume. Furthermore, all shear stress on the control volume was neglected. The no-mixing assumption implies that the structure of the flow field is determined by conditions in the immediate neighborhood of the jet slot. The x momentum balance reduces to equating the net force of a drag acting on the boundaries of the control volume to the x component of the momentum flux $\dot{m}_j v_j$ of the injectant as it leaves the control volume. Spaid and Zukoski (1968) developed the procedure for computation of these quantities.

When the jet is underexpanded, most of the jet flow passes through a normal shock before much of the turning occurs. The jet flow will be either subsonic or near-sonic relatively early in the turning process over a wide range of jet-to-free-stream pressure ratios. As shown in the top part of Fig. 6.61, the recompression shock extends much closer to the wall than a characteristic dimension of the jet; thus the majority of the jet flows through it and the jet flow must be supersonic near reattachment. Hence, it is assumed that the jet flow will be approximately sonic as it leaves the control volume. If the jet flow is also adiabatic, then the average velocity of the jet v_j leaving the control volume may be replaced by jet gas velocity for Mach number of unity a_j^*, which can be evaluated from the jet reservoir conditions and the properties of the jet gas. The drag is given by $(\bar{p}_f - \bar{p}_b)h_j$ where $\bar{p}_f$ and $\bar{p}_b$ are average pressures acting on the upstream and downstream sides of the control volume, respectively, and h_j is the jet penetration height. It may be recalled that for forward-facing step flow, drag is given by Zukoski (1967) as $(\bar{p}_f - \bar{p}_1)h$ where $\bar{p}_f$ is the average pressure on the face of the step, p_1 is ambient static pressure, and h is step height.

However, for the evaluation of $\bar{p}_f$, based on the result of an investigation of the flow over a forward facing step by Zukoski (1967), the separated flow region is considered as a constant pressure region with plateau pressure p_p which is fixed by the turning angle θ. The pressure on the face of the step is determined in part by the general pressure rise in the separated region $p_p - p_1$ (where p_1 is static pressure just upstream of separation outside of the boundary layer) and in part by the pressure rise required to turn the recirculating flow near the face of the step $\bar{p}_f - p_p$. An estimate for $\bar{p}_f - p_p$ is made from a momentum balance for Region A of the upper portion of Fig. 6.61. This leads to

$$(\bar{p}_f - p)h_j \cong \int_0^{h_j} \rho u_e^2 \, dy \cong \beta' \rho(p_p, T_{st}) u_e^2 h_j \tag{6.39}$$

Here u_e is the velocity at the outer edge of the shear layer, the quantity β' is a proportionality constant, T_{st} is stagnation temperature, and the integral is evaluated along the upstream (left) face of Region A. The value of β' has been chosen as 0.062 in agreement with jet interaction force data. This order of magnitude is consistent with measurements of maximum reverse flow speeds in the separated region by Vas and Bogdonoff (1955), and mean pressure data on forward faces of steps indicate that β' is a function of h/δ when $h/\delta < 1$ (Zukoski, 1967). Manipulation of Eq. (6.39) leads to

$$\bar{p}_f \cong \left\{1 + \beta' \frac{\gamma M_2^2}{1 + [(\gamma - 1)/2] M_2^2}\right\} p_p = (1 + \beta) p_p \tag{6.40}$$

where β is a function of M_1 through its dependence on M_2. Again subscripts 1 and 2 respectively refer to the region upstream of separation outside of the boundary layer and the region downstream of separation shock at the edge of the shear layer.

The average base pressure is given by αp_1 where α is a constant amounting to $1.2 \geqslant \alpha \geqslant 0.4$. If the x momentum balance is combined with expressions for $\bar{p}_f$ and $\bar{p}_b$, then

$$h_j = \frac{\dot{m} a_j^*}{(1 + \beta)(p_p - p_1) + (1 + \beta - \alpha) p_1} \tag{6.41}$$

The interaction force F_1 is approximated by

$$F_i = (p_p - p_1)\left(\frac{x_s}{h_j}\right) h_j \tag{6.42}$$

taking the pressure in the separated flow region as p_p, with x_s as the coordinate of the separation point parallel to the wall aligned with the external flow.

By combining Eqs. (6.41) and (6.42), the amplification factor K is defined as follows

$$K = \frac{F_i + T}{T_{s_v}} = \frac{[\gamma_j/(\gamma_j + 1)]\,(x_s/h)}{(1+\beta) + (1+\beta-\alpha)[p_1/(p_p - p_1)]} + \frac{T}{T_{s_v}} \tag{6.43}$$

Here T is the jet thrust and T_{s_v} is the vacuum thrust of a sonic jet, and their ratio is approximately 1 when $p_{0_j}/p_1 > 20$ (where p_{0_j} is the stagnation pressure of the jet).

Empirical data for separation upstream of a forward facing step was used for the evaluation of Eq. (6.43). Zukoski (1967) found that for such separation in the range of $2 \leqslant M_1 \leqslant 6$, $(p_p - p_1) \approx 0.5 M_1 p_1$, $x_s/h_j \approx 4.2$, and $M_2 \approx \frac{3}{4} M_1$. Equation (6.43) was evaluated for $\gamma_j = 1.4$ and $p_{0_j}/p_1 \geqslant 20$ for several values of α and β'. As shown in Fig. 6.62, K was only a weak function of α, β', and Mach number.

The analytical model presented is for two-dimensional flow. It may be improved and further developed by using the details of experiments by Charwat and Allegre (1964), two-dimensional flow experiments such as those of Werle et al. (1969), and the critical review of Werle (1968). Billig et al. (1970) felt it was unrealistic to use a rounded blunt body shape to represent an obstacle due to jet interaction for the Spaid and Zukoski (1968) model. However, because the compressible turbulent separated flow pressure profile is independent of the mechanism that causes separation for the modeling of pressure-profile, the Spaid and Zukoski (1968) model may still be considered

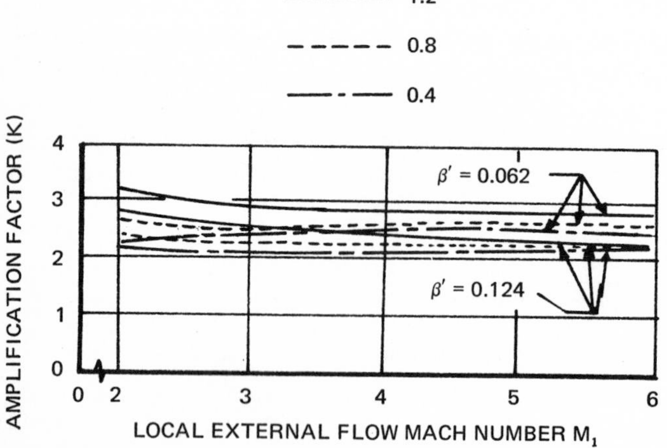

FIG. 6.62 Dependence of calculated values of amplification factor on Mach number for various values of parameters α and β' (*from Spaid & Zukoski, 1968*).

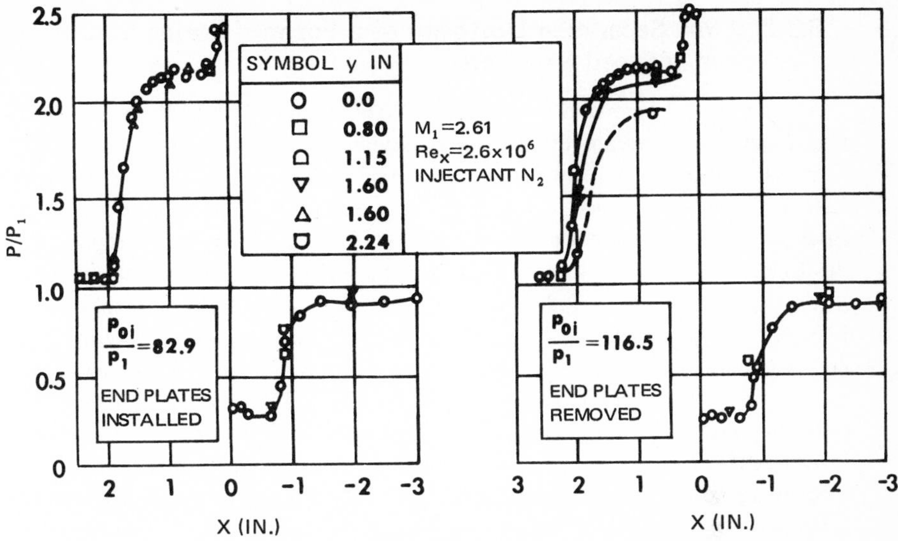

FIG. 6.63 Static pressure distributions showing effects of end plate (*from Spaid & Zukoski, 1968*).

useful. Since the jet interaction problem is generally not two-dimensional and a three-dimensional effect exists, this effect is summarized by Spaid and Zukoski (1968). The three-dimensional flow effect is illustrated by Fig. 6.63 which compares flow conditions with fixed end plates (representing the two-dimensional flow) and without them.

When the end plates are removed, the pressure distribution becomes a function of y, the distance transverse to the flow direction, as shown in Fig. 6.63. The length of the separated region decreases and there are strong lateral flows from the upstream separated region (as indicated by oil flow studies of Heyser & Maurer, 1962). However, the plateau pressure and hence the initial separation angle near the plate centerline are not affected by the removal of the end plates. When transverse flows exist, the streamline dividing the flow that passes over the jet flow entering the recirculating zone shifts outward to include a part of the boundary-layer flow upstream of the separation point. This shift moves the separation point toward the disturbance. Furthermore, the new dividing streamline is now located in a region of higher average velocity in the shear layer than that for the two-dimensional flow. This produces an increase in the drag of the effective obstacle produced by injection, a decrease in penetration height, and an additional decrease in the length of the separated region. However, if the ratio of penetration height to slot span is sufficiently large, these transverse flows will become negligible; they can be prevented by mounting end plates.

6.3.3 Flow Separation Upstream of a Forward Facing Step and Induced Side Force

As mentioned previously, Vogler (1963) found that with a subsonic mainstream, the pressure distribution upstream of the jet can be correlated with that caused by a rod or a spoiler. Spaid and Zukoski (1968) concluded that with a supersonic mainstream, the turbulent separated flow field produced by a jet is the same as that of two-dimensional turbulent separation due to a forward facing step provided (*a*) the region in the immediate vicinity of the step or jet is excluded, and (*b*) the step height and jet penetration distance are adjusted so that the lengths of the separated regions are approximately the same. This finding is very useful because the data available on separated flow upstream of forward facing steps may be utilized to correlate with the jet-induced flow separation. However, for the flow downstream of a jet, as Vogler (1963) indicated, no similarity of flow downstream of a rod may be found. Recently Kaufman and Koch (1968) attempted to correlate the reattaching flow downstream of a jet issuing into a supersonic mainstream with the flow downstream of a rearward facing step. Additional investigation is desirable on the correlation of flow downstream of a jet with the rearward facing step or the spoiler. The monograph by Chang (1970) included a presentation of separated flow for forward and rearward facing steps.

It has been demonstrated that for a fixed turbulent boundary layer and supersonic external flow, the pressure profile in the upstream separated flow region is independent of the mechanism that causes separation. Plateau pressure produced by separation in an overexpanded conical rocket nozzle (Arens & Spiegler, 1963) for the range $2.0 < M_s < 5.5$ (M_s is Mach number at separation) agree within the scatter of data with step-induced values from gas injection studies by Spaid and Zukoski (1968). Data on the pressure rise in front of a cylinder (Bernstein & Brunk, 1955) are also in agreement. Hence an attempt may be made to correlate the wall pressure distribution in the upstream separated region due to a jet and that induced by some geometric shape. The step is the simplest geometric shape and the one for which most data are available. Therefore, let us study the wall pressure distribution in the separated region upstream of a forward facing step.

Table 6.3 summarizes data from various pertinent studies over the period 1954–1966 as reviewed by Zukoski (1967). The information covers cases where the boundary layer upstream of separation is turbulent and the step height is large compared with the boundary-layer thickness. Zukoski (1967) concludes that the pressure rise in the separation region, expressed in normalized form, is independent of Mach number and Reynolds number and that the scale for the separation phenomena is the boundary-layer thickness. Moreover, for the turbulent regime, the plateau pressure rise in the separated

TABLE 6.3 Dependence of properties on step height

h	h/δ	x_s/h	p_p/p_0	α	β	F_i/p_0h*
0.10	0.42	3.0	2.50	1.00	0.87	7.3
0.15	0.63	4.2	2.65	0.85	0.79	7.5
0.20	0.83	4.3	2.76	0.86	0.86	7.9
0.25	1.04	4.2	2.90	0.90	0.90	8.6
0.30	1.25	4.3	3.02	0.89	0.89	8.8
0.35	1.45	4.4	3.05	0.98	0.83	9.0
0.40	1.67	4.5	3.02	0.89	0.89	9.1

Note:

$$\alpha = \left(\int_{x_s}^{x_s+\Delta x_s} \frac{\Delta p}{\Delta p_p} \frac{dx}{\delta}\right)$$

$$\beta = \left(\int_0^{x_s} \frac{\Delta p}{\Delta p_p} \frac{dx}{x_s}\right)$$

M = 3.85

$\delta = 0.24$ in.

From Zukoski (1967).
*Normalized side force; see Eq. (6.44).

region is independent of Reynolds number, but the induced side force increases linearly with Mach number.

6.3.3.1 Wall pressure distribution and geometric parameters

The general features of a typical steady separated flow based on the experiment of Bogdonoff (1955) is sketched in Fig. 6.64 along with the wall pressure distribution.

At the initial location of the interaction, the rise of the wall static pressure to the separation pressure takes place in about two boundary-layer thicknesses and then increases more slowly to a first plateau. At the beginning of the interaction region, the pressure gradient rises to a maximum value of $(\Delta p_p/2\delta)$ in a distance of about one boundary-layer thickness and remains constant for the distance of another boundary-layer thickness. As the step is approached, there is a further rapid pressure rise amounting to at least 50 percent more than the plateau value. The locations of the separation and the plateau pressure are about $4\frac{1}{2}$ and 2 step heights upstream of the step, respectively.

The boundaries shown in the lower sketch were obtained from measured pitot pressure profiles; these crudely approximate the profiles obtained by tangents drawn at extreme values and at inflection points. As determined by this method, the shock wave is first clearly defined at the edge of the boundary layer and at a point almost directly above the separation point. As seen from Fig. 6.64, the first noticeable movement of the lower edge of the boundary layer or shear zone away from the wall occurs roughly linearly at

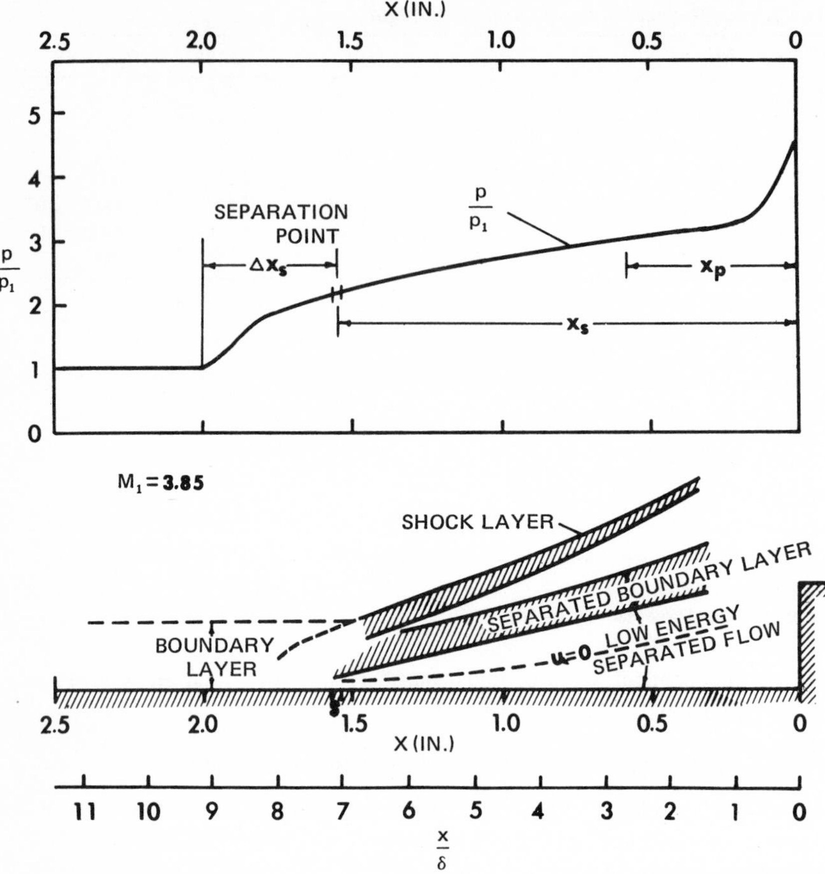

FIG. 6.64 Flow field and wall pressure distribution produced by separation of a turbulent boundary layer ahead of a step (*Bogdonoff, 1955*; pressure data *from Zukoski, 1967*).

the separation point, making an approximate angle of 13 deg with respect to the wall. The upper boundary of this shear zone makes a slightly larger angle of about 16.5 deg. Thus, the region "spreads" with an angle of about 3.5 deg. Supersonic flow above the shear layer is more or less uniform. Beneath the shear zone, a line along which the horizontal velocity is zero may be drawn close to the center of a strongly recirculating flow region. The fluid velocity in this low energy separated flow region can be as high as 35 percent of the undisturbed stream value even though the total pressure may amount to only 3 or 4 percent of that for the undisturbed stream.

Pressure distribution on the face of the step is not uniform, but is at a maximum near the top and at the bottom of the step face. It may be concluded from such pressure measurements that the maximum speed along the face is at least 40 percent of the free-stream value. Flow separation on forward facing steps is often unsteady; thus in addition to the steady flow separation illustrated in Fig. 6.64, the unsteady features should also be described. The motions of the shock system are observed to be unsteady and the separation point and accompanying region of rapid rise of pressure may move upstream and downstream (with a frequency of less than 1 kHz) and over a region of the order of one boundary-layer thickness, causing wall pressure fluctuations. Near the separation point of a turbulent boundary layer and for large steps, the ratio of pressure fluctuation (rms values) to the free-stream dynamic pressure amounts to about 0.06 and over the plateau region to 0.03 and is roughly independent of Mach number. For the average data of Fig. 6.64, the pressure rise to the plateau may amount to only 20 percent of the free-stream dynamic pressure, but it is three to seven times the rms amplitude of the fluctuations.

Now consider the geometric parameters shown in Fig. 6.64:

1. Length of region Δx_s between the initial location of the pressure disturbance and the separation point.
2. Length x_s between the separation point and the face of the step.
3. Length x_p between the first pressure peak or beginning of the first pressure plateau and the face of the step.

The normalized geometric length of the separated zone is plotted in Fig. 6.65 as a function of Mach number. The symbols x_s and x_p refer to normalized x_s and x_p by step height; their values were about 4.3 and 2.0, respectively, and almost independent of Mach number.

6.3.3.2 Induced side forces

The pressure distribution and geometric parameters thus far investigated are used to calculate the side force induced by separation upstream of a forward facing step by summation of the integration of the pressure rise over the two regions Δx_s and x_s. Thus

$$F_i = \delta\ \Delta p_p \int_{\Delta x_s} \frac{\Delta p}{\Delta p_p} \frac{dx}{\delta} + x_s\ \Delta p_p \int_{x_s} \left(\frac{\Delta p}{\Delta p_p}\right) \frac{dx}{x_s}$$

or

$$F_i = \alpha(\Delta p_p\ \delta) + \beta(\Delta p_p x_s)$$

Here α and β contain the dependence of the pressure profiles on parameters such as h/δ and on Mach number. The values of α and β are close to unity. The meanings of the other symbols are $\Delta p = p - p_\infty$ and $\Delta p_p = p_p - p_\infty$ where p_∞ is ambient static pressure. The distance Δx is measured along the

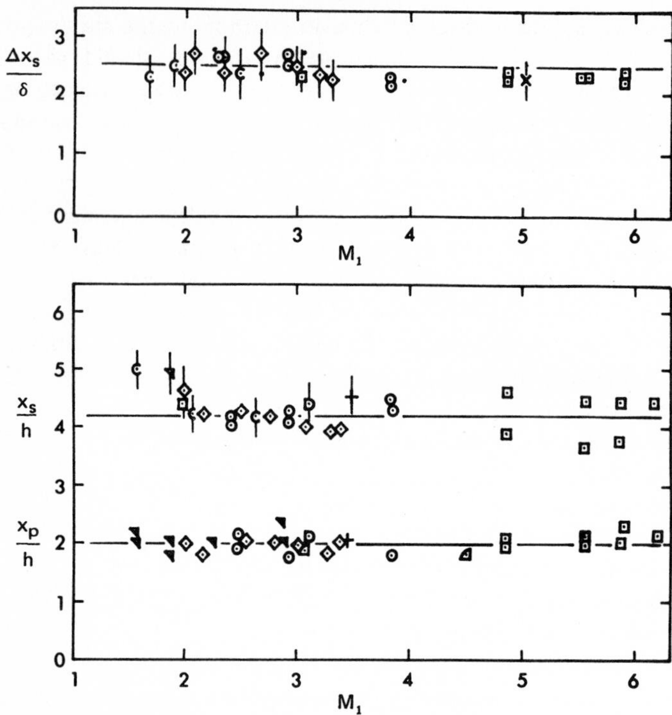

FIG. 6.65 Variation of normalized geometric parameters with Mach number (*from Zukoski, 1967*).

wall from the location of initial pressure rise, h is step height, and δ is boundary layer thickness. The normalized side force is then

$$\frac{F_i}{p_1 h} = \left(\frac{\Delta p_p}{p_1}\right)\left(\frac{x_s}{h}\right)\beta\left[1 + \frac{\alpha}{\beta}\left(\frac{\delta}{h}\right)\frac{h}{x_s}\right] \tag{6.44}$$

In this equation the dependence on h/δ appears explicitly in the bracketed term but is implicitly present through the dependence of the rest of the parameters on h/δ. It may be noted, however, that the value of h/x_s is about $\frac{1}{4}$; if α/β is near unity, the value of the bracketed term is close to 1 when $h/\delta > 1$.

The dependence of normalized side force on Mach number may be estimated by using the approximations based on the experimental data from Eq. (6.44). The approximations for the various parameters are $\Delta p_p/p_1 \approx \frac{1}{2}M_1$, $x_s/h \approx 4.2$, and $\beta \approx \alpha \approx 0.9$. Then for $h/\delta > 1.5$, Eq. (6.44) becomes

$$\frac{F_i}{p_1 h} \approx 2.1 M_1 \tag{6.45}$$

The normalized induced side force is plotted in Fig. 6.66 as a function of Mach number and compared with the experimental data. The agreement between the measured induced side force and that predicted by Eq. (6.45) is better when the constant in the equation is taken as 2.3. On the basis of the experimental data, the induced side force is reasonably estimated by

$$\frac{F_i}{p_1 h} \approx \left(\frac{\Delta p_p}{p_1}\right)\left(\frac{x_s}{h}\right) \tag{6.46}$$

Next, the drag of the step can be estimated on the basis of the following assumptions: The pressure rise from the plateau value to the average value on the face of the step is produced by the forces required to turn the recirculating flow lying beneath the dividing streamline. For an adiabatic flow, the density in this region is evaluated at $\rho = p_p/RT_{1_t}$, where subscript t refers to total condition. The average of the speed is taken to be proportional to the speed at the edge of the separated boundary layer u_e. Then

$$(\bar{p}_f - p_p) \approx \left[\frac{\gamma M_e^2}{1 + (\gamma - 1/2)M_e^2}\right]\kappa' p_p = \kappa p_p \tag{6.47}$$

where $\bar{p}_f$ is the average pressure on face of the step,
κ is assumed to be independent of Mach number, and
M_e is Mach number at the outer edge of the separated flow.

The value of κ' is evaluated approximately by 0.13 by matching the following experimental data of Mach number upstream of separation from 3 to 6, namely, $\bar{p}_f - p_p \approx 0.6\, p_p$.

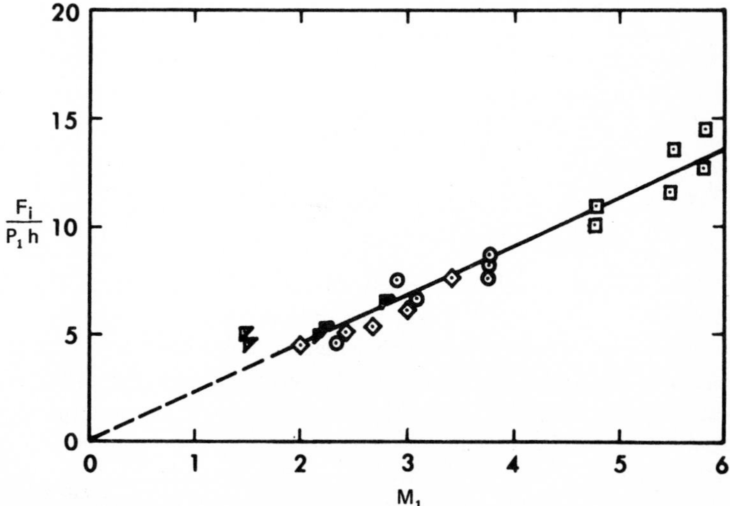

FIG. 6.66 Variation of induced side force with Mach number (*from Zukoski, 1967*).

By defining the drag of the step by $D_f = (p_f - p_1)h \approx [(1+\kappa)p_p - p_1]h$ and by using Eq. (6.47) to calculate the side force, we obtain a ratio of side force to drag as

$$\frac{F_i}{D_f} \cong \frac{(p_p - p_1)x_s/h}{(1+\kappa)(p_p - p_1) + \kappa p_1} = \frac{x_s/h}{(1+\kappa) + (p_1/\Delta p_p)} \tag{6.48}$$

If Eq. (6.45) and the approximate relations $M_s \approx \frac{3}{4} M_1$ are used, then numerical values of the ratios of side force to drag are obtained as follows:

For $M_1 = 2$: 2.6

For $M_1 = 4$: 2.2

For $M_1 = 8$: 2.1

It may thus be concluded that the ratio of side force to drag is approximately 2 and that effect of Mach number is not significant.

It is noted that the scaling of the free interaction region by the boundary layer thickness rather than the step height is intuitively satisfying because some measure of the boundary layer thickness is the most obvious dimension associated with this region. It should also be noted that the most direct scale length is the physical boundary layer thickness δ rather than momentum or displacement thickness. However for the special case of forward facing step, the normalized pressure profile $(\Delta p/\Delta p_p)$ (between the separation point and step face) scales with the step height, is independent of Reynolds number, and is only weakly dependent on Mach number when the step height is larger than the boundary layer thickness.

Zukoski (1967) has shown that the normalized profile $[(p - p_1)/(p_p - p_1)]$ versus $(\Delta x/\delta)$ is independent of Mach number and Reynolds number up to $\Delta p \approx 0.6\, p_p$ and that the maximum slope is given by $(\partial p/\partial x)_{max} \approx \frac{1}{2}\, \Delta p_p/\delta$ regardless of Mach or Reynolds numbers.

6.3.4 Analogy between Side Force Generated by Different Systems

The analogy which exists between the characteristics of flow separation provoked by a forward facing step and by a two-dimensional jet plume is presented in this section. This presentation supports the fact that the separated flow in a supersonic free stream is independent of the cause of the separation. Needless to say, such an analogy is very useful in that data available from one system may be applied to another system, thus widening the scope of practical application.

The pressure distributions within the separated flow for the two cases are shown in Figs. 6.67 and 6.68. The experimental, turbulent, separated-flow

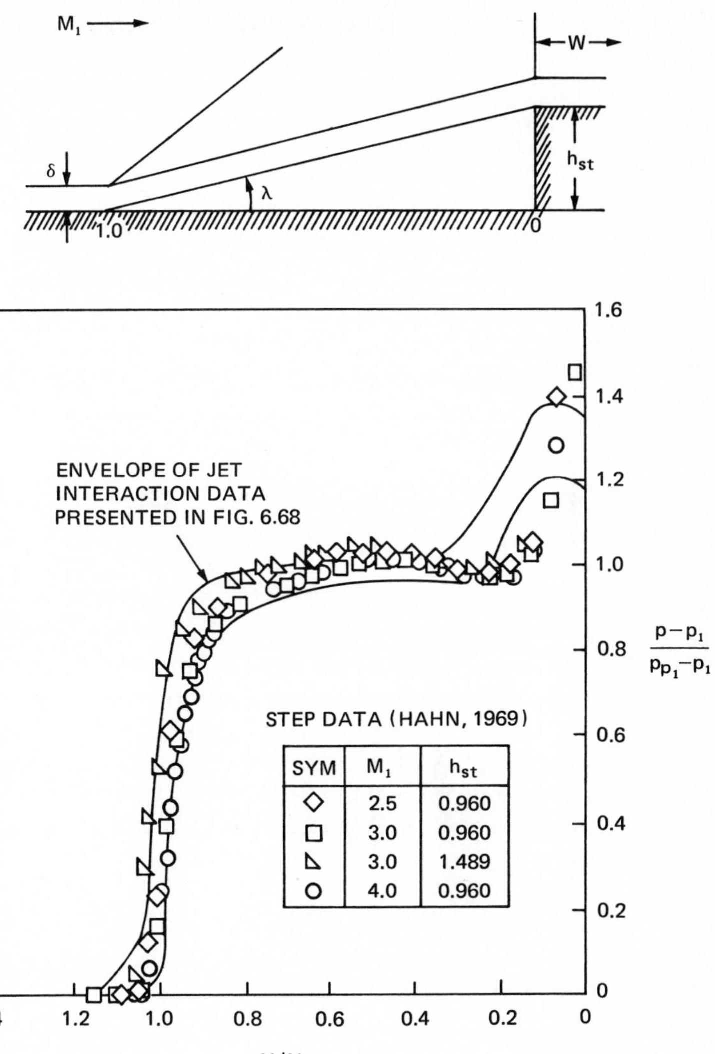

$$\frac{p_{p_1}}{p} = 1 + \frac{M_1}{2}, \quad \tan\lambda = \frac{h_{st}}{x_s} = \frac{1}{4.2} = \text{const.}$$

FIG. 6.67 Step separation, turbulent boundary layer (*from Hahn, 1969*).

pressure distribution upstream of a forward facing step as determined in an experimental investigation by Hahn (1969) is shown in Fig. 6.67. The pressure distribution upstream of jet plumes as measured by Werle et al. (1970) is indicated in Fig. 6.68*a*. It is evident from a comparison of these figures that there is a close analogy of the pressure distribution although the conditions were not the same. For the step flow, the M_1 range was 2.5 to 4.0 and the

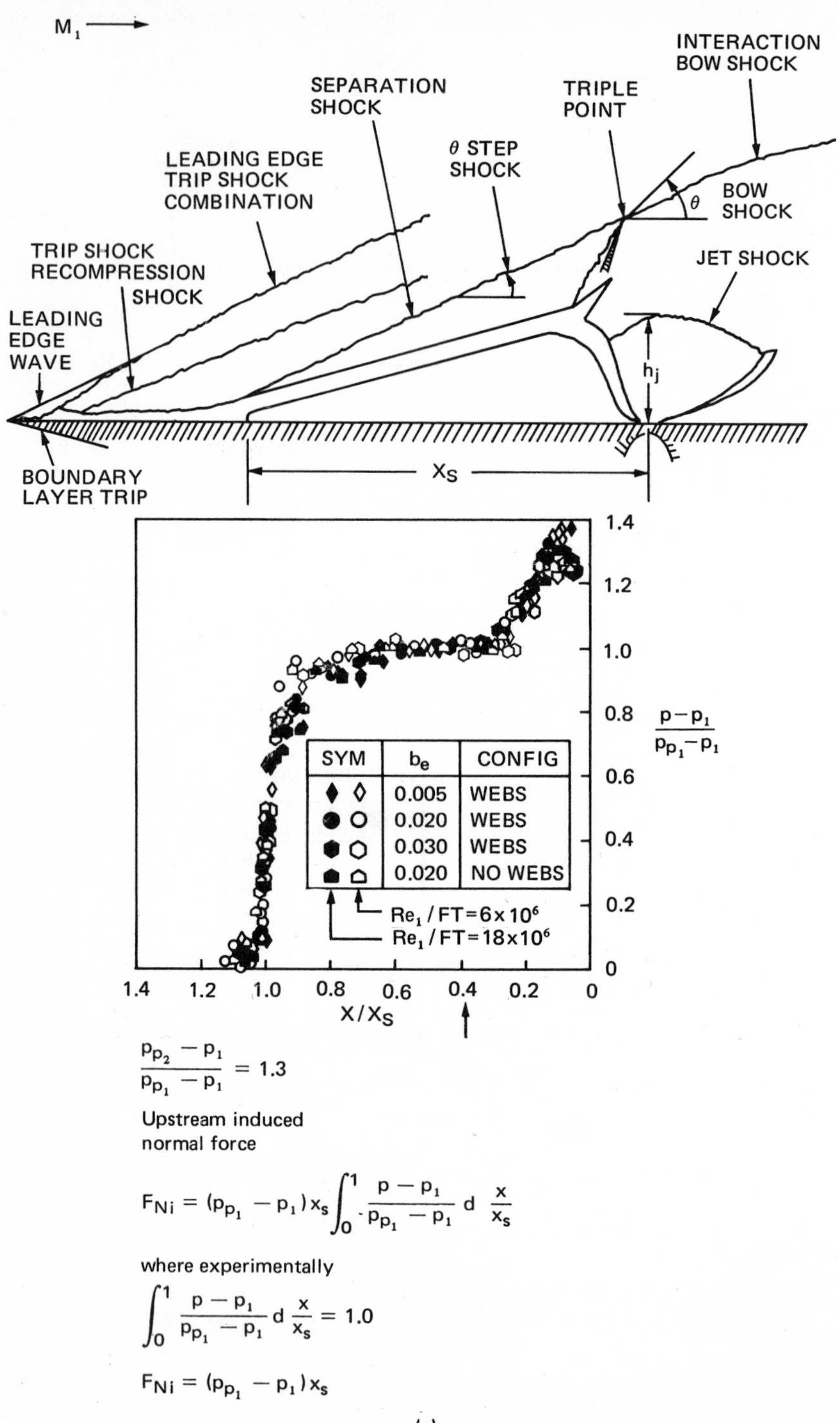

FIG. 6.68 Jet interaction, turbulent boundary layer: (*a*) pressure distribution upstream of jet (*from Werle et al., 1970*).

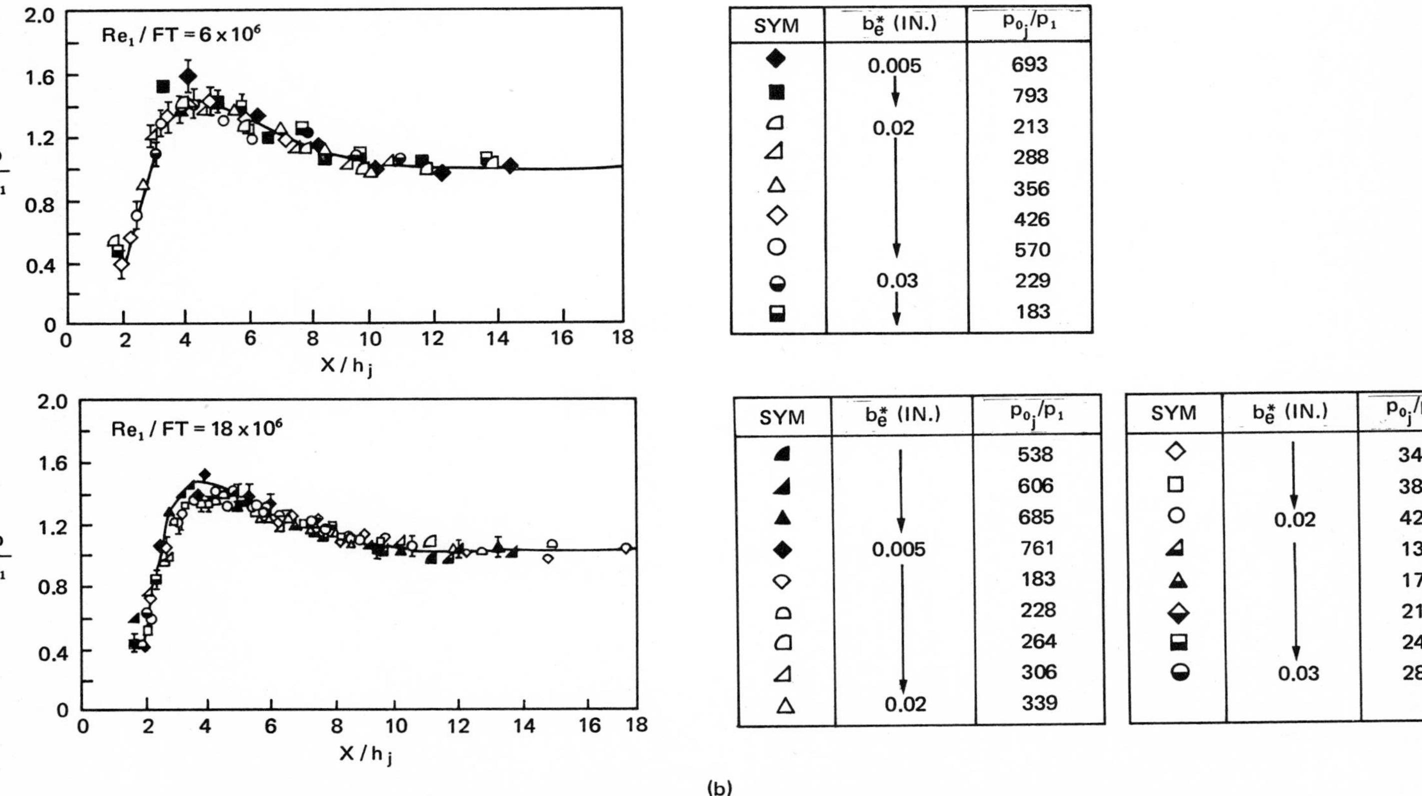

FIG. 6.68 *(continued)* Jet interaction, turbulent boundary layer: (*b*) pressure distribution downstream of jet (*from Werle et al., 1970*).

step height 0.96 to 1.489 in.; for the jet interaction case, the Re_∞/ft range was 6 to 18 × 10^6 at $M_\infty = 4$.

It is seen from the figures that the first plateau pressure p_{p_1} is the most influential pressure and contributes the major portion to the side force. Small discrepancies in the pressure distribution are observed in the neighborhood of separation, $x/x_s \to 1.0$, and also in the region approaching the face of the step (or the centerline of the jet port) involving a second peak pressure p_{p_2} at $x/x_s \to 0$. However, these discrepancies do not appreciably affect the generated side force, as will be discussed later. For the step flow, the height of the solid step h_{st} is the actual height of the obstruction, but the situation is more complicated for the jet. The external configuration of the jet plume is altered somewhat if it is exposed to a supersonic airstream, as seen in Fig. 6.68*b*. The terminal shock position along the centerline of the jet plume is equivalent to the static free jet shock height h_j, as shown in Fig. 6.69. Therefore, the normal distance between the jet exit plane at the wall surface and the terminal shock position along the centerline of the jet port, as indicated in

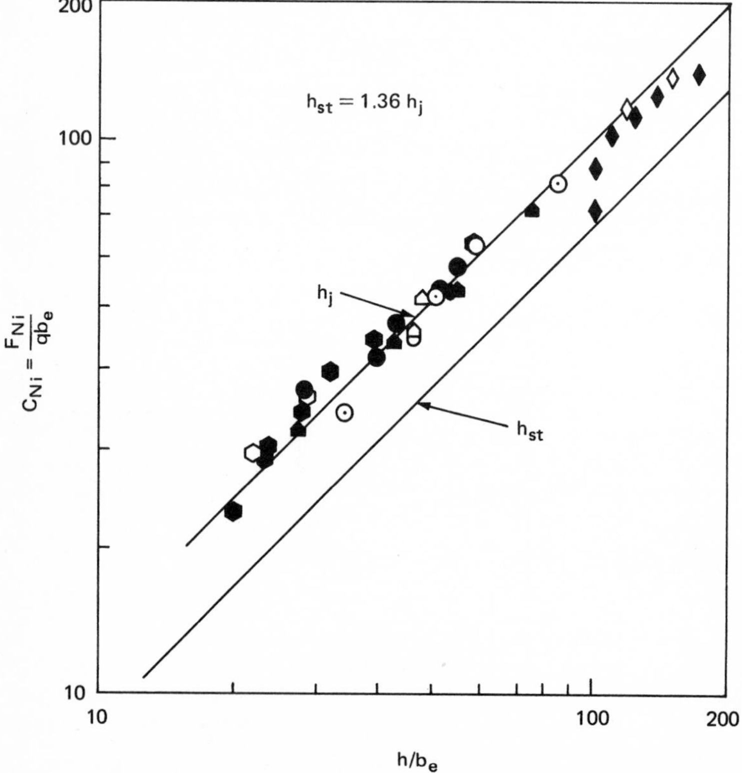

FIG. 6.69 Relationship between step height and jet height for a sonic jet (*from Werle et al., 1970*).

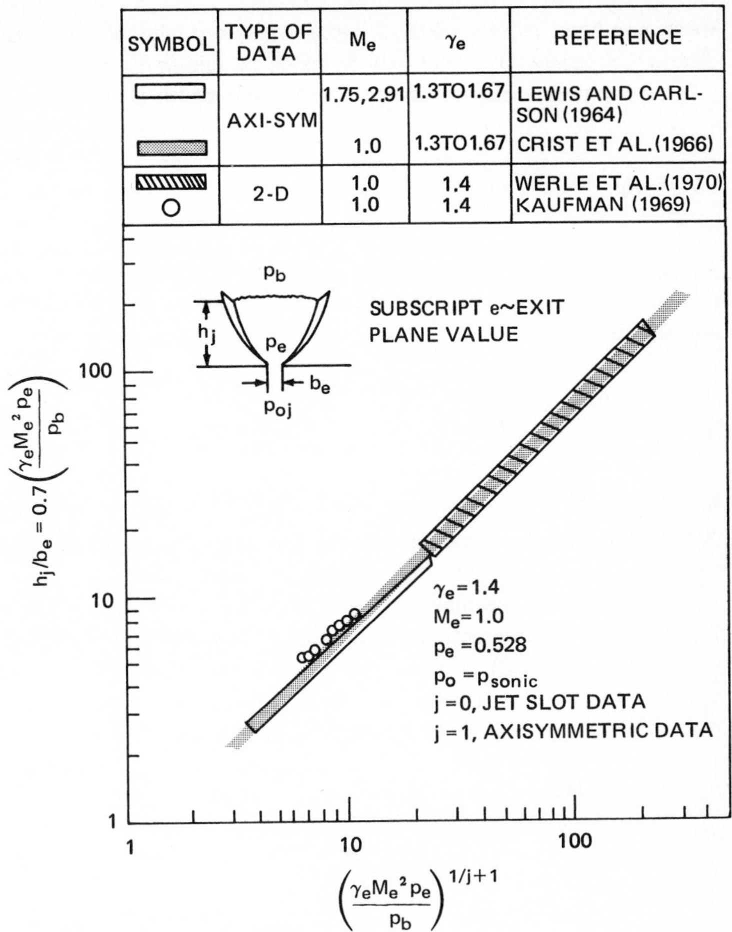

SYMBOL	TYPE OF DATA	M_e	γ_e	REFERENCE
(open bar)	AXI-SYM	1.75,2.91	1.3TO1.67	LEWIS AND CARLSON (1964)
(shaded bar)	AXI-SYM	1.0	1.3TO1.67	CRIST ET AL.(1966)
(hatched bar)	2-D	1.0	1.4	WERLE ET AL.(1970)
O	2-D	1.0	1.4	KAUFMAN (1969)

FIG. 6.70 Free jet correlation (*from Werle et al., 1970*).

Fig. 6.70 for a symmetrical jet, may be considered as the reference height to evaluate an equivalent step height for the side force to be generated. The equivalent step height is determined from Fig. 6.69 for a given jet shock height to correlate normal induced force coefficient as

$$h_{st} = 1.36 h_j \tag{6.49}$$

Note that this relation is independent of jet strength, slot width, and free-stream Reynolds number within the range of test conditions (Re_∞/ft from 6 to 18×10^6 and slot width from 0.005 to 0.030 in.) used at the U.S. Naval Ordnance Laboratory* by Werle et al. (1970). Unfortunately, it is not presently known how this relation given by Eq. (6.49) is affected by free-stream or jet

*now U.S. Naval Surface Weapon Center

Mach numbers. Nevertheless, a conservative estimate of the equivalent step obstruction height of the jet may be made by taking $h_{st} = h_j$.

Now the normal induced force F_{N_i} for unit width is given by

$$F_{N_i} = (p_{p_1} - p_1)x_s \int_0^1 \frac{p - p_1}{p_{p_1} - p_1}\, d\left(\frac{x}{x_s}\right) \tag{6.50}$$

where subscript i refers to induced force and x_s is a function of step height or jet shock height. By using the measured data shown in Fig. 6.67, it was found that the integral

$$\int_0^1 \frac{p - p_1}{p_{p_1} - p_1}\, d\left(\frac{x}{x_s}\right)$$

amounts to 1.02 for both step flow and jet interaction, indicating that the discrepancies of the pressure for the two cases are negligible. For simplicity therefore, the value of this integral may be taken as 1.0.

An attempt will be made to show how the information obtained from the step flow and jet interaction can be used to compute the generated side force. Note from Eq. (6.50) that the side force is dependent on p_{p_1} and x_s. These two quantities are evaluated from the step data of Hahn (1969), resulting in the following first plateau pressure given by Zukoski (1967):

$$\frac{p_{p_1}}{p_1} = 1 + \left(\frac{M_1}{2}\right) \tag{6.51}$$

and

$$x_s = \frac{h_{st}}{\tan\lambda} = 4.2 h_{st} \tag{6.52}$$

The value of $\tan\lambda$ (λ is the separation angle as shown in Fig. 6.67) is constant, amounting to 1/(4.2).

Next, from the experimental data upstream of a jet, the jet interaction may be computed by

$$\frac{h_j}{b_e} = 0.7 \left(\frac{\gamma_e M_e^2 p_e}{p_b}\right)^{1/j+1} \tag{6.53}$$

which is in good agreement with the experimental data shown in Fig. 6.70. In this equation, b_e is the jet exit slot width or exit diameter; $j = 0$ and $j = 1$, respectively, refer to two-dimensional and axisymmetric jets; and subscript e refers to conditions at the jet nozzle exit. The symbol p_b refers to the effective back pressure acting on the jet plume in the interaction model. From the experimental results of Werle et al. (1970) it was found that

$$p_{e_b} = p_{p_2} \tag{6.54}$$

and

$$\frac{p_{p_2} - p_1}{p_{p_1} - p_1} \approx 1.3 \tag{6.55}$$

Thus, for two-dimensional jet interaction, substitution of Eqs. (6.53) and (6.55) into (6.50) yields

$$F_{N_i} = 0.7 \frac{h_{st}}{h_j} \frac{p_{p_1} - p_1}{p_{p_2} \tan \lambda} \gamma_e \mathrm{M}_e^2 p_e b_e \tag{6.56}$$

or

$$F_{N_i} = 0.7 \frac{(p_{p_1}/p_1) - 1}{\tan \lambda} \frac{\gamma_e \mathrm{M}_e^2 p_e b_e}{1.3\,[(p_{p_1}/p_1) - 1] + 1} \tag{6.57}$$

for $h_{st} = h_j$.

Downstream of the obstruction, reattachment of the flow to the wall surface may take place both for the jet interaction and for the step flow. For the case of the step flow, however, the step may be reduced to a spoiler by shortening the running length w, as indicated in Fig. 6.67, followed by a rearward facing step surface. At present, the available experimental data on the pressure distribution downstream of a spoiler are insufficient to enable correlation with that of a jet interaction shown in the right-hand side of Fig. 6.68*b*. Although the pressure distribution downstream of the jet plume shows

$$\frac{p}{p_1} \approx 0.4 \quad \text{at} \quad \frac{x}{h_j} \approx 2 \qquad \text{(pressure rise start)}$$

$$\frac{p}{p_1} \approx 1.5 \quad \text{at} \quad \frac{x}{h_j} \approx 4.1 \qquad \text{(peak pressure)}$$

$$\frac{p}{p_1} \approx 1.0 \quad \text{at} \quad \frac{x}{h_j} \approx 12 \qquad \text{(pressure decrease to free stream value)}$$

the net side force generated there is negligible, amounting only to about 2 percent of the force generated upstream of the jet slot. Hence, for the analogy of step flow with jet interaction, the side force generated downstream of a jet plume may be neglected, and only the side force upstream of the jet plume need be considered.

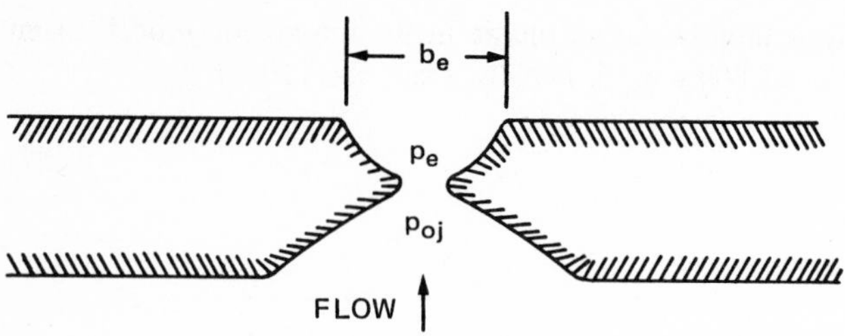

$$F_{Nj} = p_e b_e + \rho_e b_e v_e^2 \quad \text{or} \quad F_{Nj} = p_e(1 + \gamma_e M_e^2) b_e$$

TWO-DIMENSIONAL JET INTERACTION:

$$A = 1 + \frac{F_{Ni}}{F_{Nj}}; \quad A = 1 + \left(\frac{0.7}{\tan\lambda}\right)\left(\frac{1}{1 + \dfrac{1}{\gamma_e M_e^2}}\right)\left(\frac{1}{1.3 + \dfrac{1}{(p_{p_1}/p_1) - 1}}\right)\left(\frac{h_{st}}{h_j}\right)$$

FIG. 6.71 Jet vacuum normal force (*from Werle et al., 1970*).

Finally, a jet amplification factor A is presented to enable correlation of the jet interaction in a mainstream with jet issuance into vacuum. The amplification factor A is defined by

$$A = 1 + \left(\frac{F_{N_i}}{F_{N_j}}\right) \tag{6.58}$$

where F_{N_j} is the vacuum thrust force of the jet per unit width, given by $F_{N_j} = p_e b_e + \rho_e v_e^2$ or $F_{N_j} = p_e(1 + \gamma_e M_e^2) b_e$ as shown in Fig. 6.71.

Then for a sonic jet

$$A \cong 1 + \frac{0.7[(p_{p_1}/p_1) - 1]\gamma_e M_e^2 h_{st} \, ctn\, \lambda}{\{1.3[(p_{p_1}/p_1) - 1] + 1\}(1 + \gamma_e M_e^2) h_j}$$

or

$$A = 1 + \left(\frac{0.7}{\tan\lambda}\right)\left[\frac{1}{1 + (1/\gamma_e M_e^2)}\right]\left\{\frac{1}{1.3 + [1/(p_{p_1}/p_1) - 1]}\right\}\left(\frac{h_{st}}{h_j}\right)$$

By taking $\tan\lambda = 1/4.2$,

$$A = 1 + \frac{1.47 M_1}{1 + 0.65 M_1} \, \frac{\gamma_e M_e^2}{1 + \gamma_e M_e^2} \, \frac{h_{st}}{h_j} \tag{6.59}$$

The value of A is slightly dependent on the fluid used in the jet; when the value of γ_e is increased from 1.4 to 1.67, there is a 4-percent increase in the value of the jet amplification factor.

Further investigation is required to verify the amplification factor A for jets that differ from a sonic jet.

NOMENCLATURE

A	Point in the Q-plane; also cross-sectional area or jet flow amplification factor
A_f	Extent of nose bluntness in percent
B	Point in the Q-plane
b_e	Jet slot width or exit diameter
C	Point in the Q-plane
C_{LD}	Theoretical value of C_L when $\alpha = \alpha_0$
C_m	Pitching moment coefficient
$C_{\dot{m}}$	Mass flow coefficient
C_Q	$C_Q = Q/2\pi u_\infty$
C_μ	Moment coefficient
C_Γ	$C_\Gamma = \Gamma/2\pi u_\infty$
d	Point in the ζ-plane; also body diameter; also slot width
F_i	Induced side force
f	Point in the Q-plane
h	Height of step, height of jet plume
K	Constant; also $K = l/d$
L	Length of free streamline
l	Length
$\dot{m}$	Mass flow rate
p_B	Base pressure
p_{eb}	Effective back pressure
p_f	Average pressure
p_j^*	Pressure at sonic point in the jet orifice
p_p	Plateau pressure
p_{stag}	Stagnation pressure of free streamline
p_t	Total pressure behind normal shock
Q	Transformed plane of a free streamline or half sink strength
q	Flap; also velocity or momentum
R	Universal gas constant
R_j	Jet exhaust radius
r	Radial distance or radius
S	Distance along jet trajectory
s	Distance along surface measured from stagnation point
T	Thrust
T_{s_v}	Vacuum thrust of sonic jet
v	Slot velocity or jet velocity
w	$w = \phi + i\psi$, the complex velocity potential
x_1	Distance along upper plate from front of flat portion

x_2	Distance along lower plate from leading edge
α	A parameter
α_D	Incidence of lower plate of a free streamline flap
β	Angle or constant
β'	Proportionality constant
Γ	Vortex strength
γ	Slot angle
θ	Hemispherical angle; also cone angle; also direction of flow
λ	Angle of separation
τ	Angle; also transformation
ϕ	Semi apex angle of conical shock

Subscripts

e	Conditions at edge of shear layer or conditions at jet exit nozzle
j	Injected air or jet
N	Normal
n	Nozzle
p_1	First plateau
p_2	Second plateau
R	Restored value
s	Separation or stagnation
st	Step
t	Total
1	Upstream of separation outside the shear layer or ahead of Mach disk
2	Downstream of separation outside the shear layer

REFERENCES

Abbett, M. (1970). "The Mach disc in underexpanded exhaust plumes," AIAA Paper 70-231, AIAA 8th Aerospace Sciences Meeting, 19–21 Jan, New York.

Abdelhamid, A. N. and D. S. Dosanjh (1969). "Mach disc and Riemann wave in underexpanded jet flows," AIAA Paper 69-665 (Jun).

Adamson, T. C., Jr. and J. A. Nicholls (1959). "On the structure of jets from highly underexpanded nozzles into still air," *J. Aerospace Sci.*, Vol. 26, No. 1 (Jan), pp. 16–24.

Album, H. H. (1961). "Spiked blunt bodies in a supersonic flow," Air Force Office of Scientific Research Report 307 (Jun).

Arens, M. and E. Spiegler (1963). "Shock-induced boundary layer separation in overexpanded conical exhaust nozzles," *AIAA J.*, Vol. 1, No. 3 (Mar), pp. 578–581.

Bernstein, H. and W. E. Brunk (1955). "Exploratory investigation of flow in the separated region ahead of two blunt bodies at Mach number 2," NACA RM E55D07b (Jun).

Billig, F. S. et al. (1970). "A unified approach to the problem of gaseous jet penetration into a supersonic stream," AIAA Paper 70-93, AIAA 8th Aerospace Sciences Meeting, (19–21 Jan), New York.

Bogdonoff, S. M. (1955). "Some experimental studies of the separation of supersonic turbulent boundary layers," Princeton Univ. Report 336 (Jun).

Bogdonoff, S. M. and I. E. Vas (1959). "Preliminary investigations of spiked bodies at hypersonic speeds," *J. Aeronaut. Sci.*, Vol. 26, No. 2 (Feb), pp. 65–74.

Boyer, J. et al. (1963). "The flow field resulting from Mach reflection of a convergent conical shock at the axis of a supersonic axially symmetric jet," General Dynamics, Astronautics Report GDA 63-0586.

Boyer, J. et al. (1964). "Transonic aspects of hypervelocity rocket plumes," in *Supersonic flow, chemical processes and radiative transfer*, D. B. Olfe and V. Zakkay, eds., Pergamon Press, New York.

Brindle, C. C. and M. J. Malia (1963). "Longitudinal aerodynamics and heat-transfer characteristics of a hemisphere-cylinder missile configuration with an aerodynamic spike," David Taylor Model Basin Aero Lab Report 1061 (Jul).

Chang, P. K. (1970). *Separation of flow*, Pergamon Press, New York.

Chapman, D. R. et al. (1957). "Investigation of separated flows in supersonic and subsonic streams with emphasis on the effect of transition," NACA Report 1356.

Charwat, A. F. and J. Allegre (1964). "Interaction of a supersonic stream and a transverse supersonic jet," *AIAA J.*, Vol. 2, No. 11 (Nov), pp. 1965–1972.

Crawford, D. H. (1959). "Investigation of the flow over a spiked-nose hemisphere-cylinder at a Mach number of 6.8," NASA TN D-118 (Dec).

Crist, S. et al. (1966). "Study of the highly underexpanded sonic jet," *AIAA J.*, Vol. 4, No. 1 (Jan), pp. 68–71.

Cubbison, R. W. et al. (1961). "Surface pressure distribution with a sonic jet normal to adjacent flat surfaces at Mach 2.92 to 6.4," NASA TN D-580.

Daniels, L. E. and H. Yoshihara (1954). "Effects of the upstream influence of shock wave at supersonic speeds in the presence of a separated boundary layer," WADC Tech. Report 54-31 (Jan).

Eastman, D. W. and L. P. Radtke (1963). "Location of the normal shock wave in the exhaust plume of a jet," *AIAA J.*, Vol. 1, No. 4 (Apr), p. 919.

Frey, K. P. H. and N. C. Vasuki (1964). "Tests on flow development in diffusers," ASME Symposium on Fully Separated Flows, 18–20 May, Philadelphia, Pennsylvania, pp. 40–47.

Hahn, J. S. (1969). "Experimental investigation of turbulent step-induced boundary-layer separation at Mach numbers of 2.5, 3 and 4." AEDC-TR-69-1 (Mar).

Hahn, M. (1966). "Pressure distribution and mass injection effects in the transitional separated flow over a spiked body at supersonic speed," *J. Fluid Mech.*, Vol. 24, Part 2, pp. 209–223.

Hartley, R. M. (1957). "Leading-edge spikes to reduce the drag of wings at supersonic air speeds," David Taylor Model Basin Aero Lab Report 925 (Sep).

Hawk, N. E. and J. L. Amick (1965). "An experimental and theoretical investigation of two-dimensional jet-flap aerodynamic interaction at supersonic speeds," Applied Physics Laboratory, Johns Hopkins University Report CR-23 (Oct).

Hawk, N. E. and J. L. Amick (1967). "Two-dimensional secondary jet interaction with a supersonic stream," *AIAA J.*, Vol. 5, No. 4 (Apr), pp. 655–660.

Heyser, A. and F. Maurer (1962). "Experimental investigation on solid spoilers and jet spoilers at Mach numbers of 0.6 to 2.8," *Z. Flugwissenschaften*, Vol. 10, 1962, pp. 110–130. (Translation 32, California Inst. of Technology, Jet Propulsion Lab., 1964.)

Hunt, G. K. (1958). "Supersonic wind-tunnel study of reducing the drag of a bluff body at incidence by means of a spike," RAE Report Aero. 2606 (May).

Hurley, D. G. (1961). "The use of boundary layer control to establish free streamline flows," in *Boundary layer and flow control*, G. V. Lachmann, ed., Pergamon Press, New York.

Janos, J. J. (1961). "Loads induced on a flat-plate wing by an air jet exhausting perpendicularly through the wing and normal to a free-stream flow of Mach number 2.0," NASA TN D-649.

Kaufman, L. G. (1969). "Interaction between high speed flows and transverse jets: A method for predicting the resultant surface pressure distribution," Paper 35, Proc. 8th Navy Symposium, *Aeroballistics*, Vol. 3 (May), pp. 885–920.

Kaufman, L. G. and F. Koch (1968). "High-speed flows past transverse jets," Grumman Aircraft Engineering Corp. Research Dept. Report RE 348 (Oct).

Kopal, Z. (1947). "Tables of supersonic flow around cones," Massachusetts Institute of Technology.

Letko, W. (1963). "Loads induced on a flat plate at a Mach number of 4.5 with a sonic or supersonic jet exhausting normal to the surface," NASA TN D-1935 (Jul).

Lewis, C. H. and D. L. Carlson (1964). "Normal shock location in underexpanded gas and gas-particle jets," *AIAA J.*, Vol. 2, No. 4 (Apr), pp. 776–777.

Love, E. S. et al. (1959). "Experimental and theoretical studies of axisymmetric free jets," NASA TR-R-6.

Mair, W. A. (1952). "Experiments on separation of boundary layers on probes in front of blunt-nosed bodies in a supersonic air stream," *Phil. Mag.*, Vol. 43, Seventh Series (Jul), pp. 695–716.

Mandl, P. (1959). "Effect of standing vortex on flow about suction aerofoils with split flaps," NRC of Canada, Aero. Report LR-239.

Maull, D. J. (1960). "Hypersonic flow over axially symmetric spiked bodies," *J. Fluid Mech.*, Vol. 8, Part 4 (Aug), pp. 584–592.

Maurer, F. (1966). "Three-dimensional effects in shock-separated flow regions ahead of lateral control-jets issuing from slot nozzles of finite length," in *Separated flows*, Part II, Proceedings Fourth AGARD Conference (May).

Melbourne, W. H. (1960). "Experiments on a delta wing with jet-assisted lift," ARC CP 21, 968 (May).

Mitchell, J. W. (1964). "An analytical study of a two-dimensional flow field associated with some secondary injection into a supersonic flow," Vidya Research and Development Tech Note 9166-TN-2 (Mar).

Morkovin, M. V. et al. (1952). "Interaction of a side jet with a supersonic main stream," Univ. Michigan, Engineering Research Institute Bulletin 35.

Narashima, R. et al. (1967). "Leading shock in two-dimensional secondary fluid injection," *AIAA J.*, Vol. 5, No. 11 (Nov), pp. 2064–65.

Orth, R. C. and J. A. Funk (1967). "An experimental and comparative study of jet penetration in supersonic flow," *J. Spacecraft and Rockets*, Vol. 4, No. 9 (Sep), pp. 1236–1242.

Otis, J. H. Jr. (1962). "Induced interference effects on a four-jet VTOL configuration with various wing planforms in the transition speed range," NASA TN D-1400.

Owen, P. L. and C. K. Thornhill (1952). "The flow in an axially symmetric supersonic jet from a nearly sonic orifice into a vacuum," ARC Tech. Report R&M 26116.

Peters, C. E. and W. J. Phares (1970). "The structure of plumes and moderately underexpanded supersonic nozzles," AIAA Paper 70-229, AIAA 8th Aerospace Sciences Meeting, 19–21 Jan, New York.

Poisson-Quinton, P. (1956). "Einige physikalische Betrachtungen über das Ausblasen an Tragflügeln," Jahrb. d. W. G. L. Fridr. Vieweg & Sohn, Braunsweig & Akademie-Verlag, Berlin, pp. 29–51.

Povinelli, F. P. et al. (1970). "Supersonic jet penetration (up to Mach 4) into a Mach 2 airstream," AIAA Paper 70-92, AIAA 8th Aerospace Sciences Meeting, 19–21 Jan, New York.

Ringleb, F. O. (1961). "Separation control by trapped vortices," in *Boundary layer and flow control*, Vol. 1, G. V. Lachmann, ed., Pergamon Press.

Ringleb, F. O. (1964). "Discussion of problems associated with standing vortices and their applications," Symposium on Fully Separated Flows, ASME Fluid Engineering Division, 18–20 May, Philadelphia, Pennsylvania, pp. 33–39.

Romeo, D. J. (1963). "Aerodynamic interaction effects ahead of rectangular sonic jets exhausting perpendicularly from a flat plate into a Mach number 6 free stream," NASA TN D-1800 (May).

Romeo, D. J. and J. R. Sterrett (1961). "Aerodynamic interaction effects ahead of a sonic jet exhausting perpendicularly from a flat plate into a Mach number 6 free stream," NASA TN D-743 (Apr).

Ruggeri, R. S. et al. (1950). "Penetration of air jets issuing from circular, square, and elliptical orifices directed perpendicularly to an air stream," NACA TN 2019 (Feb).

Schetz, J. A. et al. (1967). "Structure of highly underexpanded transverse jets in a supersonic stream," *AIAA J.*, Vol. 5, No. 5 (May), pp. 882–884.

Schetz, J. A. et al. (1968). "Supersonic transverse injection into a supersonic stream," *AIAA J.*, Vol. 6, No. 5 (May), pp. 933–934.

Smith, A. M. O. (1956). "Experimental investigation of the snow cornice or cusp effect for control of separation over blunt fairings," Douglas Aircraft Company Report ES 26154.

Spaid, F. W. (1964). "A study of secondary injection of gases into a supersonic flow," Ph. D. Thesis, California Institute of Technology (Jun).

Spaid, F. W. and E. E. Zukoski (1968). "A study of the interaction of gaseous jets from transverse slots with supersonic external flows," *AIAA J.*, Vol. 6, No. 2 (Feb), pp. 205–212.

Spreemann, K. P. (1961). "Investigation of interference of a deflected jet with free stream and ground on aerodynamic characteristics of a semispan delta-wing VTOL model," NASA TN D-915.

Sterrett, J. R. and J. B. Barber (1966). "A theoretical and experimental investigation of secondary jets in a Mach 6 free stream with emphasis on the structure of the jet and separation ahead of the jet," in *Separated flow,* Part 2, Proceedings Fourth AGARD Conference, pp. 667–700 (May).

Vas, I. E. and S. M. Bogdonoff (1955). "Interaction of turbulent boundary layers with a step at M = 3.85," Princeton University Report 295 (Apr).

Vinson, P. W. et al. (1959). "Interaction effects produced by jet exhausting laterally near base of Ogive-cylinder model in supersonic main stream," NASA Memo 12-5-58W (Feb).

Vogler, R. D. (1963). "Surface pressure distribution induced on a flat plate by a cold air jet issuing perpendicularly from the plate and normal to a low-speed free-stream flow," NASA TN D-1629 (Mar).

Vranos, A. and J. J. Nolan (1965). "Supersonic mixing of a light gas and air," presented at AIAA Propulsion Joint Specialists Conference, 14–18 Jun, Colorado Springs, Colorado.

Walker, R. E. et al. (1963). "Secondary gas injection in a conical rocket nozzle," *AIAA J.*, Vol. 1, No. 2 (Feb), pp. 334–338.

Welzenbach, W. (1930). "Untersuchungen über die Stratigraphie der Schneeablagerungen," Wiss. Veröffentlichungen des Deutschen und Oesterreichischen Alpenvereins 9.

Werle, M. J. (1968). "A critical review of analytical methods for estimating control forces produced by secondary injection," NOLTR 68-5 (Jan).

Werle, M. J. et al. (1969). "Two-dimensional jet-interaction experiments–Results of flow-field probing and scale effects studies," 8th Navy Symposium on Aeroballistics, 6–8 May 1969, Corona, California.

Werle, M. J. et al. (1970). "Jet-interaction-induced separation of supersonic turbulent boundary layers–The two-dimensional problem," AIAA Paper 70-765, AIAA 3rd Fluid and Plasma Dynamics Conference, 29 Jun–1 Jul, Los Angeles, California.

Wood, C. J. (1961). "A study of hypersonic separated flow," Ph. D. Thesis, Univ. of London (Oct); AD 401652.

Zukoski, E. E. (1967). "Turbulent boundary-layer separation in front of a forward facing step," *AIAA J.*, Vol. 5, No. 10 (Oct), pp. 1746–1753.

Zukoski, E. E. and F. W. Spaid (1964). "Secondary injection of gases into a supersonic flow," *AIAA J.*, Vol. 2, No. 10 (Oct), pp. 1689–1696.

SUPPLEMENTARY REFERENCES

Abbett, M. (1971). "Mach disk in underexpanded exhaust plumes," *AIAA J.*, Vol. 9, No. 3 (Mar), p. 539.

Amick, J. L. (1966). "Circular-arc jet flaps at supersonic speeds, two-dimensional theory," Applied Physics Lab., Johns Hopkins Univ., CR-24 (Jul).

Amick, J. L. and P. B. Hays (1960). "Interaction effects of side jets issuing from flap plates and cylinders aligned with a supersonic stream," WADD Tech. Report 60-329 (Jun).

Bankston, L. T. and H. M. Larsen (1959). "Thrust-vectoring experiments with gas injection," U.S. Naval Ord. Test Station, NAVORD Report 6548 (May).

Barnes, J. W. et al. (1967). "Control effectiveness of transverse jets interacting with a high speed free stream," AFFDL Tech. Report 67-90, Vol. 1 (Sep).

Bestall, D. and J. Turner (1952). "The effect of a spike protruding in front of a bluff body at supersonic speeds," Admiralty Research Council Tech. Report R&M 3007 (Jan).

Beylich, A. A. (1970). "Condensation in carbon dioxide jet plume," *AIAA J.*, Vol. 8, No. 5 (May), p. 965.

Broadwell, J. E. (1963). "Analysis of the fluid mechanics of secondary injection thrust vector control," *AIAA J.*, Vol. 1, No. 5 (May), pp. 1067–1075.

Brower, W. B. (1961). "Leading-edge separation of laminar boundary layers in supersonic flow," *J. Aeronaut. Sci.*, Vol. 28, No. 12 (Dec), pp. 957–961.

Crocco, L. (1955). "Considerations on the shock-boundary layer interaction," in *High-Speed Aeronautics*, A. Ferri, N. J. Hoff, and P. Libby, eds., Polytechnic Institute of Brooklyn.

Dahm, T. J. (1967). "A comprehensive analytical procedure for the performance prediction of rocket thrust vector control with gaseous secondary injection," Aerotherm Corp. Final Report 67-11 (Jun). (Also AFRPL-TR-169.)

Dershin, H. (1965). "Forces due to gaseous slot jet boundary-layer interaction," *J. Spacecraft and Rockets*, Vol. 2, No. 4 (Jul–Aug), pp. 597–599.

Driftmyer, R. T. (1974). "Thick, two-dimensional turbulent boundary layers separated by steps and slot jets," *AIAA J.*, Vol. 12, No. 1 (Jan), pp. 21–27.

Dupuichs, G. (1963). "Action d'un jet transversal à un ecoulement supersonique," Publications Scientifiques et Techniques du Ministère de L'Air 396. (Translated as "Effect of a transverse jet on a supersonic flow," Office of the Chief of Naval Operations, Office of Naval Intelligence, Translation Section, No. 2174.)

Emery, J. C. et al. (1967). "Flow visualization of a secondary jet by means of lampblack injection techniques," *AIAA J.*, Vol. 5, No. 5 (May), pp. 1039–1040.

Erdos, J. and A. Pallone (1962). "Shock-boundary layer interaction and flow separation," Proceedings 1962 Heat Transfer and Fluid Mechanics Institute, June, Univ. Washington.

Gadd, G. E. (1953). "Interactions between wholly laminar or wholly turbulent boundary layers on shock waves strong enough to cause separation," *J. Aeronaut. Sci.*, Vol. 20, No. 11 (Nov), pp. 729–739.

Georges, H. A. (1959). "Some tests on the formation of conical flows due to separation," Australian Defense Scientific Service, Weapons Research Establishment Tech. Note, HSA 50 (Apr).

Hill, W. G. (1967). "Analysis of experiments on hypersonic flow separation ahead of flaps using a simple flow model," Grumman Aircraft Engineering Corp. Research Dept. Memo RM-383 (Nov).

Hsia, H. T. S. et al. (1966). "Perturbation of supersonic flow by secondary injection," Proceedings ICRPG/AIAA Solid Propulsion Conference, CPIA Publication 111, Vol. II, pp. 613–668. (See also *J. Spacecraft and Rockets*, 1965, Vol. 2, No. 1, pp. 67–72.)

Hunter, P. A. (1959). "An investigation of the performance of various reaction control devices," NASA Memo 2-11-596.

Jones, J. J. (1952). "Flow separation from rods ahead of blunt noses at Mach number 2.72," NASA RM L52E05a (Jul).

Karamcheti, K. and H. T. S. Hsia (1963). "Integral approach to an approximate analysis of thrust vector control by secondary injection," *AIAA J.*, Vol. 1, No. 11 (Nov), pp. 2538–2544.

Kistler, A. L. (1964). "Fluctuating wall pressure under a separated flow," *J. Acoust. Soc. Am.*, Vol. 36, (Mar), pp. 543–550.

Korst, H. H. (1965). "Dynamics and thermo-dynamics of separated flows," in *Single and multi-component flow processes*, R. L. Peskin and C. F. Chen, eds., Rutgers, The State University, Eng. Res. Publ. 45.

Kuehn, D. M. (1959). "Experimental investigation of the pressure rise required for incipient separation of turbulent boundary layers in two-dimensional supersonic flow," NASA Memo 1-21-59A.

Lange, R. H. (1954). "Present status of information relative to the prediction of shock-induced boundary layer separation," NACA TN 3065.

Lingen, A. (1957). "Jet-induced thrust-vector control applied to nozzles having large expansion ratios," United Aircraft Corp. Res. Dept. Report R-0937-33 (Mar).

Love, E. S. (1955). "Pressure rise associated with shock-induced boundary-layer separation," NACA TN 36-1.

Maurer, F. (1965). "Interference effects produced by gaseous side-jets issuing into a supersonic stream," Applied Physics Lab., TG 230-T460, No. 22. (Translated from DVL-Bericht, No. 382, Deutsche Luft. u. Raumfahrt, Forschungsbericht 65-04, Jan 1965.)

Munz, E. P. et al. (1970). "Some characteristics of exhaust plume rarefaction," *AIAA J.*, Vol. 8, No. 9 (Sept), p. 1051.

Ringleb, F. O. (1960a). "Two-dimensional flow with standing vortices in ducts and diffusers," U.S. Naval Air Materiel Center, NAEF ENG-6656 (Mar).

Ringleb, F. O. (1960b). "Vortex formation behind a flat plate in perpendicular flow," U.S. Naval Air Materiel Center, NAEF ENG-6627 (Feb).

Ringleb, F. O. (1963). "Geometrical construction of two-dimensional and axisymmetrical flow fields," *AIAA J.*, Vol. 1, No. 10 (Oct), pp. 2257–2263.

Roshko, A. (1955). "Some measurements of flow in rectangular cutout," NACA TN 3488.

Soyano, S. (1962). "Heat transfer in shock wave-turbulent boundary layer interaction regions," Douglas Aircraft Company Report SM-42567 (Nov).

Spaid, F. W. et al. (1966). "A study of secondary injection of gases into a supersonic flow," Jet Propulsion Laboratory, TR-32-934 (Aug).

Speaker, W. V. and C. M. Ailman (1966). "Spectra and space-time correlations of the fluctuating pressure at a wall beneath a supersonic turbulent boundary layer perturbed by steps and shock waves," NASA CR-486.

Sterrett, J. R. and J. C. Emery (1960). "Extension of boundary layer separation criteria to a Mach number of 6.5 by utilizing flat plates with forward facing steps," NASA TN D-618 (Dec).

Sterrett, J. R. and P. F. Holloway (1964). "On the effect of transition on parameters within a separation region at hypersonic speeds—with emphasis on heat transfer,"

Symposium on Fully Separated Flows, ASME Fluids Engineering Div. Conference, 18–20 May, Philadelphia, Pennsylvania, pp. 15–26.

Sterrett, J. R. et al. (1967). "Experimental investigation of secondary jets from two-dimensional nozzles with various exit Mach numbers for hypersonic control application," NASA TN D-3795 (Jan).

Strike, W. T. et al. (1963). "Interactions produced by sonic lateral jets located on surfaces in a supersonic stream," AEDC-TDR-63-22 (Apr).

Tani, I. (1958). "Experimental investigation of flow separation over a stop," in *Boundary Layer Research*, Springer Verlag.

Truitt, R. W. (1959). *Hypersonic Aerodynamics*, The Ronald Press, New York.

Vinson, P. W. (1965). "Prediction of reaction control effectiveness at supersonic and hypersonic speeds," Martin-Orlando Research Division Report OR 6487 (Mar).

Werle, R. T. et al. (1972). "Jet-interaction-induced separation: The two-dimensional problem," *AIAA J.*, Vol. 10, No. 2 (Feb), p. 188.

(*Supplementary References to Chapter 1, continued from p. 83*)

Westfalen, Herausgegeben von Staatssekretär Prof. Dr. h.c. Leo Brandt, No. 470, Westdeutscher Verlag, Köln un Opladen.

Wehrmann, O. (1960). "Characteristics of separated cylindrical boundary layers," Deutsche Versuchsanstalt für Luftfahrt, E. V. Bericht, No. 131.

Werle, H. and M. Gallon (1972). "Flow control by cross jet," NASA-TT-F-14548. (Translated into English from *Aeronaut. Astronaut. (Paris)*, Vol. 34, pp. 21–33.)

Werle, M. J. and V. N. Vatsa (1974). "New method for supersonic boundary-layer separations," *AIAA J.*, Vol. 12, No. 11, pp. 1491–1497.

Whitehead, A. H. (1970). "Vortices in separated flows," *AIAA J.*, Vol. 8, No. 6, p. 1173.

Whitehead, A. H. et al. (1972). "Effects of transverse outflow from a hypersonic separated region," *AIAA J.*, Vol. 10, No. 4, p. 553.

Williams, G. M. (1974). "Viscous modelling of wing-generated trailing vortices," *Aeronaut. Quart.*, Vol. XXV, Pt. 2, pp. 143–154.

Williams, J. C., III and W. D. Johnson (1974). "Note on unsteady boundary-layer separation," *AIAA J.*, Vol. 12, No. 10, pp. 1427–1429.

Williams, J. C., III and W. D. Johnson (1974). "Semisimilar solutions to unsteady boundary-layer flows including separation," *AIAA J.*, Vol. 12, No. 10, pp. 1388–1393.

Williams, R. V. and F. M. Burrows (1974). "Movement of a line vortex pair downstream of a circular cylinder in potential flow," *Aeronaut. J.*, Vol. 78, No. 768, p. 573.

Wilson, R. E. and F. Maurer (1970). "An experimental investigation of turbulent separated boundary layers at low supersonic Mach numbers," Deutsche Luft- und Raumfahrt, Forschungsbericht 70-33 (DLR FB 70-33).

Wilson, R. E. and F. Maurer (1971). "Turbulent boundary-layer separation at low supersonic Mach numbers," *AIAA J.*, Vol. 9, No. 1, p. 189.

Wortman, A. and A. F. Mills (1971). "Separating self-similar laminar boundary layers," *AIAA J.*, Vol. 9, No. 12, p. 2499.

Wrage, E. (1960). "Entwicklung und Anwendung einer allgemeiner Reihenmethode zur Berechnung laminarer, kompressibler Grenzschichten," *Dt. Versuchsanstalt f. Luftfahrtforschung, Bericht*, 134.

Wu, J. J. and W. Behrens (1972). "An experimental study of hypersonic wakes behind wedges at angle of attack," *AIAA J.*, Vol. 10, No. 12, p. 1582.

Young, W. H. and J. C. Williams (1972). "Boundary-layer separation on rotating blades in forward flight," *AIAA J.*, Vol. 10, No. 12, p. 1613.

ACKNOWLEDGMENTS AND PERMISSIONS

Reprinted with permission from **American Institute of Aeronautics and Astronautics:**

Fig. 5.2 from Ludwig, G. R. (1964). "An experimental investigation of laminar separation from a moving wall," AIAA Preprint 64-6, presented at the AIAA 2nd Aerospace Sciences Meeting, New York, January.

Figs. 5.70–5.83 from Hubbartt, J. E. and L. H. Bangert (1970). "Turbulent boundary layer control by a wall jet," AIAA Paper 70-107 presented at the AIAA 8th Aerospace Sciences Meeting, New York, January 19–21.

Fig. 6.51 from Abbett, M. (1970). "The Mach disc in under expanded exhaust plumes," AIAA Paper 70-231, presented at the AIAA 8th Aerospace Sciences Meeting, New York, January 19–21.

Figs. 6.52–6.54 from Peters, C. E. and W. J. Phares (1970). "The structure of plumes and moderately underexpanded supersonic nozzles," AIAA Paper 70-229, presented at the AIAA 8th Aerospace Sciences Meeting, New York, January 19–21.

Fig. 6.57 from F. P. Povinelli, L. A. Povinelli, and M. Hersch (1970). "Supersonic jet penetration (up to Mach 8) into a Mach 2 airstream," AIAA Paper 70-92, presented at the AIAA 8th Aerospace Sciences Meeting, New York, January 19–20.

Figs. 6.58–6.60 from Billig, F. S., R. C. Orth, and M. Lasky (1970). "A unified approach to the problem of gaseous jet penetration into a supersonic stream," AIAA Paper 70-93, presented at the AIAA 8th Aerospace Sciences Meeting, New York, January 19–21.

Fig. 2.33 from Beamish, J. K., D. M. Gibson, R. H. Sumner, S. M. Zivi, and G. H. Humberstone (1969). "Wind-tunnel diagnostics by holographic interferometry," *AIAA J.*, Vol. 7, No. 10 (October).

Figs. 5.51–5.54 from Pate, S. R. (1969). "Experiments on control of supersonic transitional flow separation using distributed suction," *AIAA J.*, Vol. 7, No. 5 (May), pp. 847–851.

Figs. 5.55-5.61 from Ball, K. O. and R. H. Korkegi (1968). "An investigation of the effect of suction on hypersonic laminar boundary-layer separation," *AIAA J.*, Vol. 6, No. 2 (February), pp. 239-243.

Figs. 6.61-6.63 from Spaid, F. W. and E. E. Zukoski (1968). "A study of the interaction of gaseous jets from transverse slots with supersonic external flows," *AIAA J.*, Vol. 6, No. 2 (February), pp. 205-212.

Figs. 6.64-6.66 from Zukoski, E. E. (1967). "Turbulent boundary layer separation in front of a forward facing step," *AIAA J.*, Vol. 5, No. 10 (October), pp. 1746-1753.

Figs. 1.27-1.29 from Hazen, D. C. (1967). "The rebirth of subsonic aerodynamics," *Astronaut. and Aeronaut.*, November, pp. 24-39.

Fig. 4.43 from Cornish, J. J. (1965). "Some aerodynamic and operational problems of STOL aircraft with boundary layer control," *J. Aircraft*, Vol. 2, No. 2 (March-April).

Figs. 4.73-4.87 from Brown, A., H. F. Nawrocki, and P. N. Paley (1968). "Subsonic diffusers designed integrally with vortex generators," *J. Aircraft*, Vol. 5, No. 8 (May-June), pp. 221-229.

Reprinted with permission from **the authors**:

Figs. 4.56, 4.61-4.63, and 4.88 from Frey, K. P. H. in collaboration with N. C. Vasuki (1966). *Detached flow and control*, published by the authors.

Reprinted with permission from **Butterworths, London**:

Fig. 2.4 from Vallentine, H. R. (1959). *Applied hydrodynamics*, Butterworths, London.

Reprinted with permission from **The Clarendon Press, Oxford**:

Figs. 1.30, 4.33, and 5.1 from Thwaites, B. (1960). *Incompressible aerodynamics*, Clarendon, Oxford.

Fig. 2.38 from Rosenhead, L. (1963). "Laminar boundary layers," in *Fluid motion memoirs*, Ch. 10, Clarendon, Oxford.

Reprinted with permission from **The Franklin Institute**:

Fig. 2.26 from Gawthrop, D. W., W. C. Shepherd, and St. J. G. Perott (1931). "The photograph of waves and vortices produced by the discharge of an explosive," *J. Franklin Institute*, Vol. 211.

Reprinted with permission from **Interavia**:

Figs. 4.22 and 4.48 from Liebe, W. (1952). "The boundary layer fence," *Interavia*, Vol. 7, No. 4, pp. 215-217.

Reprinted with permission from **McGraw-Hill Book Company**:

Figs. 1.9-1.10 from Schlichting, H. (1968). *Boundary layer theory*, sixth ed., McGraw-Hill, New York.

Fig. 1.24 from Prandtl, L. and O. G. Tietjens (1934). *Applied hydro and aero mechanics*, McGraw-Hill, New York.

Reprinted with permission from **The New York Academy of Sciences and the authors**:

Figs. 5.87-5.91 from Kikuhara, S. and M. Kasu (1968). "Design of blowing type BLC system on the Japanese STOL seaplane," presented at the International Congress on Subsonic Aeronautics sponsored by the New York Academy of Sciences, New York, Vol. 154, pp. 397-424.

Reprinted with permission from **Pergamon Press Ltd.**:

Figs. 1.7, 1.43, 4.51, and 4.55 from Pearcey, H. H. (1961). "Shock-induced separation and its prevention by design and boundary layer control," in *Boundary layer and flow control*, Vol. 2, G. V. Lachmann, ed., Pergamon, New York.

Figs. 1.23 and 6.1-6.19 from Hurley, D. G. (1961). "The use of boundary layer control to establish free streamline flows," in *Boundary layer and flow control*, Vol. 1, G. V. Lachmann, ed., Pergamon, New York.

Figs. 2.13, 2.24, and 2.31 from Bradshaw, P. (1964). *Experimental fluid mechanics*, Pergamon, New York.

Figs. 2.34-2.35 from Ower, E. and R. C. Pankhurst (1966). *The measurement of air flow*, Pergamon, New York.

Figs. 4.36-4.40 from Wortmann, F. X. (1961). "Progress in the design of low drag airfoils," in *Boundary layer and flow control*, Vol. 2, G. V. Lachmann, ed., Pergamon, New York.

Figs. 5.18-5.22 from Head, M. R. (1961). "Approximate methods of calculating the two dimensional laminar boundary layer with suction," in *Boundary layer and flow control*, Vol. 2, G. V. Lachmann, ed., Pergamon, New York.

Fig. 5.26 in Wuest, W. (1961). "Theory of boundary suction to prevent separation," in *Boundary layer and flow control*, Vol. 1, G. V. Lachmann, ed., Pergamon, New York.

Figs. 5.63-5.66 from Attinello, J. S. (1961). "Design and engineering features of flap blowing installations," in *Boundary layer and flow control*, Vol. 1, G. V. Lachmann, ed., Pergamon, New York.

Figs. 5.95-5.100 from Wagner, F. G. (1961). "Design and engineering features for flap suction and combined blowing and suction," in *Boundary layer and flow control*, Vol. 1, G. V. Lachmann, ed., Pergamon, New York.

Figs. 6.20-6.33 from Ringleb, F. O. (1961). "Separation control by trapped vortices," in *Boundary layer and flow control*, Vol. 1, G. V. Lachmann, ed., Pergamon, New York.

Figs. 6.45 and 6.48 from Bogdonoff, S. M. and I. E. Vas (1959). "Preliminary investigations of spiked bodies at hypersonic speeds," *J. Aero. Sci.*, Vol. 26, No. 2, pp. 65–74.

Fig. 6.50 from Adamson, T. C., Jr. and J. A. Nicholls (1959). "On the structure of jets from highly under expanded nozzles into still air," *J. Aero. Sci.*, Vol. 26, No. 1 (January), pp. 16–24.

Reprinted with permission from **Pitman Publishing, London**:

Figs. 2.1-2.3 and 2.8 from Pankhurst, R. C. and D. W. Holder (1952). *Wind-tunnel techniques*, Pitman, London.

Reprinted with permission from **Prentice-Hall, Inc.**:

Fig. 1.25 from Heinke, P. E. (1946). *Elementary applied aerodynamics*, Prentice-Hall, New York.

Reprinted with permission from **Princeton University**:

Fig. 5.96 from Malavard, L., P. Poisson-Quinton, and P. Jousserandot (1956). "Theoretical and experimental investigation of circulation control," Translation Report No. 358, Princeton University.

Reprinted with permission from **Princeton University Press**:

Fig. 1.1 from *Theory of laminar flows*, edited by F. K. Moore, Vol. 4 *High speed aerodynamics and jet propulsion* (copyright © 1964 by Princeton University Press): Fig. C.18a in A. Mager "Three-dimensional laminar boundary layers," p. 389.

Reprinted with permission from the **Royal Aeronautical Society**:

Figs. 4.11–4.20 from Chappell, P. C. (1968). "Flow separation and stall characteristics of plane, constant-section wings in subcritical flow," *Aeronaut. J.*, Vol. 22, No. 685 (January), pp. 82–90.

Reprinted with permission from **The Royal Society**:

Fig. 2.39 from Gadd, G. E., D. W. Holder, and J. D. Regan (1954). "Interaction between shock waves and boundary layers," *Proc. Roy. Soc. (London)*, Vol. 226.

Reprinted with permission from **Sawell Publications Limited (Bunhill Publications):**

Fig. 4.34 from Nonweiler, T. (1956). "The design of wing sections," *Aircraft Eng.*, Vol. 28, p. 221.

Figs. 5.4-5.11 from Alvarez-Calderon, A. A. (1964). "Rotating cylinder flaps of V/STOL aircraft," *Aircraft Eng.*, Vol. 36, No. 10 (October), pp. 304-309.

Reprinted with permission from **Springer-Verlag:**

Fig. 4.23 from Schlichting, H. and E. Truckenbrodt (1967). *Aerodynamik des Flugzeuges*, Vol. 1, Springer-Verlag, Berlin.

Fig. 4.50 from Weinig, F. (1940). *Aerodynamik der Luftschraube*, Verlag von Julius Springer, Berlin.

Reprinted with permission from **Stanford University Press:**

Figs. 1.44 and 1.54-1.55 from Erdos, J. and A. Pallone (1962). "Shock-boundary layer interaction and flow separation," *Proceedings*, 1962 Heat Transfer and Fluid Mechanics Institute (June, 1962, University of Washington), Stanford Univ. Press, Stanford, California.

Reprinted with permission from **Taylor & Francis, Ltd.:**

Figs. 1.20, 6.40-6.41 and 6.44 from Mair, W. A. (1952). "Experiments on separation of boundary layers on probes in front of blunt-nosed bodies in a supersonic air stream," *Phil. Mag.*, Vol. 43, 7th Series, July, pp. 695-716.

Reprinted with permission from **Friedr. Vieweg & Sohn:**

Figs. 5.62 and 5.67-5.68 from Poisson-Quinton, P. (1956). "Einige Physikalische Bertrachtungen uber das Ausblasen an Tragflugeln," *Jahrb. d.W.G.L.*, Friedr. Vieweg & Sohn, Braunschweig.

Reprinted with permission from **John Wiley & Sons:**

Fig. 2.23 from Liepmann, H. W. and A. Roshko (1957). *Elements of gas dynamics*, Wiley, New York.

AUTHOR INDEX

Numbers in italics refer to the pages on which the complete references are cited.

SUBJECT INDEX